## eBook Bonus.

With the purchase of this book, you also receive access to its electronic version. Use the following personal code to download the electronic pages.

aoqug-l26r8-
01800-dw2l3

Take advantage of the eBook Bonus on your computer, smartphone, or tablet. Go to **www.hanserpublications.com/ebookbonus** and register with your personal access code.

Download of this book as an eBook is subject to legal regulations, which can be found at www.hanserpublications.com/ebookbonus. Requirements: Internet connection, Adobe® Reader®, and account registration.

Hopmann, Michaeli

**Extrusion Dies for Plastics and Rubber**

Christian Hopmann
Walter Michaeli

# Extrusion Dies for Plastics and Rubber

## Design and Engineering Computations

4th Edition

With Contributions by

Dr.-Ing. Ulrich Dombrowski • Dr. Ulrich Hüsgen • Dr.-Ing. Matthias Kalwa • Dr.-Ing. Stefan Kaul • Dr.-Ing. Michael Meier-Kaiser • Dr.-Ing. Boris Rotter • Dr.-Ing. Micha Scharf • Dr.-Ing. Claus Schwenzer • Dr.-Ing. Christian Windeck • Nafi Yesildag, M.Sc.

HANSER

Hanser Publishers, Munich    Hanser Publications, Cincinnati

The Authors:

*Prof. Dr.-Ing. Christian Hopmann,*
Head of the Institute of Plastics Processing (IKV) at RWTH Aachen University, Aachen, Germany

*Prof. Dr.-Ing. Dr.-Ing. E.h. Walter Michaeli,*
former Head of the Institute of Plastics Processing (IKV) at RWTH Aachen University, Aachen, Germany

MIX
Papier aus verantwortungsvollen Quellen
FSC® C016439
FSC www.fsc.org

Distributed in the Americas by:
Hanser Publications
6915 Valley Avenue, Cincinnati, Ohio 45244-3029, USA
Fax: (513) 527-8801
Phone: (513) 527-8977
www.hanserpublications.com

Distributed in all other countries by:
Carl Hanser Verlag
Postfach 86 04 20, 81631 München, Germany
Fax: +49 (89) 98 48 09
www.hanser-fachbuch.de



Cataloging-in-Publication Data is on file with the Library of Congress


Editor: Mark Smith
Production Management: Jörg Strohbach
Coverconcept: Marc Müller-Bremer, www.rebranding.de, München
Coverdesign: Stephan Rönigk
Typesetting: Kösel Media GmbH, Krugzell
Printed and bound by Hubert & Co GmbH, Göttingen
Printed in Germany

ISBN: 978-1-56990-623-1
E-Book ISBN: 978-1-56990-624-8

# Preface

In January 2003, this book was published in its 3rd edition in English. Since then, an unrelenting demand for the book has been observed, both for the German and English versions. In order to meet this demand, it is our pleasure that Hanser now publishes this 4th edition. With this edition, "Extrusion Dies" has for the first time two editors: In April 2011 Prof. Dr.-Ing. Christian Hopmann succeeded Prof. Dr.-Ing. Dr.-Ing. E. h. Walter Michaeli as holder of the Chair of Plastics Processing and Head of the Institute of Plastics Processing (IKV) at RWTH Aachen University, Aachen, Germany. We are very pleased that this book with its long history with Hanser is continued into the next IKV generation.

This update will continue to help you in your work and life while hopefully also providing pleasure in reading. We have retained the structure of the book, which has proven itself over many years and received much positive resonance from readers.

When we say "we", we particularly refer to *Dr.-Ing. Christian Windeck,* former head of the IKV extrusion department, and his successor *Nafi Yesildag, M.Sc.,* who have critically analyzed, checked, and supplemented the contents, equations, and reference lists. We would like to express our special thanks to both of them.

We further thank *Mark Smith* and *Jörg Strohbach* of Hanser for their support in the publication of our work.

Once again, suggestions obtained from the plastics and rubber industry were taken up and addressed in this fourth edition. We thank all those who provided their suggestions and help. Many research and development efforts of the IKV form the fundament of some of the facts described in this book. Against this background, we thank the Federal Ministry for Economic Affairs and Energy (BMWi), Berlin, for the promotion of many industrial research projects through the German Federation of Industrial Research Associations (AIF e. V.), Cologne, the Deutsche Forschungsgemeinschaft (DFG), Bonn-Bad Godesberg, the Federal Ministry of Education and Research (BMBF), Bonn, and the European Commission, Brussels, with respect to extrusion dies.

*Walter Michaeli*

*Christian Hopmann*

# Preface to the Third Edition

During my last visit to Medellin, Colombia, on occasion of the 10th anniversary of the ICIPC, a thriving plastics and rubber research institute, I met many young and eager students who knew my name because they had studied the book on extrusion die design. They asked many questions and I could not say good-bye without being photographed showing me at the center of their group. I enjoyed that for two reasons: first, this event showed that the book has reached acceptance even far away from my hometown. Second, it was important to learn that those young men and women enjoyed studying the book on their way to qualify for their professional life. But this book has also been written for the people who need daily support in their practical work applications in industry and science. Twelve years have gone by since the second edition of this book was published, years with visible changes and innovations in the field of extrusion and die design. For example, spiral mandrel dies have existed for more than three decades, but some functionalities have changed. Today, we place the spiral on a flat surface and feed it from the side. And when we pile a couple of those systems on top of each other, the result is the so-called stack die, which provides several advantages over classical coextrusion dies with annular slits. We incorporated this new development in Chapter 5.

It may be the dream of an extrusion die designer to process all of the material, processing, and geometrical data of the final product with a computer and end up with a fully designed flow channel which facilitates the optimum flow distribution. In studying this book, the reader will realize that finite element analysis is a key to fulfilling this dream, but the proper description of the viscoelastic properties of the extruded materials is still a challenge for rheologists and engineers. Nevertheless, significant steps have been made in this direction. For viscous flow, that goal has almost been reached. Therefore, a new chapter on optimization of extrusion die performance with computers was integrated into this third edition.

I would like to thank my co-workers, Dr.-Ing. Boris Rotter, head of the IKV extrusion department, and Dipl.-Ing. Stefan Kaul, research engineer in this department, for their support and active help in reviewing and improving this book with their expertise. Many of the results presented in this book were produced by students at the Institute during the research for their diplomas.

Much gratitude also goes to those who provided many suggestions and help; in particular, the members of the IKV advisory committees: Extrusion, Blow Molding, and Rubber Technology. Many research and development projects of IKV form the basis of some of the relations documented in the book. They were made possible by the cooperative research between the industry and the IKV, by the support and funding of the Arbeitsgemeinschaft industrieller Forschungsvereinigungen Otto von Guericke e. V. (AiF) in Cologne, the Deutsche Forschungsgemeinschaft (DFG), Bonn-Bad Godesberg, and the Federal Ministry for Education, Research and Technology (BMBF) in Berlin, respectively Bonn and the European Commission, Brussels.

Last but not least, I woud like to thank Dr. Wolfgang Glenz of Hanser, Munich, for so many years of excellent and active cooperation and for his vital insight. Such insight is appreciated by technical authors like myself, who have a rather challenging job, which at least nourishes our families, parallel to writing books. All of these contributing factors make things easier to write a book like this.

*Walter Michaeli*

# Preface to the Second Edition

Ten years after the publication of the first edition of this book it is appropriate to start anew by reviewing and documenting the new developments and applications in the area of designing and manufacturing of extrusion dies. That is the purpose of this new, revised edition. Although the basic principles pertaining to extrusion dies are the same, there have been, in the meantime, many developments and refinements in this area due to continuously growing demands for improved quality and increased productivity, as well as emerging new polymers and novel products. For example, coextrusion has gained importance recently and the polymers based on liquid crystals represent an entirely new class of materials which will, without doubt, require new concepts in extrusion die design. That means development will continue and, therefore, this second edition can summarize the current state of technology. Particular attention is given here to the theoretical tools, such as the finite element method, which have been greatly developed in the last decade and which can provide solutions to many current problems.

The basic goal of this book, as already stated in the preface to the first edition, will not change under any circumstances; it is written for the practitioner, to help him in his daily work and for the student, to introduce him to the complex world of extrusion dies and provide him with an extensive orientation and thorough education. The response to the first edition of this book was very positive. Nevertheless, as with everything, it can be further improved and this is what we are attempting with this second edition. The chapter about the design of dies for the extrusion of elastomers was added; the area of coextrusion dies was expanded considerably; and all other chapters were subjected to substantial revisions.

When I say "we", I refer to my coworkers at the Institut fuer Kunststoffverarbeitung (IKV) at the Rheinisch-Westfaelische Technische Hochschule in Aachen. Those are Messrs. Dr. *U. Dombrowski*, Dr. *U. Huesgen*, Dr. *M. Kalwa*, Dr. M. *Meier* and Dr. *C. Schwenzer*. They took part in the work on this book and dedicated many hours of their personal time. This is also true for Mrs. *N. Petter* and Mrs. *D. Reichelt*, who transcribed the text and for Mrs. *G. Zabbai* and Mr. *M. Cosier* who assured the good quality of the illustrations. Many special and personal thanks to all of them.

Many of the results presented in this book were produced by students at the Institute during their studies and research leading to diplomas. They also deserve sincere thanks.

Suggestions obtained from the plastics and rubber industry were taken up and dealt with in this second edition. Many thanks go also to those who provided the suggestions and help, in particular the members of the advisory committees Extrusion, Blow Molding and Rubber Technology of the IKV.

Many research and development efforts of the IKV form the basis of some of the relations described in this book. They were made possible by the cooperative research between the industry and IKV, by the support of the Arbeitsgemeinschaft Industrieller Forschungsvereinigungen (AIF) in Cologne, of the Deutsche Forschungsgemeinschaft (DFG), Bonn-Bad Godesberg and the Federal Ministry for Research and Technology (BMFT) in Bonn as well as by the Volks wagen werk Foundation in Hannover.

Aachen, in November 1991

*Walter Michaeli*

# Preface to the First Edition

In this book an attempt is made to present to the practitioner and to the student a broad picture of all extrusion dies for plastics. In pursuing that objective the various types of dies and their specific features are discussed, guidelines for their design given and approaches to computational engineering analyses and its limitations demonstrated. This is even more important in view of the increasing efforts made by the industry as well as academia, starting in the recent past and continuing in the present, to model the transport phenomena (flow and heat transfer) in the extrusion die mathematically. These important projects are motivated primarily by the demand for higher productivity accompanied by better product quality (i.e. dimensional accuracy, surface quality) of the extruded semifinished goods. Purely empirical engineering methods for extrusion dies are becoming unacceptable at an increasing rate because of economical considerations.

The design of the flow channel takes a focal position in the engineering process of extruder dies. This book starts by identifying and explaining the necessary material data for designing the flow channel.

The derivation of basic equations permits estimates to be made of pressure losses, forces acting on the flow channel walls, velocity profiles, average velocities etc. in the flow channel. The simple equations that are useful for practical applications are summarized in tables. For the majority of extrusion dies these equations are sufficient to arrive at a realistic design based upon rheological considerations.

Approaches to calculating the velocity and temperature fields using finite difference and finite element methods (FEM) are also discussed because of their increasing importance in the design of extrusion dies.

The various types of single and multiple layer extrusion dies and their specific features are highlighted in detail in Chapters 5 and 6, followed by a review of the thermal and mechanical design considerations, and comments pertinent to the selection of material for extrusion dies and to their manufacture. A discussion of handling, cleaning and maintenance of extrusion dies as well as of devices for sizing of pipes and profiles concludes the book. At the end is a comprehensive list of references.

The book was written during my activity as head of the Extrusion and Injection Molding Section at the Institut fuer Kunststoffverarbeitung (IKV) at the Rheinisch-Westfaelische Technische Hochschule Aachen (Institute for Plastics Processing at the Aachen Technical University, Aachen, West Germany, Director: Prof. Dr.-Ing. G. *Menges*). I had access to all important results of the research at the IKV in the field of engineering of extrusion dies. I wish to extend my thanks to my former and present colleagues at the IKV, in particular Messrs. *J. Wortberg, A. Dierkes, U. Masberg, B. Franzkoch, H. Bangert, L. Schmidt, W. Predoehl, P.B. Junk, H. Cordes, R. Schulze-Kadelbach, P. Geisbuesch, P. Thienel, E. Haberstroh, G. Wuebken, U. Thebing, K. Beiss* and *U. Vogt* whose research work was essential in the preparation of the text and also to all other colleagues who contributed and to the students and graduate students of the Institute. But foremost, I wish to thank Prof. Dr.-Ing G. *Menges* for encouraging me to prepare this book and for his ceaseless help, promotion and support which made it possible for me to complete it.

Further thanks are extended to a number of representatives of the plastics industry, in particular to the members of the Section Extrusion and Extrusion Blow Molding of the Advisory Board of the Foerdervereinigung (Sponsors Society) of the IKV.

Many of the research and development projects of the IKV which are referred to in this book and which became the basis for some of the facts presented in it, were only made possible financially by the joint research between industry and the IKV, support by the Arbeitsgemeinschaft Industrieller Forschungsvereinigungen (AIF), Cologne, the Deutsche Forschungsgemeinschaft (DFG), Bonn-Bad Godesberg and the Ministry for Research and Technology (BMFT), Bonn.

This book was first published in German in 1979. The book in your hands is the first English translation based on this slightly revised 1979 edition. We have added an alphabetic index and checked the list of references to make sure that the most important references in English are easily identified.

'Life goes on' – also in extrusion tooling – so the list of references is completed by publications since 1979. I wish to thank all who made the English version possible: The Society of Plastics Engineers (SPE) for sponsoring this book, Dr. Herzberg for translating, Dr. Immergut and Dr. Glenz of Hanser for coordinating, Dr. Hold of Polymer Processing Institute – Stevens Institute of Technology, Hoboken, New Jersey, for being the technical editor, and Hanser for publishing.

Heppenheim, W. Germany August 1983

*Walter Michaeli*

# Contents

# 1 Introduction

In the extrusion of thermoplastics into semifinished products, two units occupy a central position: the extrusion die (also known as the extruder head), which shapes the melt, and the former or calibrator, usually mounted adjacent to the extrusion die, whose function is to guide the molten semifinished product while maintaining the desired dimensions and providing the specified degree of cooling (Figs. 1.1 and 1.2).

Conversely, when extruding elastomeric materials, the dimensions of the product are essentially determined by the geometry of the extrusion die. Only when vulcanization follows the extrusion some geometrical changes occur, mainly due to the crosslinking of the material, especially when the extrudate is allowed to shrink freely.

A sufficiently large, pulsation-free, reproducible, and thermally and mechanically homogeneous melt stream is expected from the extruder. Subsequently, the extrusion die and the calibration unit determine the dimensions of the semifinished products. In this connection, it must be taken into consideration that the rheological and thermodynamic processes in the die and in the calibrator, as well as any stretching processes that may be present between the die and the calibrator or in connection to both, have a decisive effect on the quality of the extruded semifinished products (such as the surface and the characteristic mechanical properties). In order to design the extrusion die and the calibration unit in a manner that is appropriate from a process engineering point of view, it is necessary to take into consideration the flow, deformation, and temperature relationships in both parts of the production line. If an analytical description of the physical processes is selected, the empirical portion in the design of the die and of the forming stage can be reduced, because changes, such as in the geometry of the channels of the die, in the operating conditions, or in the rheological and the thermodynamic material values of the polymer being processed, can be evaluated directly with regard to shaping and cooling of the extruded semifinished products. This leads to a more reliable design of extrusion die and of calibration unit.

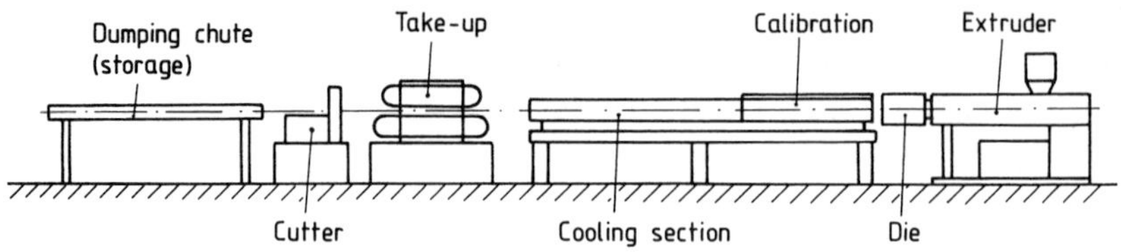


**Figure 1.1** Diagram of an extrusion line for profiles, left to right: dumping chute (storage), cutter, take-up, cooling section, calibration, die, and extruder

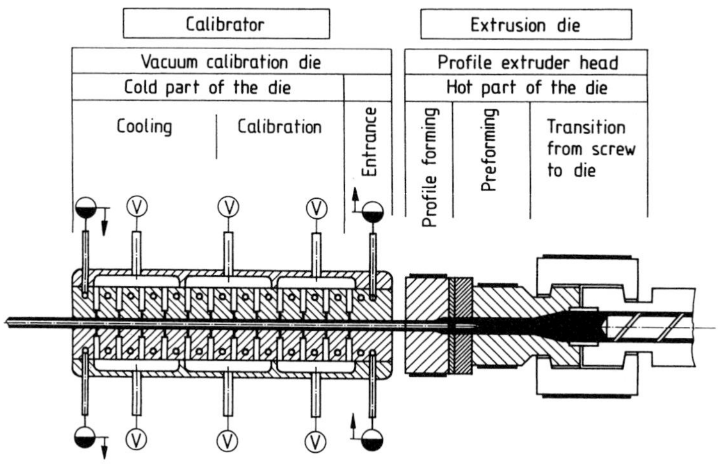


**Figure 1.2** Functional sections of a die and a calibrator [Reifenhäuser]

It is therefore an objective of this book to provide a comprehensive description of the processing and engineering methods used in extrusion dies and calibration units, with attention being focused on a description of the extrusion die. Rules are derived for their design, and simple mathematical aids are given that conform to practical needs. Moreover, reference is made to the special features of different designs for the die and the calibration unit. Specific differences between dies used for the extrusion of elastomeric materials and those used for thermoplastics are pointed out.

Rheological, thermodynamic, mechanical, manufacturing, and operational points of view arise in the design and engineering of extrusion dies and calibration units [1].

The *operational aspects* include, for example, an adequate mechanical stiffness of the extrusion die, in order to keep changes in the cross section of the outlet that are due to the action of melt pressure to a minimum; the ease of installation and dismantling of the die and calibration unit; and the ease of cleaning the die. In addition to that, as few as possible well-sealed surfaces in the dies and a readily detachable and tight connection between the extruder and the die are important [1].

*Manufacturing points of view* must be given consideration in the design of the individual die and calibration unit components to achieve the lowest manufacturing costs, such as using die materials that lend themselves to machining, polishing and, if necessary, tempering and employing established manufacturing methods.

When considering *rheological aspects,* the question must be asked [1]: How should the dimension of the flow channel in the die be selected so that

- a specified throughput is achieved for a given extrusion pressure? (This question may also be reversed.)
- the melt emerges at the same average rate from the whole of the outlet cross section?
- the desired extrudate geometry is achieved for semifinished products without axial symmetry? (This is, among other things, on the basis of visco-elastic effects.)
- the surface of the extrudate or the interfaces of different melt layers remain smooth even at high throughputs? (At high shear rates, melt fracture may arise.)
- stagnations and decompositions of the extruded material, which are partly associated with stagnations, are avoided? (This is a question of the residence times of the material in the die as well as of the temperatures existing there.)

In the *thermodynamic consideration* of the problem, which is closely associated with the rheological aspects, information must be obtained concerning the maximum temperatures occurring in the melt stream in the die on the basis of existing heat transmission and dissipation relationships, especially in view of the heat-sensitive polymeric materials. In addition, the homogeneity of temperatures within the melt is of importance because local temperature variations directly change the materials' viscosity, and throughput deviations may arise. This topic also includes the realization of a uniform and controllable temperature in the die and the calibrator.

As a rule, not all of the subjects addressed here can be realized simultaneously with the same success when designing an extrusion die as well as a calibrator. For this reason, priorities must be established. For example, in designing a pelletizer die (perforated plate for pelletizing), every effort is made to obtain as high a throughput as possible, while in wire coating a smooth surface has a high priority [1].

In these considerations, however, it is necessary to pay attention to the fact that the extruder and the extrusion die interact in their operating behavior. This is the case particularly when using a conventional single-screw extruder with a smooth cylinder and an extruder screw with three sections. As can be seen from Fig. 1.3, increased extrusion die resistance at a constant screw speed can lead to a clear reduction in the throughput in this case. The pressure drop in the extrusion die is therefore important in the mechanical design of the die body and of the bolts that hold the die together, as well as for the throughput that can be achieved. (Note: In many cases, the throughput of an extrusion line is not limited by the die or by the extruder that has been selected, but rather by the attainable rate of cooling in the region of the calibrating and cooling stages. For instance, thick-walled rods are manufactured with extruders having a small screw diameter.) Moreover, under adiabatic conditions, the temperature increases in the extrusion die, which results from internal friction of the melt (dissipation) correlated with this pressure drop according to the well-known equation

$$\Delta T = \frac{\Delta p}{\varrho c_{\mathrm{p}}}. \tag{1.1}$$

$\Delta T$ Temperature difference

$\Delta p$ Pressure loss

$\varrho$ Density

$c_{\mathrm{p}}$ Specific heat capacity

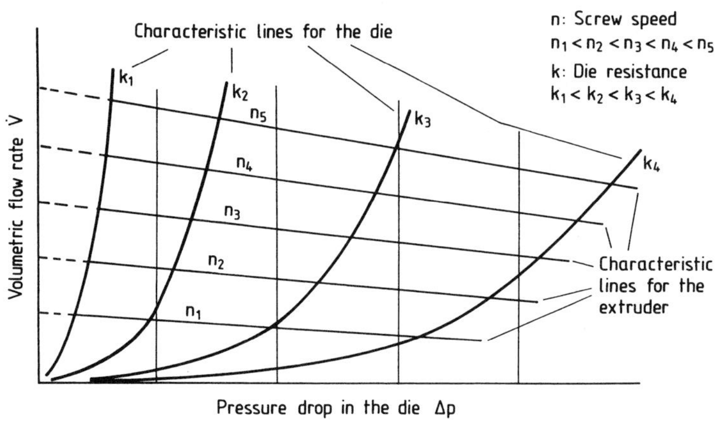


**Figure 1.3** Performance graph of a conventional extruder

With regard to the design of the extrusion die, the pressure drop is of primary importance.

The aspects addressed, which must be taken into consideration in designing an extrusion die, are presented in the following diagram (Fig. 1.4). The input data for designing an extrusion die (Fig. 1.4, Step I) are as follows:

- the geometry of the semifinished product to be extruded (e.g., pipe, flat film, any profile) and whether this is also designed in conformity with the needs of the polymer processed;
- the manner in which the die is fed and whether several semifinished products are to be manufactured simultaneously (Fig. 1.5);

- the material or the combination of materials to be processed in the case of coextrusion; and
- the operating point (or the operating region) of the extrusion die (the operating point is understood to be the throughput and the temperature in the die).

This is followed in Step II by the selection and design of the flow channel, as well as by the calculation of the pressure drop on the basis of the information provided in Step I. Moreover, the position of the heaters relative to the flow channel can be given, with minimum clearances being taken into consideration.

In Step III, the basic dimensions of the die are established. However, the sequence of the steps to be considered can be changed.

The exact design of the die takes place in Step IV. If necessary, control calculations can be carried out for the structural details.

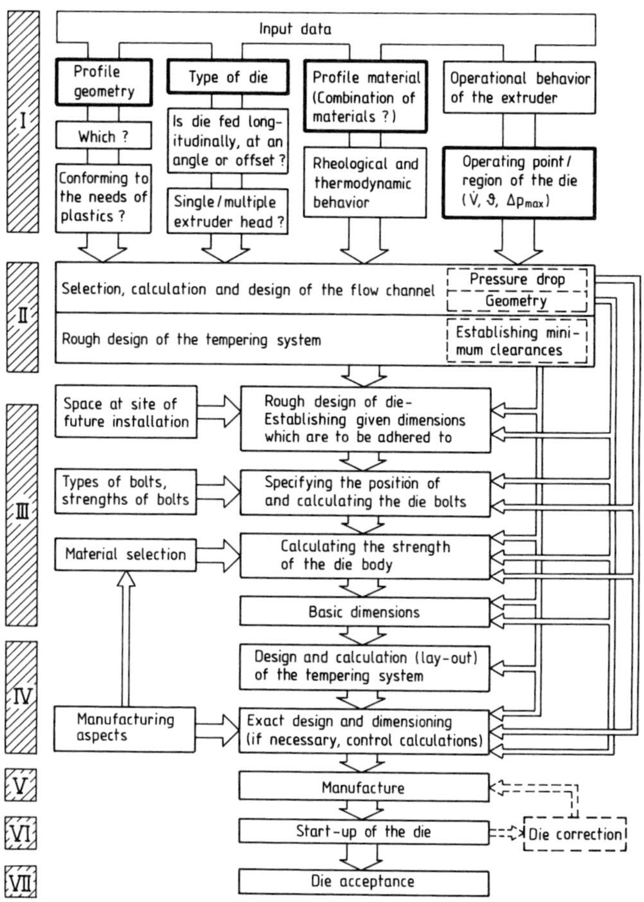


**Figure 1.4** Flow diagram of the development of an extrusion die from design to acceptance

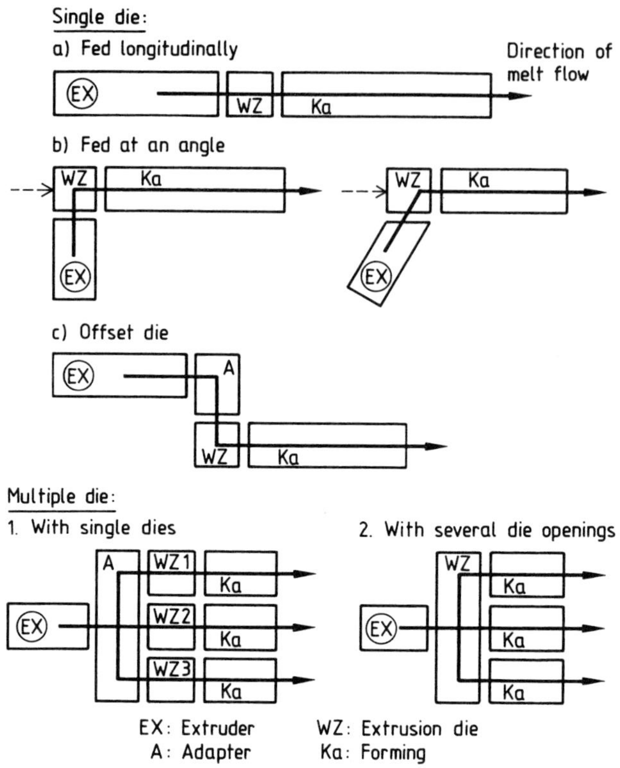


**Figure 1.5** Types of extrusion dies

The manufacture (Step V) is followed by the initial operation of the die (Step VI) with the material (or material combination) selected for the design and under the operating conditions aimed for in later use (see Step I). Several die corrections can also become necessary here.

If the result is satisfactory, the die is then finally accepted (Step VII) and is ready for production.

## Symbols and Abbreviations

| | |
|---|---|
| $c_p$ | Specific heat capacity |
| $k$ | Die resistance |
| $n$ | Screw speed |
| $p$ | Pressure |
| $\Delta p$ | Pressure difference |
| $T$ | Temperature |
| $\Delta T$ | Temperature difference |
| $\dot{V}$ | Volume flow |
| $\varrho$ | Density |

## 1.1 Reference of Chapter 1

1. Röthemeyer, F.: Bemessung von Extrusionswerkzeugen. Maschinenmarkt 76 (1970) 32, pp. 679–685

# 2 Properties of Polymeric Melts

When we choose a theoretical description of the process correlations in the extrusion die and calibration unit for a reliable design of those systems, there are two things in particular to be considered:

- Simplifications and boundary conditions based on the physical models always have to be analyzed critically with regard to the problem at hand.
- Data pertaining to the processed material and that are being entered into the models become of key importance. These are data that characterize flow, deformation, and relaxation behaviors and heat transfer; in other words, its rheological and thermodynamic data [1].

## 2.1 Rheological Behavior

A general flow is fully described by the law of conservation of mass, impulse, and energy, as well as by the rheological and thermodynamic equations of state. The rheological state equation, often referred to as the material law, describes the correlation between the flow velocity field and the resulting stress field. All the flow properties of the given polymer enter this equation. The description, explanation, and measurement of the flow properties are at the core of the science of deformation and flow called rheology [2].

Rheology will be introduced in this chapter to the extent to which it is needed for the design of extrusion dies. Polymeric melts do not behave as purely viscous liquids; they also exhibit a substantial elasticity. Their properties therefore lie between ideal Newtonian (viscous) fluids and ideal Hookean (elastic) solids. This is referred to as viscoelastic behavior or viscoelasticity. When describing rheological material behavior, a clear distinction is made between purely viscous behavior and the combination of viscous and time-dependent elastic behavior.

### 2.1.1 Viscous Properties of Melts

During the process of flow as it occurs in extrusion dies, the melt is subjected to shear deformation. This shearing flow is caused by the fact that melts adhere to the die walls. This is called Stokean adhesion. A change in flow velocity through the flow channel area is the result of this, and it is represented by the following equation:

$$\dot{\gamma} = -\frac{dv}{dy}, \tag{2.1}$$

$v$ Flow velocity

$y$ Direction of shear

During the steady-state shear flow, a shear stress $\tau$ occurs between two layers of the fluid at any point. In the simplest case of a *Newtonian fluid,* this shear stress $\tau$ is proportional to the shear rate $\dot{\gamma}$:

$$\tau = \eta \cdot \dot{\gamma}. \tag{2.2}$$

The constant of proportionality $\eta$ is called the dynamic shear viscosity or simply *viscosity.* Its dimension is Pa·s. The viscosity is the measure of the internal resistance to flow in the fluid under shear.

Generally, polymeric melts do not behave in a Newtonian fashion. Their viscosity is not constant but is dependent on the shear rate. In reference to Equation (2.2) valid for Newtonian fluids, this can be expressed in the following manner:

$$\tau = \eta(\dot{\gamma}) \cdot \dot{\gamma} \tag{2.3}$$

or

$$\eta(\dot{\gamma}) = \frac{\tau}{\dot{\gamma}} \neq \text{const.} \tag{2.4}$$

*Note:* Many polymers exhibit more or less pronounced time-dependent viscosity (thixotropy, rheopexy, lag in viscosity at sudden onset of shear or elongation [2,3]). This time dependence is usually not considered in the design of dies; hence it will be ignored in the following sections.

#### 2.1.1.1 Viscosity and Flow Functions

When plotting the viscosity $\eta$ in dependence on the shear rate $\dot{\gamma}$ in a log-log graph, we obtain a function shown in Fig. 2.1 valid for polymers at constant temperature. It can be seen that for low shear rates the viscosity remains constant; however, with increasing shear rate at a certain point it changes linearly over a relatively broad range of shear rates in a log-log graph.

This, the reduction of viscosity with increased shear rate, is referred to as pseudoplastic or shear-thinning behavior. The constant viscosity at low shear rates is called zero-shear viscosity, $\eta_0$.

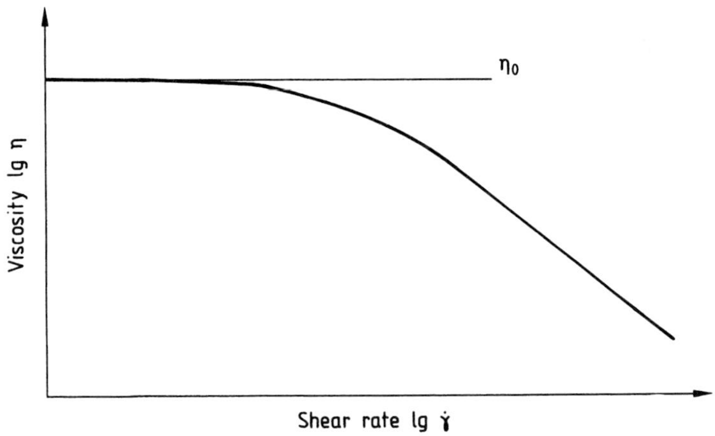


**Figure 2.1** Representation of the dependence of viscosity on the shear rate by a viscosity curve

Besides the graphic representation of viscosity vs. shear rate, the so-called *viscosity curve,* the relationship between shear stress and shear rate (also in a log-log graph) is referred to as a *flow curve* (Fig. 2.2). For a Newtonian fluid, the shear rate is directly proportional to the shear stress. A log-log graph therefore is a straight line with a slope of 1, which means that the angle between the abscissa and the flow curve is 45°. Any deviation from this slope directly indicates a non-Newtonian behavior.

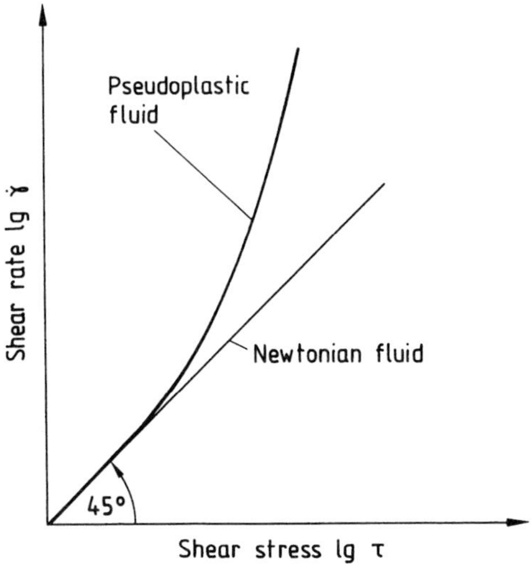


**Figure 2.2** Representation of the dependence of the shear rate on the shear stress by a flow curve

For a pseudoplastic fluid, the slope is greater than 1, meaning that the shear rate increases progressively with increasing shear stress. Conversely, the shear stress increases with the shear rate in a less-than-proportional relationship (see also Chapter 3).

#### 2.1.1.2 Mathematical Description of the Pseudoplastic Behavior of Melts

Various models describing the viscosity and flow curves were developed mathematically. They differ in the mathematical methods used on one hand and in the adaptability and hence accuracy on the other. An overview and examples are given in the literature [2,4]. The most widely used models for thermoplastics and rubbers will be discussed in the following section.

***Power Law of Ostwald and de Waele* [5,6]**

When plotting the flow curves of different polymers in a log-log graph, curves are obtained that consist of two approximately linear sections and one transition region (Fig. 2.3). In many cases we can operate in one of those two regions, so these sections of the curve can be mathematically represented in the following general form:

$$\dot{\gamma} = \phi \cdot \tau^{\mathrm{m}} \tag{2.5}$$

Equation (2.5) is called the *power law of Ostwald and de Waele*. The parameters are $m$, the flow exponent, and $\phi$, the fluidity. Characteristic for the ability of a material to flow and its deviation from Newtonian behavior is the flow exponent $m$. It can be expressed by the following relation:

$$m = \frac{\Delta \lg \dot{\gamma}}{\Delta \lg \tau} \tag{2.6}$$

Note that $m$ is also the slope of the flow curve in the given sections of the log-log diagram (Fig. 2.3).

The value of $m$ for polymeric melts lies between 1 and 6; for the range of shear rates between approximately $10^0$ and $10^4$ $s^{-1}$ applicable to the design of extrusion dies, the corresponding values of $m$ are between 2 and 4. For $m = 1$, $\phi = 1/\eta$, which is the case of a Newtonian flow.

Since

$$\eta = \frac{\tau}{\dot{\gamma}}$$

we obtain from Equation (2.5):

$$\eta = \phi^{-1} \cdot \tau^{1-m} = \phi^{-\frac{1}{m}} \cdot \dot{\gamma}^{\frac{1}{m}-1}. \tag{2.7}$$

By substituting $k = \phi^{-\frac{1}{m}}$ and $n = \frac{1}{m}$ we obtain the usual representation of the viscosity function:

$$\eta = k \cdot \dot{\gamma}^{n-1}. \tag{2.8}$$

The factor $k$ is called the consistency factor. It represents the viscosity at a shear rate of $\dot{\gamma} = 1/s$. The viscosity exponent $n$ is equal to 1 for Newtonian behavior, and its value for most polymers is between 0.2 and 0.7. It represents the slope of the viscosity curve in the observed range.

The power law is very simple mathematically: it allows an analytical treatment of almost all simple flow problems that can be solved for Newtonian fluids (see Chapter 3). The disadvantage of the power law is that when the shear rate drops to zero, the viscosity value becomes infinity, and therefore the shear-rate-independent Newtonian region cannot be depicted. Another disadvantage is that the flow exponent $m$ enters into the dimension of the fluidity.

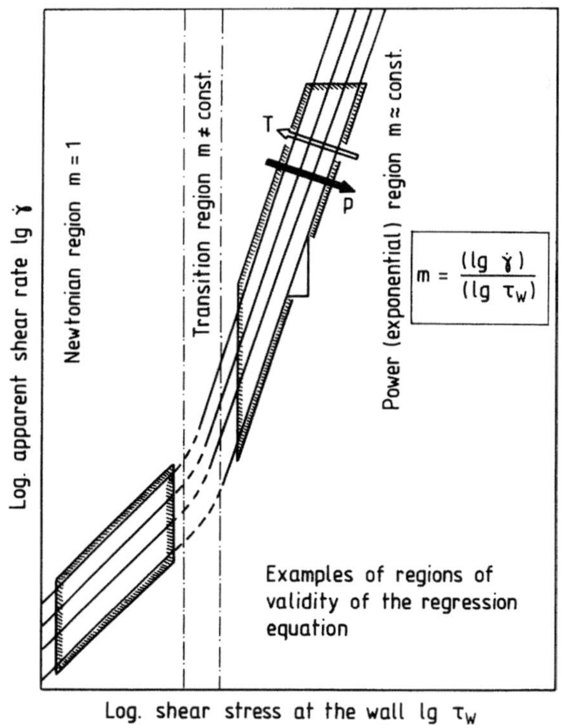


**Figure 2.3** Approximation of the flow curve by the power law

Generally, the power law can be used to represent a flow or viscosity curve with an acceptable accuracy over only a certain range of shear rates. The size of this range at a given accuracy depends on the curvature of the graph's representation of this function.

If a flow curve has to be described by the power law over a large range, it has to be divided into segments, each with its own values of $\phi$ and $m$ to be determined [7]. Therefore, in the collection of standard rheological material data [8,9], there will be different values of $\phi$ and $m$ corresponding to different ranges of shear rates.

### *Prandtl-Eyring Constitutive (sinh) Equation* [7,10–12]

This model was developed by Prandtl and Eyring from observation of the place-exchange processes of molecules during flow. It takes the following form:

$$\dot{\gamma} = C \cdot \sinh\left(\frac{\tau}{A}\right) \tag{2.9}$$

with material constants $C$ in [$s^{-1}$] and $A$ in [$N/m^2$].

The advantage of the Prandtl-Eyring model is that it describes a finite viscosity at small shear rates (zero-shear viscosity) and that it is readily applicable in dimensional analysis [13,14]. Its mathematical application is somewhat difficult, however, because of its unwieldiness.

### *Carreau Constitutive Equation* [9,15,16]

This model, which is gaining increasing importance in the design of extrusion dies, is represented by the following equation:

$$\eta(\dot{\gamma}) = \frac{A}{\left(1 + B \cdot \dot{\gamma}\right)^C} \tag{2.10}$$

where $A$ describes zero-shear viscosity in [Pa·s], $B$ the so-called reciprocal transition rate in [s], and $C$ [-] the slope of the viscosity curve in the pseudoplastic region at $\dot{\gamma} \to \infty$ (Fig. 2.4).

This model by Carreau has an advantage in that it represents the actual behavior of the material over a much broader range of shear rates than the power law, and it produces reasonable viscosity values at $\dot{\gamma} \to 0$.

In addition, it is applicable for the calculation of the correlation between pressure and throughput in a consistent analytical form for both a capillary and a slit die [9,16]. As a result, this model allows rough calculations by means of a pocket calculator. This is particularly useful when a convenient approximate calculation rather than exact analytical solution is required [9,16].

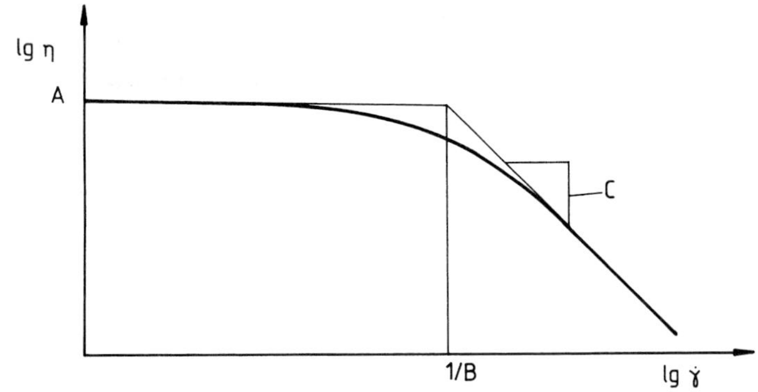


**Figure 2.4** Approximation of the viscosity curve by the Carreau constitutive equation

***Universal Viscosity Function by Vinogradov and Malkin*** **[17,18]**

Vinogradov and Malkin [17] found that, in a temperature-invariant representation (see Section 2.1.1.3), the viscosity functions of the following materials fall within the scatter range shown in Fig. 2.5: polyethylene, polypropylene, polystyrene, polyisobutylene, polyvinylbutyrate, natural rubber, butadiene-styrene rubber, as well as cellulose acetate.

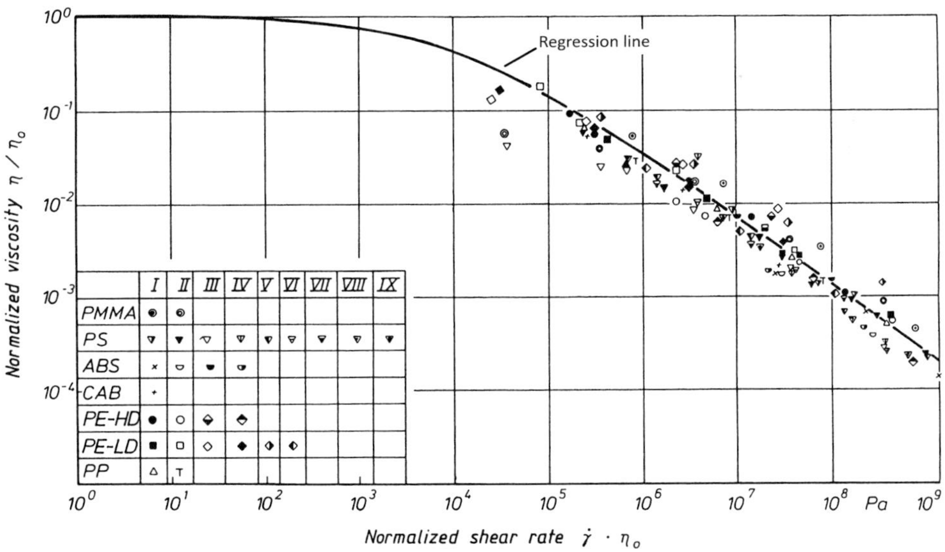


**Figure 2.5** Universal viscosity curve according to *Vinogradov* and *Malkin*

The regression line can be considered, at least for the purpose of estimation, to be apparently a universal viscosity function, independent of temperature and pressure. This function allows the estimation of the viscosity behavior over a wide range of shear rates when only one point is known, while the zero-shear viscosity is determined by iteration.

The graphic representation of this universal viscosity function is given by the following regression formula [17]:

$$\eta(\dot{\gamma}) = \frac{\eta_0}{1 + A_1 \cdot (\eta_0 \cdot \dot{\gamma})^{\alpha} + A_2 \cdot (\eta_0 \cdot \dot{\gamma})^{2\alpha}} \tag{2.11}$$

where

$\eta_0$ Zero-shear viscosity, i.e., the limiting value of viscosity for $\dot{\gamma} \to 0$

$A_1$ $1.386 \times 10^{-2}$

$A_2$ $1.462 \times 10^{-3}$

$\alpha$ 0.355

Here, $A_1$ and $A_2$ depend on the units of viscosity and shear rate. The values shown here are valid for the following units: $[\eta] = \text{Pa} \cdot \text{s}$ and $[\dot{\gamma}] = \text{s}^{-1}$.

The advantage of the universal Vinogradov function is that it only contains one free parameter, namely zero-shear viscosity $\eta_0$, which can be readily determined by the measurements of viscosity. When keeping the regression coefficients $A_1$, $A_2$, and $\alpha$ constant, the accuracy of the relation becomes limited. For $\dot{\gamma} \to 0$ the Vinogradov function approaches the limiting value, namely $\eta_0$.

In the following section, it will be shown briefly how to calculate, by a simple iteration, the zero-shear viscosity from a measured point $[\dot{\gamma}_p; \eta(\dot{\gamma}_p)]$ with shear rate $\dot{\gamma}_p$ and viscosity $\eta(\dot{\gamma}_p)$. However, the viscosity function obtained by this procedure is only an estimation, and it cannot replace the viscosity measurement in the entire relevant range of shear rates.

The deviations from the actual function are increasing with increasing distance from the known point on the curve $[\dot{\gamma}_p; \eta(\dot{\gamma}_p)]$.

First, the known values are put into Equation (2.11), which is then rearranged as follows:

$$\eta_0 = \eta(\dot{\gamma}_p) \cdot \left[1 + A_1 \cdot (\eta_0 \cdot \dot{\gamma}_p)^{\alpha} + A_2 \cdot (\eta_0 \cdot \dot{\gamma}_p)^{2\alpha}\right]. \tag{2.12}$$

Equation (2.12) contains $\eta_0$ on both sides. An explicit solution for $\eta_0$ is not possible. Therefore it is subjected to an iteration procedure. It follows

$$\eta_{0_{n+1}} = \eta(\dot{\gamma}_p) \cdot \left[1 + A_1 \cdot (\eta_{0_n} \cdot \dot{\gamma}_p)^{\alpha} + A_2 \cdot (\eta_{0_n} \cdot \dot{\gamma}_p)^{2\alpha}\right]. \tag{2.13}$$

From Equation (2.13) results, with the estimated value of $\eta_{0_n}$, an improved estimated value of the zero-shear viscosity $\eta_{0_{n+1}}$ in the $n$th iteration step. The value of $\eta_{0_{n+1}}$ is then put into the $(n + 1)$th iteration step using Equation (2.13). The following iteration process results:

Step 0: Set $\eta_{0_0}$ equal to the known value of viscosity, $\eta(\dot{\gamma}_p)$.

Step 1: Calculate the new estimated value for $\eta_0$ by putting the previous estimated value into Equation (2.13).

Step 2: Decision: If the difference of the two subsequent estimated values is small enough, the iteration is stopped. The last estimated value for $\eta_0$ is the desired result. If the difference is not small enough, return to step 1

A sufficiently accurate result is usually found after 5 to 10 iterations. The iteration pattern can be easily programmed on a pocket calculator because very few programming steps are required.

Of course, the Vinogradov model in its general form can also be used for the description of the viscosity function. In this case $A_1$, $A_2$, and $\alpha$ are free parameters, which can be determined by regression analysis. By this a more accurate approximation is possible than with the parameters of the universal function.

On the other hand, as a universal function defined by the regression line drawn through the data points (Fig. 2.5), any model that approximates the curve with a satisfactory accuracy can be used here instead of Equation (2.11).

### *Herschel-Bulkley Model* [2,12,19]

With many polymers, especially with rubbers, a so-called yield stress is observed. Such fluids start to flow only when a finite shear stress is exceeded (yield stress). Fluids exhibiting this behavior are called Bingham fluids.

The flow curve of a Bingham fluid is shown schematically in Fig. 2.6. It is clearly seen that the shear rate is equal to zero up to the yield stress, $\tau_0$, which means no flow occurs. Only beyond $\tau_0$ will there be flow. This means that the viscosity below the yield stress is infinite [2].

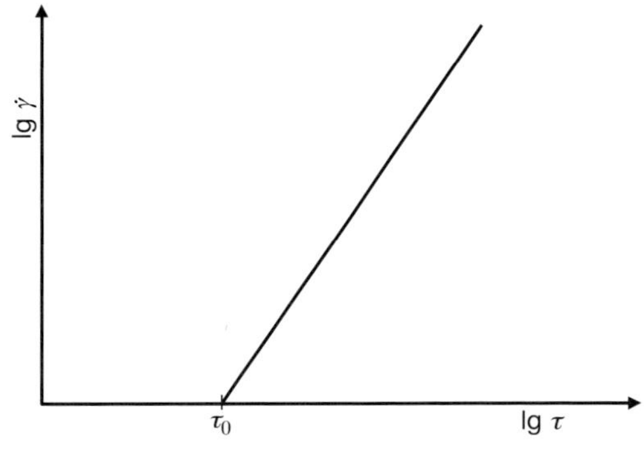


**Figure 2.6** Schematic representation of the flow curve of a Bingham fluid

In a developed flow rate profile of a Bingham fluid, there will be one range of shear flow in which the shear stress $\tau$ is larger than $\tau_0$ and another one in which $\tau$ is smaller than $\tau_0$ (Fig. 2.7 [20]). Figure 2.7 also shows that the proportion of the so-called plug flow diminishes with the increase in the ratio of the shear stress at the wall and the yield stress. Therefore, the plug-shear flow model is valid when the shear stress at the wall is low, that is, when there is a small volumetric flow rate or a large die cross section.

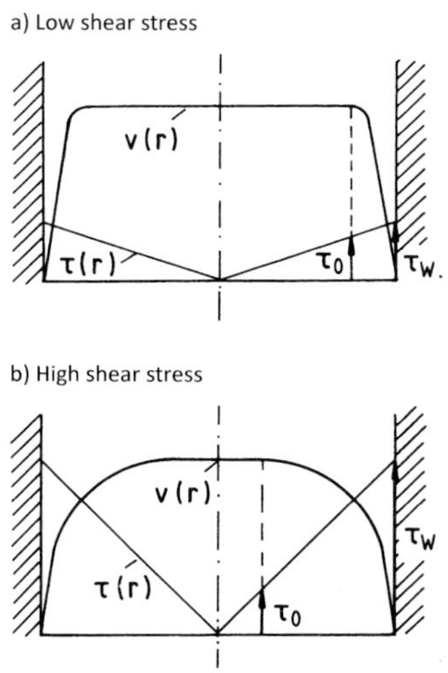


**Figure 2.7** Velocity profile of a Bingham fluid in dependence on the shear stress at the wall and the yield stress [20]: (a) low shear stress, (b) high shear stress

The Herschel-Bulkley model [19] has been successful in describing the flow behavior of polymers with a yield stress. This model results from the combination of a simplified Bingham model (with $\eta = \text{const.}$ for $\tau > \tau_0$ [2]) and the power law, yielding

$$\dot{\gamma} = \phi \cdot (\tau - \tau_0)^{m}. \tag{2.14}$$

For $\tau_0 = 0$ the relation becomes the power law (Equation (2.5)) and for $m = 1$ the simple Bingham model.

When rearranging Equation (2.14) the following expression for the shear stress is obtained:

$$\tau - \tau_0 = k \cdot \dot{\gamma}^{n-1} \cdot \dot{\gamma} \tag{2.15}$$

where

$$k = \phi^{-\frac{1}{m}} \quad \text{and} \quad n = \frac{1}{m}.$$

with

$$\eta = \frac{\tau - \tau_0}{\dot{\gamma}} \tag{2.16}$$

A relation analogous to the power law (Equation (2.8)) is derived from Equation (2.15) for $\tau > \tau_0$:

$$\eta = k \cdot \dot{\gamma}^{n-1}$$

#### 2.1.1.3 Influence of Temperature and Pressure on the Flow Behavior

Factors determining the flow of melts besides shear rate $\dot{\gamma}$ and shear stress $\tau$ for a specific polymer melt are the melt temperature $T$, the hydrostatic pressure in the melt $p_{hyd}$, the molecular weight, and the molecular weight distribution, as well as additives, such as fillers and lubricants. For a given polymer formulation, the only free variables having an effect are $\dot{\gamma}$ or $\tau$, $p_{hyd}$, and $T$.

Figure 2.8 (from [21]) illustrates a quantitative effect of the changes in temperature and pressure on the shear viscosity: an increase in pressure of approximately 550 bar for an observed sample of PMMA (polymethylmethacrylate) resulted in a tenfold increase in viscosity. Or, in order to keep viscosity constant, in this case the temperature would have to be increased by approximately 23 °C.

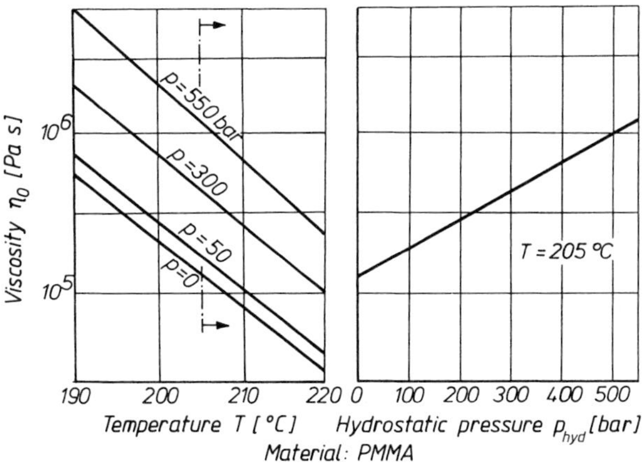


**Figure 2.8** Viscosity as a function of temperature and of hydrostatic pressure (according to [21])

Figure 2.9 [22] provides a picture of the behavior of viscosity with the change of temperature for various polymers. It can be clearly seen that semicrystalline polymers, which have a low $T_g$ when compared to amorphous polymers, exhibit a considerably lesser temperature dependence of their viscosity than the latter. This influence on the ability of polymers to flow can be essentially caused by two factors [23,24]:

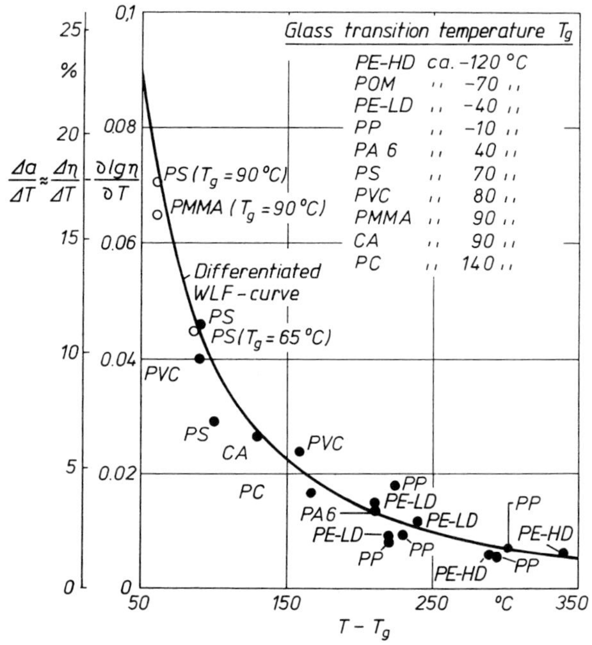


**Figure 2.9** Change in viscosity with temperature for different polymers [22]

- A thermally activated process causing the mobility of segments of a macromolecular chain (i. e., the intramolecular mobility)
- The probability that there is enough free volume between the macromolecular chains allowing their place exchange to occur

### *The Influence of Temperature*

When plotting viscosity curves at varied temperatures for identical polymer melts in a log-log graph (Fig. 2.10), the following can be established:

- First, the effect of temperature on the viscosity is considerably more pronounced at low shear rates, particularly in the range of the zero-shear viscosity, when compared to that at high shear rates.
- Second, the viscosity curves in the diagram are shifted with the temperature, but their shape remains the same.

It can be shown that for almost all polymeric melts (so-called thermorheologically simple fluids [25]) the viscosity curves can be transformed into a single master curve that is independent of temperature. This is done by dividing the viscosity by the temperature corresponding to $\eta_0$ and multiplying the shear rate with $\eta_0$ [1,2,25,26]. Graphically, this means that the curves are shifted along a straight line with a slope of −1, i.e., along a line $\log(\eta_0(T))$ to the right and simultaneously downward and thus transformed into a single curve (Fig. 2.10). This is referred to as the *time-temperature superposition principle.*

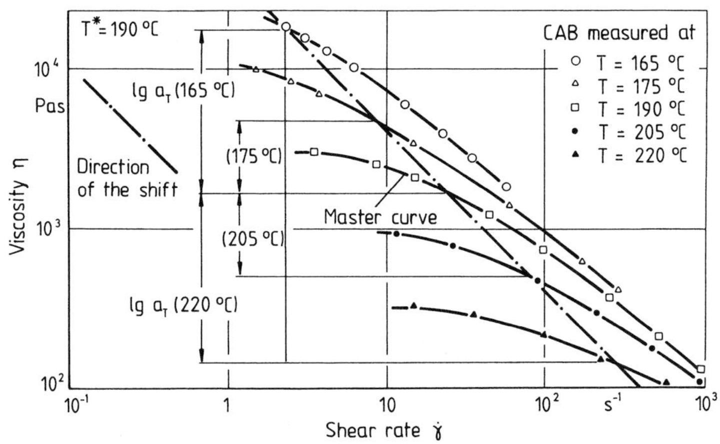


**Figure 2.10** Viscosity curves for cellulose acetate butyrate (CAB) at various temperatures

This time-temperature superposition leads to the plotting of the reduced viscosity $\eta/\eta_0$ against $\eta_0\dot{\gamma}$. In this way a single characteristic function for the polymer is obtained:

$$\frac{\eta(\dot{\gamma},T)}{\eta_0(T)} = f\left(\eta_0(T)\cdot\dot{\gamma}\right). \tag{2.17}$$

Here, $T$ as the reference temperature can be chosen freely.

When seeking the viscosity function for a certain temperature $T$ with only the master curve or the viscosity curve at a certain other temperature $T_0$ given, a temperature shift is necessary to obtain the required function. First, it is not known how much the curve has to be shifted. The shift factor $a_T$ required here can be found as follows:

$$a_T = \frac{\eta_0(T)}{\eta_0(T_0)} \quad \text{or} \quad \lg a_T = \lg\frac{\eta_0(T)}{\eta_0(T_0)}. \tag{2.18}$$

The quantity lg $a_T$ is the distance that the viscosity curve at the reference temperature $T_0$ has to be shifted in the direction of the respective axes (Fig. 2.11).

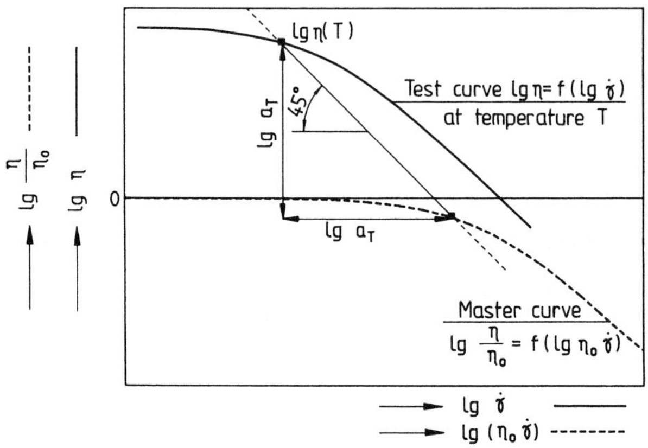


**Figure 2.11** Time-temperature superposition principle for a viscosity function

There are several formulas for the calculation of the temperature shift factor. Two of them are the most important and should be mentioned, namely the Arrhenius law and the WLF equation.

The Arrhenius law can be derived from the study of a purely thermally activated process of the interchange of places of molecules:

$$\lg a_T = \lg \frac{\eta_0(T)}{\eta_0(T_0)} = \frac{E_0}{R}\left(\frac{1}{T} - \frac{1}{T_0}\right). \tag{2.19}$$

where $E_0$ is the flow activation energy in J/mol specific for the given material, and $R$ is the universal gas constant equal to 8.314 J/(mol·K).

The Arrhenius law is suitable particularly for the description of the temperature dependence of the viscosity of semicrystalline thermoplastics [9,25].

For small temperature shifts or rough calculations, $a_T$ can be approximated from an empirical formula, which is not physically proven and which takes the following form [1,9]:

$$\lg a_T = -\alpha \cdot (T - T_0) \tag{2.20}$$

where α is the temperature coefficient of viscosity specific to the given material.

Another approach based on the free volume, i.e., the probability of the place exchange, was developed by *Williams, Landel,* and *Ferry* [27]. It was originally applied to the temperature dependence of relaxation spectra and later was applied to viscosity. The relationship (also known as the WLF equation) in its most usual form is

$$\lg a_T = \lg \frac{\eta(T)}{\eta(T_S)} = -\frac{C_1 \cdot (T - T_S)}{C_2 + (T - T_S)}, \tag{2.21}$$

which relates the viscosity $\eta(T)$ at the desired temperature $T$ to the viscosity $\eta(T_S)$ at the standard temperature $T_S$ with shear stress being constant. For $T_S$ equal approximately to $T_g + 50$ K [27] (i.e., 50 K above the glass transition temperature), $C_1 = -8.86$ and $C_2 = 101.6$ K.

The glass transition temperatures of several polymers are shown in Fig. 2.9, and additional values are in [28]. The measurement of $T_g$ of amorphous polymers can be done in accordance with DIN EN ISO 75-2, Procedure A, which is a test for the deflection temperature of plastics under load; in the USA the corresponding ASTM standard is ASTM D 648 ISO 75. The softening temperature determined by this test can be set equal to $T_g$ [7].

A more accurate description is possible when $T_S$ (and, if necessary, $C_1$ and $C_2$, which also can be considered as almost material independent) is determined from regression of the viscosity curves, measured at different temperatures. Although the WLF equation pertains by definition to amorphous polymers only and is superior to the Arrhenius law [9,24,25], it still can be used for semicrystalline polymers with an acceptable accuracy [22,29–32].

Figure 2.12 compares the determination of the shift factor $a_T$ obtained from the Arrhenius law to that from the WLF equation [30]. When operating within the temperature range ±30 K from the reference temperature, which is often sufficient for practical purposes, both relations are satisfactory.

There are basically two reasons for favoring the WLF equation, however:

- The standard temperature $T_S$ is related to the known $T_g$ for the given material with a high enough accuracy ($T_S \approx T_g + 50$ K).
- The effect of pressure on the viscosity can be easily determined when operating above the standard temperature (this will be explained further at a later point).

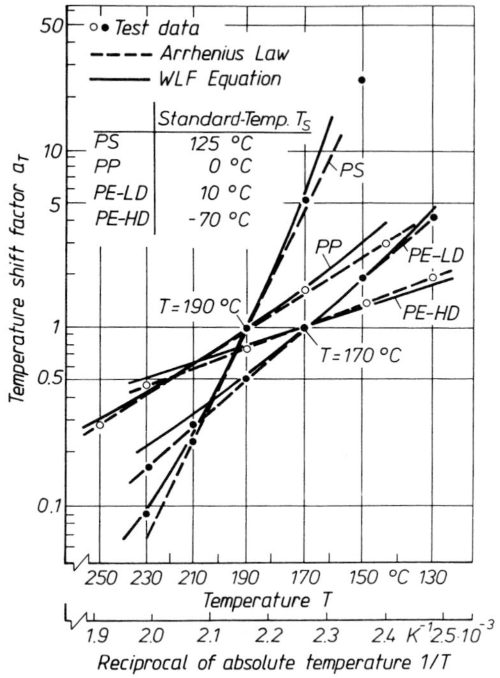


**Figure 2.12** Temperature shift factor $a_T$ for different polymers

When the shift of a viscosity curve from one arbitrary temperature $T_0$ to the desired temperature $T$ is performed using the WLF equation, Equation (2.21) is used twofold:

$$\begin{aligned}\lg a_T &= \lg \frac{\eta(T)}{\eta(T_0)} = \lg\left(\frac{\eta(T)}{\eta(T_S)}\cdot\frac{\eta(T_S)}{\eta(T_0)}\right) = \lg\left(\frac{\eta(T)}{\eta(T_S)}\right) + \lg\left(\frac{\eta(T_S)}{\eta(T_0)}\right) = \lg\left(\frac{\eta(T)}{\eta(T_S)}\right) - \lg\left(\frac{\eta(T_0)}{\eta(T_S)}\right) \\ &= \frac{C_1\left(T_0 - T_S\right)}{C_2 + \left(T_0 - T_S\right)} - \frac{C_1\left(T - T_S\right)}{C_2 + \left(T - T_S\right)},\end{aligned} \tag{2.22}$$

with $C_1 = 8.86$ and $C_2 = 101.6\ \mathrm{K}$, and $T_0$ is the reference temperature at which the viscosity is known.

### *The Influence of Pressure*

The effect of pressure on the flow behavior can be determined along with the expression for the temperature dependence from the WLF equation [29]. It turns out that the standard temperature $T_S$, which lies at approximately $T_g$ + 50 K, used in the WLF equation at 1 bar, increases with pressure. This shift corresponds in turn to the shift in $T_g$, which can be determined directly from a $p$-$\nu$-$T$ diagram [22,33].

The pressure dependence of the glass transition temperature can be assumed to be linear up to pressures of about 1000 bar (Fig. 2.13 [34]), thus:

$$T_g(p) = T_g(p = 1\,\text{bar}) + \xi \cdot p. \tag{2.23}$$

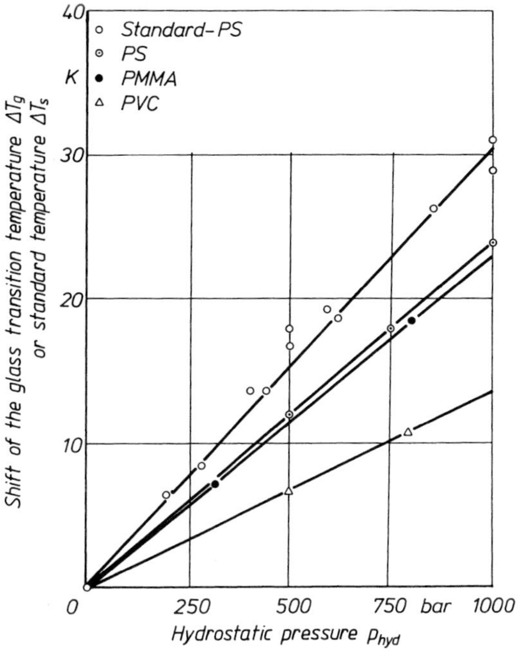


**Figure 2.13** Effect of pressure on the glass transition and the standard temperature

The resulting shifts in $T_g$ are of the order of 10 to 30 K per 1000 bar. At pressures higher than 1000 bar, the glass transition temperature increases with increasing pressure at a much smaller rate.

If there is no $p$-$\upsilon$-$T$ diagram available for the given polymer from which the pressure dependence of its $T_g$ could be determined, it can be estimated by the following relation:

$$T_g \approx T_g(1\ \text{bar}) + (10\ \text{to}\ 30) \cdot 10^{-3}\ \text{K/bar} \cdot p \tag{2.24}$$

where $p$ is the desired pressure in bar. (Generally, it is known that the pressure affects the flow properties of amorphous polymers stronger than the flow of semi-crystalline polymers.)

Then, the WLF shift from a viscosity curve at the temperature $T_1$ and the pressure $p_1$ to another one at $T_2$ and $p_2$ respectively can be performed. The shift factor $a_T$ can be calculated as follows:

$$\begin{aligned} \lg a_T &= \lg\left(\frac{\eta\,(T_2, p_2)}{\eta\,(T_1, p_1)}\right) \\ &= \frac{C_1 \cdot (T_2 - T_S(p_2))}{C_2 + (T_2 - T_S(p_2))} - \frac{C_1 \cdot (T_1 - T_S(p_1))}{C_2 + (T_1 - T_S(p_1))}. \end{aligned} \tag{2.25}$$

### Computational Application

When a point $P_1(\dot{\gamma},\eta(T_1))$ on a viscosity curve, constructed at temperature $T_1$, is shifted to a viscosity curve valid at the temperature $T_2$ (Fig. 2.11), the following applies:

$$\eta_2 = \eta_1 \cdot a_T$$

$$\dot{\gamma}_2 = \frac{\dot{\gamma}_1}{a_T}.$$

where $a_T$ is computed from one of the superposition equations. If the WLF equation is used, then the $a_T$ so determined can also contain the shift due to the pressure, provided $p_1$ and $p_2$ are not equal. Accordingly, the viscosity functions can be formulated to become independent of temperature and pressure. Thus, the Carreau law can become

$$\eta(\dot{\gamma},T,p) = \frac{a_T(T,p) \cdot A}{\left(1 + a_T(T,p) \cdot B \cdot \dot{\gamma}\right)^C}. \tag{2.26}$$

The analog for the universal viscosity function by Vinogradov and Malkin then is

$$\eta(\dot{\gamma},T,p) = \frac{a_T(T,p) \cdot \eta_0}{1 + A_1\left[a_T(T,p) \cdot \eta_0 \dot{\gamma}\right]^{\alpha} + A_2\left[a_T(T,p) \cdot \eta_0 \cdot \dot{\gamma}\right]^{2\alpha}}. \tag{2.27}$$

where

$$\left.\begin{aligned} A_1 &= 1.386 \cdot 10^{-2} \\ A_2 &= 1.462 \cdot 10^{-3} \\ \alpha &= 0.355 \end{aligned}\right\} \quad \text{for} \quad \begin{aligned} [\eta] &= \text{Pa s} \\ [\dot{\gamma}] &= \text{s}^{-1} \end{aligned}$$

As the Vinogradov-Malkin function is virtually independent of material [17], Equation (2.27) represents a universal material-, temperature-, and pressure-invariant viscosity function.

### 2.1.2 Determination of Viscous Flow Behavior

For the measurement of the flow function $\dot{\gamma} = f(\tau)$ or the viscosity function $\eta = f(\dot{\gamma},T)$ within the range of shear rates relevant to the design of extrusion dies (from 10 $s^{-1}$ to $10^4$ $s^{-1}$), the capillary viscometer (or a laboratory extruder provided with a viscometer nozzle and the proper instrumentation for pressure and temperature measurement) is suitable. The cross section of the capillary die can be circular, annular, or in slit form.

The principle of measurement is to determine the pressure loss $\Delta p$ in a capillary with an exactly defined and constant geometry and a known volumetric flow rate $\dot{V}$ at a constant melt temperature.

If the capillary has a circular cross section, the pressure gauge is not placed in the capillary because of its small diameter (mostly 1–3 mm) but in a much larger cylinder. The second pressure measuring point is at the capillary exit (ambient pressure) (Fig. 2.14).

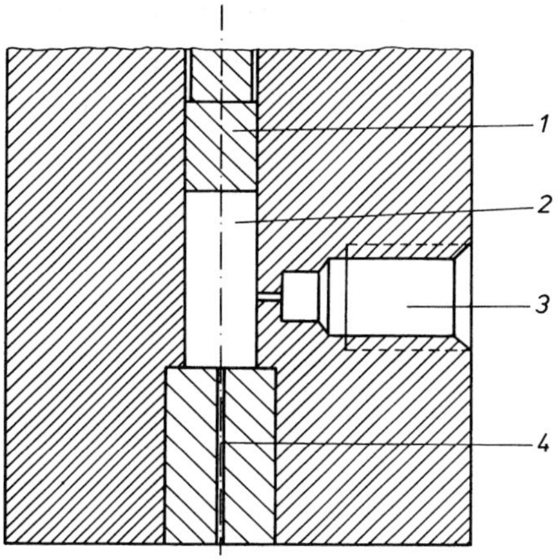


**Figure 2.14** Principle for the capillary viscometer: (1) piston (ram), (2) cylinder, (3) pressure transducer well, (4) capillary

Assuming a laminar, steady-state isothermal flow with wall adhesion, the shear stress at the wall of a round capillary with the radius $R$ is

$$\tau_W = \frac{\Delta p \cdot R}{2 \cdot L}, \tag{2.28}$$

where $L$ is the length of the capillary and $\Delta p$ the pressure loss.

With the assumption of Newtonian flow, the shear rate at the wall $\dot{\gamma}_W$ is

$$\dot{\gamma}_W = \frac{4\dot{V}}{\pi R^3}. \tag{2.29}$$

where $\dot{V}$ is the known volumetric flow rate (note that Equations (2.28) and (2.29) are valid only for a round capillary, but similar relations are arrived at for annular and slit-shaped capillary dies; see Chapter 3). From the definition of viscosity, we get

$$\eta = \frac{\tau_W}{\dot{\gamma}_W}. \tag{2.30}$$

From Equations (2.28), (2.29), and (2.30) a point $(\dot{\gamma}, \eta)$ on the viscosity curve can be determined from the known geometric parameters $R$ and $L$, the given volumetric flow rate $\dot{V}$, and the measured pressure drop $\Delta p$. By variation of $\dot{V}$ and with it the corresponding $\dot{\gamma}$, the viscosity can be determined over a broad range of shear rates.

### Correction for Pseudoplastic Flow

In the above considerations it was assumed that the melt behaves like a Newtonian fluid. Since this is not often the case with polymers, the resulting viscosity is not the true viscosity but the so-called *apparent* viscosity (e.g., [2]).

While Equation (2.28) is valid for both Newtonian and pseudoplastic flow (see Chapter 3), Equation (2.29) is applicable for the shear rate in the Newtonian region only. Hence for non-Newtonian materials the so-called apparent shear rate $D_S$ is obtained, which is equal to the shear rate at the wall for a Newtonian fluid.

To obtain a *true viscosity curve,* two correction methods may be used.

When applying the *Weissenberg-Rabinowitsch correction* [1,2,35], the true shear rate is determined from the differentiation of the apparent flow curve. The relation for the true shear rate $\dot{\gamma}$ is

$$\dot{\gamma} = D_S \frac{3+s}{4} \tag{2.31}$$

where $s$ is the slope of the log-log graph for the apparent flow curve:

$$s = \frac{d\left(\lg D_S\right)}{d\left(\lg \tau_W\right)}. \tag{2.32}$$

Subsequently, the division of the values of shear stress obtained from Equation (2.28) by the values of the corresponding shear rates yields the values of the true viscosities. This method is quite time consuming because it involves a point-by-point graphic differentiation of the flow curve.

The above shortcoming was eliminated by *Chmiel* and *Schümmer* [36] for the first time and described as the *concept of the representative viscosity* [30,36–38]. It requires considerably less effort and still is equally accurate.

The thinking behind this correction method is that, during a laminar and isothermal flow under pressure as it occurs in a channel of a rheometer or in the extrusion die (assuming wall adhesion and neglecting elastic effects), there will be a point in the flow channel at which the shear rates of both the Newtonian and the pseudoplastic materials are equal (assuming an equal volumetric flow rate $\dot{V}$). Once the shear stress and the shear rate of a Newtonian fluid are determined, the true viscosity of corresponding representative shear rate and shear stress result

(Fig. 2.15). The position of the point of intersection and hence of the representative radius in the pipe depends in principle on the flow properties of the material [38]. However, for most polymeric melts it can be considered constant. The representative distance $e$ from the center of the channel in a circular pipe is [37]

$$e_{\bigcirc} = \frac{r_S}{R} = \frac{\pi}{4} \approx 0.785$$

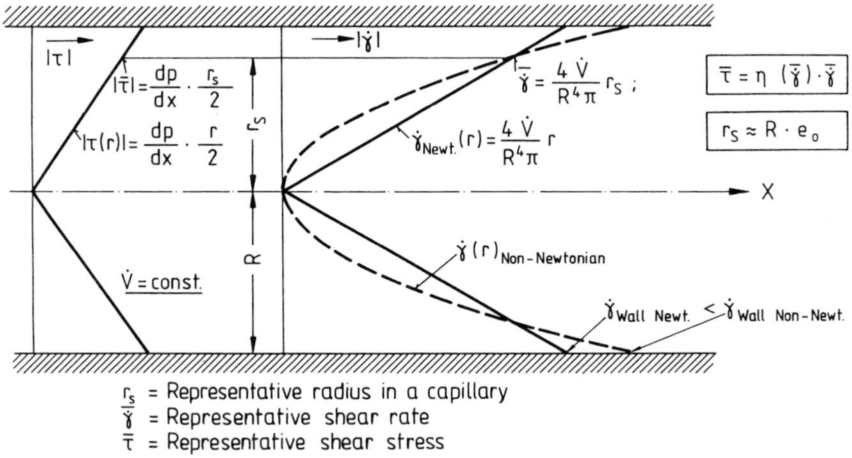


**Figure 2.15** Determination of the representative shear rate in a circular capillary

and for a rectangular slit

$$e_{\square} = \frac{h_S}{H/2} = \frac{\pi\sqrt{\frac{3}{2}}}{6} \approx 0.64$$

The representative distances from the center generated from the application of the *Ostwald-de* Waele Power Law are published in [38].

As Wortberg shows [30], it is obvious that, for the range of flow exponents $m$ between 2 to 4, corresponding to practical applications, the corresponding representative distances $e_{\bigcirc}$ and $e_{\square}$ can be considered constant. The respective errors associated with this assumption are less than 1.8% and less than 2.5%.

The true shear rate at the point of intersection then can be calculated from the apparent shear rate $D_S$ simply as

$$\dot{\gamma}_{rep} = D_S \cdot e \tag{2.33}$$

where the values for $e$ can be obtained from Fig 2.16, or the calculation can be done by the mean values contained therein.

The true viscosity (which is equal to the representative viscosity) can be calculated from the following equation:

$$\eta_{\text{rep}} = \frac{\tau_W}{D_S} = \frac{\tau_W \cdot e}{\dot{\gamma}} \tag{2.34}$$

which is valid for the representative shear rate $\dot{\gamma}_{\text{rep}}$.

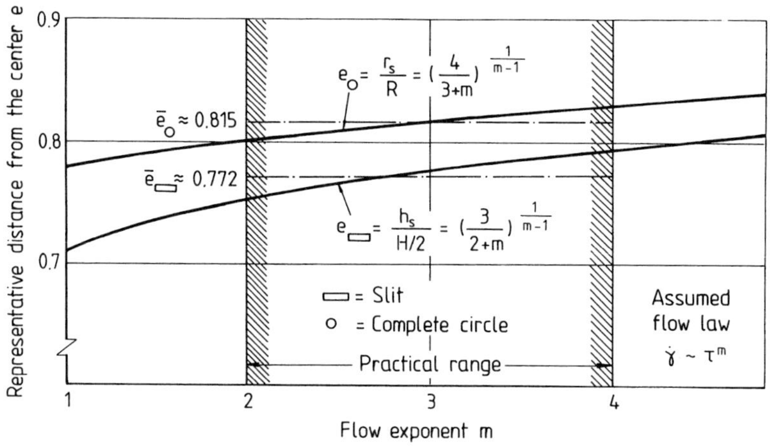


**Figure 2.16** Representative distance from the center of the channel as a function of the flow exponent (power-law exponent)

When plotting the viscosity function in a log-log graph as usual, the correction is the result of a shift in the abscissa by the value of log *e* (Fig. 2.17).

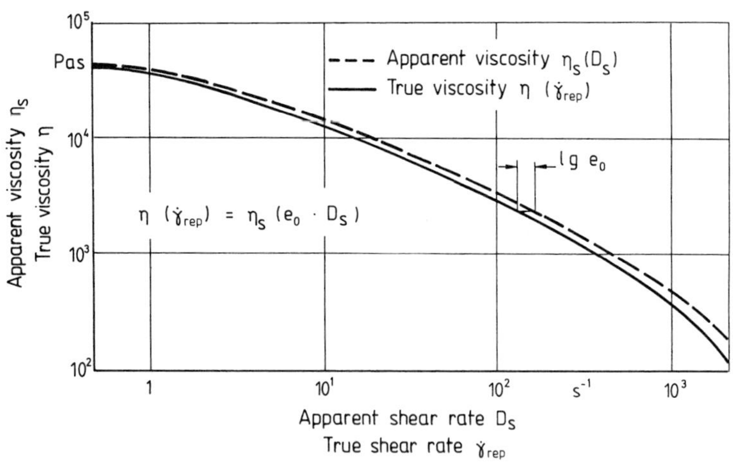


**Figure 2.17** Correction of the viscosity function according to the concept of the representative viscosity

### *Consideration of the Entry Pressure Loss*

The pressure drop as measured by the capillary viscometer is often obtained not as the difference between two pressure transducers in the capillaries, but as the pressure in front of them (Fig. 2.14). This way the pressure drop is not resulting only from the friction of the fluid in the capillary (which is the only correct quantity to be used in the appropriate equations for the calculation of viscosity), but it also includes the so-called *entry pressure drop,* which is the result of the elastic deformation of the melt at the entry to the capillary. In the area of contraction of the flow channel, the melt particles are stretched (elongated) in the direction of the flow. This deformation is transported through the capillary in the form of elastically stored energy and then partly released at the exit by the swelling of the extrudate (see Section 2.1.3). This elastically stored energy has to be first generated by an additional pressure drop. This pressure drop is picked up by the pressure gauge in addition to the shear flow pressure drop and therefore must be eliminated from the result. This is done by the so-called *Bagley correction* [2,39].

The Bagley correction [39] is based on the following two assumptions:

1. The entry pressure drop occurs only at the entry to the capillary and is therefore independent of the length of the capillary.
2. The pressure in the capillary drops because of viscous flow, and the pressure gradient is constant.

When determining the pressure drop in capillaries of different $L/D$ ratios (usually the diameter, $D$, is kept constant and the length, $L$, is varied) and at a constant shear rate, a $\Delta p$ - $L/D$ diagram is obtained (Fig. 2.18). The points in this diagram lie on a straight line. The slope of this line corresponds to the pressure gradient in the capillaries. When changing the shear rate, a multitude of straight lines is obtained with the shear rate as parameter.

When extrapolating from this line to the zero length of the capillary ($L/D = 0$), the value of the entry pressure loss, independent of the capillary length, is found as the intersection of the line with the ordinate. This entry pressure drop increases with the increase of the shear rate, the same as for the pressure gradient in the capillary. For the calculation of the viscosity curve, the difference between the measured pressure $\Delta p_{\mathrm{m}}$ and the entry pressure drop $\Delta p_{\mathrm{e}}$ must be obtained simply as

$$\Delta p_{\mathrm{v}} = \Delta p_{\mathrm{m}} - \Delta p_{\mathrm{e}}. \tag{2.35}$$

Only this difference can be used in the Equation (2.28) for the shear stress.

Note that $\Delta p_{\mathrm{e}}$ is dependent not only on the shear rate but also on the type of material: the higher the elasticity of the melt, the higher the entry pressure drop.

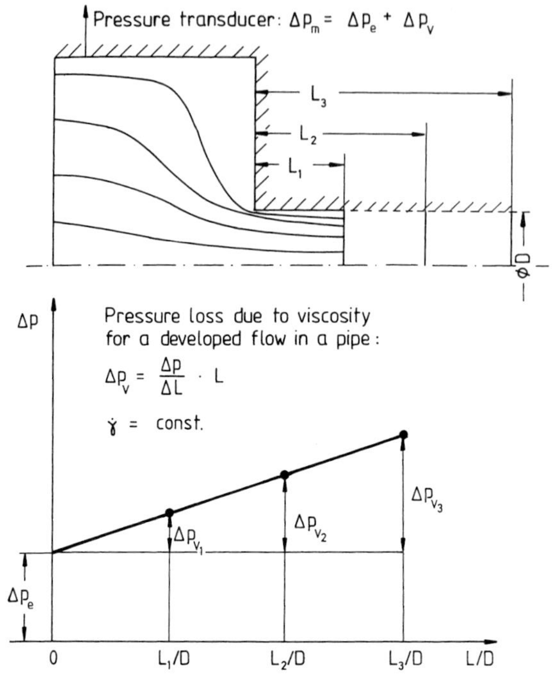


**Figure 2.18**
Principle of the Bagley correction

### 2.1.3 Viscoelastic Properties of Melts

Polymeric melts exhibit not only viscous but also elastic behavior (viscoelasticity). This manifests itself in the behavior of the melt when entering an extrusion die with a pronounced variation in cross section, for example. The result of that is the entry pressure drop. This happens because the melt is deformed in the transitions between the different cross sections, and part of this deformation is stored elastically. At the outlet of the die, these elastic deformations are released, and the emerging extrudate swells (the so-called die swell). The released elastic energy, generated by the processing machine (e.g., extruder), is lost completely for the system.

This ability of reversible deformation is caused by the so-called entropic elasticity of polymer melts. This means that the initially disorderly entangled macromolecules (the state of the most possible disorder equals the state of the largest possible entropy) can be oriented to a large degree along the direction of the deformation. A state of increased order, which is a state of lower entropy, is formed. However, as soon as this happens, the oriented network will strive to move back toward the state of the most possible disorder (following the second law of thermodynamics). The material adapts to the imposed state of deformation; however, some rearrangements of the macromolecules occur gradually with time (so-called relaxation).

Since this reduction in orientation is relatively slow, it is not finished when the material emerges from the die. As a result, further relaxation occurs after the material leaves the die. The time required for the oriented state to relax shortens with increased temperature because of the increased mobility of the molecules and free volume with increased temperature.

Since the melt seemingly remembers the deformations to which it was subjected in the die, we refer to this as the *memory effect* or *memory fluid.*

Another typical characteristic of viscoelastic fluids is the occurrence of differences in the normal stress during the flow on curved paths as well as a phase shift in periodic deformation, which manifests itself as a shift in the maxima of shear stress and shear rate. These maxima do not coincide [e.g., 2–4,12,25,26]. Both these effects are used for the measurement of viscoelastic behavior and the characterization of polymers.

When a polymeric melt flows through an extrusion die, it is exposed to two types of deformation (Fig. 2.19 [30]):

- Deformation due to elongation or compression in divergent or convergent sections of the channel, respectively (the extrudate can also be subjected to a free stretching outside the die)
- Deformation due to shear generated by the velocity profiles in the flow channel

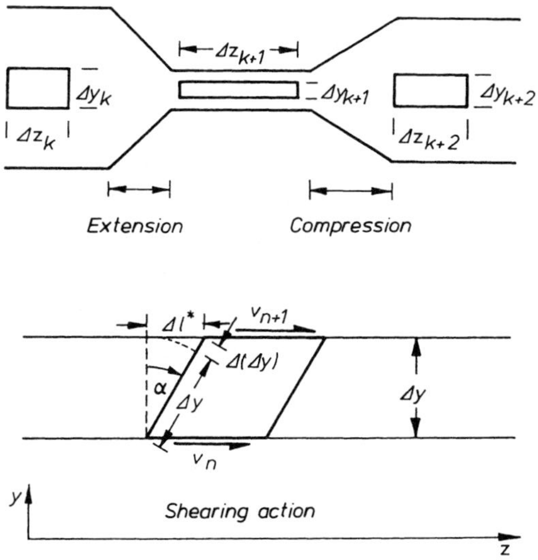


**Figure 2.19** Tensile and shear deformations due to velocity differences in the flow and shear directions [30]

In the first case, the elongation rate is defined as

$$\dot{\varepsilon}_D = \frac{dv}{dz}. \tag{2.36}$$

Fig. 2.19 shows that a volume unit is deformed by an angle $\alpha$ per a unit of time due to the differences in velocity existing on its edges. With continuing deformation $\alpha$ approaches 90° while $\Delta l^*$ per unit of time moves toward $\Delta(\Delta y)$. That means that a shear deformation, too, can be seen as a way to orient the macromolecular chains in the direction of the flow. This is described by the following:

$$\dot{\varepsilon}_{\mathrm{E}} = \frac{\Delta l^*}{\Delta t} = \dot{\gamma}. \tag{2.37}$$

This offers a possibility to measure the reversible deformation in a uniaxial extension experiment, using an extension rheometer. With such an instrument, a polymeric strand is, for example, stretched at a defined temperature and kept in this stretched state for different periods of time. Subsequently, it is allowed to shrink freely [30,31]. Reversible deformation can be defined by applying the definition of natural extension according to Henky:

$$\varepsilon_{\mathrm{R}}(t) = \ln\left(\frac{l(t)}{l_\infty}\right) = \ln\left(\lambda_{\mathrm{R}}(t)\right) \tag{2.38}$$

where

- $\lambda$ Extension ratio
- $l\,(t)$ Sample length at time $t$
- $l_\infty$ Sample length after a complete shrinkage

The relation is in principle depicted in Fig. 2.20. The decay of the reversible deformation is also defined as relaxation. It is assumed that the total deformation is completely reversible immediately after the deformation took place. The viscous portion of the deformation develops subsequently as a function of time through the decay of the elastic component of the deformation, in other words, by relaxation [30,31]. This time-dependent decay of reversible deformation is essentially a function of the material, temperature, and the level of deformation initially present [30].

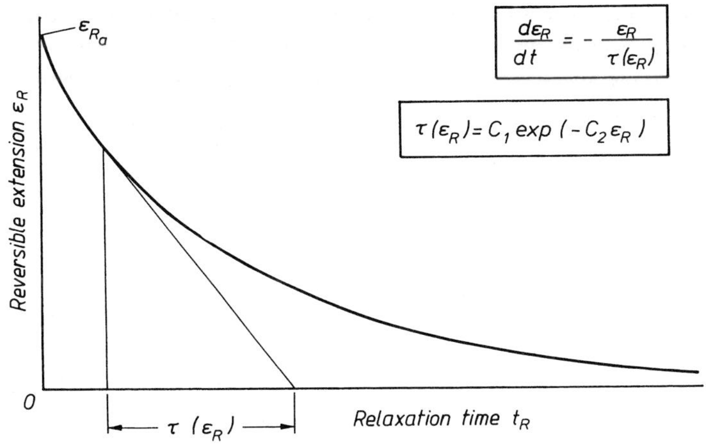


**Figure 2.20** Reversible extension as a function of the relaxation time at constant temperature

This process can also be well explained by a spring and dashpot connected in series to characterize the elastic and viscous properties of a melt (Fig. 2.21). A melt filament (spring and dashpot) is stretched at time $t = 0$ s by the amount $\Delta l$. While the external dimensions are maintained, relaxation processes take place in the melt and are symbolized by the extension of the dashpot. At a very long time ($t \to \infty$), the spring once again attains the initial length; that is, the energy introduced, which is stored fully elastically at $t = 0$ s, has been converted totally into viscous, that is, irreversible deformation in the dashpot. This process and the shape of the curve shown in Fig. 2.20 have the characteristic features of an exponential function, which can be expressed by the following equation [30,31]:

$$\frac{d\varepsilon_R}{dt} = -\frac{\varepsilon_R}{\tau(\varepsilon_R)}. \tag{2.39}$$

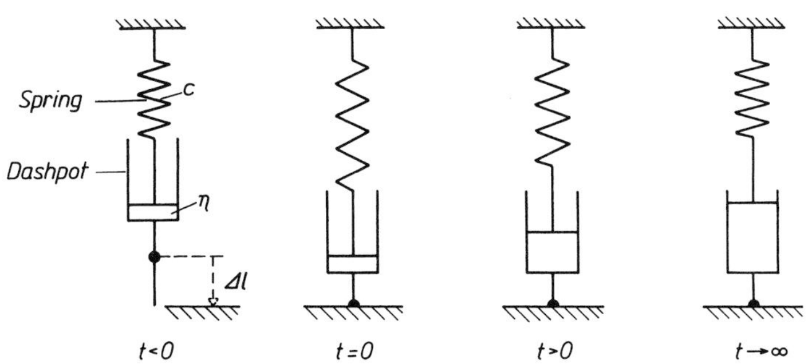


**Figure 2.21** The model representing the decay of elastic deformations (spring and dashpot model)

In this relationship, $\tau(\varepsilon_R)$ is the relaxation time, which is a function of the momentary state of deformation $\varepsilon_R$.

Measurements have shown that this characteristic relaxation time $\tau(\varepsilon_R)$ can be approximated very well for various thermoplastics at different temperatures and initial deformation $\varepsilon_{R,in}$ by the relationship

$$\tau\left(\varepsilon_R\right)=C_1\cdot\exp\left(-C_2\cdot\varepsilon_R\right). \tag{2.40}$$

where $C_1$ and $C_2$ are material-specific constants [30,31].

Figure 2.22 shows a normalized relaxation curve; this distribution of the characteristic relaxation time $\tau(\varepsilon_R)$ derived from this curve is shown in Fig. 2.23.

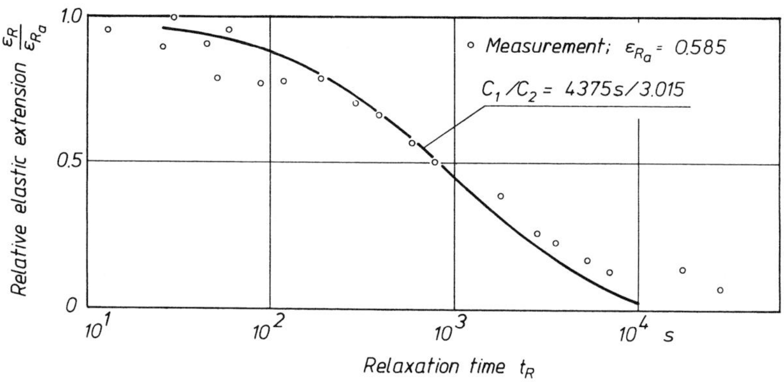


**Figure 2.22** Relaxation of polystyrene (PS) at 130 °C

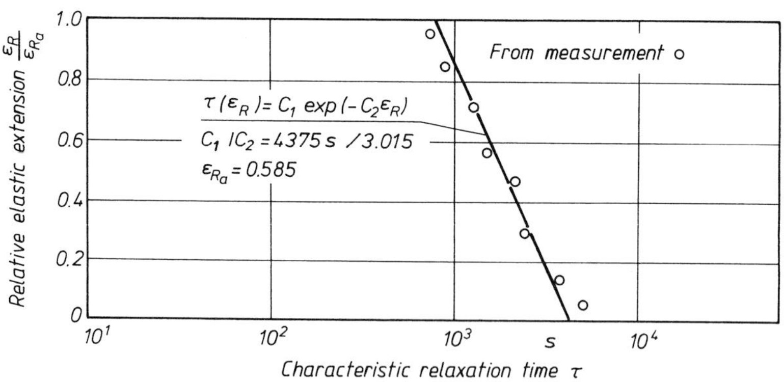


**Figure 2.23** Characteristic relaxation time as a function of the relative elastic extension for PS at 130 °C

The quantity $C_1$ contains the temperature effect, and, if known for a particular temperature, it can be converted to other temperatures in the melt range by a simple temperature-shift law, such as the WLF equation (since relaxation like flow is based on molecular site exchange processes) [30]):

$$\tau\left(\varepsilon_{\mathrm{R}}, T\right)=C_1\left(T^*\right) \cdot a_{\mathrm{T}}(T) \cdot \exp\left(-C_2 \varepsilon_{\mathrm{R}}\right), \qquad (2.41)$$

where $a_T$ is the shift factor for the selected time-temperature superposition equation, which can be calculated from one of the following equations: Equations (2.19) through (2.22).

Simultaneously with the buildup of reversible deformations, as a result of shearing or elongation, relaxation phenomena exist in a polymeric material. A simple form of describing this buildup of reversible deformations is given in [30,31], with this buildup being treated as a superposition of relaxing deformations already present onto the new deformations being added at the same time (Fig. 2.24). During the interval of time $\Delta t_i$ under consideration, the external shape of a melt particle is retained. The initially present, reversible deformation decreases during this time through relaxation from the value $\varepsilon_{Ri-1}$ to $\varepsilon^*_{Ri-1}$. At the end of the time interval $\Delta t_i$ the additional deformation $\Delta\varepsilon_i = \dot{\varepsilon}\Delta t_{i+1}$ is reached through a step change and is therefore completely reversible. The relationships describing this concept are shown in Fig. 2.24.

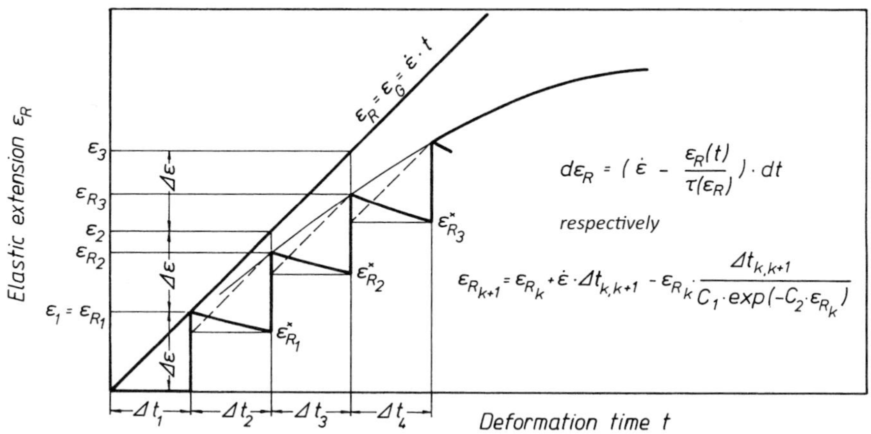


**Figure 2.24** Model of the deformation buildup at a constant rate of deformation

Experiments with the extension rheometer show that the buildup of reversible deformations can be approximated well by this description, provided that the step width $\Delta t_i$ is sufficiently small. The model can be used not only for deformation at various temperatures, but also for any other types of deformations. In this connection, an appropriate step width $\Delta t_i$ should be selected, and the effect of temperature on the relaxation behavior should be taken into consideration through the quantity $C_1$.

The model developed by Wortberg and Junk [30,31] was presented above because of its clarity and relatively simple application. There are many other models dealing with this subject. Some of them are described in the referenced literature [4,12,18,40,41].

All models have their advantages and disadvantages, which become particularly clear when they are applied in computer programs used for the computation of flow. At present, there is no model that describes the behavior under different deformations (shear, uniaxial, or multiaxial extension, as well as any superposition of the above) for a great multitude of materials accurately enough. For that reason, all conditions under which the material models are developed or the material parameters are determined must be known so that the analysis is done within the limitations of the given method.

## 2.2 Thermodynamic Behavior

In order to compute heat transfer in the extrusion die (nonisothermal flow), in cooling (for example in a calibration line), or in heating in a cross-linking line or to determine the necessary heating and cooling capacities to be installed for an extrusion die, thermodynamic data are required. These are both pressure and temperature dependent.

The dependence on pressure can be generally neglected in the view of relatively low pressure in the extrusion dies (usually less than 300 bar). Whether the material data can be taken as constant depends on the temperature range specified by the system under consideration. If the temperature does not pass through a transition and stays, for example, only in the melt range, as is the case in an extrusion die, some material values can be taken to be constant. This is frequently possible above the glass transition temperature ($T_g$) for amorphous polymers and above the crystalline melting temperature ($T_m$) for semicrystalline polymers. If the simplification of using constant material data is not permissible, temperature-dependent values must be employed. This can be done with one of the known polynomial equations or with a table of data for the numeric, stepwise solution of the problem at hand. Detailed explanations of thermodynamic material data of thermoplastics as well as of their measurements are in [42]. A standard work pertaining to thermodynamic material data for thermoplastics is [43].

### 2.2.1 Density

The densities of polymers lie considerably below those of metals. For unfilled polymers they are around 1 g/cm³ at room temperature and normal pressure. The value of the density is dependent on temperature and pressure.

The reciprocal of density $\rho$ is the specific volume $v$:

$$v = \frac{1}{\rho}. \tag{2.42}$$

Figure 2.25 depicts the dependence of the specific volume on the temperature at a pressure $p = 1$ bar [44]. For semicrystalline polymers, there is a steeper slope of the specific volume in the area around the crystalline melting temperature ($T_m$) because the crystalline areas are highly packed and melting enables the expansion of the macromolecules.

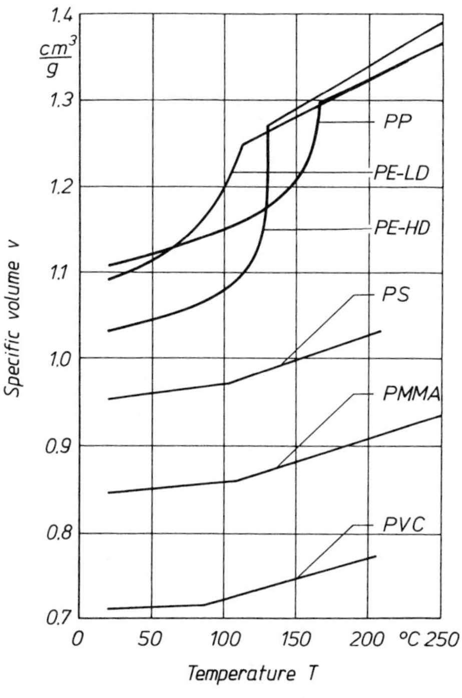


**Figure 2.25** Specific volume as a function of temperature

The correlation between specific volume, temperature, and pressure is frequently represented by a $p$-$v$-$T$ diagram. An example of this for a semicrystalline polymer is shown in Fig. 2.26.

For the computation of flow in extrusion dies, the melt is frequently considered to be incompressible. This is acceptable because the pressures are relatively low and the temperature variation within the melt is relatively small.

A collection of *p-v-T* diagrams is given in [43].

For the conversion of density from a reference temperature $T_0$ to a different temperature $T$, the following simple relation is used:

$$\rho(T) = \rho(T_0) \cdot \frac{1}{1+\alpha(T-T_0)} \tag{2.43}$$

where

$\alpha$ Linear coefficient of thermal expansion

$\rho(T_0)$ Density at the reference temperature $T_0$

$\rho(T)$ Density at temperature $T$

The range of validity is limited to the linear sections only: under or over $T_g$ for amorphous polymers or under or over $T_m$ for semicrystalline polymers.

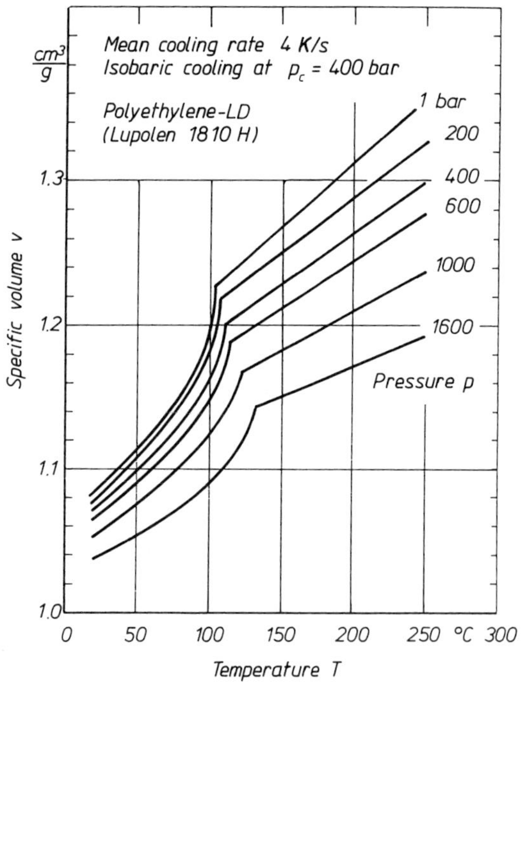


**Figure 2.26** *p-ν-T* diagram of PE-LD

### 2.2.2 Thermal Conductivity

The thermal conductivity λ of plastics is very low. Its values are around 0.12 W/(mK) and thus two to three orders of magnitude less than those of metals.

As shown in [45] the thermal conductivity increases with pressure. The increase is approximately 5% at pressures usual during the extrusion (less than 300 bar), but it is mostly hidden below the anisotropy occurring due to the molecular orientation resulting from flow [46,47]. Because of that, the dependence of thermal conductivity on pressure is frequently neglected.

The thermal conductivity of semicrystalline thermoplatics particularly is dependent on temperature (Fig. 2.27 [48]). It is higher than that of amorphous polymers and drops to approximately their level when in the molten state [42]. It turns out that generally the thermal conductivity of molten polymers can be considered constant with a good accuracy.

Diagrams depicting the temperature dependence of thermal conductivity for numerous thermoplastics are in [43].

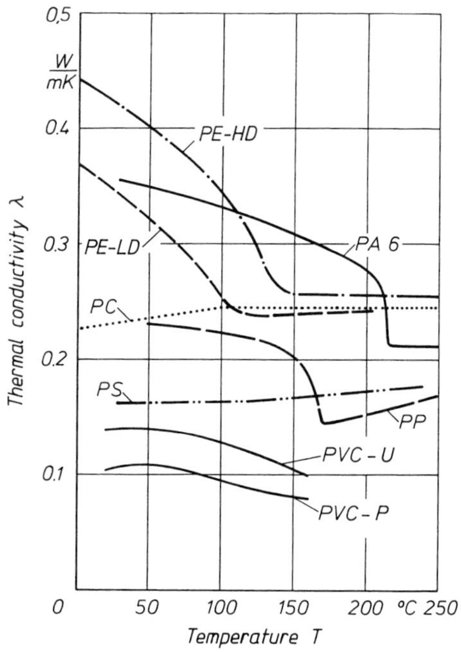


**Figure 2.27** Thermal conductivity as a function of temperature for different thermoplastics [48]

### 2.2.3 Specific Heat Capacity

The specific heat capacity $c_p$, often called the "specific heat," is the amount of heat necessary to increase the temperature of 1 g of substance by 1 K at a constant pressure [43]. From this definition, it is clear that $c_p$ is relevant to only nonstationary processes such as heating or cooling. The values of $c_p$ for plastics at room temperature are around 1.5 J/(g·K) and hence approximately three times higher than that for steel, but only slightly over one-third of that for water ($c_p$ for water at 20 °C is 4.18 J/(g·K)). The pressure dependence of $c_p$ can be neglected [7].

For amorphous polymers, $c_p$ below and above the glass transition temperature increases almost linearly with temperature (Fig. 2.28) with a step change at $T_g$ [43].

For semicrystalline polymers, this step is less pronounced or missing altogether; however, at the crystalline melt temperature ($T_m$) there is a steep maximum (Fig. 2.28) that results from the heat required for the melting of the crystalline portion of the polymer [42].

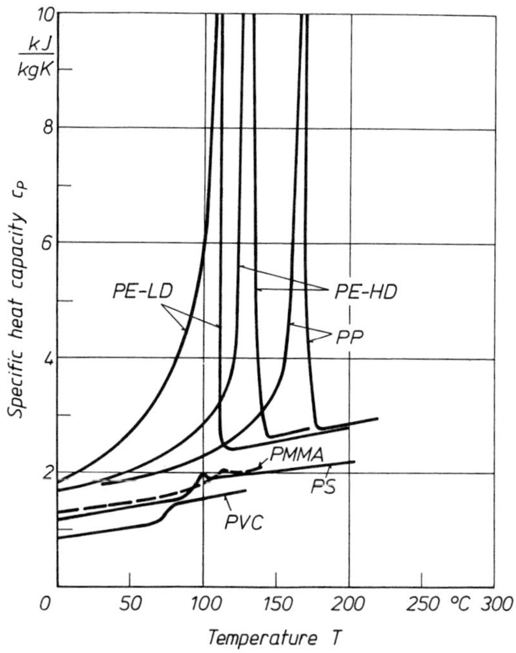


**Figure 2.28** Specific heat capacity as a function of temperature for different thermoplastics

### 2.2.4 Thermal Diffusivity

Thermal diffusivity $a$ is defined as

$$a = \frac{\lambda}{\rho \cdot c_p}. \tag{2.44}$$

It is depicted for several polymers in Fig. 2.29. Since it is calculated from heat conductivity, density, and specific heat capacity, its temperature and pressure dependence is based on that of the above-mentioned components.

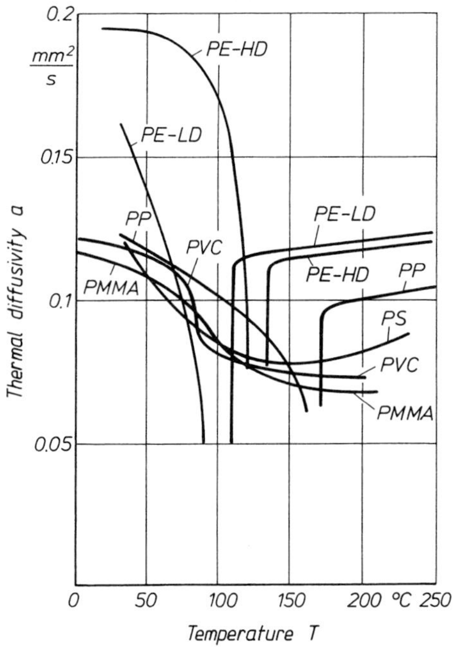


**Figure 2.29** Thermal diffusivity as a function of temperature for different thermoplastics

### 2.2.5 Specific Enthalpy

Specific enthalpy $h$ is defined by the following equation:

$$\Delta h = \int_{T_1}^{T_2} c_p(T)\,\mathrm{d}T. \tag{2.45}$$

It results from the integration of the expressions for $c_p$ in the temperature range $T_1$ to $T_2$. Here, $\Delta h$ is the difference of heat content of a substance at the temperature $T_1$ and its heat content at some reference temperature (usually 0 °C or 20 °C).

Enthalpy is necessary for the computation of required capacities for heating or cooling of polymers: $\dot{Q} = \dot{m} \cdot \Delta h \left(\dot{Q} = \text{heating or cooling power}, \dot{m} = \text{mass flow}\right)$.

Figure 2.30 depicts specific enthalpies of a variety of polymers. Additional graphs are found in [43].

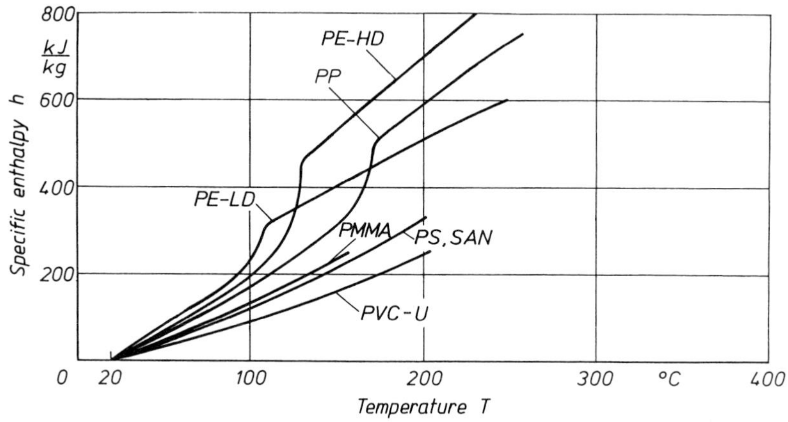


**Figure 2.30** Specific enthalpy as a function of temperature for different thermoplastics

## Symbols and Abbreviations

| | |
|---|---|
| $A$ | Material constant in the Prandtl-Eyring model |
| $A$ | Material constant in the Carreau model (zero viscosity) |
| $A_1$, $A_2$ | Coefficients of the Vinogradov-Malkin function |
| $a$ | Thermal diffusivity |
| $a_T$ | Temperature shift factor |
| $B$ | Material constant in the Carreau model (reciprocal transition shear rate) |
| $C$ | Material constant in the Prandtl-Eyring model |
| $C$ | Material constant in the Carreau model (slope of the viscosity curve) |
| $C_1$, $C_2$ | (a) coefficients of the WLF equation<br>(b) constants in Equation (2.40) |
| $c$ | Spring constant |
| $c_p$ | Specific heat capacity |
| $D$ | Diameter |
| $D_s$ | Apparent shear rate |
| $E_0$ | Flow activation energy |
| $e$ | Representative mean distance |
| $e_\bigcirc$ | Representative mean distance in a circular channel |
| $e_\square$ | Representative mean distance in a rectangular slit |

| | |
|---|---|
| $H$ | Height |
| $h$ | Specific enthalpy |
| $h_s$ | Representative height in a rectangular slit |
| $k$ | Consistency factor |
| $L$ | Length (length of a capillary) |
| $l(t)$ | Specimen length at time $t$ |
| $l_\infty$ | Specimen length after a complete shrinkage |
| $m$ | Flow exponent |
| $\dot{m}$ | Mass throughput |
| $n$ | Viscosity exponent |
| $p$ | Pressure |
| $p_{hyd}$ | Hydrostatic pressure in the melt |
| $p_m$ | Measured pressure |
| $\Delta p$ | Pressure loss |
| $\Delta p_E$ | Entry pressure loss |
| $\Delta p_V$ | Viscous pressure drop in a capillary |
| $\dot{Q}$ | Heating (or cooling) power |
| $R$ | Universal gas constant |
| $R$ | Radius |
| $r_s$ | Representative radius in a circular channel |
| $s$ | Slope |
| $T$ | Temperature |
| $T_0$ | Reference temperature |
| $T_g$ | Glass transition temperature |
| $T_s$ | Standard temperature |
| $t$ | Time |
| $t_R$ | Relaxation time |
| $\dot{V}$ | Volumetric flow rate |
| $v$ | Flow velocity |
| $y$ | Coordinate vertical to flow direction respectively in direction of flow channel height |
| $z$ | Coordinate in flow direction |
| $\alpha$ | Coefficient of the Vinogradov-Malkin function |
| $\alpha$ | Linear coefficient of thermal expansion |
| $\alpha$ | Temperature coefficient of viscosity |
| $\alpha$ | Angle |
| $\dot{\gamma}$ | Shear rate |
| $\dot{\gamma}_W$ | Shear rate at the wall |

| | |
|---|---|
| $\dot{\gamma}_{rep}$ | Representative shear rate |
| $\dot{\varepsilon}$ | Elongation rate |
| $\varepsilon_{Ra}$ | Initial deformation |
| $\varepsilon_R$ | Reversible elongation |
| $\varepsilon_{Ri}$ | Reversible elongation in step *i* |
| $\varepsilon^*_{Ri}$ | Reversible deformation in step *i* after relaxation |
| $\dot{\varepsilon}_D$ | Elongation rate |
| $\dot{\varepsilon}_S$ | Elongation rate due to shear |
| $\eta$ | Dynamic shear viscosity, viscosity |
| $\eta_0$ | Zero-shear (rate) viscosity |
| $\lambda$ | Thermal conductivity |
| $\lambda$ | Stretch ratio |
| $\xi$ | Coefficient of pressure dependence of glass transition temperature |
| $v$ | Specific volume ($p$-$v$-$T$ diagram) |
| $\rho$ | Density |
| $\tau$ | Shear stress |
| $\tau$ | Characteristic relaxation time |
| $\tau_0$ | Yield stress |
| $\tau_W$ | Shear stress at the wall |
| $\phi$ | Fluidity |

## 2.3 References of Chapter 2


1. Münstedt, H.: Die Bedeutung physikalischer Kenndaten für die Kunststoffverarbeitung In: Berechnen von Extrudierwerkzeugen. VDI-Verl., Düsseldorf (1978)
2. Pähl, M.: Praktische Rheologie der Kunststoffschmelzen und Lösungen. VDI-Verl., Düsseldorf (1982)
3. Gleißle, W.: Kurzzeitmessungen zur Ermittlung der Fließeigenschaften von Kunststoffen bis zu höchsten Schergeschwindigkeiten. In: Praktische Rheologie der Kunststoffe. VDI-Verl., Düsseldorf (1978)
4. Bird, R. B., Armstrong, R. C. and Hassager, O.: Dynamics of PolymerLiquids. Vol. 1: Fluid Mechanics. Wiley, New York (1977)
5. Ostwald, W.: Über die Geschwindigkeitsfunktion der Viskosität disperser Systeme I. Kolloid-Z. 36 (1925) pp. 99–117
6. Waele, A. de: J. Oil Colour Chem. Assoc. 6 (1923) p. 33
7. Schulze-Kadelbach, R., Thienel, P., Michaeli, W., Haberstroh, E., Dierkes, A., Wortberg, J. and Wübken, G.: Praktische Stoffdaten für die Verarbeitung von Piastomeren. Proceedings of 9. Kunststofftechnisches Kolloquium of IKV, Aachen (1978)

8. N.N.: Kenndaten für die Verarbeitung thermoplastischer Kunststoffe. VDMA (Ed.). T. 2: Rheologie. Hanser, München (1982)
9. N.N.: Kenndaten für die Verarbeitung thermoplastischer Kunststoffe. VDMA (Ed.). T. 4: Rheologie II. Hanser, München (1986)
10. Prandtl, L.: Ein Gedankenmodell zur kinetischen Theorie der festen Körper. Z. Ang. Math. Mech. 8 (1928) pp. 86–106
11. Eyring, H.: Viscosity, plasticity and diffusion as examples of absolute reaction rates. J. Chem. Phys. 4 (1936) pp. 283–291
12. Ebert, F.: Strömung nicht-newtonscher Medien. Vieweg, Braunschweig (1980)
13. Rautenbach, R.: Kennzeichnung nicht-newtonscher Flüssigkeiten durch zwei Stoffkonstanten. Chem. Ing. Techn. 36 (1964) p. 277
14. Schultz-Grunow, F.: Exakte Viskosimetrie und Modelltheorie für die Rheologie. Kolloid- Z. 138 (1954) p. 167
15. Carreau, P.J.: Rheological Equations from Molecular Network Theories. Ph. D. Thesis, University of Wisconsin (1968)
16. Geiger, K. and Kühnle, H.: Analytische Berechnung einfacher Scherströmungen aufgrund eines Fließgesetzes vom Carreauschen Typ. Rheol. Acta 23 (1984) pp. 355–367
17. Vinogradov, G.V. and Malkin, A.Y.: Rheological properties of polymer melts. J. Polym. Sci., Polym. Chem. Ed. 4 (1966) pp. 135–154
18. Pearson, J.R.A.: Mechanics of Polymer Processing. Elsevier, London (1985)
19. Herschel, W.H. and Bulkley, R.: Kolloid-Z. 39 (1926) p. 291
20. Limper, A.: Methoden zur Abschätzung der Betriebsparameter bei der Kautschukextrusion. Thesis at the RWTH Aachen (1985)
21. Knappe, W. and Schönewald, H.: Anwendung der temperaturinvarianten Auftragung Theologischer Daten für die Auslegung von Düsen. Kunststoffe 60 (1970) 9, pp. 657–665
22. Menges, G. et al.: Eine Abschätzmethode für die Relaxation von Molekülorientierungen in Kunststoffen. Kunststoffe 66 (1976) 1, pp. 42–48
23. Lenk, R.S.: Rheologie der Kunststoffe. Hanser, München (1971)
24. Ramsteiner, F.: Abhängigkeit der Viskosität einer Polymerschmelze von Temperatur, hydrostatischem Druck und niedermolekularen Zusätzen. Rheol. Acta 9 (1970) 3, pp. 374–381
25. Laun, H.M.: Rheologie von Kunststoffschmelzen mit unterschiedlichem molekularem Aufbau. Kautsch. Gummi Kunstst. 40 (1987) 6, pp. 554–562
26. Meissner, J: Rheologisches Verhalten von Kunststoffschmelzen und Lösungen. In: Praktische Rheologie der Kunststoffe. VDI-Verl., Düsseldorf (1978)
27. Williams, M.L. et al.: The temperature dependence of relaxation mechanisms in amorphous polymers and other glass-forming liquids. J. Am. Chem. Soc. 77 (1955) 7, pp. 3701–3706
28. Lee, W.A. and Knight, G.J.: Ratio of the glass transition temperature to the melting point in polymers. Br. Polym. J. 2 (1970) 1, pp. 73
29. Wübken, G.: Einfluß der Verarbeitungsbedingungen auf die innere Struktur thermoplastischer Spritzgußteile unter besonderer Berücksichtigung der Abkühlverhältnisse. Thesis at the RWTH Aachen (1974)
30. Wortberg, J.: Werkzeugauslegung für die Ein- und Mehrschichtextrusion. Thesis at the RWTH Aachen (1978)
31. Junk, P.B.: Betrachtungen zum Schmelzeverhalten beim kontinuierlichen Blasformen. Thesis at the RWTH Aachen (1978)

32. Kühnle, H.: Evaluation of the viscoelastic temperature and pressure shift factor over the full range of shear rates. P. I: Int. Polym. Process. 1 (1987) 2, pp. 89–97; P. II: Int. Polym. Process. 1 (1987) 3, pp. 116–122
33. Hellwege, K. H. et al.: Die isotherme Kompressibilität amorpher und teilkristalliner Hochpolymerer. Kolloid-Z. 183 (1962) 2, pp. 110–119
34. Menges, G. et al.: Abschätzung der Viskositätsfunktion über den Schmelzindex. Kunststoffe 68 (1978) 1, pp. 47–50
35. Rabinowitsch, B.: Über die Viskosität und Elastizität von Solen. Phys. Chem. 145 (1929) pp. 1–26
36. Chmiel, H. and Schümmer, P.: Eine neue Methode zur Auswertung von Rohrrheometer-Daten. Chem. Ing. Techn. 43 (1971) 23, pp. 1257–1259
37. Schümmer, P. and Worthoff, R. H.: An elementary method for the evaluation of a flow curve. Chem. Eng. Sci. 38 (1978) pp. 759–763
38. Giesekus, H. and Langer, G.: Die Bestimmung der wahren Fließkurven nicht-newtonscher Flüssigkeiten und plastischer Stoffe mit der Methode der repräsentativen Viskosität. Rheol. Acta 16 (1977) pp. 1–22
39. Bagley, E. B.: End corrections in the capillary flows of polyethylene. J. Appl. Phys. 28 (1957) 5, pp. 624–627
40. Carreau, P. J. and De Kee, D.: Review of some useful rheological equations. Can. J. Chem. Eng. 57 (1979) pp. 3–15
41. Astarita, G. and Marrocci, G.: Principles of Non-Newtonian Fluid Mechanics. McGraw Hill, London (1974)
42. Menges, G.: Werkstoffkunde der Kunststoffe. 3. Edition Hanser, München (1990)
43. N. N.: Kenndaten für die Verarbeitung thermoplastischer Kunststoffe. VDMA (Ed.). T. 1: Thermodynamik. Hanser, München (1979)
44. Michaeli, W.: Berechnen von Kühlprozessen bei der Extrusion. In: Kühlen von Extrudaten. VDI-Verl., Düsseldorf (1978)
45. Dietz, W.: Bestimmung der Wärme- und Temperaturleitfahigkeit von Kunststoffen bei hohen Drücken. Kunststoffe 66 (1976) 3, pp. 161–167
46. Picot, J. J.C. and Debeauvais, F.: Molecular orientation effects on thermal conductivity of polydimethylsiloxane under shearing strain. Polym. Eng. Sci. 15 (1975) 5, pp. 373–380
47. Retting, W.: Orientierung, Orientierbarkeit und mechanische Eigenschaften von thermoplastischen Kunststoffen. Colloid Polym. Sci. (1975) 253, pp. 852–874
48. Haberstroh, E.: Analyse von Kühlstrecken in Extrusionsanlagen. Thesis at RWTH Aachen (1981)

# 3 Fundamental Equations for Simple Flows

In this chapter, simple fundamental equations for the flow in channels with circular, rectangular and annular cross-sections will be derived. These allow in many cases at least an estimate of such factors as pressure consumption or throughput. The basis for the computations pertaining to the flow processes are the conservation laws of mass, momentum, and energy (e.g., [1–3]).

The following assumptions have to be made to derive these easily applicable equations:

- Steady state flow: no transient change in the flow at any point in the flow channel.
- Slow moving flow: the forces of inertia should be negligible when compared to the friction forces; the flow is to be laminar based on low Reynolds numbers.
- Isothermal flow: all particles of the fluid have the same temperature.
- Hydrodynamically fully developed flow.
- Incompressible fluid: the density is constant.
- No external forces, e.g., the effect of gravity is neglected.
- The melt adheres to a solid wall, meaning the velocity of the fluid at the wall is equal to the velocity of the wall (no slip boundary condition).

When all these conditions are met, the simple fundamental equations can be derived from the following momentum flux balance:

$$\begin{Bmatrix} \text{Entering (initial)} \\ \text{momentum / time} \\ \text{(momentum flux)} \end{Bmatrix} - \begin{Bmatrix} \text{Exiting (final)} \\ \text{momentum / time} \\ \text{(momentum flux)} \end{Bmatrix} + \begin{Bmatrix} \text{Sum of all} \\ \text{forces acting} \\ \text{on the system} \end{Bmatrix} = 0 \tag{3.1}$$

$$\vec{I}_1 - \vec{I}_2 + \sum \vec{F} = \vec{0}.$$

In the system under consideration, momentum flux can occur, for example, as a result of the general motion of the flowing melt or as a result of the forces acting on the medium.

However, if the velocity in the equilibrium space under consideration does not change in magnitude or direction or if the flowing medium is regarded as incompressible (i. e., the density is constant), $\vec{I}_1 = \vec{I}_2$ ; and then the problem is reduced to the balance of forces $\sum \vec{F} = \vec{0}$ to be equal to zero. Pressure and shear forces are those that can act on the system under consideration in the balance space.

If the additional assumption can be made that the flow lines are straight as, for example, in the case of flow through a rectangular channel, the general procedure, according to [1], for solving simple, viscous flow problems is as follows:

1. Set up a momentum flux balance on a fluid element of finite wall thickness (taking the continuity equation into consideration).
2. Let the wall thickness approach zero, as a result the momentum flux balance converts into a differential equation (i. e., differential momentum flux balance).
3. Introduce a material law

$$\dot{\gamma} = f(\tau). \tag{3.2}$$

4. On introducing boundary conditions, which make unambiguous physical statements at defined values of the independent variables, the following can be determined from Nos. 2 and 3 (above) after integration: velocity profile $v = f(r)$ or $f(x)$, average velocity profile $\overline{v}$, maximum velocity $v_{max}$, flow rate $\dot{V}$, forces $F$ occurring at the channel walls, pressure drop $\Delta p$, the residence time spectrum $t(r)$ or $t(x)$, and the average residence time $\overline{t}$.

This procedure is explained with examples in Sections 3.1 through 3.3. More detailed and partially further developed considerations are in [4–6].

## 3.1 Flow through a Pipe

In a circular channel (radius $R$, length $l$) in which entry and exit effects can be neglected, the balance of forces acting on a cylindrical element of mass with wall thickness d$r$ (velocity $v_z$) is set up (Fig. 3.1); as explained earlier, the momentum flux balance is reduced to a force balance, as a result of the incompressibility of the fluid and the fact that it flows on straight, parallel paths with a uniform velocity. It follows:

$$\begin{aligned} &2\pi r\,\mathrm{d}r \cdot \left[p(z) - p(z + \mathrm{d}z)\right] + \tau(r) \cdot 2\pi r\,\mathrm{d}z \\ &- \tau(r + \mathrm{d}r) \cdot 2\pi \cdot (r + \mathrm{d}r) \cdot \mathrm{d}z = 0. \end{aligned} \tag{3.3}$$

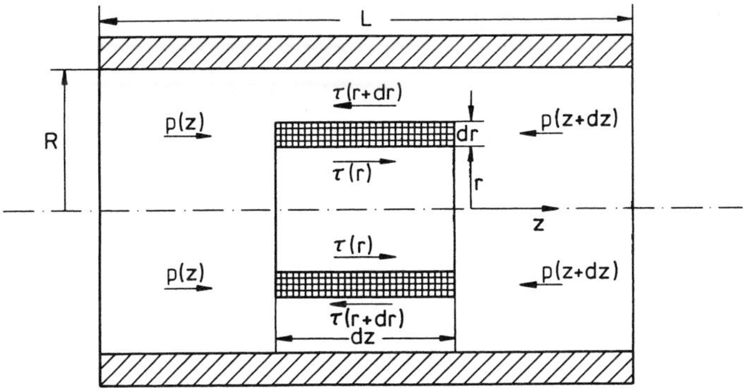


**Figure 3.1** Balance of forces acting on an element of mass during flow through a pipe

Developing the above into Taylor series and keeping only the first term:

$$p(z+\mathrm{d}z)=p(z)+\frac{\partial p}{\partial z}\mathrm{d}z$$
$$\tau(r+\mathrm{d}r)=\tau(r)+\frac{\partial \tau}{\partial r}\mathrm{d}r. \tag{3.4}$$

Because the flow is developed, the pressure gradient can be taken as constant:

$$\frac{\partial p}{\partial z}=-\frac{\Delta p}{L},\ \textit{with}\ \Delta p>0\ \textit{bar}. \tag{3.5}$$

After neglecting all terms of the higher order, the following differential equation is obtained:

$$\frac{\Delta p}{L}=\frac{\tau(r)}{r}+\frac{\mathrm{d}\tau}{\mathrm{d}r}=\frac{1}{r}\frac{\partial}{\partial r}(\tau\cdot r). \tag{3.6}$$

After integration, the result is:

$$\tau(r)=\frac{\Delta p}{2L}r+\frac{C_1}{r}. \tag{3.7}$$

For $r=0$, all forces have to disappear, so the first boundary condition follows:

For $\tau(r=0)=0 \Rightarrow C_1=0$

$$\tau(r)=\frac{\Delta p\cdot r}{2L}. \tag{3.8}$$

This results in a shear stress at the wall $\tau_W = \tau(r = R)$ of:

$$\tau_W = \tau(r = R) = \frac{\Delta p \cdot R}{2 \cdot L} \tag{3.9}$$

Equations (3.8) and (3.9) result strictly from the balance of forces; no assumptions were made as to the material law. This means that the linear dependence of the shear stress in Equation (3.8) is independent of the behavior of the material. Now the material law can be introduced into the Equation (3.8):

**Case A: Newtonian Flow**

In the case of a Newtonian Flow the correlation between shear stress and shear rate is the following:

$$\tau(r) = -\eta \frac{\mathrm{d}v_z(r)}{\mathrm{d}r} = \eta \cdot \dot{\gamma}(r). \tag{3.10}$$

When introducing this expression into Equation (3.8)

$$-\eta \frac{\mathrm{d}v_z(r)}{\mathrm{d}r} = \frac{\Delta p \cdot r}{2 \cdot L} \tag{3.11}$$

$$-\dot{\gamma} = \frac{\mathrm{d}v_z(r)}{\mathrm{d}r} = -\frac{\Delta p \cdot r}{2 \cdot L \cdot \eta}. \tag{3.12}$$

The negative sign means that $v_z$ in the direction of $r$ decreases. From Equation (3.12) we obtain:

$$\mathrm{d}v_z(r) = -\frac{\Delta p \cdot r}{2 \cdot L \cdot \eta} \cdot \mathrm{d}r. \tag{3.13}$$

After integration and considering the boundary condition $v_z\ (r = R) = 0$ (no slip boundary condition) the velocity distribution over the cross section of the channel is given by the following equation:

$$v_z(r) = \frac{\Delta p \cdot R^2}{4 \cdot L \cdot \eta} \left[ 1 - \left( \frac{r}{R} \right)^2 \right]. \tag{3.14}$$

The maximum velocity is at $r = 0$. From this it follows:

$$v_{z\,\max} = v_z\,(r = 0) = \frac{\Delta p \cdot R^2}{4 \cdot L \cdot \eta}. \tag{3.15}$$

For the mean velocity $\overline{v}_z$:

$$\overline{v}_z = \frac{1}{A}\int v_z(r)\mathrm{d}A \quad \text{with} \quad A = \pi \cdot R^2 \quad \text{and} \quad \mathrm{d}A = 2 \cdot \pi \cdot r \cdot \mathrm{d}r. \tag{3.16}$$

From the above:

$$\overline{v}_z = \frac{\Delta p \cdot R^2}{8 \cdot L \cdot \eta}. \tag{3.17}$$

A comparison with Equation (3.15) shows that the mean velocity is equal to half of the maximum velocity:

$$\overline{v}_z = \frac{1}{2} v_{z\,\max} \tag{3.18}$$

The relation for the flow rate $\dot{V}$ is:

$$\dot{V} = \overline{v}_z \cdot A \tag{3.19}$$

where the pipe cross-sectional area $A = \pi \cdot R^2$.

The Hagen-Poiseuille Law follows:

$$\dot{V} = \frac{\pi R^4}{8L} \cdot \frac{1}{\eta} \cdot \Delta p, \text{ with } K = \frac{\pi \cdot R^4}{8 \cdot L} = \text{const. (die conductance)} \tag{3.20}$$

The pressure drop, $\Delta p$, can be calculated from Equation (3.20) when the volumetric flow rate $\dot{V}$ is known.

The residence time of a particle on the radius $r$ in a pipe with the length $l$ can be calculated from the following equation:

$$t(r) = \frac{L}{v_z(r)} = \frac{4 \cdot L^2 \cdot \eta}{\Delta p \cdot R^2 \cdot \left[1 - \left(\frac{r}{R}\right)^2\right]}. \tag{3.21}$$

The mean residence time $\overline{t}$ is:

$$\overline{t} = \frac{L}{\overline{v}_z} = \frac{8 \cdot \eta \cdot L^2}{R^2 \cdot \Delta p}. \tag{3.22}$$

Because the residence time is inversely proportional to the velocity, its mean value is twice as long as the shortest time, at which the particle is at $v_{z\,\max} = v_z\,(r = 0)$.

The dependence of the shear rate on the flow rate is given by combining Equations (3.12) and (3.20):

$$\dot{\gamma}(r) = \frac{4 \cdot \dot{V}}{\pi \cdot R^4} \cdot r \qquad (3.23)$$

At the pipe wall the shear rate $\dot{\gamma}_W$ is:

$$\dot{\gamma}_W = \dot{\gamma}\left(r = R\right) = \frac{4 \cdot \dot{V}}{\pi \cdot R^3}. \qquad (3.24)$$

The above equation is particularly important for the evaluation of results from the measurements with the capillary viscometer (see Section 2.1.2).

The force $F_z$ acting on the surface of the flow channel results from the multiplication of the surface area and the shear stress at the wall:

$$F_z = 2\pi \cdot R \cdot L \cdot \tau\left(r = R\right) = \tau \cdot R^2 \cdot \Delta p. \qquad (3.25)$$

The validity of Equation (3.25) is independent of the flow behavior.

**Case B: Pseudoplastic Flow Following the Power Law**

From the definition of the Power Law (Equation (2.5)), the following relation can be derived:

$$\tau(r) = \left(\frac{\dot{\gamma}(r)}{\phi}\right)^{\frac{1}{m}} = \left(\frac{1}{\phi}\right)^{\frac{1}{m}} \cdot \left(-\frac{dv_z(r)}{dr}\right)^{\frac{1}{m}}. \qquad (3.26)$$

Applying Equation (3.8) with a negative sign for $\dot{\gamma}$ (because $v_z$ decreases in the direction of $r$):

$$\dot{\gamma}(r) = -\frac{dv_z(r)}{dr} = \phi\left(\frac{\Delta p}{2 \cdot L} \cdot r\right)^{m}. \qquad (3.27)$$

After integration:

$$v_z(r) = -\phi\left(\frac{\Delta p}{2 \cdot L}\right)^{m} \frac{r^{m+1}}{m+1} + C_1 = -\frac{\phi \cdot L}{m+1} \cdot \left(\frac{\Delta p}{2}\right)^{m} \cdot \left(\frac{r}{L}\right)^{m+1} + C_1. \qquad (3.28)$$

For a boundary condition $v_z\ (r = R) = 0$ (no slip condition) it follows:

$$C_1 = \frac{\phi \cdot L}{m+1} \cdot \left(\frac{\Delta p}{2}\right)^{m} \cdot \left(\frac{R}{L}\right)^{m+1}. \qquad (3.29)$$

Then:

$$v_z(r)=\frac{\phi\cdot L}{m+1}\cdot\left(\frac{\Delta p}{2}\right)^{m}\cdot\left[\left(\frac{R}{L}\right)^{m+1}-\left(\frac{r}{L}\right)^{m+1}\right]. \tag{3.30}$$

Analogically to Equations (3.15) to (3.24) the following relations can be derived:

$$v_{z\,max}=v_z\,(r=0)=\frac{\phi\cdot R}{m+1}\cdot\left(\frac{\Delta p\cdot R}{2\cdot L}\right)^{m} \tag{3.31}$$

$$\bar{v}_z=\frac{\phi\cdot R}{m+3}\left(\frac{\Delta p\cdot R}{2\cdot L}\right)^{m} \tag{3.32}$$

$$\dot{V}=\phi\cdot\Delta p^{m}\cdot\frac{\pi\cdot R^3}{m+3}\cdot\left(\frac{R}{2\cdot L}\right)^{m},\ \text{with}\ K'=\frac{\pi\cdot R^3}{m+3}\cdot\left(\frac{R}{2\cdot L}\right)^{m} \tag{3.33}$$

$$\bar{t}=\frac{L}{\bar{v}_z}=\frac{m+3}{\phi}\cdot\left(\frac{2}{\Delta p}\right)^{m}\cdot\left(\frac{L}{R}\right)^{m+1}. \tag{3.34}$$

$$\dot{\gamma}(r)=\frac{\dot{V}}{\pi\cdot R}\cdot(m+3)\cdot\left(\frac{r}{R}\right)^{m} \tag{3.35}$$

$$\dot{\gamma}_W=\dot{\gamma}(r=R)=\frac{\dot{V}}{\pi\cdot R^3}\cdot(m+3) \tag{3.36}$$

As stated before, the action of the shear force $F_z$ upon the surface of the flow channel is independent of the flow behavior and is described by the following equation:

$$F_z=\pi\cdot R^2\cdot\Delta p. \tag{3.37}$$

If $m=1$ and $\eta=\frac{1}{\phi}$, Equations (3.26) to (3.34) are transformed into corresponding equations for Newtonian fluids.

## 3.2 Flow through a Slit

When taking a momentum balance for the flow between two parallel plates in the same way as in Section 3.1, the analysis is performed on a rectangular volume element with the height d$x$, width $B$ and length $L$ (Fig. 3.2) and the resulting force balance leads to a differential equation analogous to Equation (3.6):

$$\frac{\Delta p}{L} = \frac{\mathrm{d}\tau}{\mathrm{d}x}. \tag{3.38}$$

It is important to bear in mind the assumption that all quantities in the direction of the width remain constant (which would mean neglecting the edge effect which is equivalent to the assumption of an infinite width of the flow channel).

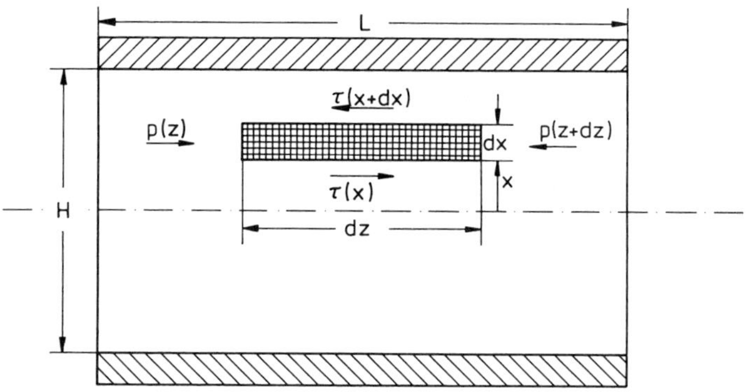


**Figure 3.2** Balance of forces acting on an element of mass during flow through a slit

When applying the boundary condition $\tau\ (x = 0) = 0$ after integration, we obtain:

$$\tau(x) - \frac{\Delta p}{L} \cdot x. \tag{3.39}$$

For the shear stress at the wall follows:

$$\tau_W = \tau\left(r = \frac{H}{2}\right) = \frac{\Delta p \cdot H}{2 \cdot L} \tag{3.40}$$

Again, this correlation is independent of the flow behavior.

### Case A: Newtonian Flow

Assuming Newtonian flow behavior, the following equation results:

$$\tau(x) = \dot{\gamma} \cdot \eta = -\eta \frac{dv_z(x)}{dx}. \tag{3.41}$$

Substituting for $\tau$ into Equation (3.39), we obtain:

$$-\eta \frac{dv_z(x)}{dx} \cdot B = \frac{\Delta p}{L} \cdot B \cdot dx. \tag{3.42}$$

Integrating over $x$ and taking into consideration the boundary conditions $v_z$ $(x = -H/2) = v_z\ (x = H/2) = 0$ yields the velocity distribution for a slit-shaped flow channel of an infinite width:

$$v_z(x) = \frac{\Delta p}{2 \cdot \eta \cdot L}\left[\left(\frac{H}{2}\right)^2 - x^2\right] = \frac{\Delta p \cdot H^2}{8 \cdot L \cdot \eta} \cdot \left[1 - \left(\frac{2 \cdot x}{H}\right)^2\right]. \tag{3.43}$$

The maximum flow velocity $v_{z\,max}$ occurs at $x = 0$:

$$v_{z\max} = v_z\,(x = 0) = \frac{\Delta p \cdot H^2}{8 \cdot \eta \cdot L}. \tag{3.44}$$

The mean velocity $\bar{v}_z$ is:

$$\bar{v}_z = \frac{1}{H} \int_{-\frac{H}{2}}^{\frac{H}{2}} v_z(x)\,dx = \frac{\Delta p \cdot H^2}{12 \cdot \eta \cdot L}. \tag{3.45}$$

By comparing Equations (3.43) and (3.44) we find that

$$\bar{v}_z = \frac{2}{3} v_{z\max} \tag{3.46}$$

The volumetric flow rate is obtained by integration over the entire cross-section:

$$\dot{V} = \int_{-\frac{H}{2}}^{\frac{H}{2}} \int_{0}^{B} v_z(x)\,dy\,dx = \frac{\Delta p \cdot B \cdot H^3}{12 \cdot L \cdot \eta} \tag{3.47}$$

where $B$ is the channel width. Equation (3.47) is valid only when $B$ is much greater than $H$, that means when edge effects can be neglected (this can be done with a satisfactory accuracy from $B > 10H$ and up).

The mean residence time $\bar{t}$ is:

$$\bar{t} = \frac{L}{\bar{v}_z} = \frac{12 \cdot \eta \cdot L^2}{\Delta p \cdot H^2}. \tag{3.48}$$

For the shear rate $\dot{\gamma}(x)$ follows from Equations (3.39), (3.41), and (3.47):

$$\dot{\gamma}(x) = \frac{\Delta p}{L \cdot \eta} \cdot x = \frac{12 \cdot \dot{V} \cdot x}{\Delta p \cdot H^3} \tag{3.49}$$

The shear rate at the wall $\dot{\gamma}_W$ therefore is:

$$\dot{\gamma}_W = \dot{\gamma}\left(x = \frac{H}{2}\right) = \frac{\Delta p \cdot H}{2 \cdot \eta \cdot L} = \frac{6 \cdot V}{H^2 \cdot B}. \tag{3.50}$$

The shear force acting upon the surface of the plate $F_z$ is obtained from Equation (3.40).

The following relation results:

$$F_z = B \cdot L \cdot \tau\left(x = \frac{H}{2}\right) = \Delta p \cdot B \cdot \frac{H}{2}. \tag{3.51}$$

Again, the validity of Equation (3.51) is independent of the material law.

**Case B: Pseudoplastic Flow Following the Power Law**

From the definition of the Power Law (Equation (2.5)), the expression for the shear stress is as follows:

$$\tau(x) = \left(\frac{\dot{\gamma}(x)}{\phi}\right) = \left(\frac{1}{\phi}\right)^{\frac{1}{m}} \cdot \left(-\frac{dv_z(x)}{dx}\right)^{\frac{1}{m}}. \tag{3.52}$$

Combining the above with Equation (3.39), we obtain:

$$\left(\frac{1}{\phi}\right)^{\frac{1}{m}} \cdot \left(-\frac{dv_z(x)}{dx}\right)^{\frac{1}{m}} = \frac{\Delta p}{L} \cdot x. \tag{3.53}$$

After rearranging the above, we obtain for the shear rate:

$$\dot{\gamma}(x) = -\frac{dv_z(x)}{dx} = \phi\left(\frac{\Delta p}{L}\cdot x\right)^m. \tag{3.54}$$

After integration and solving for the boundary condition $v_z\ (x = H/2) = 0$, we obtain the velocity distribution:

$$v_z(x) = \phi\left(\frac{\Delta p}{L}\right)^m \cdot \frac{\left(\frac{H}{2}\right)^{m+1} - x^{m+1}}{m+1} = \frac{\phi \cdot L}{m+1}\cdot \Delta p^m \cdot \left[\left(\frac{H}{2\cdot L}\right)^{m+1} - \left(\frac{x}{L}\right)^{m+1}\right]. \tag{3.55}$$

Analogous to Equations (3.44) and (3.45), we obtain:

$$v_{z\,max} = v_z(x=0) = \frac{\phi \cdot H}{2\cdot(m+1)}\cdot\left(\frac{\Delta p\cdot H}{2\cdot L}\right)^m \tag{3.56}$$

$$\bar{v}_z = \frac{\phi\cdot H}{2\cdot(m+2)}\left(\frac{\Delta p\cdot H}{2\cdot L}\right)^m. \tag{3.57}$$

The ratio of the mean and maximum velocities is obtained by dividing Equation (3.57) by Equation (3.56):

$$\frac{\bar{v}_z}{v_{z\,max}} = \frac{m+1}{m+2}. \tag{3.58}$$

This ratio for the Newtonian behavior ($m = 1$) is also 2/3 of what was found previously. With increasing $m$, that is, with increasing pseudoplasticity, the velocity profile becomes increasingly block-like (the maximum velocity decreases and the velocity ratio approaches 1); see Fig. 3.3.

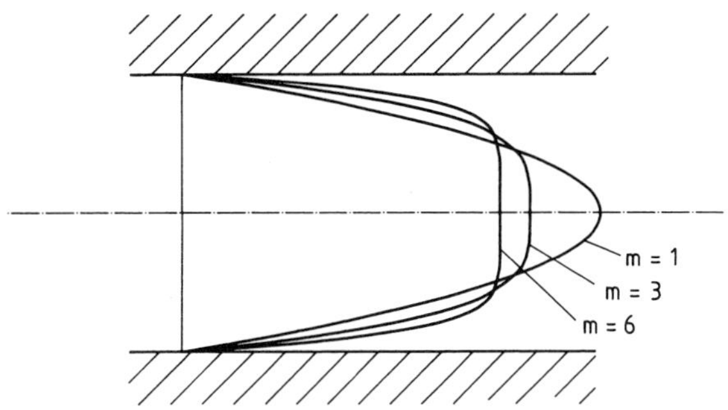


**Figure 3.3** Velocity profiles for different flow exponents

The flow rate again is obtained by the integration of Equation (3.55) over the entire cross-section:

$$\dot{V} = \frac{\phi \cdot B \cdot H^2}{2 \cdot (m+2)} \left( \frac{\Delta p \cdot H}{2 \cdot L} \right)^{\mathrm{m}} = B \cdot H \cdot \overline{v}_{\mathrm{z}}. \tag{3.59}$$

The mean residence time is obtained from Equation (3.57):

$$\overline{t} = \frac{L}{\overline{v}_{\mathrm{z}}} = \frac{m+2}{\phi \cdot \Delta p^{\mathrm{m}}} \left( \frac{2 \cdot L}{H} \right)^{\mathrm{m+1}}. \tag{3.60}$$

For the shear rate $\dot{\gamma}(x)$ is following from Equations (3.39), (3.52), and (3.59):

$$\dot{\gamma}(x) = \frac{2 \cdot (m+2) \cdot \dot{V}}{B \cdot H^2} \cdot \left( \frac{2x}{H} \right)^{\mathrm{m}} \tag{3.61}$$

Hence, the shear stress at the wall $\dot{\gamma}_{\mathrm{W}}$ is:

$$\dot{\gamma}_{\mathrm{W}} = \dot{\gamma} \left( x = \frac{H}{2} \right) = \phi \cdot \left( \frac{\Delta p \cdot H}{2 \cdot L} \right)^{\mathrm{m}} = \frac{2 \cdot (m+2) \cdot \dot{V}}{B \cdot H^2} \tag{3.62}$$

The shear force acting upon the surface of the plates $F_{\mathrm{z}}$ is independent of the flow behavior as per Equation (3.51):

$$F_{\mathrm{z}} = B \cdot L \cdot \tau_{\mathrm{W}} = p \cdot B \cdot \frac{H}{2} \tag{3.63}$$

Again, with $m = 1$ and $\eta = \frac{1}{\phi}$, we obtain equations valid for Newtonian flow.

## 3.3 Flow through an Annular Gap

Considering the flow through an annular gap, we can apply the same assumptions and conditions as we did for the flow through a pipe (Section 3.1). The momentum balance Equation (3.1) here too leads to Equation (3.7), which is given here again as Equation (3.64) [1]:

$$\tau(r) = \frac{\Delta p}{2L} \cdot r + \frac{C_1}{r}. \tag{3.64}$$

To solve this, the assumption has to be made that the shear stress $\tau$ becomes zero at the point $r = \lambda R$ (see Fig. 3.4), where the maximum velocity $v_{z\,max}$ occurs; $R$ is the outside radius of the annular gap. It follows:

$$C_1 = -\frac{\Delta p}{2L} \cdot (\lambda \cdot R)^2. \tag{3.65}$$

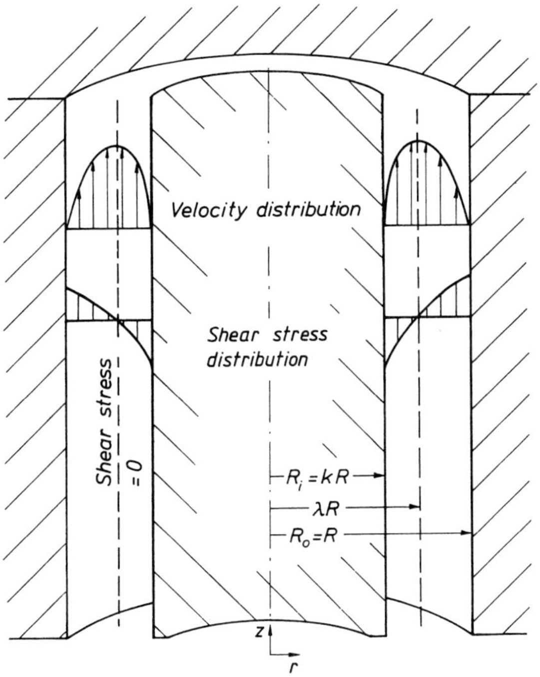


**Figure 3.4** Flow through a cylindrical annulus

However, $\lambda$ is still unknown at this point. With Equation (3.65) in (3.7), we obtain:

$$\tau(r) = \frac{\Delta p \cdot R}{2L}\left[\frac{r}{R} - \lambda^2 \cdot \frac{R}{r}\right]. \tag{3.66}$$

This correlation is independent of the material behavior. In Equation (3.7) a material law is introduced now.

**Case A: Newtonian Flow**

Assuming Newtonian flow behavior it follows:

$$\tau(r) = \eta \cdot \dot{\gamma}(r) = -\eta \cdot \frac{dv_z(r)}{dr} \tag{3.67}$$

For a Newtonian flow the shear rate will be calculated as follows with Equation (3.10):

$$\dot{\gamma}(r) = -\frac{dv_z(r)}{dr} = -\frac{\Delta p \cdot R}{2 \cdot \eta \cdot L}\left[\frac{r}{R} - \lambda^2 \cdot \frac{R}{r}\right] \tag{3.68}$$

From Equations (3.67) and (3.68), it follows for the shear rate:

$$\dot{\gamma}(r) = -\frac{p \cdot R}{2\eta \cdot L} \cdot \left(\frac{r}{R} - \lambda^2 \frac{R}{r}\right) \tag{3.69}$$

and after integration of *r*:

$$v_z(r) = -\frac{\Delta p \cdot R^2}{4 \cdot \eta \cdot L}\left[\left(\frac{r}{R}\right)^2 - 2\lambda^2 \cdot \ln\left(\frac{r}{R}\right) + C_2\right]. \tag{3.70}$$

The two unknowns, $\lambda$ and $C_{2,}$ can be determined using the following boundary conditions:

1. For $r = kR$, $v_z\ (r = kR) = 0$
2. For $r = R$, $v_z\ (r = R) = 0$

The symbol *k* is the ratio of the inside radius to the outside radius, thus

$$k = \frac{R_i}{R} \tag{3.71}$$

substituting this into Equation (3.70), we determine:

$$2\lambda^2 = \frac{1 - k^2}{\ln\left(\frac{1}{k}\right)} \qquad C_2 = -1. \tag{3.72}$$

Finally, we obtain for the velocity distribution

$$v_z(r) = \frac{R^2 \cdot \Delta p}{4 \cdot \eta \cdot L}\left[1 - \left(\frac{r}{R}\right)^2 + \frac{1 - k^2}{\ln\left(\frac{1}{k}\right)} \cdot \ln\left(\frac{r}{R}\right)\right]. \tag{3.73}$$

When $r = \lambda R$, we obtain the expression for $v_{z\,max}$:

$$v_{z\,max} = v_z\,(r = \lambda \cdot R) = \frac{R^2 \cdot \Delta p}{4 \cdot \eta \cdot L}\left\{1 - \frac{1-k^2}{2 \cdot \ln(\frac{1}{k})}\left[1 - \ln\left(\frac{1-k^2}{2 \cdot \ln(\frac{1}{k})}\right)\right]\right\}. \tag{3.74}$$

We get the mean velocity $\bar{v}_z$ by the integration of Equation (3.73):

$$\bar{v}_z = \frac{R^2 \cdot \Delta p}{8 \cdot \eta \cdot L}\left(\frac{1-k^4}{1-k^2} - \frac{1-k^2}{\ln\left(\frac{1}{k}\right)}\right) = \frac{\Delta p \cdot R^2}{8 \cdot L \cdot \eta} \cdot \left(1 + k^2 - \frac{1-k^2}{\ln\left(\frac{1}{k}\right)}\right). \tag{3.75}$$

Multiplication of the expression for the mean velocity by the area of the annular cross section produces the volumetric flow rate, $\dot{V}$:

$$\dot{V} = \pi R^2 \cdot \left(1-k^2\right) \cdot \bar{v}_z = \frac{\pi \cdot R^4}{8 \cdot L} \cdot \left[\left(1-k^4\right) - \frac{\left(1-k^2\right)^2}{\ln(\frac{1}{k})}\right] \cdot \frac{1}{\eta} \cdot \Delta p,$$

$$\text{with} \quad K'' = \text{die conductance} = \frac{\pi \cdot R^4}{8 \cdot L} \cdot \left[\left(1-k^4\right) - \frac{\left(1-k^2\right)^2}{\ln\left(\frac{1}{k}\right)}\right]. \tag{3.76}$$

The mean residence time $\bar{t}$ is inversely proportional to the mean velocity $\bar{v}_z$:

$$\bar{t} = \frac{L}{\bar{v}_z} = \frac{8 \cdot \eta \cdot L^2}{R^2 \cdot \Delta p} \cdot \frac{1}{\left(\frac{1-k^4}{1-k^2} - \frac{1-k^2}{\ln\left(\frac{1}{k}\right)}\right)} \tag{3.77}$$

The expression for the shear forces $F_z$ acting upon the walls of the flow channel is as follows:

$$\begin{aligned} F_{z\,total} &= F_{z,outer} + F_{z,inner} = \tau\,|_{r=R} \cdot 2\pi RL - \tau\,|_{r=kR} \cdot 2\pi kRL \\ &= \pi R^2 \cdot \left(1-k^2\right) \cdot \Delta p \end{aligned} \tag{3.78}$$

$$F_{z\,inner} = -\tau\,|_{r=kR} \cdot 2\pi kRL = -\Delta p \cdot \pi R^2 \cdot \left(\frac{1-k^2}{2 \cdot \ln\left(\frac{1}{k}\right)} - k^2\right) \tag{3.79}$$

$$F_{z\ outer} = \tau\,|_{r=R}\; 2\pi RL = \Delta p\pi R^2 \left(1 - \frac{1-k^2}{2\ln\frac{1}{k}}\right). \tag{3.80}$$

**Case B: Pseudoplastic Flow Following the Power Law**

By definition of Power Law (Equation (2.5)) it follows for the shear stress $\tau(r)$ in a flow through an annular slit:

$$\tau(r) = \left(\frac{\dot{\gamma}(r)}{\phi}\right)^{\frac{1}{m}} = \left(\frac{1}{\phi}\right)^{\frac{1}{m}} \cdot \left(\frac{-dv_z(r)}{dr}\right)^{\frac{1}{m}} \tag{3.81}$$

For the shear rate $\dot{\gamma}(r)$ it follows from Equations (3.66) and (3.81):

$$\dot{\gamma}(r) = -\frac{dv_z(r)}{dr} = \phi \cdot \left(\frac{\Delta p \cdot R}{2L}\right) \cdot \left[\frac{\left(\frac{r}{R}\right)^2 - \lambda^2}{\frac{r}{R}}\right]^m \tag{3.82}$$

To determine the velocity profile the following integral has to be solved:

$$v_z(r) = -\int \phi \cdot \left(\frac{\Delta p \cdot R}{2L}\right) \cdot \left[\frac{\left(\frac{r}{R}\right)^2 - \lambda^2}{\frac{r}{R}}\right] dr \tag{3.83}$$

However, this integral can only be solved for whole-number flow exponents $m$. For the flow behavior in an annular slit no general equations can be derived, while the Power Law is applied.

## 3.4 Summary of Simple Equations for Dies

The equations for shear stress, shear rate, axial velocity profile, and pressure loss for different simple flow problems (simple geometries) are concluded here. As simple geometries the flow channel geometries of a circle, a slit, and an annular gap are constituted. Beyond that, approximation equations for flow channels with irregular cross section, as well as conical bores, are given. The geometries and their dimensions are depicted in Fig. 3.5.

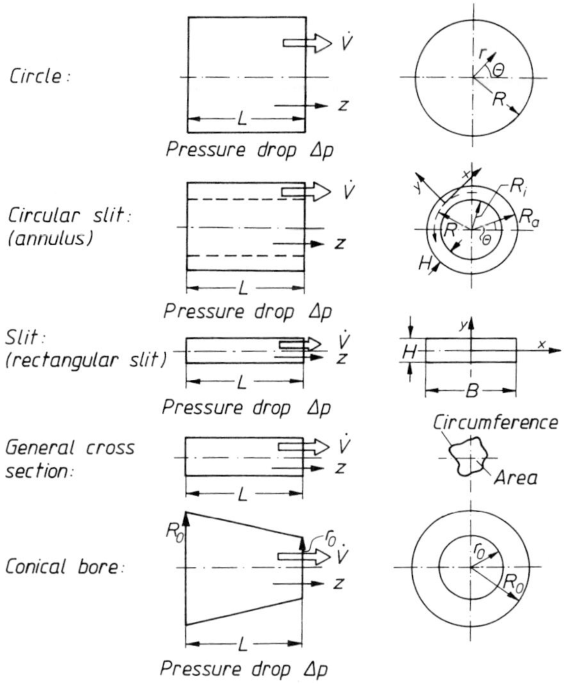


**Figure 3.5** Geometric definition of the basic shape elements

It needs to be reminded that the following assumptions and simplifications are necessary:

1. *Isothermal flow* (that mean that all particles of the melt stream have the same temperature; the wall of the die has the same temperature as the melt, i.e., there is no heat flow between the melt and the die wall. This condition can usually be assumed to exist approximately in the steady-state extrusion process).
2. *Steady-state flow,* i.e., no change in flow profile with time (valid for the steady state extrusion process).
3. *Laminar flow* for Reynolds numbers, $Re = 2 \cdot r \cdot v_z \rho / \eta$, smaller than 2100 (applies practically always to plastics melts).
4. *Incompressible Fluid* (constant density) (may be applied approximately to polymeric melts, see Section 2.2.1).
5. *Disregarding the inlet and outlet effects* (This is applicable as a rule, since the equation below [20] gives the inlet length, $L_e$, for Newtonian fluid for building up a parabolic velocity profile, so-called hydrodynamic inlet distance).

   $$L_e = 0.035 \cdot D \cdot Re$$

   In [21] this inlet distance for plastic melts is shown to be generally shorter than $R$. This can also be proven by somewhat more involved numerical flow simulations (e.g., [22]).

Viscoelastic liquids require an additional rheological inlet distance in which the material remembers all previous deformations. This inlet distance for plastic melts is longer than for the hydrodynamic one [23]. Its length depends on the rheological behavior of the melt and cannot be explicitly accessible by simple means.

6. *No-slip boundary condition* (This condition is mostly applicable to polymer melts. Exceptions possibly are polyethylene at higher shear rates [e. g., 24,25], several rigid PVC formulations and as a rule rubber compounds [26]; in those cases there will be a finite slip velocity at the die walls (see Section 3.5).

Analysis of the equations for volumetric flow rate $\dot{V}$, derived in Sections 3.1 through 3.3, shows that they can be written in the following general form:

*For Newtonian fluids*:

$$\dot{V} = \frac{K}{\eta} \Delta p = \frac{1}{W\eta} \Delta p. \tag{3.84}$$

Here $K$ is the so-called die conductance or alternately, its reciprocal, W, the so-called die resistance. Both quantities are only functions of the die geometry.

*For pseudoplastic fluids:*

Applying the Power Law according to *Ostwald-de Waele* (Equation (2.5)), the flow rate can be written in the following general form:

$$\dot{V} = K' \phi \Delta p^{\mathrm{m}} = G^{\mathrm{m}} \phi \Delta p^{\mathrm{m}} \tag{3.85}$$

where $K'$ is the die conductance applicable to the Power Law; $G$, appearing in the alternative and less frequently used form of this equation, is referred to as the die constant. Both $K'$ and $G$ are functions of the die geometry and the flow exponent $m$.

Equations (3.84) and (3.85) can be regarded as fundamental, simple relationships describing the flow through a die. The values of coefficients, $K$ and $K'$ for dies of simple geometry.

In the following, the most important equations for different simple geometries are summarized for Newtonian and, as far as it can be derived, pseudoplastic flow.

In Table 3.1 the equations for the shear stress in different simple cross-sections are summarized. In Table 3.2 the equations for the die conductances of different simple cross-sections are given. Table 3.3 summarizes the equations for shear rate of different simple cross-sections. For the rectangular cross-section (slit), of width $B$, height $H$, and length $L$, an infinite width is assumed in the derivations, unless otherwise stated, so that effects at the boundaries can be neglected. These values are therefore only valid for flow channels with width much larger than height. For circular/annular cross-sections with a width much smaller than the average diameter, these less practical equations can be replaced by the equations for a rectangular slit by approximation. In this case it is approximated for slit height $H$ and slit width $B$:

$$H \approx R_o - R_i \quad \text{and} \quad B \approx \pi \cdot (R_o + R_i) \quad \text{for} \quad \frac{R_o - R_i}{R_o + R_i} \gg 1. \tag{3.86}$$

In Table 3.4 the the velocity equations of different cross-sections are summarized. These equations are rules of thumb, based on an approximation of the irregular cross section by annular or rectangular cross-sections including a correction factor.

**Table 3.1** Shear Stress $\tau$, Shear Stress at the Wall $\tau_w$

| *Geometry* | Shear stress |
|---|---|
| Circle (Pipe) | $\tau = \frac{\Delta p}{2L} r$<br>$\tau_W = \frac{\Delta p}{2L} R \quad \tau = \tau_W \left(\frac{r}{R}\right)$ |
| Circular slit (Annular slit) | $\tau = \frac{\Delta p}{L} x$<br>$\tau_W = \frac{\Delta p}{2L} H \quad \tau = \tau_W \left(\frac{2x}{H}\right)$<br>$\tau = \frac{R}{2}\frac{\Delta p}{L}\left[\left(\frac{r}{R}\right) - \frac{1-k^2}{2\ln 1/k}\left(\frac{R}{r}\right)\right]$<br>$k = \frac{R_i}{R_a}$ |
| Rectangular slit | $\tau = \frac{\Delta p}{L} \cdot x$<br>$\tau = \frac{\Delta p}{2L} H \quad \tau = \tau_W \left(\frac{2x}{H}\right)$ |
| General | $\tau_W = \frac{\Delta p A}{U L}$ |

Correction factors $f_p$ (Fig. 3.6) [7,8], as applicable to the flow of Newtonian materials through a rectangular and other selected cross-sections, are introduced in Table 3.1. These factors take into consideration also the larger ratio between the surface area of the flow channels in contact with the molten polymer and their cross-sectional areas (see also [17,18]).

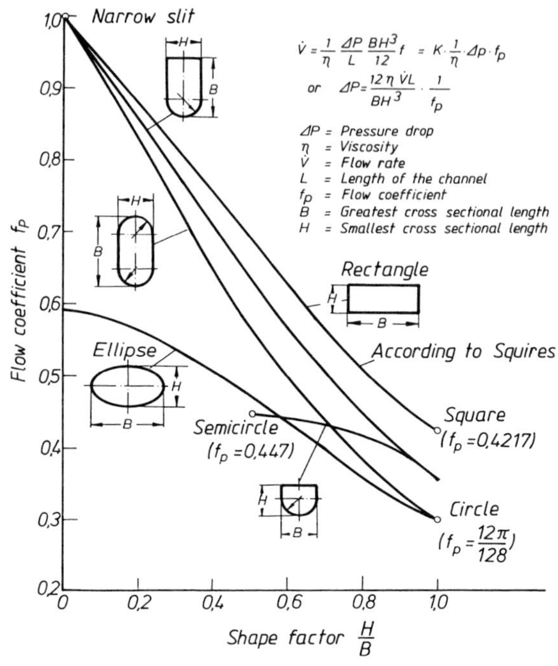


**Figure 3.6** Flow coefficient as a function of the shape factor for different shapes of cross sections

If a flow channel of an extrusion die consists of a series of successively arranged basic geometries, as is usually the case, then the respective pressure losses, for example can be calculated and then simply added up to the total pressure loss for the entire die. This method also lends itself to the computations involving conical and divergent flow channels. The channels have to be subdivided into sections with a constant geometry (Fig. 3.7). Subsequently the respective pressure losses are calculated separately and then the total pressure loss is determined by a simple addition of the single contributions. An estimate of a pressure loss with errors between 10 to 30% can be calculated from simple equations and tables published in [19].

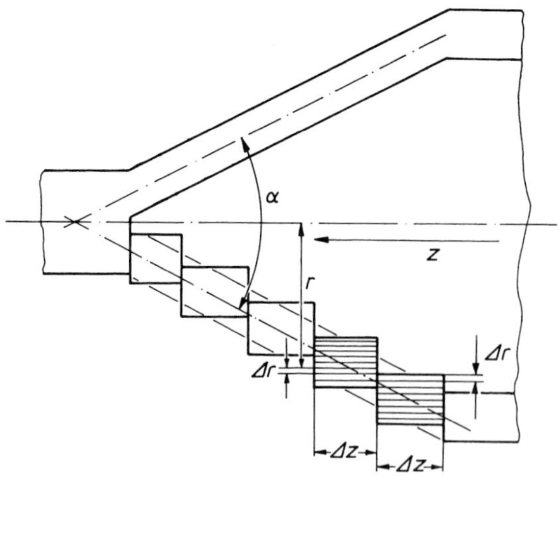

**Figure 3.7** Approximation of a flow channel with a changing geometry by sections of constant geometry

**Table 3.2** Die Conductances $K$, $K'$

| Geometry | Newtonian $\left(\tau = \eta \cdot \dot{\gamma}\right)$ | Pseudoplastic $\left(\tau^{m} = \frac{1}{\phi} \cdot \dot{\gamma}\right)$ |
|---|---|---|
| Circle (Pipe) | $K = \frac{\pi R^4}{8L}\left(L/R \gg 1\right)$ | $K' = \frac{\pi R^{m+3}}{2^{m}\left(m+3\right)}\left(\frac{1}{L}\right)^{m}$ |
| Circular slit (Annular slit) | $K = \frac{2\pi R + H}{12L} H^3$<br>$\left(H/R > 0.1\right)$<br>(With inlet correction)<br>$K = \frac{\pi R H^3}{6L}\left(H/R \ll 0.1\right)$<br>(Considered as slit)<br>$K = \frac{\pi D H^3}{12L}\left(H/R \ll 0.1\right)$<br>(Considered as slit)<br>$K = \frac{\pi R^4}{8L}\left[\left(1-k^4\right) - \frac{\left(1-k^2\right)^2}{\ln 1/k}\right]\left(k = \frac{R_i}{R_a}\right)$ | $K' = \frac{\pi\left(R_a + R_i\right)\left(R_a - R_i\right)^{m+2}}{2^{m+1}\left(m+2\right)}\left(\frac{1}{L}\right)^{m}$<br>$= \frac{\pi D H^{m+2}}{2^{m+1}\left(m+2\right)}\left(\frac{1}{L}\right)^{m}$ |
| Rectangular slit | $K = \frac{B \cdot H^3}{12L}\left(B \gg H\right)$<br>Introduction of a correction factor $f_p$ for $B/H \leq 20$ [111] (see Fig. 3.6) | $K' = \frac{B \cdot H^{m+2}}{2^{m+1}\left(m+2\right)}\left(\frac{1}{L}\right)^{m}$<br>$\left(B/H > 20\right)$ |
| Cone (Conical bore) | $K = \frac{\pi r^4}{8L} \cdot \frac{3\left(\frac{R_0}{r_0} - 1\right)}{1 - \left(\frac{r_0}{R_0}\right)^3}$ | $K' = \frac{\pi}{2^{m}\left(m+3\right)}\left(\frac{3\left(\frac{R_0}{r_0} - 1\right)}{m \cdot L\left(1 - \left(\frac{r_0}{R_0}\right)^{\frac{3}{m}}\right)}\right)^{m}$<br>$\cdot r^{m+3}$ |
| Irregular cross-section (General) | a) $K = \frac{B \cdot H^3}{12L} \cdot f_p$<br>$f_p$ see Fig. 3.6<br>b) $K = \frac{A^3}{2L \cdot U^2}$<br>(Approximation formula!) | $K' = \frac{2 \cdot A^{m+2}}{\left(m+3\right) L \cdot U^{m+1}}$<br>(Approximation formula!) |

**Table 3.3** Shear rates $\dot{\gamma}$, Shear Rates at the Wall $\dot{\gamma}_W$

| Geometry | Newtonian $(\tau = \eta \cdot \dot{\gamma})$ | Pseudoplastic $\left(\tau^m = \frac{1}{\phi}\dot{\gamma}\right)$ |
|---|---|---|
| Circle (Pipe) | $\dot{\gamma} = \frac{4\bar{v}_z \cdot r}{R^2} = \frac{4 \cdot \dot{V} \cdot r}{R^4 \pi}$ <br> $\dot{\gamma}_W = \frac{4 \cdot \dot{V}}{\pi R^3}$ | $\dot{\gamma} = (m+3)\frac{\bar{v}_z}{R}\left(\frac{r}{R}\right)^m$ <br> $\dot{\gamma}_W = \frac{(m+3) \cdot \dot{V}}{\pi R^3}$ |
| Circular slit (Annular slit) | $\dot{\gamma}_W = \frac{6 \cdot \dot{V}}{\pi D H^2}$ <br> $\dot{\gamma}_W = \frac{6 \cdot \dot{V}}{\pi (R_a + R_i)(R_a - R_i)^2}$ | $\dot{\gamma}_W = \frac{2(m+2)\dot{V}}{\pi D H^2}$ <br> $\dot{\gamma}_W = \frac{2(m+2)\dot{V}}{\pi (R_a + R_i)(R_a - R_i)^2}$ |
| Rectangular slit | $\dot{\gamma}_W = \frac{6 \cdot \dot{V}}{B \cdot H^2}$ | $\dot{\gamma} = (m+2)\frac{2\bar{v}_z}{H}\left(\frac{2x}{H}\right)^m$ <br> $\dot{\gamma}_W = \frac{2(m+2)\dot{V}}{B H^2}$ |
| Cone (Conical bore) | $\dot{\gamma} = \frac{4\dot{V}}{\pi r^3}\left[\frac{1-\left(\frac{r_0}{R_0}\right)^3}{3\left(\frac{R_0}{r_0}-1\right)}\right]^{\frac{3}{4}}$ | |
| Irregular cross section (General) | a) $\dot{\gamma} = \frac{2 \cdot \dot{V} \cdot U}{A^2}$ <br> b) $R_{eq} = \sqrt{\frac{A}{\pi}}$ <br> $\dot{\gamma}_W = \frac{4 \cdot \dot{V}}{\pi R_{eq}^3}$ <br> (Approximation formula!) | |

**Table 3.4** Velocities $v_z$, Average Velocities $\bar{v}_z$, Maximum Velocity $v_{z_{max}}$

| Geometry | Newtonian $(\tau = \eta \cdot \dot{\gamma})$ | Pseudoplastic $\left(\tau^m = \frac{1}{\phi} \cdot \dot{\gamma}\right)$ |
|---|---|---|
| Circle (Pipe) | $v_z = \frac{R^2}{4\eta}\frac{\Delta p}{L}\left[1-\left(\frac{r}{R}\right)^2\right]$<br>$v_z = v_{z_{max}}\left(1-\left(\frac{r}{R}\right)^2\right)$<br>$v_{z_{max}} = \frac{\Delta p R^2}{4\eta L}$<br>$\bar{v}_z = \frac{\Delta p R^2}{8\eta L}$<br>$\left(v_{z_{max}} = 2\bar{v}_z\right)$ | $v_z = \phi\left(\frac{\Delta p}{2L}\right)^m \frac{R^{m+1}-r^{m+1}}{m+1}$<br>$v_z = v_{z_{max}}\left(1-\left(\frac{r}{R}\right)^{m+1}\right)$<br>$v_{z_{max}} = \phi\left(\frac{\Delta p}{2L}\right)^m \frac{R^{m+1}}{m+1}$<br>$\bar{v}_z = \frac{\phi}{m+3}\left(\frac{\Delta p}{2L}\right)^m R^{m+1}$ |
| Circular slit (Annular slit) | $v_z = \frac{R^2}{4\eta}\frac{\Delta p}{L}\left[1-\left(\frac{r}{R}\right)^2+\left(\frac{1-k^2}{\ln 1/k}\right)\ln\left(\frac{r}{R}\right)\right] \quad k = \frac{R_i}{R_a}$<br>$v_{z_{max}} = \frac{R^2}{4\eta}\frac{\Delta p}{L}\left\{1-\left(\frac{1-k^2}{2\ln 1/k}\right)\cdot\left[1-\ln\left(\frac{1-k^2}{2\ln 1/k}\right)\right]\right\}$<br>$\bar{v}_z = \frac{R^2\Delta p}{8\eta L}\left(\frac{1-k^4}{1-k^2}-\frac{1-k^2}{\ln 1/k}\right)$ | |
| Rectangular slit | $v_z = \frac{H^2\Delta p}{8\eta L}\left\{1-\left(\frac{2x}{H}\right)^2\right\}$<br>$v_{z_{max}} = \frac{H^2\Delta p}{8\eta L}$<br>$\bar{v}_z = \frac{H^2\Delta p}{12\eta L}$<br>$\left(v_{z_{max}} = \frac{3}{2}\bar{v}_z\right)$ | $v_z = \phi\left(\frac{\Delta p}{L}\right)^m \frac{\left(\frac{H}{2}\right)^{m+1}-x^{m+1}}{m+1}$<br>$v_z = v_{z_{max}}\left\{1-\left(\frac{2x}{H}\right)^{m+1}\right\}$<br>$v_{z_{max}} = \phi\left(\frac{\Delta p}{L}\right)^m \frac{\left(\frac{H}{2}\right)^{m+1}}{m+1}$<br>$\bar{v}_z = \frac{\phi}{m+2}\left(\frac{\Delta p}{L}\right)^m\left(\frac{H}{2}\right)^{m+1}$ |

Pressure losses for the flow of Newtonian materials through eccentric annular channels and diamond shaped channels can be calculated by using methods outlined in [5]. The flow of Newtonian materials through elliptic and elliptically convergent channels is described in [5,28] and [28], respectively. Die resistances for conical channels are published in [5,27,28,32–35]; these are only valid for Newtonian materials, however. Triangular cross-sections are treated in [18,36].

**Table 3.5** Pressure Drop $\Delta p/L$

| Geometry | Newtonian $(\tau = \eta \cdot \dot{\gamma})$ | Pseudoplastic $\left(\tau^{m} = \frac{1}{\phi}\dot{\gamma}\right)$ | Pseudoplastic: Representative data |
|---|---|---|---|
| Circle (Pipe) | $\frac{\Delta p}{L} = \frac{8\eta \cdot \dot{V}}{\pi R^4}$ | $\frac{\Delta p}{L} = \left[\frac{2^{m}(m+3) \cdot \dot{V}}{\phi \pi R^{m+3}}\right]^{\frac{1}{m}}$ | $\frac{\Delta p}{L} = \frac{8\bar{\eta} \cdot \dot{V}}{\pi R^4} = \frac{8\bar{\eta} \cdot \bar{\dot{\gamma}}}{\pi R} = \frac{8\bar{\eta} \cdot \bar{V}_z}{R^2}$ |
| Circular slit (Annular slit) | $\frac{\Delta p}{L} = \frac{12\eta \cdot \dot{V}}{\pi D H^3};\quad H \ll D$ | $\frac{\Delta p}{L} = \left[\frac{2^{m+1}(m+2)\dot{V}}{\phi \pi D H^{m+2}}\right]^{\frac{1}{m}}$ | $\frac{\Delta p}{L} = \frac{12\bar{\eta}\dot{V}}{\pi D H^3}$<br>$\frac{\Delta p}{L} = \frac{8\bar{\eta}\dot{V}}{\pi\left(R_a^2 - R_i^2\right) \cdot \bar{R}^2}$<br>$\bar{R} = R_a\left(1 + k^2 + \frac{1-k^2}{\ln k}\right)^{\frac{1}{2}};k = \frac{R_i}{R_a}$ |
| Rectangular slit | $\frac{\Delta p}{L} = \frac{12\eta \cdot \dot{V}}{B \cdot H^3}(\mathrm{B} \gg \mathrm{H})$<br>$\frac{\Delta p}{L} = \frac{12\eta \cdot \dot{V}}{B \cdot H^3 \cdot f_p}\left(B / H \leq 20\right)$<br>$f_p$ see Fig. 3.6 | $\frac{\Delta p}{L} = \left[\frac{2^{m+1}(m+2)\dot{V}}{\phi B \cdot H^{m+2}}\right]^{\frac{1}{m}}$ | $\frac{\Delta p}{L} = \frac{12\bar{\eta}\dot{V}}{B H^3} = \frac{12\overline{\eta \upsilon}}{H^2}$ |
| Irregular cross section (General) | a) $\frac{\Delta p}{L} = \frac{12\eta \cdot \dot{V}}{B \cdot H^3 \cdot f_p}$<br>$f_p$ see Fig. 3.6<br>b) $\frac{\Delta p}{L} = \frac{2\eta U^2 \cdot \dot{V}}{A^3}$<br>(Approximation formula) | $\frac{\Delta p}{L} = \left[\frac{(m+3)U^{m+1}\dot{V}}{2\phi A^{m+2}}\right]^{\frac{1}{m}}$<br>(Approximation formula!) | ($\bar{\eta}$ There is no generally valid formula for $\bar{\eta}$ ) |

Often it is possible to compute the pressure loss in complicated channel cross-sections by subdividing the channel cross section into individual subsystems which can be easily calculated or by replacing it by a substitute pipe with a radius with an equivalent friction, $R_{eq}$ (an analog to hydraulic diameter). Values of this radius for irregular cross-sections are given in Table 3.3 [10]. The so obtained are only an approximation.

**Table 3.6** Representative Shear Rate $\bar{\dot{\gamma}}$

| Geometry | Pseudoplastic: Representative quantities |
|---|---|
| Circle (Pipe) | $\bar{\dot{\gamma}} = \frac{4 \cdot \dot{V}}{\pi R^3} \cdot \bar{e}_{\bigcirc} \quad \bar{e}_{\bigcirc} \approx 0.815$ |
| Circular slit (Annular slit) | $\bar{\dot{\gamma}} = \frac{\dot{V}}{\left(R_a^2 - R_i^2\right)\bar{R}} = \pi \frac{\bar{v}_z}{R}$ <br> $\bar{R} = R_a \left[1 + k^2 + \frac{1-k^2}{\ln k}\right]^{\frac{1}{2}} \quad k = \frac{R_i}{R_a}$ |
| Rectangular slit | $\bar{\dot{\gamma}} = \frac{6 \cdot \dot{V}}{BH^2} \cdot \bar{e}_{\square} \quad \bar{e}_{\square} \approx 0.772$ <br> The $\bar{e}_{\bigcirc,\square}$ values are valid for flow exponent $2 \le m \le 4$ as average (compare Fig. 2.16) |

*Calculations Using the Concept of the Representative Viscosity*

The relationships for pseudoplastic materials shown in Tables 3.1 to 3.5 are mathematically unwieldy when using the Power Law. When using the Prandtl-Eyring or Carreau constitutive laws they become even more so. The relationships for Newtonian materials are much simpler, however. Therefore it is advantageous to apply the Method of Representative Viscosity described in Section 2.1.2. This method allows the application of relationships developed for Newtonian materials to the pseudoplastic materials by introducing representative quantities.

If a representative point in the flow channel is known (Fig. 2.15), by the application of this method, the representative shear rate $\dot{\gamma}$ (Table 3.6) at that location can be calculated from the flow rate $\dot{V}$ and the representative viscosity $\bar{\eta}_{\text{rep}}$ from the known viscosity curve. The obtained value is then introduced into the equations derived for Newtonian materials for the determination of the pressure loss (Table 3.5).

Thus an equation analogous to Equations (3.84) and (3.85) is obtained which is a simple die equation employing the representative parameters:

$$\dot{V} = \frac{K}{\bar{\eta}_{\text{rep}}} \Delta p. \tag{3.87}$$

The die conductivities $K$ can be found in Table 3.2.

This procedure should be used as often as possible because of its simplicity.

## 3.5 Phenomenon of Wall Slip

In the derivation of the applicable equations to this point, the assumption was made that the melt flowing through the die adheres to its surface and hence has a velocity $v_z = 0$ at that point (Fig. 3.8(a)) [37]. This assumption is not correct for some rigid PVC formulations, high molecular weight polyethylene, and for rubber compounds after a critical shear stress is reached [e.g., 25,26,38]. Beyond that point the melt can slip along the wall (sometimes with friction) and have a finite velocity $v_g$ (Fig. 3.8(c)). [25] publishes a range of critical shear rates for polyethylene between 0.1 and 0.14 N/mm$^2$, above which a wall slip generally occurs. Between the two extremes, namely wall adhesion and wall slippage, there is the formation of a low viscosity slip film (Fig. 3.8(b)). This happens frequently when extruding a compound containing an external lubricant resulting in a lower pressure drop at constant throughput.

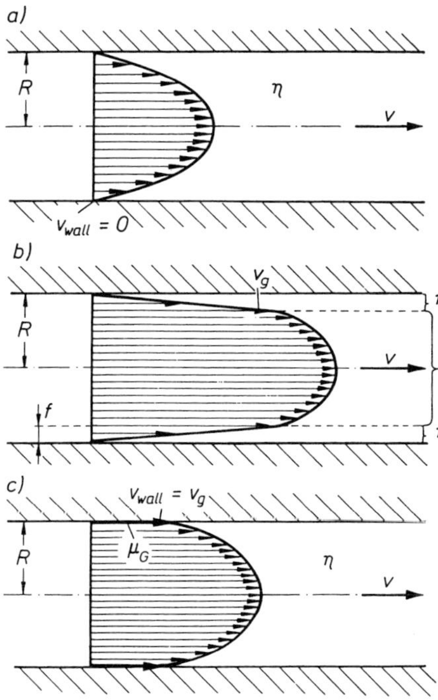


**Figure 3.8** Velocity profile at various boundary conditions, a) Adhesion to the wall, b) Sliding film, c) Slippage along the wall

### 3.5.1 Model Considering the Wall Slip

An analytical mathematical formulation considering the wall slip is that by Uhland [37], a brief outline of which is presented here. It starts out from Coulomb's Law of the friction of a solid body on a wall and formulates the equilibrium between the viscosity and frictional forces for a flow through a tube on an element of volume (Fig. 3.9):

$$\tau_W = \frac{R}{2}\frac{dp}{dz} = -p\mu_G = \frac{F_R}{A} \tag{3.88}$$

$\big($with $\mu_G$ = coefficient of sliding friction and

$p$ = pressure existing in the element under consideration$\big)$

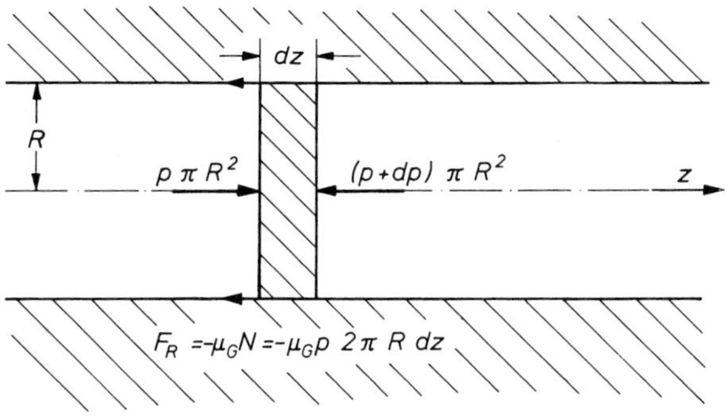


**Figure 3.9** Force equilibrium on a volume element of a viscous material with friction at the wall

Integration leads to the variation of pressure along the tube with $p(z{=}L) = p_L$ ($p_L$: pressure at the die exit):

$$p(z) = p_L \cdot \exp\left[\frac{2\mu_G}{R}(L-z)\right]. \tag{3.89}$$

It follows from Equation (3.88), that:

$$\tau_W = -p\mu_G \exp\left[\frac{2\mu_G}{R}(L-z)\right]. \tag{3.90}$$

This equation states that in contrast to the case of wall adhesion, the wall shear stress is not constant along the flow channel for wall slippage (see Table 3.1 and Fig. 3.10).

It can be seen from Equations (3.88) and (3.90) that the frictional force $F_R$ increases with the distance from the end of the die. It is conceivable that the frictional force $F_R$ becomes so large that there no longer can be any slip ($z < z_1$). In this case, the wall shear stress $\tau_W$, which brings about shear flow, is less than the shear stress required for overcoming frictional forces. Initially, the fluid adheres in a flow channel ($0 < z < z_1$). Subsequently (for $z_1 < z < L$), and there is wall slippage accompanied by a shear flow of the fluid (Fig. 3.10). This is made clear from Equation (3.88) and from Table 3.5 using the Power Law (Equation (2.5)):

$$z_1 = L - \frac{R}{2\mu_G} \ln \left[ \frac{1}{p_L \mu_G} \left\{ \frac{(m+3)\dot{V}}{\phi \pi R^3} \right\}^{\frac{1}{m}} \right]. \tag{3.91}$$

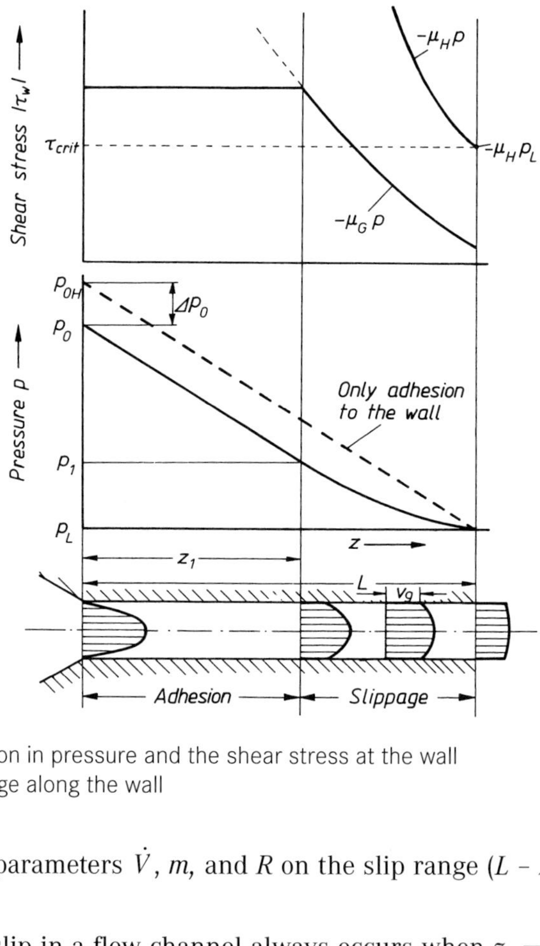


**Figure 3.10** Variation in pressure and the shear stress at the wall in the case of slippage along the wall

The effect of the parameters $\dot{V}$, $m$, and $R$ on the slip range ($L - z_1$) is discussed in detail in [37].

A complete wall slip in a flow channel always occurs when $z_1 = 0$. The necessary volumetric flow rate $\dot{V}$ for this case can be then obtained from the following equation:

$$\dot{V} \geq \frac{\phi \pi R^3}{m+3} \left[ p_L \mu_G \exp\left( \frac{2\mu_G L}{R} \right) \right]^m. \tag{3.92}$$

The pressure at the point of separation $z_1$ is calculated from Equations (3.89) and (3.91):

$$p_1 = \frac{1}{\mu_G}\left[\frac{(m+3)\cdot \dot{V}}{\phi\cdot\pi\cdot R^3}\right]^{\frac{1}{m}}. \tag{3.93}$$

The pressure at the capillary die inlet $p_0$ follows from:

$$\frac{p_0 - p_1}{z_1} = -\frac{dp}{dz}. \tag{3.94}$$

With $\frac{dp}{dz} = \frac{-\Delta p}{L}$ from Table 3.5 and Equation (3.92) it follows that:

$$p_0 = \left[\frac{(m+3)\cdot \dot{V}}{\phi\cdot\pi\cdot R^3}\right]^{\frac{1}{m}} \cdot \left(\frac{2L}{R} + \frac{1}{\mu_G}\left[1 - \ln\left[\frac{1}{p_L \mu_G}\left(\frac{(m+3)\cdot \dot{V}}{\phi\cdot\pi\cdot R^3}\right)^{\frac{1}{m}}\right]\right]\right). \tag{3.95}$$

The pressure in the case of a complete wall adhesion along the capillary $p_{0H}$ is (compare Table 3.5):

$$p_{0H} = \frac{2L}{R}\left[\frac{(m+3)\dot{V}}{\phi\pi R^3}\right]^{\frac{1}{m}}. \tag{3.96}$$

This value is greater than that for the case of a partial wall slip (see Fig. 3.10). From Equation (3.90), the material function (Equation (2.5)), and $\tau = \tau_W \cdot \frac{r}{R}$ (see Table 3.2), it follows:

$$\dot{\gamma}(r,z) = -\frac{dv_z(r,z)}{dr} = \phi\cdot\left(p_L\cdot\mu_G\cdot\frac{r}{R}\right)^{m} \exp\left[\frac{2\mu_G}{R}(L-z)\right]. \tag{3.97}$$

After integration and the boundary condition $v_z = v_g$ for $r = R$, the velocity profile is given by:

$$v_z(z) = v_g(z) + \phi\cdot\left[\frac{p_L\cdot\mu_G}{R}\right]^{m}\frac{R^{m+1} - r^{m+1}}{m+1}\exp\left[\frac{2\mu_G\cdot m}{R}(L-z)\right], \tag{3.98}$$

in which $v_g$, as well as the velocity portion of the viscous flow, are functions of $z$.

After further integration we obtain the expression for $v_g(z)$

$$v_g(z) = \frac{\dot{V}}{\pi R^2} - \phi \cdot (p_L \cdot \mu_G)^m \cdot \frac{R}{m+3} \cdot \exp\left[\frac{2\mu_G \cdot m}{R}(L-z)\right]. \tag{3.99}$$

From this equation, it is evident that $v_g(z)$ has its maximum value at the die end and it turns out that $v_g(z = z_1) = 0$ for the point of flow separation $z_1$ (transition from adhesion to slip, Equation (3.91)).

Figure 3.11 shows the development of a velocity profile in the die for $z_1 \le z \le L$ [37]. For $(L - Z) \ge 2.36$ cm in the example considered here, there is pure shear flow with $v_z = v_g = 0$ at the die wall, while plug flow sets in at the die exit (where $v_g \approx v_z$). Because of the rearrangement of the velocity profiles, there is also a velocity component in $r$-direction. This component is disregarded ($v_r \ll v_z$) [37]. It is evident that the wall slip is most pronounced at the die exit and decreases in the direction opposite to the direction of flow. This means that wall slip also commences at the die exit. This phenomenon was also observed in rheological investigations of the wall slip behavior of plastic melts.

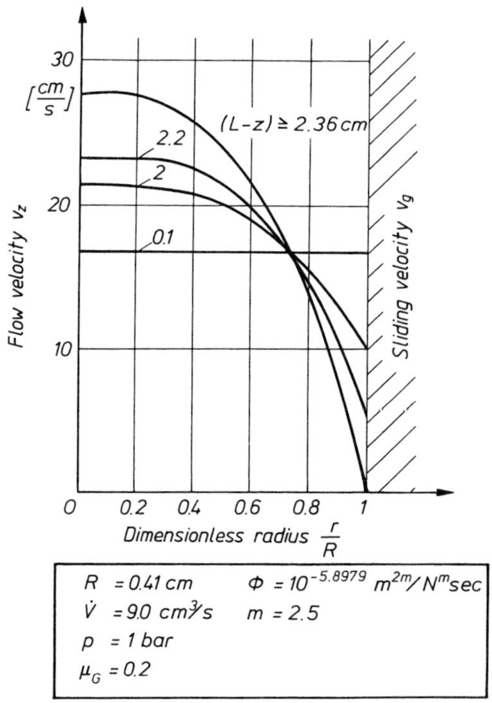


**Figure 3.11** Change of the velocity profile in the slippage area

### 3.5.2 Instability in the Flow Function – Melt Fracture

Although the method presented here correctly describes the wall slip trend of polymers, the greatest obstacle to its application is the unavailability of the coefficient of friction $\mu_G$ for polymers in the molten state or the choice of the correct slip velocity $v_g$ on the die wall. Technical difficulties with the measurements prevent access to these data [39], which depend both on the polymeric melt and the material from which the wall is made [25]. Therefore, the expressions derived in this section unfortunately have more of theoretical than practical importance at this point.

Therefore, when designing extrusion dies for melts that exhibit wall slip, the only alternatives remaining are either an empirical way of determining the pressure loss in a similar prototype die or the use of mathematical formulations, which are selected in a nonspecific manner and which start out from total wall adhesion and therefore lead to calculated pressure losses in the die being too high.

During rheological experiments to determine flow behavior, especially of PE-HD melts, it was frequently found that when the volumetric flow rate $\dot{V}$ exceeded a critical value, there was a discontinuous change in the shear stress or, when the shear stress (at constant pressure in the experiment) exceeded a critical value, there was a discontinuous change in the volume throughput (e. g., [40]). Both cases necessarily lead to an instability in the flow function (Fig. 3.12).

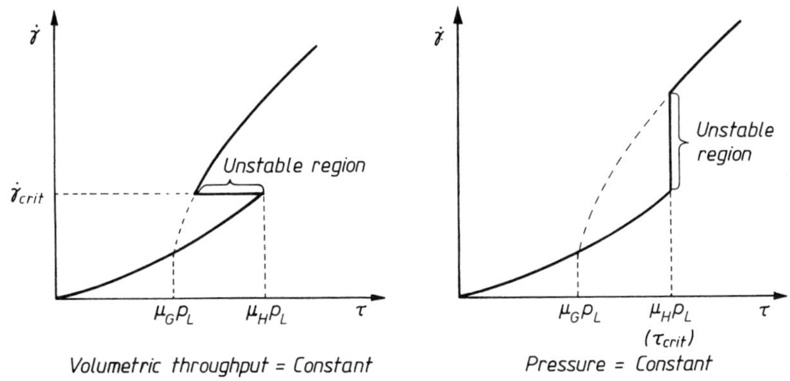


**Figure 3.12** Variation of the flow curve with an unstable region as a function of the method of operating the capillary rheometer

This instability is interpreted using the approach of Uhland [37], described in Section 3.5.1. The wall adhesion of the melt can be described by means of a coefficient of friction, namely by introducing the coefficient of static friction $\mu_H$, and since wall slip, as shown, starts at the die end, a critical wall shear stress $\tau_{crit}$ can be defined below which there can be no wall slip:

$$\tau_{crit} = \mu_H \cdot p_L \, . \tag{3.100}$$

If the wall shear stress is larger than $\tau_{crit}$, an unstable region arises in the vicinity of the die orifice, in which the flowing melt changes from adhering to slipping with the coefficient of friction changing discontinuously ($\mu_H \rightarrow \mu_G$). The shear stress also changes discontinuously here (see Fig. 3.10) and therefore also the pressure. An unstable region is formed, in which adhesion, as well as slipping, are possible. This results in the so-called stick-slip effect, which leads to periodically rough surfaces on the extrudate.

The so-called melt fracture, which manifests itself by a rough and periodically deformed surface, is also partly interpreted by the effect described above [40]. However, opinions differ considerably on this point. Melt fracture is also frequently ascribed to the elastic properties of the melt [40–43]. According to this theory, the molecules are no longer able to align themselves during the flow process, when the shear rate or shear stress exceed the critical value, which depends on the geometry (inlet geometry) and on the molecular parameters of the polymer. This phenomenon leads to disturbances in the stream lines. As a result, there are different retardation behaviors of the melt at the die orifice which, in turn, manifest themselves in the extrudate deformation described above.

It should be stated, however, that, for designing extrusion dies, the operating points of those dies must not lie in the described regions of instability of the flow function. Accordingly, critical shear rates and shear stresses, which are made evident by the flow function (instability or melt fracture), have to be taken into consideration during this design work.

## Symbols and Abbreviations

| | |
|---|---|
| $A$ | Cross section |
| $B$ | Width |
| $C_1$ | Constant of integration |
| $C_2$ | Constant of integration |
| $d$ | Diameter |
| $e_{\bigcirc}$ | Ratio of representative radius to outer radius of a circular flow channel |
| $e_{\square}$ | Ratio of representative height to height of a rectangular channel |
| $\vec{F}$ | Force |
| $F_R$ | Force of friction |
| $F_z$ | Force of shear |
| $f_p$ | Correction factor |
| $G$ | Die constant |
| $H$ | Height |
| $\vec{I}$ | Momentum flux |
| $K$ | Die conductance |

| | |
|---|---|
| $K'$ | Die conductance for fluids described by the Power Law |
| $k$ | Ratio of inner radius to outer radius |
| $L$ | Die length |
| $L_e$ | Entry length |
| $m$ | Flow exponent (material parameter of Power Law) |
| $N$ | Force in normal direction |
| $p$ | Pressure |
| $p_0$ | Pressure at die inlet |
| $p_{0H}$ | Pressure with total wall adhesion |
| $p_1$ | Pressure at point of separation |
| $p_L$ | Pressure at die outlet (die length $L$) |
| $\Delta p$ | Pressure loss |
| $R$ | Radius |
| $R_o$ | Outer radius |
| $R_{eq}$ | Equivalent radius (considering friction) |
| $R_i$ | Inner radius |
| $r$ | Radius |
| $Re$ | Reynolds number |
| $t$ | Residence time |
| $\bar{t}$ | Mean residence time |
| $U$ | Circumference |
| $\dot{V}$ | Volume flow |
| $v$ | Flow velocity |
| $\bar{v}$ | Mean flow velocity |
| $v_g$ | Slip velocity |
| $v_{max}$ | Maximum flow velocity |
| $v_r$ | Flow velocity in radial direction |
| $v_z$ | Flow velocity in $z$-direction |
| $v_{z\,max}$ | Maximum flow velocity in $z$-direction |
| $\bar{v}_z$ | Mean velocity in $z$-direction |
| $W$ | Die resistance |
| $x, y, z$ | Coordinates |
| $z_1$ | $z$-coordinate at the point of separation |
| $\dot{\gamma}$ | Shear velocity |
| $\dot{\gamma}_W$ | Shear velocity at flow channel wall |
| $\dot{\gamma}_{rep}$ | Representative shear velocity |
| $\eta$ | Shear viscosity |

| | |
|---|---|
| $\overline{\eta}$ | Representative shear viscosity |
| $\lambda$ | Ratio of radius of maximum flow velocity at annular slit to outer radius of an annular slit |
| $\mu_G$ | Friction coefficient in slippage |
| $\mu_H$ | Friction coefficient in adhesion (coefficient of static friction) |
| $\rho$ | Density |
| $\tau$ | Shear stress |
| $\tau_{crit}$ | Critical wall shear stress |
| $\tau_W$ | Wall shear stress |
| $\phi$ | Fluidity (material parameter of Power Law) |

## 3.5 References of Chapter 3

1. Bird, R. B., Stewart, W. E., and Lightfoot, E. N.: Transport Phenomena. Wiley, London (1960)
2. Schade, H. and Kunz, E.: Strömungslehre, de Gruyter, Berlin (1980)
3. Böhme, G.: Strömungsmechanik nicht-newtonscher Fluide. Teubner, Stuttgart (1981)
4. Winter, H. H.: Ingenieurmäßige Berechnung von Geschwindigkeits- und Temperatur-feldern in strömenden Kunststoffschmelzen. In: Berechnen von Extrudierwerkzeugen. VDI-Verlag, Düsseldorf (1978)
5. Pahl, M. H.: Rheologie der Rohrströmung. In: Praktische Rheologie der Kunststoffe. VDI-Verl, Düsseldorf (1978)
6. Pahl, M. H.: Praktische Rheologie der Kunststoffschmelzen und Lösungen. VDI-Verl, Düsseldorf (1982)
7. Squires, P. H.: Screw-Extruder Pumping Efficiency. SPE-J. 14 (1958) 5, pp. 24–30
8. Lahti, G. P.: Calculation of Pressure Drops and Outputs. SPE-J. 19 (1963) 7, pp. 619–620
9. Röthemeyer, F.: Bemessung von Extrusionswerkzeugen. Maschinenmarkt 76 (1970) 32, pp. 679–685
10. Rao, N.: Strömungswiderstand von Extrudierwerkzeugen verschiedener geometrischer Querschnitte. In: Berechnen von Extrudierwerkzeugen. VDI-Verl., Düsseldorf (1978)
11. Carley, J. F.: Rheology and Die Design. SPE-J. 19 (1963) 9, pp. 977–983
12. Carley, J. F.: Problems of Flow in Extrusion Dies. SPE-J. 39 (1983) 12, pp. 1263–1266
13. Weeks, D. J.: Berechnungsgrundlagen für den Entwurf von Breitschlitz- und Ringdüsen. Br. Plast. 31 (1958) 4, pp. 156–160; Brit. Plast. 31 (1958) 5, pp. 201–205 and Kunststoffe 49 (1959) 9, pp. 463–467
14. Worth, R. A. and Parnaby, J: The Design of Dies for Polymer Processing Machinery. Trans. Inst. Chem. Eng. London 52 (1974) 4, pp. 368–378
15. Jacobi, H. R.: Berechnung und Entwurf von Schlitzdüsen. Kunststoffe 17 (1957) 11, pp. 647–650
16. Sors, L.: Planung und Dimensionierung von Extruderwerkzeugen. Kunststoffe 64 (1974) 6, pp. 287–291
17. Schenkel, G.: Zur Extrusion von Kunststoffen aus Rechteck-Kanälen. Kunststoffe 74 (1981) 8, pp. 479–484

18. Schenkel, G. and Kühnle, H.: Zur Bemessung der Bügellängenverhältnisse bei Mehrkanal-Extrudierwerkzeugen für Kunststoffe. Kunststoffe 73 (1983) 1, pp. 17-22
19. Lenk, R. S.: Auslegung von Extrusionswerkzeugen für Kunststoffschmelzen. Kunststoffe 75 (1985) 4, pp. 239-243
20. Perry, J. H.: Chemical Engineers Handbook. McGraw Hill, New York (1950) pp. 388-389
21. Winter, H. H.: Temperaturänderung beim Durchströmen von Rohren. In: Praktische Rheologie der Kunststoffe. VDI-Verl., Düsseldorf (1978)
22. Grajewski, F.et al.: Von der Mischung bis zum Profil, Hilfsmittel für die Extrusion von Elastomerprofilen. Kautsch. Gummi Kunstst. 39 (1986) 12, pp. 1198-1214
23. Retting, W.: Orientierung, Orientierbarkeit und mechanische Eigenschaften von thermoplastischen Kunststoffen. Colloid Polym. Sci. 253 (1975), pp. 852-874
24. Uhland, E.: Das anormale Fließ verhalten von Polyäthylen hoher Dichte. Rheol. Acta 18 (1979) 1, pp. 1-24
25. Ramamurthy, A. V.: Wall Slip in Viscous Fluids and Influence of Materials of Construction. J. Rheol. 30 (1986) 2, pp. 337-357
26. Limper, A.: Methoden zur Abschätzung der Betriebsparameter bei der Kautschukextrusion. Thesis at the RWTH Aachen (1985)
27. Plajer, O.: Praktische Rheologie für Extrudierwerkzeuge. Plastverarbeiter 20 (1969) 10, pp. 693-703; Plastverarbeiter 20 (1969) 11, pp. 803-807
28. Plajer, O.: Schlauchkopfgestaltung beim Extrusionsblasformen. (Teil I-III). Kunststofftechnik 11 (1972) 11, pp. 297-301; Kunststofftechnik 11 (1972) 12, pp. 336-340; Kunststofftechnik 12 (1973) 1/2, pp. 18-23
29. Starke, L.: Zur analytischen Ermittlung des Druckverlaufs, der Fließgeschwindigkeit und Durchflußmengen bei der Füllung von Spritzgießwerkzeugen I. Plaste Kautsch. 22 (1975) 12, pp. 975-981
30. Losson, J.-M.: A Mathematical Model is used in Predicting Die Performance. Japan Plast. (1974) 1, pp. 22-28
31. McKelvey, J. M.: Polymer Processing. Wiley, New York (1962)
32. Plajer, O.: Praktische Rheologie für Kunststoffschmelzen. Plastverarbeiter 23 (1972) 6, pp. 407-412
33. Plajer, O.: Vergleichende Viskositätsmessungen und Druckberechnungen. Plastverarbeiter 20 (1969) 1, pp. 1-6; Plastverarbeiter 20 (1969) 2, pp. 89-94
34. Plajer, O.: Praktische Rheologie für Extrudierwerkzeuge. Plastverarbeiter 21 (1970) 8, pp. 731-734
35. Plajer, O.: Angewandte Rheologie. Plastverarbeiter 29 (1978) 5, pp. 249-252
36. Ramsteiner, F.: Fließverhalten von Kunststoffschmelzen durch Düsen mit kreisförmigem, quadratischem, rechteckigem oder dreieckigem Querschnitt. Kunststoffe 61 (1971) 12, pp. 943-947
37. Uhland, E.: Modell zur Beschreibung des Fließens wandgleitender Substanzen durch Düsen. Rheol. Acta 15 (1976) pp. 30-39
38. Offermann, H.: Die Rheometrie von wandgleitenden Kunststoffschmelzen untersucht am Beispiel von Hart-PVC. Thesis at the RWTH Aachen (1972)
39. Westover, R. F.: The Significance of Slip in Polymer Melt Flow. Polym. Eng. Sci. 6 (1966) 1, pp. 83-89
40. Pearson, J. R.A.: Mechanics of Polymer Processing. Elsevier, London (1985)
41. Fiedler, P. and Braun, D.: Fließverhalten und Schmelzbruch von Äthylen-Vinylazetat- Kopolymeren im Vergleich zu Hochdruckäthylen bei der Extrusion durch Kapillaren. Plaste Kautsch. 17 (1970) 4, pp. 246-250
42. Han, C. D.: Rheology in Polymer Processing. Academic Pr., New York (1976)
43. Den Otter, J. L.: Mechanisms of melt fracture. Plast. Polym. 38 (1970) pp. 155-168

# 4 Computation of Velocity and Temperature Distributions in Extrusion Dies

## 4.1 Conservation Equations

The basis of the general mathematical treatment of flow processes is the conservation equations for mass, momentum, and energy. The flow can be fully described only when the velocity vector and the thermodynamic data as pressure, density, and temperature are known at any time and at any point of the area of the flow.

To determine these quantities, the conservation equations are combined with the constitutive (material) equations that describe the correlations between parameters relating to motion and kinetics on the one hand and between the individual thermodynamic parameters on the other (e.g., [1–5]).

To describe an isothermal flow fully, the following are required:

- the law of conservation of mass (continuity equation)
- the law of conservation of momentum (equation of motion)
- a rheological material law

If there are events involving heat transfer during the flow, it is no longer an isothermal process, and it requires a full description of the following relations in addition to the above:

- the law of conservation of energy (energy equation)
- thermodynamic state and material equations [5] (e.g., Fourier law of heat conductivity)

In the following section, the so-called conservation equations will be briefly explained. For detailed derivations and representations in different coordinate systems, see the pertinent literature (e.g., [1–5]). The equations treated below are interpreted strictly in Cartesian coordinates.

### 4.1.1 Continuity Equation

Considering a mass balance on a small control volume $\Delta x \cdot \Delta y \cdot \Delta z$ in a space, we obtain [1]

$$\begin{Bmatrix} \text{Stored mass} \\ \text{per unit time} \end{Bmatrix} = \begin{Bmatrix} \text{Entering mass} \\ \text{per unit time} \end{Bmatrix} - \begin{Bmatrix} \text{Leaving mass} \\ \text{per unit time} \end{Bmatrix}. \tag{4.1}$$

For Cartesian coordinates, it follows that

$$\frac{\partial \rho}{\partial t} = -\left( \frac{\partial}{\partial x} \rho \cdot v_x + \frac{\partial}{\partial y} \rho \cdot v_y + \frac{\partial}{\partial z} \rho \cdot v_z \right). \tag{4.2}$$

This relationship describes the change in density with time at a fixed reference point as a function of the mass flow vector $\rho \vec{v}$ $\left(v_x, v_y, v_z\right.$ are the components of the velocity vector $\left.\vec{v}\right)$. Written in vector symbolism, Equation (4.2) becomes

$$\frac{\partial \rho}{\partial t} = -\rho(\nabla \vec{v}). \tag{4.3}$$

By using the multiplication rule of differential calculus, Equation (4.2) can be reformulated as follows:

$$\frac{\partial \rho}{\partial t} + v_x \frac{\partial \rho}{\partial x} + v_y \frac{\partial \rho}{\partial y} + v_z \frac{\partial \rho}{\partial z} = -\rho \left( \frac{\partial v_x}{\partial x} + \frac{\partial v_y}{\partial y} + \frac{\partial v_z}{\partial z} \right). \tag{4.4}$$

or in vector symbolism:

$$\frac{\mathrm{D}\rho}{\mathrm{D}t} = -\rho(\nabla \vec{v}). \tag{4.5}$$

The expression $\frac{\mathrm{D}\rho}{\mathrm{D}t}$ is called the substantial derivative of the density. This means that the time-dependent change in density is described by Equations (4.4) and (4.5) as seen by an observer who follows the motion of the flow (i.e., the substance). Equations (4.3) and (4.5) contain the same physical statements, however. The assumption of incompressibility of polymeric melts (i.e., of constant density) is permissible in most cases. In this case, the left side of Equation (4.4) becomes zero, and the continuity equation is simplified thus:

$$\nabla \vec{v} = \frac{\partial v_x}{\partial x} + \frac{\partial v_y}{\partial y} + \frac{\partial v_z}{\partial z} = 0. \tag{4.6}$$

### 4.1.2 Momentum Equations

If a momentum flux balance is made on a fixed-space Cartesian volume element $\Delta x \cdot \Delta y \cdot \Delta z$, then

$$\begin{Bmatrix} \text{Change in the} \\ \text{momentum per unit time} \\ (\text{Momentum Flux}) \end{Bmatrix} = \begin{Bmatrix} \text{Entering} \\ \text{momentum per unit time} \\ (\text{Momentum Flux}) \end{Bmatrix} - \begin{Bmatrix} \text{Leaving} \\ \text{momentum per unit time} \\ (\text{Momentum Flux}) \end{Bmatrix} + \begin{Bmatrix} \text{Forces} \\ \text{acting on the} \\ \text{system} \end{Bmatrix}. \tag{4.7}$$

The surface forces, as well as the momentum, transported with the flow in the balance space and the pressure and gravitational forces must be taken into consideration. The flow component in the $z$ direction, for example, can thus be represented with the following equation [1]:

$$\frac{\partial}{\partial t}\rho v_z = -\left(\frac{\partial}{\partial x}\rho v_x v_z + \frac{\partial}{\partial y}\rho v_y v_z + \frac{\partial}{\partial z}\rho v_z v_z\right) - \frac{\partial p}{\partial z} - \left(\frac{\partial}{\partial x}\tau_{xz} + \frac{\partial}{\partial y}\tau_{yz} + \frac{\partial}{\partial z}\tau_{zz}\right) + \rho g_z. \tag{4.8}$$

For definition of the components of the always symmetric, so-called extra-stress tensor, see Fig. 4.1. The first index specifies the axis of the coordinate system, which cuts through the plane in which the stress is acting. The second index indicates the direction of the stress.

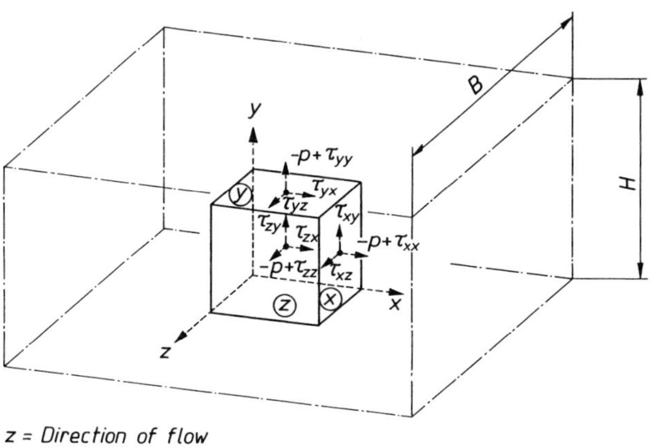


**Figure 4.1** Definition of the stress components acting on a fluid particle in a rectangular channel

In a generally valid vector and tensor symbolism (i.e., independent of the coordinate system) the momentum equation becomes

$$\underbrace{\frac{\partial}{\partial t}\rho\vec{v}}_{\substack{\text{Momentum change}\\ \text{per unit time and}\\ \text{unit of volume}}} = \underbrace{-(\nabla\rho\cdot\vec{v}\cdot\vec{v})}_{\substack{\text{Change in the motion}\\ \text{(momentum flux)}\\ \text{per unit time and}\\ \text{unit volume}}} \underbrace{-\nabla p}_{\substack{\text{Pressure forces}\\ \text{acting on the element}\\ \text{under consideration}\\ \text{per unit volume}}} \underbrace{-\nabla\tau}_{\substack{\text{Change in surface}\\ \text{forces per unit time and}\\ \text{volume unit}}} \underbrace{+\rho\vec{g}}_{\substack{\text{Gravitational force}\\ \text{on the element under}\\ \text{consideration per}\\ \text{unit volume}}} \tag{4.9}$$

where $\vec{g}$ is the vector of acceleration due to gravity.

Equation (4.8) can be rearranged so that it appears in a substantial notation:

$$\rho\frac{\mathrm{D}v_z}{\mathrm{D}t} = -\frac{\partial p}{\partial z} - \left(\frac{\partial\tau_{xz}}{\partial x} + \frac{\partial\tau_{yz}}{\partial y} + \frac{\partial\tau_{zz}}{\partial z}\right) + \rho g_z. \tag{4.10}$$

The expressions for the $x$ and $y$ directions can be derived in the same fashion. The general formulation of the equation is

$$\rho\frac{\mathrm{D}\vec{v}}{\mathrm{D}t} = -\nabla p - \nabla\tau + \rho\vec{g}. \tag{4.11}$$

This equation states that an element of mass, moving with the flow, is accelerated because of the forces acting on it. This means that it complies with the well-known Newton's second law: mass times acceleration equals the sum of the forces acting.

Equations (4.3) and (4.5) are equivalent just as are Equations (4.9) and (4.11); however, the placement of the observer is different in each case.

### 4.1.3 Energy Equation

The law of conservation of energy for a volume element through which there is flow can be formulated in a similar fashion as the momentum equation [1]:

$$
\begin{aligned}
\left\{\begin{array}{l}\text{Change in internal}\\ \text{and kinetic energy}\\ \text{per unit time}\end{array}\right\} = & \left\{\begin{array}{l}\text{Increase in internal}\\ \text{and kinetic energy}\\ \text{through convection}\\ \text{per unit time}\end{array}\right\} - \left\{\begin{array}{l}\text{Decrease in internal}\\ \text{and kinetic energy}\\ \text{through convection}\\ \text{per unit time}\end{array}\right\} \\
& + \left\{\begin{array}{l}\text{Change in internal}\\ \text{energy through}\\ \text{conduction per}\\ \text{unit time}\end{array}\right\} - \left\{\begin{array}{l}\text{Work which the}\\ \text{system carries out}\\ \text{on the environment}\\ \text{per unit time}\end{array}\right\} \\
& + \left\{\begin{array}{l}\text{Change in internal}\\ \text{energy due to}\\ \text{heat sources per unit time}\end{array}\right\}.
\end{aligned}
\tag{4.12}
$$

Kinetic energy is understood to be the energy that is coupled directly to the motion of the fluid. Internal energy is understood to be the energy that is coupled directly to molecular motion and interactions; it depends on the local temperature and on the density of the flowing medium.

The energy equation for Cartesian coordinates is derived, for example, in [1]:

$$
\begin{aligned}
\frac{\partial}{\partial t}\left(\rho U+\frac{1}{2}\rho v^2\right)= & -\left(\frac{\partial}{\partial x}v_x\left(\rho U+\frac{1}{2}\rho v^2\right)+\frac{\partial}{\partial y}v_y\left(\rho U+\frac{1}{2}\rho v^2\right)+\frac{\partial}{\partial z}v_z\left(\rho U+\frac{1}{2}\rho v^2\right)\right) \\
& -\left(\frac{\partial \dot{q}_x}{\partial x}+\frac{\partial \dot{q}_y}{\partial y}+\frac{\partial \dot{q}_z}{\partial z}\right)+\rho\left(v_x g_x+v_y g_y+v_z g_z\right) \\
& -\left(\frac{\partial}{\partial x}pv_x+\frac{\partial}{\partial y}pv_y+\frac{\partial}{\partial z}pv_z\right)-\left(\frac{\partial}{\partial x}\left(\tau_{xx}v_x+\tau_{xz}v_y+\tau_{xy}v_z\right)\right. \\
& \left.+\frac{\partial}{\partial y}\left(\tau_{yx}v_x+\tau_{yy}\tau_y+\tau_{yz}v_z\right)+\frac{\partial}{\partial z}\left(\tau_{zx}v_x+\tau_{zy}v_y+\tau_{zz}v_z\right)\right)+\phi.
\end{aligned}
\tag{4.13}
$$

Or in general vector symbolism:

$$
\underbrace{\frac{\partial}{\partial t}\rho\left(U+\frac{1}{2}v^2\right)}_{\substack{\text{Change in energy}\\ \text{per unit time and}\\ \text{volume}}}
=\underbrace{-\left[\nabla\cdot\rho\vec{v}\left(U+\frac{1}{2}v^2\right)\right]}_{\substack{\text{Change in energy per unit}\\ \text{time and volume through}\\ \text{convection}}}
\underbrace{-\nabla\cdot\vec{q}}_{\substack{\text{Change in energy}\\ \text{per unit time and}\\ \text{volume}\\ \text{due to heat}\\ \text{conduction}}}
\underbrace{+\rho\left(\vec{v}\cdot\vec{g}\right)}_{\substack{\text{Work per unit}\\ \text{time and}\\ \text{volume due}\\ \text{to gravitational}\\ \text{forces}}}
\underbrace{-\nabla\cdot p\vec{v}}_{\substack{\text{Work per unit}\\ \text{time and}\\ \text{volume due}\\ \text{to pressure}\\ \text{forces}}}
\underbrace{-\nabla\cdot\left[\tau\cdot\vec{v}\right]}_{\substack{\text{Work per unit}\\ \text{time and}\\ \text{volume due}\\ \text{to viscous}\\ \text{forces}}}
\underbrace{+\phi}_{\substack{\text{Change in}\\ \text{internal}\\ \text{energy}\\ \text{due to heat}\\ \text{sources}}}
\tag{4.14}
$$

where $\vec{q}$ is the heat flux vector (components $\dot{q}_x, \dot{q}_y, \dot{q}_z$) and $v^2 = v_x^2 + v_y^2 + v_z^2$. After casting in substantial form, Equation (4.14) becomes [1]

$$\rho = \frac{D}{Dt}\left(U + \frac{1}{2}v^2\right) = -\nabla \cdot \vec{q} + \rho(\vec{v} \cdot \vec{g}) - \nabla p \cdot \vec{v} - \nabla \cdot (\tau \cdot \vec{v}) + \phi. \tag{4.15}$$

As also shown in [1], the following is obtained by subtracting an expression for $\rho \frac{D}{Dt}\left(\frac{1}{2}v^2\right)$:

| $\rho \frac{Du}{Dt}$ | $= -\nabla \cdot \vec{q}$ | $-p(\nabla \cdot \vec{v})$ | $-(\tau : \nabla \vec{v})$ | $+\phi.$ | (4.16) |
|---|---|---|---|---|---|
| Change in internal energy per unit time and volume | Change in energy per unit time and volume due to heat conduction | Recoverable portion of the work per unit time and volume because of compression | Nonrecoverable portion of the work per unit time and volume because of viscous dissipation (work of friction in the flowing medium) | Change in energy per unit time and volume due to heat sources | |

The dissipation term is

$$\tau : \nabla \vec{v} = \tau_{xx}\frac{\partial v_x}{\partial x} + \tau_{yy}\frac{\partial v_y}{\partial y} + \tau_{zz}\frac{\partial v_t}{\partial z} + \tau_{xy}\left(\frac{\partial v_x}{\partial y} + \frac{\partial v_y}{\partial x}\right) + \tau_{yz}\left(\frac{\partial v_y}{\partial z} + \frac{\partial v_z}{\partial y}\right) + \tau_{zx}\left(\frac{\partial v_z}{\partial x} + \frac{\partial v_x}{\partial z}\right). \tag{4.17}$$

For many applications, it is more meaningful to formulate the energy Equation (4.16) in terms of temperature $T$ and the specific heat capacity $c_p$, instead of in terms of the internal energy.

At constant pressure $p$ [1]:

$$dU = c_p \, dT - p \, dv_p \tag{4.18}$$

where $v_p$ is specific volume.

It follows that

$$\rho\frac{\mathrm{D}U}{\mathrm{D}t}=\rho c_{\mathrm{p}}\frac{\mathrm{D}T}{\mathrm{D}t}-\rho p\frac{\mathrm{D}v_{\mathrm{p}}}{\mathrm{D}t}. \tag{4.19}$$

with

$$\frac{\mathrm{D}T}{\mathrm{D}t}=\frac{\partial T}{\partial t}+\left(\vec{v}\cdot\nabla\right)T \tag{4.20}$$

from the definition of a substantial derivative

$$\rho\frac{\mathrm{D}v_{\mathrm{p}}}{\mathrm{D}t}=\rho\underbrace{\frac{\partial v_{\mathrm{p}}}{\partial t}}_{=0}+\rho(\vec{v}\cdot\nabla)v_{\mathrm{p}} \tag{4.21}$$

and

$$\rho=\frac{1}{v_{\mathrm{p}}} \tag{4.22}$$

It follows from Equation (4.16) that

$$\rho c_{\mathrm{p}}\left[\frac{\partial T}{\partial t}+\left(\vec{v}\cdot\nabla\right)T\right]=-\nabla\vec{q}-(\tau:\nabla\vec{v})+\phi. \tag{4.23}$$

According to the Fourier law of heat conduction:

$$\vec{q}=-\lambda\nabla T, \tag{4.24}$$

where $\lambda$ is the thermal conductivity.

It therefore follows that

$$\rho c_{\mathrm{p}}\left[\frac{\partial T}{\partial t}+\left(\vec{v}\cdot\nabla\right)T\right]=\lambda\cdot\nabla^{2}T-(\tau:\nabla\vec{v})+\phi. \tag{4.25}$$

This equation, suitable for calculating temperature profiles in extrusion dies, contains—apart from the dissipative term—an additional heat source $\phi$. This term must be taken into consideration, for example, if there are heats of chemical reaction. This term is dropped when processing thermoplastic materials, but must possibly be taken into consideration for cross-linkable compounds.

## 4.2 Restrictive Assumptions and Boundary Conditions

The system of the conservation equations cannot be solved in its general form, so, for the computation of velocity and temperature fields, appropriate assumptions for its simplification must be made. Which simplifications must be made depends on the choice of computational method. The accuracy of the results from a computation depends strongly on how close the restrictive assumptions are to the reality. Therefore, in each case a critical evaluation of the validity of the assumptions is necessary.

The most important assumptions made for the design of extrusion dies are summarized below:

1. Steady-state laminar flow $\left(\frac{\partial}{\partial t}(\ldots)=0\right)$: For continuous extrusion processes, the assumption of a steady-state flow is permissible; this is not valid for start-ups or discontinuous processes such as blow-molding with so-called accumulator heads.
2. Inertial and gravitational forces can be neglected when compared to frictional forces and pressure. This assumption is always a good approximation for polymeric melts.
3. Wall adhesion: This condition is under certain circumstances not met locally (see Section 3.5). Since the generally valid models and material data describing the wall slip are not available, wall adhesion is almost always assumed.
4. Reduction of the variable direction of coordinates (fixation of neutral direction of coordinates): For example, for circular channels $\frac{\partial}{\partial \theta}(\ldots)=0$, which means a concentric distribution of the variables of state (rotational symmetry).

   For the flow through a slit, the effect of the confining channel side walls can be neglected when $B/H$ is greater than 10, which means that there is no change in the direction of width (plane flow).
5. The velocity gradient in the direction of flow is much smaller than in the transverse direction. If the cross section of the flow channel in the direction of the flow remains constant, the borderline case of the plane viscometric flow [3,4] is obtained, as occurs, for example, in the land region of dies for profile extrusion.
6. Constant pressure gradient in the channel cross section, meaning no consideration of the effects of normal stress is necessary.
7. Hydrodynamically fully developed flow in every cross section of the channel with one direction of shear perpendicular to the axis of the channel. The gra-

dients of temperature and shear rate in the direction of flow are so small that the shear stress can be expressed in terms of the viscosity corresponding to the local shear rate and temperature (compare to Assumption 5).

8. The convective heat transfer in the direction of the flow is greater than the conductive heat transfer.
9. The heat transfer perpendicular to the flow direction occurs strictly due to heat conduction.
10. Constant density (incompressibility), constant heat transfer, and constant thermal diffusivity of the melts.

    The constancy of these material parameters can be applied only when the thermodynamic state in the given process zone changes only very little, when no points of transition occur, and when the material parameters can be taken as average values within this zone with a good approximation (compare to Chapter 2).

The following thermal boundary conditions are often set:

11. The heat flux in the melt layer at the wall of a die for a melt adhering to the wall is considered by the following:

$$\dot{q}_{\text{wall}} = -\lambda \frac{\partial T}{\partial y} = -k \cdot \left(T_{\text{R}} - T_{\text{W}}\right) \tag{4.26}$$

where

$\lambda$ Heat conductivity

$T_{\text{W}}$ Temperature of the melt at the wall

$T_{\text{R}}$ Known, controlled temperature in the body of the die

$k$ Coefficient of heat transmission for the wall between the contact layer of melt/die and the point of the temperature $T_{\text{R}}$ at distance $s$ from the wall

It follows from Equation (4.26) that

$$\left.\frac{\partial T}{\partial y}\right|_{\text{wall}} = Bi \frac{T_{\text{R}} - T_{\text{W}}}{s}; \quad Bi = \frac{k \cdot s}{\lambda}. \tag{4.27}$$

The dimensionless Biot number *Bi*, which characterizes the heat transfer through the wall, varies for extrusion dies between 1 and 100 [6,7]. It pertains only to the heat flux perpendicular to the contact surface between melt and die.

12. The temperature of the melt at the wall is equal to the wall temperature: $T_{\text{R}} = T_{\text{W}}$. This is referred to as an isothermal wall ($Bi \rightarrow \infty$). However, with $Bi =$ 100, the wall can be considered isothermal already [8].

13. If $(\partial T/\partial y)_{wall}$ equals zero, *Bi* equals zero, too. Here, then, the wall is considered to be adiabatic: it takes up the temperature of the melt closest to it. (This assumption is permissible for long spider legs, breaker plates, or melt strainers. This adiabatic assumption is in principle valid on the axis of rotation for axially symmetric computations of flow and temperature.)

    For *Bi* less than one, the adiabatic condition can be assumed with a good approximation [8].

The following two items are chosen partially as initial conditions:

14. Temperature profiles at the die or calibrator inlet.
15. Velocity profiles at the die inlet.

## 4.3 Analytical Formulas for Solution of the Conservation Equations

Analytical solutions of the conservation equations, which form a linked system of nonlinear (Fig. 4.2) partial differential equations, exist only for simple, special cases; some of them are demonstrated in Chapter 3.

In the following paragraphs, the derivation of an analytical solution for the velocity and temperature fields on a simple case will be explained.

Consider a channel with a shape of a slit (Fig. 4.1) with height *H* and width *B*. Here, *H* is much smaller than *B*, so the edge effects can be neglected and an assumption of a uniform flow can be made.

A pseudoplastic melt flows through that channel $\left(\tau^{m} = \frac{1}{\phi} \cdot \dot{\gamma}\right)$, while the flow is laminar and steady state (the entry effects can be neglected). The melt exhibits wall adhesion $\left(v_z = 0 \text{ for } y = \pm \frac{H}{2}\right)$ and has only the flow components in the *z* direction.

Find the velocity profile $v_z(y) \neq f(z)$ as well as the temperature profile $T(y) \neq f(z)$ in the melt (one-dimensional solution since all parameters are only dependent on the *z* coordinate). That means that the convective heat flux in the *z* direction is not taken into consideration. (However, in reality it is quite high because of the low thermal diffusivity of the polymeric melts.) The temperature of the die wall is equal to $T_W$; all thermodynamic material parameters are constant.

The solution presented in the following sections is characteristic for the description of flows with constant material parameters and shows the coupling between the velocity and temperature in the flow channel:

1. Fixing a Coordinate System

   Generally, Cartesian and cylindrical coordinates are available for the design of dies (see Fig. 4.1 for the coordinate system chosen for the example).

2. Selection of the Conservation Equations and Their Simplification

   Often not all of the three components of the continuity, motion, and energy equations are required, but one or two coordinate directions can be considered indifferent. In the case under consideration, there are only velocity components in the $z$ direction, whereas gradients of velocity and temperature are only nonzero in the $y$ direction. Therefore, all the parts of the conservation equations that contain $v_x$ and $v_y$ as well as derivatives in the $x$ and $z$ directions are omitted.

   The conservation equations as well as the possible simplifications are presented in detail below:

   *Motion Equation* (see Equations (4.10) and (4.11), e.g., [1]):

$$\rho\left(\underset{1}{\frac{\partial v_z}{\partial t}}+\underset{2}{v_x\frac{\partial v_z}{\partial x}}+\underset{3}{v_y\frac{\partial v_z}{\partial y}}+v_z\frac{\partial v_z}{\partial z}\right)=-\frac{\partial p}{\partial z}-\left(\underset{4}{\frac{\partial \tau_{xz}}{\partial x}}+\frac{\partial \tau_{yz}}{\partial y}+\underset{5}{\frac{\partial \tau_{zz}}{\partial z}}\right)+\underset{6}{\rho g_z} \tag{4.28}$$

   *1* = 0: Since it is steady state

   *2* = 0: Since $v_x = 0$

   *3* = 0: Since $v_y = 0$

   *4* = 0: Since $B \gg H$, the effect of shear stresses at the surfaces laterally limiting the channel ($\tau_{xz}$) can be disregarded

   *5* = 0: Normal stresses due to deformation can be disregarded

   *6* = 0: Gravitational forces can generally be disregarded for polymer melt flows

   After simplification, it follows from Equation (4.28) that

$$\rho\left(v_z\frac{\partial v_z}{\partial z}\right)=-\frac{\partial p}{\partial z}-\frac{\partial \tau_{yz}}{\partial y} \tag{4.29}$$

   *Continuity Equation* (see Equations (4.2) and (4.3), e.g., [1]):

$$\underset{1}{\frac{\partial \rho}{\partial t}}+\underset{2}{\frac{\partial}{\partial x}(\rho v_x)}+\underset{3}{\frac{\partial}{\partial y}(\rho v_y)}+\frac{\partial}{\partial z}(\rho v_z)=0 \tag{4.30}$$

   *1* = 0: Density is constant, so there is no change with time

   *2* = 0: Since $v_x = 0$

   *3* = 0: Since $v_y = 0$

It follows from Equation (4.30) that

$$\frac{\partial}{\partial z}(\rho v_z) = 0 \triangleq \rho \frac{\partial v_z}{\partial z} + \underbrace{v_z \frac{\partial \rho}{\partial z}}_{=0 \text{ since } \rho = \text{constant}} = 0 \tag{4.31}$$

$$\text{It follows that} \quad \rho \frac{\partial v_z}{\partial z} = 0 \quad \text{and therefore} \quad \frac{\partial v_z}{\partial z} = 0 \tag{4.32}$$

Combining Equation (4.32) (continuity equation) with Equation (4.29) (equation of motion) and simplifying:

$$0 = -\frac{\partial p}{\partial z} - \frac{\partial \tau_{yz}}{\partial y} \tag{4.33}$$

*Energy Equation* (see Equations (4.16) and (4.23), e. g., [1]):

$$\begin{aligned}
\rho c_p & \left( \underset{1}{\frac{\partial T}{\partial t}} + \underset{2}{v_x \frac{\partial T}{\partial x}} + \underset{3}{v_y \frac{\partial T}{\partial y}} + \underset{4}{v_z \frac{\partial T}{\partial z}} \right) = \\
& - \underset{5}{\left[ \frac{\partial q_x}{\partial x} + \frac{\partial q_y}{\partial y} + \underset{6}{\frac{\partial q_z}{\partial z}} \right]} - \underbrace{\left( \frac{\partial v_x}{\partial x} + \frac{\partial v_y}{\partial y} + \frac{\partial v_z}{\partial z} \right)}_{7} \cdot p \\
& - \left\{ \underset{7}{\tau_{xx} \frac{\partial v_x}{\partial x}} + \underset{7}{\tau_{yy} \frac{\partial v_y}{\partial y}} + \underset{7}{\tau_{zz} \frac{\partial v_z}{\partial z}} \right\} \\
& - \left\{ \tau_{xy} \left( \frac{\partial v_x}{\partial y} + \underset{7}{\frac{\partial v_y}{\partial x}} \right) + \tau_{xz} \left( \underset{7}{\frac{\partial v_x}{\partial z}} + \underset{7}{\frac{\partial v_z}{\partial x}} \right) + \tau_{yz} \left( \underset{7}{\frac{\partial v_y}{\partial z}} + \underset{7}{\frac{\partial v_z}{\partial y}} \right) \right\}
\end{aligned} \tag{4.34}$$

*1* = 0: Since steady-state conditions

*2* = 0: Since $v_x = 0$

*3* = 0: Since $v_y = 0$

*4* = 0: Since *T*(*y*) as well as $v_z \neq f(z)$

*5* = 0: Temperature gradient in the *z* direction was assumed to be 0

*6* = 0: Since $T \neq f(z) \rightarrow q_z = 0$

*7* = 0: See above

After simplifying, it follows from Equation (4.34) that

$$0 = -\frac{\partial q_y}{\partial y} - \tau_{yz}\frac{\partial v_z}{\partial y} \tag{4.35}$$

and with $\tau_{yz} = \tau_{zy}$ and $q_y = -\lambda\frac{\partial T}{\partial y}$ to *Fourier* that

$$\lambda\frac{\partial}{\partial y}\left(\frac{\partial T}{\partial y}\right) = \tau_{yz}\frac{\partial v_z}{\partial y} \tag{4.36}$$

To solve the problem, use the

*Material Law* (see Equation (2.3)):

$$\tau_{yz}^{m} = \frac{1}{\phi}\frac{\partial v_z}{\partial y} \tag{4.37}$$

It follows that

$$\tau_{yz} = \left(\frac{1}{\phi}\right)^{\frac{1}{m}}\left(\frac{\partial v_z}{\partial y}\right)^{\frac{1}{m}} \tag{4.38}$$

3. Computation of the Velocity Profile

Assuming $\phi$ and $m$ to be approximately independent of temperature, the momentum equation can be decoupled from the energy equation. In this case, only Equations (4.32) and (4.29) suffice for the computation of the velocity profile $v(y)$:

From Equation (4.33) it follows that

$$\tau_{yz} = -\left(\frac{\partial \rho}{\partial z}\right)\cdot y + C_1. \tag{4.39}$$

With $\tau_{yx} = 0$ for $y = 0 \Rightarrow C_1 = 0$:

$$\tau_{yz} = -\frac{\partial \rho}{\partial z}\cdot y. \tag{4.40}$$

It follows from Equation (4.39) and with a negative algebraic sign for $\dot{\gamma}$, since $v_z$ decreases in the $y$ direction:

$$\frac{\partial v_z}{\partial y} = \phi\left(\frac{\partial p}{\partial z}\right)^{m} y^{m}. \tag{4.41}$$

Integration leads to

$$v_z = \phi \left( \frac{\partial p}{\partial z} \right)^m \frac{y^{m+1}}{m+1} + C_2. \tag{4.42}$$

With $v_z = 0$ for $y = \frac{H}{2}$:

$$C_2 = -\phi \left( \frac{\partial p}{\partial z} \right)^m \frac{\left( \frac{H}{2} \right)^{m+1}}{m+1} \tag{4.43}$$

$$v_x(y) = -\phi \left( \frac{\partial p}{\partial z} \right)^m \frac{\left( \frac{H}{2} \right)^{m+1} - y^{m+1}}{m+1}. \tag{4.44}$$

The negative algebraic sign illustrates that flow takes place in the negative pressure gradient, that is, in the direction of lower pressure. Equation (4.44) corresponds to the relationships given in Table 3.4 $\left( -\frac{\partial p}{\partial z} = \frac{\Delta p}{L} \right)$.

As in Section 3.1, the maximum flow velocity $v_{max}$, the average flow velocity $\bar{v}$, and the volumetric flow rate $\dot{V}$ can be calculated from Equation (4.44).

4. *Computation of the Temperature Profile*

Assuming that $\phi$, $m$, and $\lambda$ are independent of the local melt temperature and pressure (i. e., $\phi$, $m$, $\lambda$ = const.) the temperature profile $T$(y) follows from Equations (4.36), (4.38), and (4.42):

$$\frac{\partial}{\partial y} \left( \frac{\partial T}{\partial y} \right) = -\frac{\phi}{\lambda} \left( \frac{\partial p}{\partial z} \right)^{m+1} y^{m+1}. \tag{4.45}$$

After integration:

$$\frac{\partial T}{\partial y} = -\frac{\phi}{\lambda} \left( \frac{\partial p}{\partial z} \right)^{m+1} \frac{y^{m+2}}{m+2} + C_1. \tag{4.46}$$

With $\frac{\partial T}{\partial y} = 0$ at $y = 0$ (since the temperature profile in the flow channel is symmetrical), it follows for $C_1 = 0$:

$$T = -\frac{\phi}{\lambda} \left( \frac{\partial p}{\partial z} \right)^{m+1} \frac{y^{m+3}}{(m+2)(m+3)} + C_2. \tag{4.47}$$

With $T = T_w$ for $y = H/2$:

$$\Rightarrow C_2 = T_W + \frac{\phi}{\lambda}\left(\frac{\partial p}{\partial z}\right)^{m+1} \frac{\left(\frac{H}{2}\right)^{m+3}}{(m+2)(m+3)}. \tag{4.48}$$

The resulting temperature profile $T(y)$ is

$$T(y) = T_W + \frac{\phi}{\lambda}\left(\frac{\partial p}{\partial z}\right)^{m+1} \frac{\left(\frac{H}{2}\right)^{m+3} - y^{m+3}}{(m+2)(m+3)}. \tag{4.49}$$

This solution for the temperature profile is valid under the assumption that the total energy dissipated in the flow is removed by the heat conduction in the direction perpendicular to the direction of flow, that is, through the flow channel wall. The heat development due to dissipation and the heat removed by conduction are therefore equal at any point (see Equation (4.35)).

This generally does not happen in extrusion dies at typical flow rates, so the heat removed by the flow, the convective energy transport, has to be considered. This means that the term $\rho \cdot c_p \cdot v_z \cdot \frac{\partial T}{\partial z}$ in Equation (4.34) does not become zero. Hence it follows (compare Equation (4.35)) that

$$\rho c_p v_z \frac{\partial T}{\partial z} = -\frac{\partial q_y}{\partial y} - \tau_{yz} \frac{\partial v_z}{\partial y} \tag{4.50}$$

with the Fourier law of heat conduction

$$\frac{\partial q}{\partial y} = -\lambda \frac{\partial T}{\partial y} \tag{4.51}$$

and

$$\tau_{yz} = -\eta(y) \frac{\partial v_z}{\partial y} \tag{4.52}$$

with

$$\eta(y) = \left(\frac{1}{\phi}\right)^{\frac{1}{m}} \left(\frac{\partial v_z}{\partial y}\right)^{\frac{1}{m}-1} \tag{4.53}$$

It follows from Equation (4.40) that

$$\underbrace{\rho c_p v_z \frac{\partial T}{\partial z}}_{\mathrm{I}} = \underbrace{\lambda \frac{\partial^2 T}{\partial y^2}}_{\mathrm{II}} + \underbrace{\eta(y)\left(\frac{\partial v_z}{\partial y}\right)^2}_{\mathrm{III}} . \tag{4.54}$$

- I is the convective heat transfer in the $z$ direction
- II is the conductive heat transfer in the $y$ direction
- III is the heat generation by dissipation due to the velocity gradient in the $y$ direction

Equation (4.54) makes clear that during a flow process the temperature field is coupled with the velocity field. The coupling is illustrated by the viscosity $\eta$, which throughout the process is a function of temperature $T$, hydrostatic pressure $p$, and the shear rate $\dot{\gamma}\left(\dot{\gamma} = -\frac{\partial v_z}{\partial y}\right)$ (see Chapter 2). The velocity field affects the temperature field since heat is generated in the melt by internal friction, which in turn depends on the velocity gradient $(\dot{\gamma})$. Moreover, the term in Equation (4.44) describing the convective heat transfer depends directly on the velocity field.

The coupled system from the momentum and energy equations in which the convective heat transport is considered cannot be solved analytically even for the simple flow discussed here. Numerical approximation methods are used for this purpose. Their advantage is that they are applicable to other than very simplified geometries and limited boundary conditions. Moreover, their application today is very simple since they can be applied in many ways in the form of relatively simple computer programs suitable for the inexpensive personal computer (e.g., [9–13]).

## 4.4 Numerical Solution of Conservation Equations

In the numerical solution of differential equations, the closed-form solution, a function that satisfies the differential equation over the entire domain (e.g., flow channel), is no longer the objective. Instead, values of the function at discrete points are computed or the function is approximated section-wise. The finer this division, the more accurate the result.

Among the computational methods used in engineering science today, the two most important are the finite difference method (FDM) and the finite element method (FEM).

Both these methods will be explained briefly theoretically and in examples in the next chapters. Detailed introductions into both are given, for example, in [14,15].

### 4.4.1 Finite Difference Method

When applying FDM, the subject under consideration, for example the flow channel, is covered by a grid (Fig. 4.2). The points of the grid lie always within the area or exactly on its edges [14]. In the simplest case, the grid consists of squares and its spacing is constant.

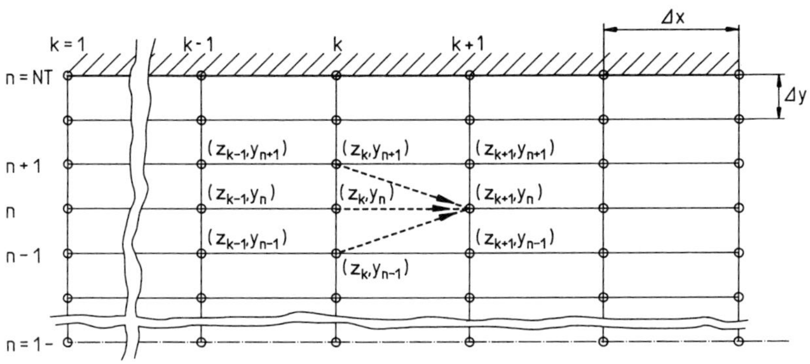


**Figure 4.2** Difference grid in a two-dimensional plane or axially symmetrical flow channel

To solve the differential equations describing a given problem, they are converted into difference equations by replacing the differential quotients by the difference quotients. For example, for the derivative of function $u$ in the $z$ coordinate at the point $z_n$ (Fig. 4.3), we obtain

$$\frac{\partial u}{\partial z}\Big|_{z_n} \approx \frac{u(z_n + \Delta z) - u(z_n)}{2 \cdot \Delta z} \tag{4.55}$$

This means that the tangent at point $(z_n, u(z_n))$ is replaced by a chord running through both points $(z_{n-1}, u(z_{n-1}))$ and $(z_{n+1}, u(z_{n+1}))$ (Fig. 4.3). The slope of the tangent that corresponds to the differential quotient is approximated by the slope of the chord, that is, by the difference quotient. Equation (4.55) corresponds to the so-called central difference quotient. This one approximates the derivative more accurately than the forward difference quotient or backward difference quotient (Fig. 4.3). The central difference quotient can be constructed only if there are points of the grid on both its sides. If the grid point $z_n$ lies on the edge and so close to the edge of the area that there is no grid point on the other side, the derivative is approximated by a forward difference quotient

$$\frac{\partial u}{\partial z}\Big|_{z_n} \approx \frac{u(z_n + \Delta z) - u(z_n)}{\Delta z} \tag{4.56}$$

or by a backward difference quotient

$$\frac{\partial u}{\partial z}\Big|_{z_n} \approx \frac{u(z_n) - u(z_n - \Delta z)}{\Delta z} \tag{4.57}$$

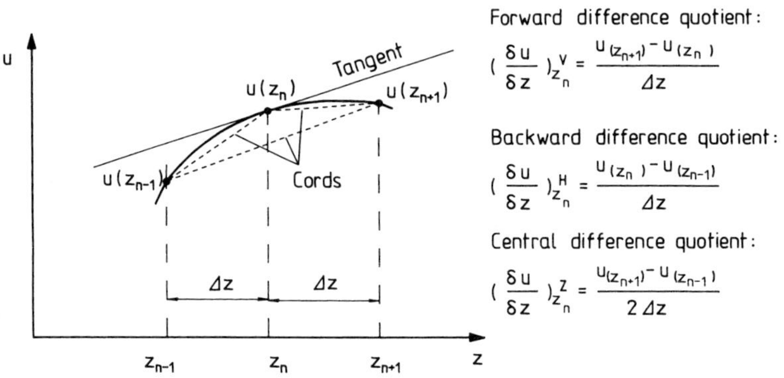


**Figure 4.3** Different ways to develop the difference quotient

As was shown schematically in Fig. 4.3 above, this approximation is less accurate than the one with the central quotient.

### *Explicit and Implicit Difference Method*

In the FDM two different variants exist: the implicit and the explicit method. When arranging the difference quotients for all the grid points simultaneous, a system of equations is obtained that contains all the points on the grid and therefore as many equations as the number of unknowns in the system. This system, which, if necessary, has to be first linearized, must be solved with a special equation-solving algorithm (some of them are described, e. g., in [14]). The values on the grid points are the solution. The system of equations therefore contains the solution implicitly; an explicit formulation that supplies the solution explicitly is not possible. Therefore, it is referred to as the implicit difference method. In the explicit difference method, the values of the function at the grid point $k + 1$ are computed explicitly from the values at the point $k$ and sometimes even from points farther back in the grid, such as $k - 1$ and $k - 2$. Thus the equation for the value of the function $u_{k+1}$ at the grid point $k + 1$ of the following form is obtained:

$$u_{k+1} = f(u_k, u_{k-1}, \ldots). \tag{4.58}$$

Using an example of a flow channel, this means that the flow in the channel is calculated successively through from the inlet to the outlet. The equations are not solved simultaneously, as is the case with the implicit method.

Explicit methods are often utilized for the computation of time-dependent problems (transient processes): here the state at the time point $t + 1$ is computed from the past states $t$, $t - 1$, ... and so on.

The advantage of the explicit difference method is that, instead of a system of equations, only a single equation has to be solved. The procedure is therefore simpler from the point of view of the programming technique and considerably faster than the implicit FDM. However, the accuracy, the stability, and the rate of convergence of the explicit FDM are smaller. It should also be noted that, in an explicit FDM analysis, retroactive effects of downstream conditions, such as from a converging channel, cannot be evaluated. Provided that certain criteria of stability and convergence are met, results with acceptable accuracy are still obtained using the explicit difference method [11,45,46]. This is particularly true when retroactive effects of downstream conditions are less important. Because of their speed and low storage requirements, the explicit difference method is particularly good for use on personal computers. Many examples are given in [9–12]. Therefore, the following sections will deal with the explicit difference method exclusively.

#### *Transformation of the Energy Equation into a Discrete Form by Explicit FDM*

In the following section, the energy equation (Equation (4.54)) is transformed into an explicit difference equation as an example of the use of this method.

To accomplish this, first, the differential quotients in Equation (4.54) are replaced by the difference quotients:

$$\frac{\partial T}{\partial z} = \frac{T_{k+1,n} - T_{k,n}}{\Delta z} \tag{4.59}$$

and

$$\frac{\partial^2 T}{\partial y^2} = \frac{T_{k,n+1} - 2T_{k,n} + T_{k,n-1}}{(\Delta y)^2}. \tag{4.60}$$

By substituting this into Equation (4.54) and solving for $T_{k+1,n}$ [16], we obtain

$$\begin{aligned} T_{k+1,n} &= T_{k,n} + \frac{\lambda \Delta z}{\rho c_p \bar{v}(n) \Delta y^2} \left( T_{k,n+1} - 2T_{k,n} + T_{k,n-1} \right) \\ &+ \frac{\Delta z \eta(n)}{\rho c_p \bar{v}(n)} \left( \frac{v_{k,n} - v_{k,n+1}}{\Delta y} \right)^2 . \end{aligned} \tag{4.61}$$

For the value of $v_z$ here the average velocity in the given time step is set:

$$v_z \sim \bar{v}(n) = \frac{1}{2} \cdot (v_{k,n} + v_{k+1,n}). \quad (4.62)$$

Equation (4.61) makes clear that the temperature distribution at the grid point $k + 1$, that is, in the direction of flow, is computed from the temperature distribution at the grid points $k$ farther back in the grid (Fig. 4.2).

When dividing into layers, as in Fig. 4.2, it has to be kept in mind that Equation (4.61), as it is, is only valid for $2 \le n \le NT - 1$. If the die temperature $T_W$ is given, the temperature in the layer $NT$ is computed by applying Equation (4.61) and introducing the new step width $\Delta y/2$ in the proximity of the die wall (compare Fig. 4.2):

$$T_{k,NT-1/2} = \frac{T_{k,NT} - T_{k,NT-1}}{2}. \quad (4.63)$$

Setting with the step width $\Delta y/2$ into Equation (4.61): $T_{k,n+1} = T_W$, $T_{k,n} = T_{k,NT}$, and $T_{k,n-1} = T_{k,NT-1/2}$

For the reason of symmetry in the flow through a slit, the layer $n = 0$ has to have the same temperature as the layer $n = 1$ so that Equation (4.61) can be set up: for $n = 1$, $T_{k,n} = T_{k,n-1}$ [16,17].

For channels with axial symmetry the procedure is similar, as is shown in [17], for example.

### 4.4.2 Finite Element Method

The finite element method (FEM) is an approximation procedure to solve differential equations. Originally, it was developed for structural analysis, that is, for the computation of forces, stresses, and deformations in solid bodies. In the past few years, the area of applications of FEM has steadily expanded; today, FEM is considered to be a universal procedure for the numerical solution of a wide range of engineering problems.

In the following section, the fundamentals of FEM will be explained only to the extent necessary for users of FEM programs. For a more detailed treatment of this subject, there are many textbooks, e.g., [14,18–22].

#### *General Formulation of the Finite Element Equations*

The FEM is based in principle on the approximation of a continuum with an infinite number of degrees of freedom by a sum of contiguous subregions with a finite number of unknowns. These subregions are the finite elements. The shape of the function that satisfies the differential equation is approximated. The approximation is done by interpolation functions and their derivatives; the points used in the interpolation are the nodal points of the finite elements. The equations are first set up for each element (local equations). Subsequently they are combined, and the result of this is a system of equations (global equations) describing the entire problem.

The system then is solved with the consideration of the pertinent initial and boundary conditions. The values of the desired function on the nodal points are the solution. In the following section, it will be explained how to obtain the local finite element equations for each element from a given differential equation by means of weighted residuals.

Consider a differential equation (DE) of a function dependent on position $u(z)$:

$$\mathrm{DE}\left(u\right)=0 \tag{4.64}$$

(the type of the DE does not matter here because the general procedure does not depend on it).

In the first step, an approximation of the form

$$u\left(\vec{z}\right)=\sum_{\mathrm{N}}\phi_{\mathrm{N}}\left(\vec{z}\right)\cdot u_{\mathrm{N}} \tag{4.65}$$

is made for the shape of the function $u$ in the (finite) element. Here, $N$ is the number of nodes in the element, $\phi_{\mathrm{N}}$ the interpolation functions, and $u_{\mathrm{N}}$ are the (constant) values at the nodes.

The interpolation functions $\phi_{\mathrm{N}}$ are referred to as shape functions. They have a central significance in the FEM, as will be explained later.

The values $u_{\mathrm{N}}$ of the approximation function (Equation (4.65)) at the interpolation points are constant; only the shape functions are dependent on the coordinate. Thus, the first derivative of the function $u$ in the direction of the $z$ coordinate, for example, is

$$\frac{\partial u\left(\vec{z}\right)}{\partial z}=\sum_{\mathrm{N}}\frac{\partial\phi_{\mathrm{N}}\left(\vec{z}\right)}{\partial z}\cdot u_{\mathrm{N}}. \tag{4.66}$$

Thus, the first local derivative is obtained by multiplying the values at the nodes by the derivatives of the shape functions and subsequently summing these $N$ products. When inserting Equations (4.65) and (4.66) into DE (Equation (4.64)), generally a nonzero value is obtained on the right side of the equation (as an error), which is called the residual, $\varepsilon$:

$$\mathrm{DE}\left(\phi_{\mathrm{N}}\left(\vec{z}\right),\frac{\partial\phi_{\mathrm{N}}\left(\vec{z}\right)}{\partial z},u_{\mathrm{N}}\right)=\varepsilon. \tag{4.67}$$

To minimize this residual, the scalar product of the residual and the weighting functions $W_i$ (where $i = 1, 2, \dots N$) is used; this product is then integrated over element area $\Omega$, and the resulting integral is set to zero:

$$\int_{\Omega} \varepsilon \cdot W_i \, d\Omega = 0. \tag{4.68}$$

This way the error of the approximate solution of the differential equation is forced to be zero-averaged over the element (the so-called weak formulation). In this general form this procedure is called the method of weighted residuals. Depending on the choice of the weighting functions $W_i$ different expressions are obtained; one of them, the Galerkin method, has become the most important since it provides the best approximate solution [19].

In the Galerkin method, the $N$ weight functions $W_i$ are set equal to the shape functions $\phi_i$. Thus Equation (4.68) becomes

$$\int_{\Omega} \varepsilon \cdot \phi_i \, d\Omega = 0. \tag{4.69}$$

This way a system of $N$ equations is obtained for each element.

When derivatives of a higher order than one appear in the DE (Equation (4.64)), Equation (4.69) must be converted by means of the Green-Gauss theorem into one volume integral with the highest order derivative reduced by one and one surface integral. This becomes important since higher than first-order derivatives of the shape functions generally are not defined, as will be shown in the following section [18,19].

#### *Shape Functions and Accuracy*

The stability and accuracy of a finite element solution depends on the choice of the shape functions. They are characterized by the shape of the finite elements as well as by the order of the approximation. When the domain has two dimensions, a choice can be made between rectangular and triangular elements.

For three-dimensional elements, the choice is increased by including, for example, tetrahedron, hexahedron (cube), or prismatic elements. Linear, quadratic, and (less frequently) cubic polynomials are used as shape functions almost exclusively. The shape functions are utilized not only for the approximation of the desired quantity but also of the area being studied (such as a profile cross section and flow channels), hence their name. When computing areas with curved borderlines, it is clear that linear shape functions are less suitable for that purpose.

More important, though, is the choice of the appropriate shape functions for the approximation of the desired quantity. These can be equal to shape functions for the approximation of the area, but it is not imperative. The residual defined by Equation (4.67) becomes smaller when the interpolation Equation (4.65) describes the actual course of the desired function in the element with increasing accuracy. This is clearly demonstrated by Fig. 4.4 [23]. It shows a velocity profile for a pseudoplastic fluid in a pipe. A solution using FEM with one element over the radius provided the function course marked with 1 when a linear shape function was selected. Although the interval under the straight line (the volumetric flow rate) can come very close to the actual volumetric flow rate (the residual is forced to zero in the mean, Equation (4.69)), there are very large deviations at discrete points. This can be markedly improved by the choice of a quadratic shape function (curve 2). Figure 4.4 shows also that an increase in the density of the discrete points by increasing the number of elements leads to a reduction in a local error. In order to obtain equal accuracy for linear and quadratic interpolation, considerably more linear elements would be necessary.

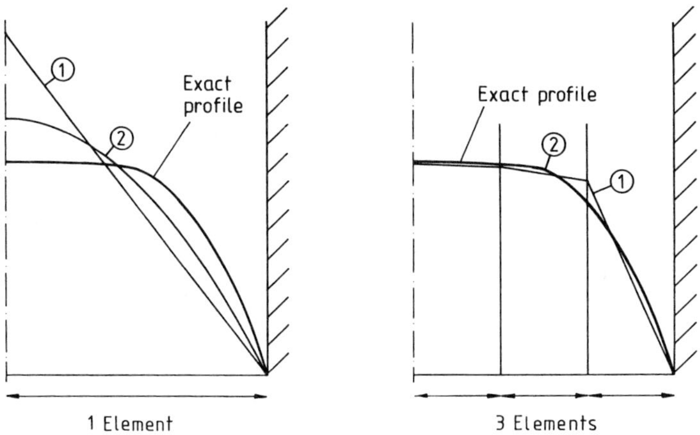


**Figure 4.4** Dependence of the accuracy of approximation on the degree of the polynomial and the number of elements

However, it cannot be generally stated that the required computation effort decreases with an increased order of the polynomial. The reason is that the size of the systems of equations for the elements increases with the increasing number of nodes; additionally, with the increase in the band width of the ordinary narrow-banded total equation systems, the amount of computation required increases at an exceedingly fast rate (the band width here is the width of a band along the main diagonal of the coefficient matrix, outside of which there are no longer nonzero coefficients). Another disadvantage of polynomials of higher order is that they lead to oscillations in the solutions and also to a decrease in the smoothness of the results. For the computations of flow and temperature problems, parabolic shape functions are used most frequently.

Finally, the question of the description of the first derivative (from here on, the derivative is the local derivative) by Equations (4.66) or (4.67) should be discussed. This is important since, in the differential equations, there are derivatives at least of the first order; furthermore, the parameters derived from the velocity field (e.g., shear rates) are needed for the computation of the flow. The shape functions guarantee the continuity of the interpolated quantities not only within the element, but also at their borderlines [18–22]. This is generally not the case for the first derivative, as explained in Fig. 4.5 [23] with an example of two elements with linear shape functions. Because of a curvature in the function being approximated, a bend in the shape of the approximating function appears at the borderline of the element and, thus, a discontinuity in the first derivative. The second derivative is here no longer defined; therefore such occurrences are not permissible in the DE while applying the usual shape functions. For this reason, as mentioned before, the order of the DE is reduced by the Green-Gauss theorem. It is also clear from Fig. 4.5 that the quality of the definition of the first derivative near the borderline of the element is low. This has to be considered when computing, for example, stresses or shear and extension rates from the velocity fields.

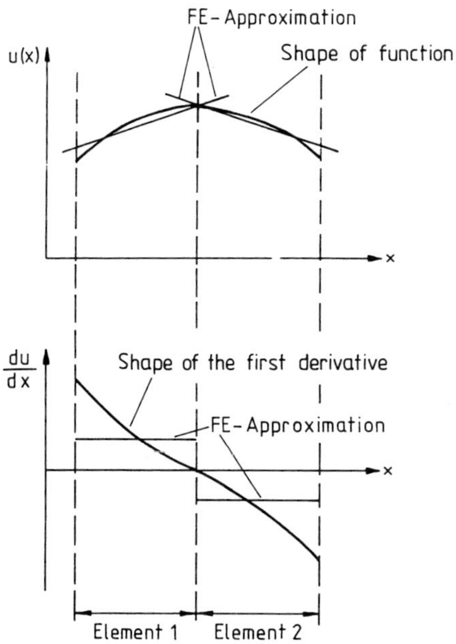


**Figure 4.5** Discontinuity of the first derivative of the quantity $u(x)$ interpolated by the means of the shape function

However, methods can be designed, using subsequent computations, that allow an improvement of the quality of parameters computed from local derivatives and a smooth shape in the entire domain (the so-called smoothing procedure [20,23–25]).

### 4.4.3 Comparison of FDM and FEM

For the user, the question arises as to which type of computation is the most suitable for the given situation. As mentioned already, the analytical methods and the various numerical methods differ in the amount of computation required as well as in the accuracy and utility of the results [26]. In order to make a general classification of the numerical methods of solution, the simple methods of computation based on analytical solutions as are described in Chapters 5 and 7 will be included in the comparison.

Simple computational methods are based on a one-dimensional analytical solution of the differential equations describing the problem (see Chapter 3 and Section 4.3). Hereby, the real process is approximated by a suitable superposition of one-dimensional procedures. Advantages of analytical methods are [26] as follows:

- They are simple and fast to use since the desired parameters are connected with the given boundary conditions by simple equations. Therefore, they can be performed on a pocket calculator. A personal computer is even more effective since it provides the desired answers even faster.
- The main advantage of the simple computational methods is, however, that the relationships are reversible so that they can be differentiated, integrated, and solved with regard to any parameter. Consequently, a suitable geometry for a (melt) manifold, for example, can be computed in one step from the requirement of uniform velocity distribution at the outlet (see Chapter 5).

On the other hand, the simple methods have considerable disadvantages:

- Their application is limited to dies that can be approximated by basic geometries defined easily by a simple analysis.
- Interactions and transition effects cannot be described in their entirety.
- Results can be often obtained only integrated over the entire cross section (e.g., pressure loss, volumetric flow rate, cooling time).

When more complex geometries are computed or more accurate simulations performed, numerical solutions of differential equations describing the problem, such as FDM or FEM, have to be used.

In the difference method, as explained previously, the differential equations are replaced by difference quotients, which are formed by the points of a rectangular grid constructed over the geometry under consideration.

| | Superposition of analytical results | Numerical methods | |
|---|---|---|---|
| | | FDM | FEM |
| Geometry | 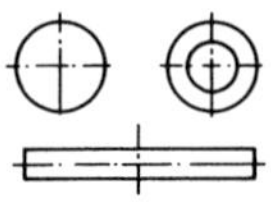 | 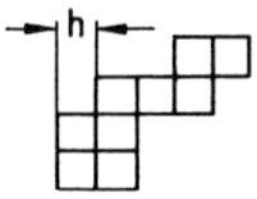 | 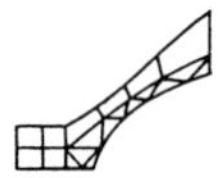 |
| Result | $\Delta p$, $\dot{\gamma}$, $\dot{V}$<br>$t_k$ | 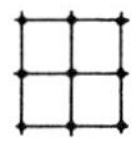 | 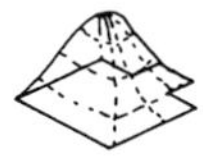 |
| Accuracy | Low | Average | High |
| Retroactions Superpositions | No | Yes | Yes |
| Determination of geometry | Direct<br>$x, y = f(\Delta p, \eta)$ | Iterative | Iterative |
| Hardware | Pocket calculator<br>PC | PC | PC<br>Servers<br>Super computer |
| Amount of time required | Small | Average | Great |

**Figure 4.6** Comparison of ananalytical methods, FDM, and FEM

The development of the differences as well as the subsequent solution of the obtained equation system and often even the construction of the grid are done by the computer program itself. Compared to simple computing techniques, FDM has the following advantages (see [26]):

- Geometries that normally are not accessible to an analytical solution can be computed.
- As a result, values on all grid points are obtained and not only ones integrated over the entire cross section.
- Almost any material law can be inserted into the differential equations, thus increasing the accuracy of the results.
- Coupling of differential equations (e.g., the momentum and energy equations, temperature and shear rate dependence of the viscosity) can be taken into consideration.
- Since the system of equations is solved for the entire domain simultaneously, all interaction effects are reflected in the results (this is not true for explicit FDM).

Besides the above-mentioned advantages of the difference method over the simple ones, it has some inherent disadvantages:

- The determination of a geometry or a operating point is possible only by iteration since the required equations are not reversible.
- The computations require at least a personal computer: they can no longer be done by a pocket calculator or by hand.
- The times required for computations for the FDM programs are considerably longer than those for simpler methods.

Like FDM, there are both advantages and disadvantages for FEM when compared to simple computing methods.

The comparison between FDM and FEM is truly difficult since the results depend strongly on the individual cases and methods used. For rectangular areas and quadratic finite difference grids or finite elements, special cases show that implicit FDM and FEM (while using linear shape functions) lead to the same equations and therefore are identical [19,27].

When working with curvilinear irregular areas, it is possible to take advantage of FEM, which permits the construction of finite elements with edges that do not have to be perpendicular to each other and can be curvilinear. When applying parabolic shape functions, the edges of the elements can take up a parabolic shape. As mentioned previously, these shape functions are at the core of the FEM: they serve not only the purpose of interpolating the geometry but also obtaining the desired quantities (velocity, pressure, temperature on each node). Thus, they directly affect the accuracy of the results. The introduction of the shape functions is at the same time the main difference between FEM and FDM: while in the latter only the values on the nodal points are known, in the former the value of the desired parameter is clearly determined at each point of the domain by the interpolation permitted by the shape function. Since this fact is taken into consideration when deriving the FE equations, FEM is by definition more accurate than FDM. Besides a better approximation of the geometry and a more accurate description of the course of the desired parameters, FEM offers the following advantages over FDM:

- The geometry under consideration can be of any shape since this application is defined independently of the computer program. This means that FE programs work independent of the geometry.
- Results can be determined and displayed at any point of the domain.
- Since the FE equations are always solved simultaneously, all system interactions are taken into consideration, resulting in high accuracy and flexibility.

But FEM has some disadvantages:

- Computing times, requirements on the core, and mass storage are multiples of those required by FDM. The lowest class of computer to be used for this purpose is a high-performance 32- to 64-bit personal computer.

- Since the geometry as well as the boundary and initial conditions are set by the user himself, the time required for the computations is longer than that required by an FDM program in which these data are more or less fixed.
- The great flexibility with regard to geometry, mesh density, and the choice of types of elements and boundary conditions demand a considerably greater understanding of the method by the user if he or she wants to obtain reliable results.

In summary, the following conclusions can be drawn from these observations: a more demanding method is advantageous only when the simpler method reaches its limitations. When the areas are regular and allow the construction of a difference grid for their approximation, then the advantages of the FDM prevail [27,28]. However, when various geometries exist that differ strongly from each other, then the application of the FEM, which is independent of geometry, is justified. This method has advantages in the presentation of the geometry, in the mesh generation and definition of boundary conditions, and in the presentation and evaluation of results. As extrusion dies are generally very complex, the current state of research focuses increasingly on the usage of FEM.

### 4.4.4 Examples of Computations of Extrusion Dies

In this section, examples of computations applying numerical methods will be presented in order to clarify possible applications. The discussion is limited in the case of FDM to computations of flow by the means of explicit difference procedures since these are the most important for polymer processing.

#### *Convergent Slit* [16]

When studying velocities and temperatures in a slit with a varying gap height, a strong influence of the channel geometry and a weak influence of the temperature on the velocity profile are observed (see Fig. 4.7).

The temperature distribution shows a maximum near the channel wall because a high shear rate causes increased dissipation at that point. With a constriction of the channel, this maximum becomes more distinct due to the increase in dissipation. The dissipated heat can be transferred only slowly to the isothermal wall ($T_W = 200$ °C) or to the cooler melt in the middle of the channel because of the low thermal conductivity of the melt.

In the section of the channel with parallel walls, the melt temperature in the middle of the channel rises faster, while the equilibrium state between friction heating (dissipation) and heat removal is reached. The temperature maximum travels toward the middle of the channel.

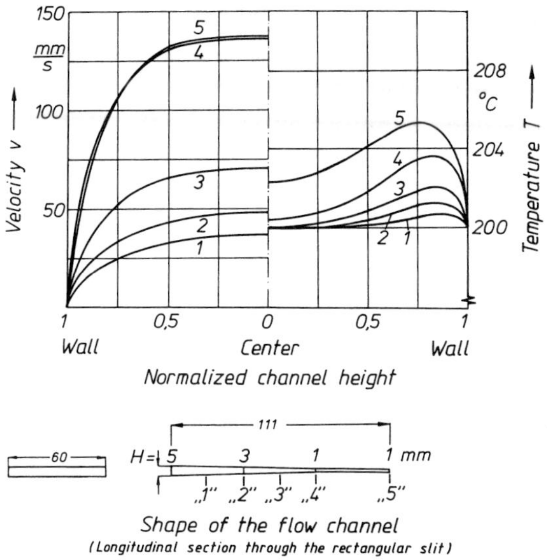


**Figure 4.7** Distribution of velocities and temperatures across the cross section of a convergent slit at a constant wall temperature of 200 °C

Figure 4.8 depicts the influence of different die wall temperatures on the temperature profile in a relatively short rectangular channel and a convergent inlet. At the entrance ($x = 0$), a constant mass temperature $T_M = 170$ °C is assumed.

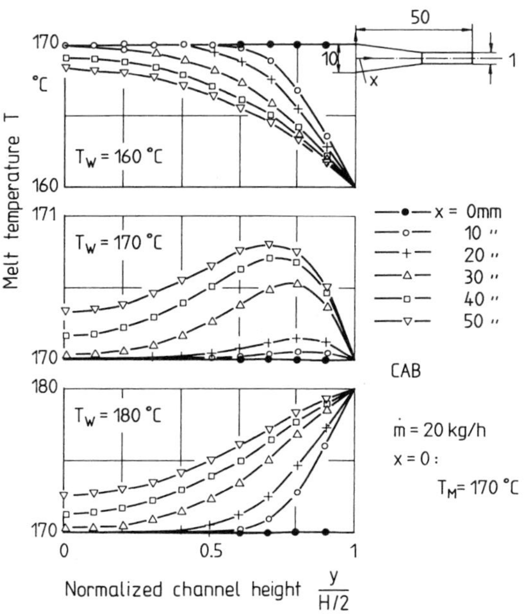


**Figure 4.8** Temperature profiles across the channel height at different wall temperatures

It can be concluded from the illustration that a uniform increase or decrease in mass temperature in the flow channel by changing the wall temperature is not possible. Therefore, one should try to deliver an extrudate with the highest possible thermal homogeneity (with the minimum temperature difference over the cross section of the melt) from the extruder to the die.

### *Crosshead for Blow-Molding* [17]

In [17] the velocity and temperature distributions in a crosshead for blow-molding are analyzed: Figure 4.9 shows a cut through the flow channel of the crosshead and the computed velocity, shear rate, and temperature profiles. A constant mass temperature at the inlet (position $x = 0$) was set at 200 °C; the mandrel wall is assumed to be adiabatic.

With the help of computations, it is possible to determine if and how the temperature differences in the melt stream in a die equalize and what effect they have on the temperature profile of the emerging extrudate. Figure 4.10 shows computed temperature profiles at the exit for two operating conditions (with extreme differences in dissipation). Two different temperature profiles at the inlet were assumed, each with the same mean temperature. In this example, it is seen that even extreme temperatures at the die inlet alone have an insignificant effect on the temperature at the outlet. This trend can be confirmed by measurements (see [17]). Even when computing with a mean, constant mass temperature at the inlet, only small temperature differences are found. It has to be taken into consideration that such results always depend on the operating conditions.

### *Plane Coextrusion Flow with Constant Cross Section* [11]

Not only the flow of one melt but also of multilayers can be subject to computation. Here the position of the layer interface (or interfaces) of two melts in addition to unknown pressure, velocity, and temperature must be determined by iteration.

Consider an asymmetrical PE-LD/PE-HD/PE-LD three-layer flow in a simple rectangular channel with the cross section $H \times B = 12$ mm × 200 mm. The mass temperature at the channel entrance ($x = 0$ mm) is 235 °C for all three layers, and the die wall temperature throughout the entire channel is 260 °C. The description of the dependence of the viscosity on the shear rate is by the Carreau model (see Section 2.1.1.2), and the temperature shift is according to the WLF equation (see Section 2.1.1.3). The relevant material data are shown in Table 4.1. The computation is performed with constant heat conductivity and constant specific heat capacity.

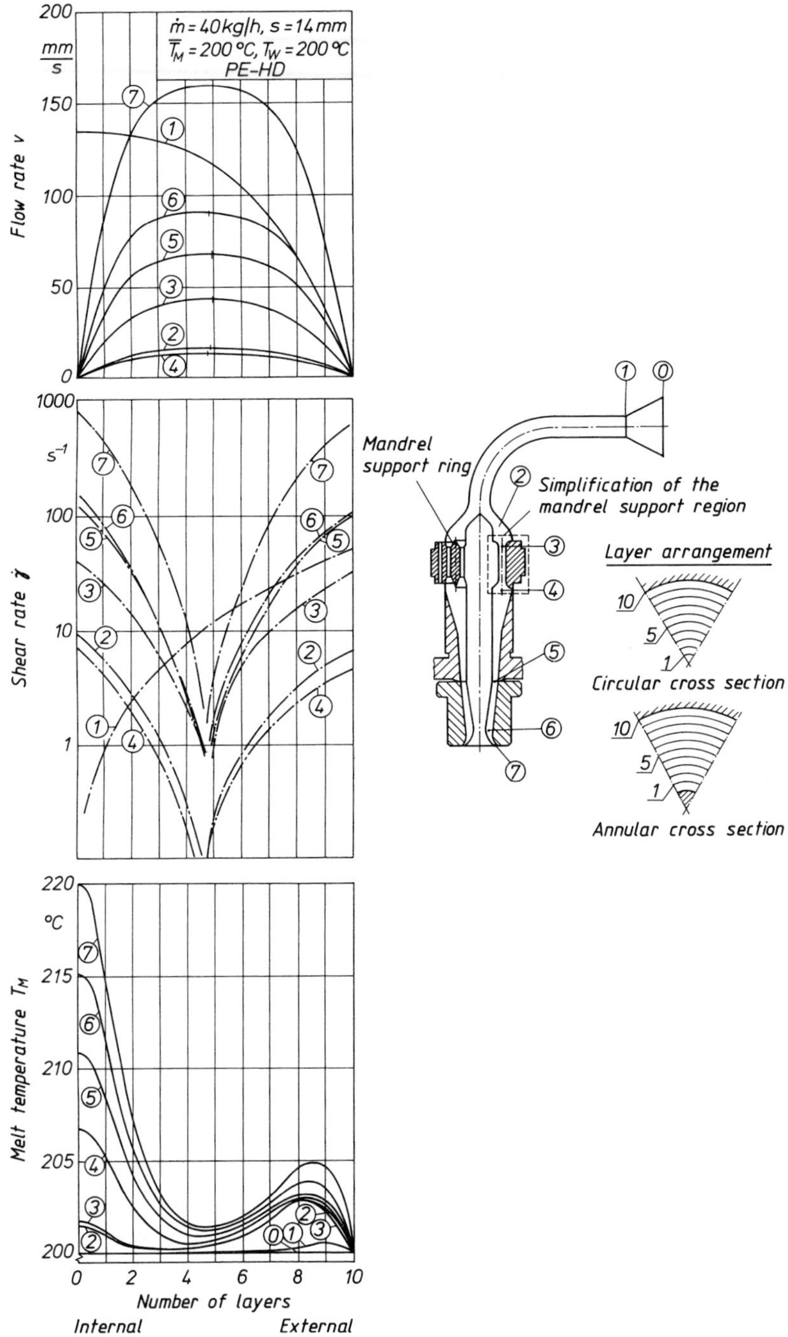


**Figure 4.9** Computed distributions of velocities, shear rates, and temperatures in a blow-molding crosshead die

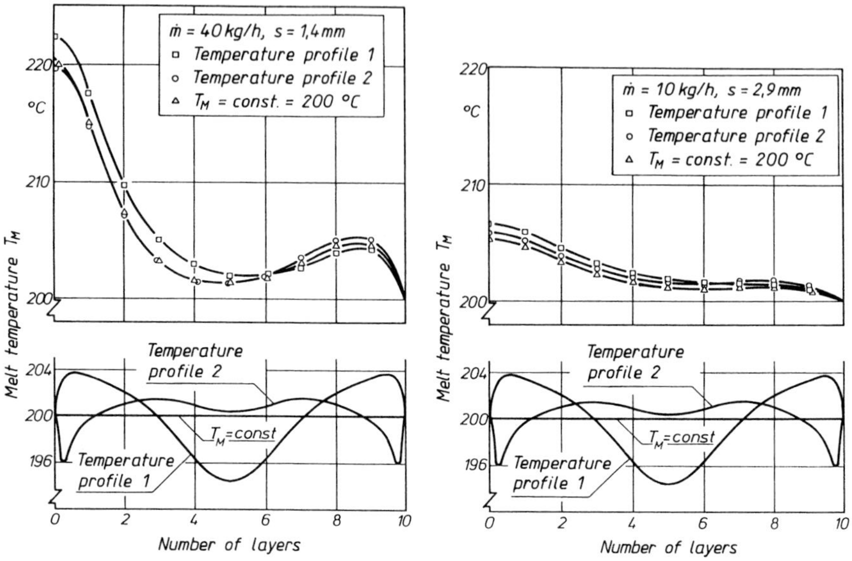


**Figure 4.10** Temperature profiles at the die inlet (bottom) and outlet (top)

**Table 4.1** Material Data

| Material | | PE-LD | PE-HD |
|---|---|---|---|
| *Thermodynamic Material Data:* | | | |
| Density | (kg/m³) | 801 | 850 |
| Reference temperature | (°C) | 115 | 115 |
| Linear coefficient of thermal expansion | (1/°C) | 0.001 | 0.001 |
| Thermal conductivity | (W/m K) | 0.257 | 0.257 |
| Specific heat capacity | (J/kg K) | 2620 | 2620 |
| *Data for WLF Equation:* | | | |
| Standard temperature | (°C) | 10 | -70 |
| Reference temperature | (°C) | 235 | 235 |
| *Carreau Coefficients:* | | | |
| *A* | (Pa·s) | 5733 | 39033 |
| *B* | (1/s) | 0.418 | 1.052 |
| *C* | | 0.938 | 1.101 |

Figure 4.11 and Figure 4.12 depict the computed velocity and temperature profiles both at the entry region of the channel ($x = 1$ mm) and after different distances in the channel ($x = 50$ mm and $x = 500$ mm). While the higher temperature of the channel wall (Fig. 4.11), depending on the flow distance, has not affected the temperature in the middle of the channel, the nonuniformities in the layer temperatures can result in additional undesirable temperature profiles in the co-

extruded product (Fig. 4.12). The effect of the change of material on the velocity and temperature profiles is shown in Fig. 4.13. Here the PE-LD/PE-HD/PE-LD melt stream was replaced by PE-HD/PE-LD/PE-HD while all the remaining parameters (throughput and temperature) were held constant with regard to the operating point in Fig. 4.11.

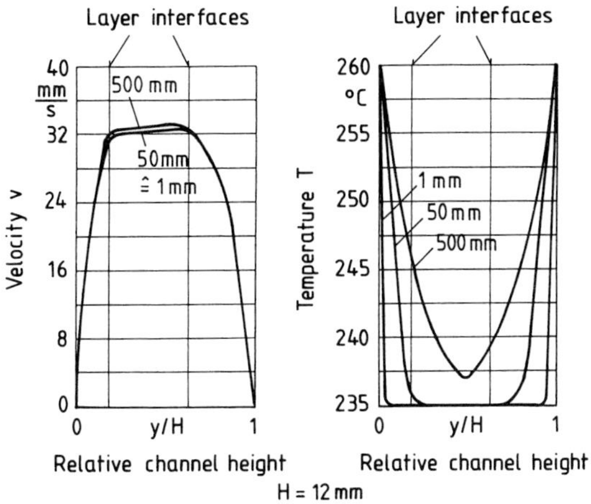


| Material | ṁ [kg/h] | $T_M$ [°C] | $T_W$ [°C] |
|---|---|---|---|
| PE-LD | 25 | 235 | 260 |
| PE-HD | 100 | 235 | - |
| PE-LD | 55 | 235 | 260 |

**Figure 4.11** Development of velocity and temperature profile in a flow with uniform entry melt temperature and constant wall temperature

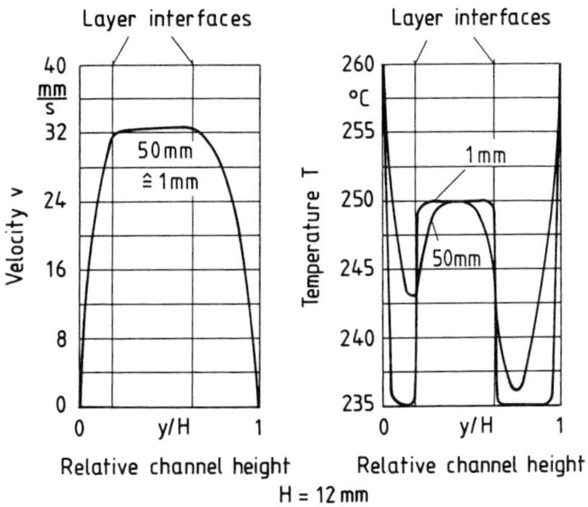


| Material | ṁ [kg/h] | $T_M$ [°C] | $T_W$ [°C] |
|---|---|---|---|
| PE-LD | 25 | 235 | 260 |
| PE-HD | 100 | 250 | - |
| PE-LD | 55 | 235 | 260 |

**Figure 4.12** Effect of the differences in the melt temperatures of the cover layers and the middle layer on temperature and velocity

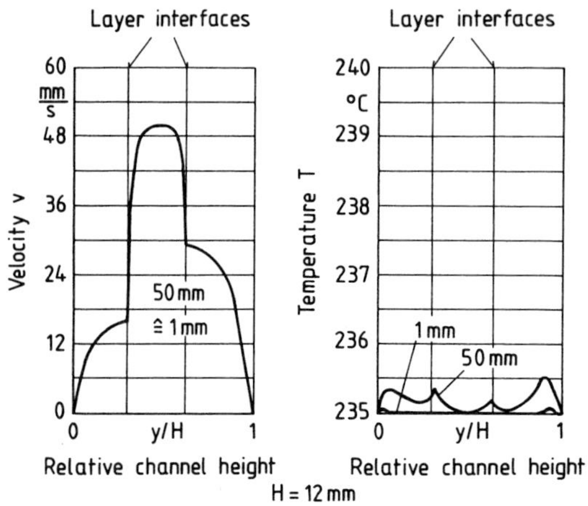


| Material | $\dot{m}$ [kg/h] | $T_M$ [°C] | $T_W$ [°C] |
|---|---|---|---|
| PE-HD | 25 | 235 | 235 |
| PE-LD | 100 | 235 | - |
| PE-HD | 55 | 235 | 235 |

**Figure 4.13** Effect of the change of the material on temperature and velocity distributions

The result of the above is a drastic change in the velocity and temperature profiles when compared to those in Fig. 4.11; this leads to a change in the layer thicknesses. A computer program for the determination of the temperature profile, velocity profiles (and hence of the residence time), shear rate profile, pressure loss, and thicknesses of the layers in coextrusion allows a simulation of the process as it responds to the changes in processing and geometric parameters. This way the effect of the individual process parameters on the process can be evaluated.

### *Flow in the Die Land of a Profile Die* [23,30]

A simple and fast way to obtain information about the velocity and the wall shear stress distribution in the die land of a profile die of any cross section is to compute the distribution of the one velocity component assuming an isothermal, planar, viscometric flow. This velocity component is always perpendicular to the cross section under evaluation. Hence, each cross section corresponds to an isobaric area, and all streamlines are straight lines. The computed distributions are valid strictly for fully developed flows in infinitely long die lands. In spite of that, the reality can also be approximated quite accurately for short zones, provided their length is the same for all cross-sectional regions.

Figure 4.14 [23,30] shows the cross section of the die land of a truly complicated die for a sealing profile and the velocity distribution calculated by FEM and represented by a shape of a mountain range. Each node is assigned a height over the flat FE network, and then the network is stretched over these points. The pseudoplastic melt behavior is approximated by the Carreau model with coefficients $A = 11{,}216$ Pa·s, $B = 0.14\ \mathrm{s}^{-1}$, and $C = 0.644$. The takeoff speed is 22 m/min (367 mm/s), and the pressure gradient is 3.9 bar/mm. The computed velocity distribution shows clearly that the melt lags behind in the nose (tab) and in the side piece that is sticking out to the left and therefore is stretched excessively in the takeoff. The mean velocity in the primary profile (including the nose) is 422 mm/s and in the relatively narrow side piece only 69 mm/s, about one-fifth of the takeup speed. Here the cross section has to be increased upstream to increase the melt velocity (see also Section 7.4.3).

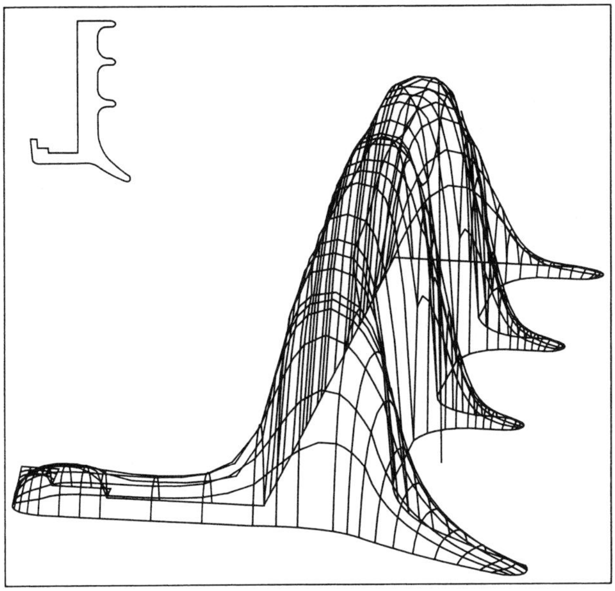

**Figure 4.14** Die cross section and velocity distribution in the parallel zone of the die for a sealing profile

### *Pipe Bend*

Figure 4.15 and Figure 4.16 show the results of computation using three-dimensional FEM for pipe bends of different ratios of pipe bend radius to pipe radius, $R_b/R_p$ (using a two-dimensional computation, only conditions in an arched slit with very large width in which the wall influence may be neglected can be simulated).

The results of the computation show that in the inside area of the pipe bend there are higher velocity gradients than in its outside area: the maximum velocity shifts from the center to the inside of the bend.

The velocity difference between the inside and the outside area increases with the decreasing radius of the bend. The effect is depicted in Fig. 4.17. Here the ratio of the shear stress at the wall (respectively inside and outside the bend) to that of a straight pipe ($R_b \rightarrow \infty$), $\tau_{R'}$ is plotted against the ratio $R_b/R_p$.

From Fig. 4.17 the following can be concluded. The shear stresses increase in the inner area, while in the outside they decrease with the decreasing radius. This can be the reason for deposits (plate-out) in this zone as they are found occasionally. From the ratio $R_b/R_p = 10$ and up, the differences become so small that they can be neglected; the velocity maximum lies approximately in the middle of the pipe (Fig. 4.15). Above this limit, the curve of the pipe has a negligible effect on the velocity distribution and hence on the shear stress difference. The smaller the ratio $R_b/R_p$, the more the flow at the outlet from the bend deviates from the developed straight pipe flow. Therefore, in every case there should be a straight pipe attached to the outlet of a bend that would act as a compensator. This ensures a well-developed symmetrical flow into other areas (such as the mandrel area).

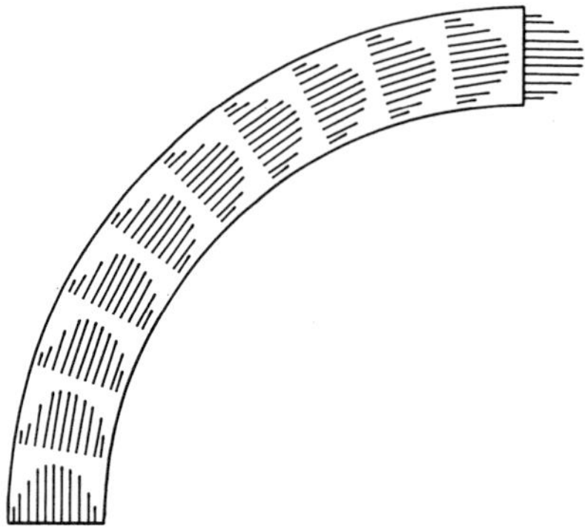

**Figure 4.15** Velocity distribution in a pipe bend of 90 degrees ($R_b/R_p = 10$)

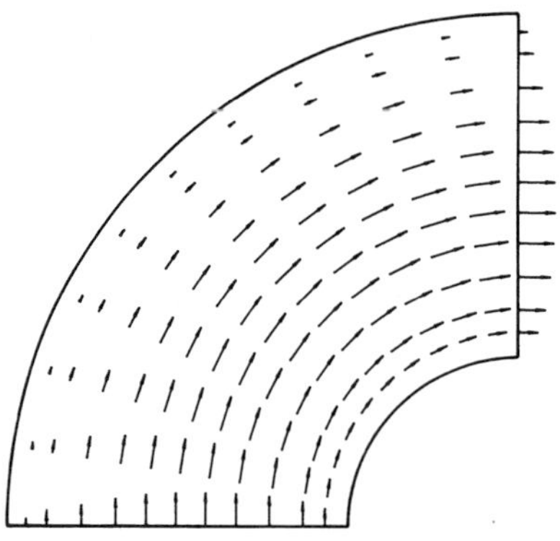

**Figure 4.16** Velocity distribution in a pipe bend of 90 degrees ($R_b/R_p = 2$)

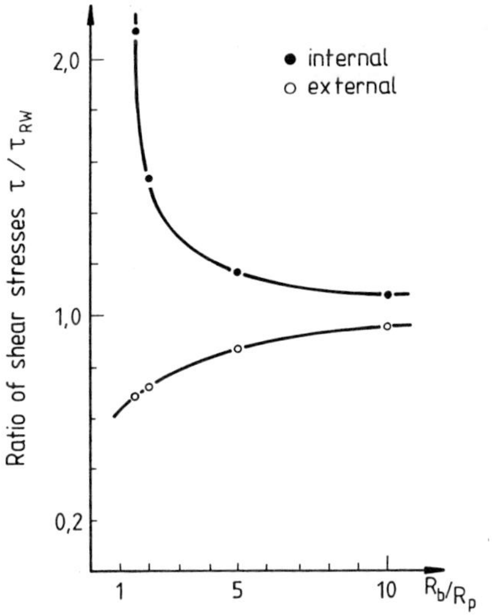


**Figure 4.17** Shear stress at the wall as a function of the ratio of radii

### *Streamlines in a Planar Two-Layer Flow* [23]

Multilayer flows are particularly critical in practical applications because they can become unstable under certain conditions, depending on the geometry of the flow channel, the behavior of the material, and the operating point. When this occurs, the layers are no longer smooth, and each of them exhibits a thickness fluctuation. In addition to that, when there are differences in viscosity of the individual layers, the problem of melt rearrangement occurs: the lower-viscosity melt tries to surround the higher-viscosity melt, since this minimizes the energy dissipation. Besides that, at the point where the layers flow together, other irregularities can occur, such as dead spots or rupturing of one layer. This is particularly true when one of the layers is very thin (e.g., the tie layer) or when the layers differ in viscosity or in velocity.

Such effects are undesirable, and they have to be eliminated by proper die design or process control. These can be determined effectively by a computational simulation.

Figure 4.18 and Figure 4.19 depict an isothermal planar two-layer flow at the point where the two streams merge. The description of the viscosity functions was done by the Carreau model. The results for a flow rate ratio of 2.5:1 is shown in Fig. 4.18. The boundary between the layers shifts downward due to the higher volumetric flow rate in the upper part of the channel. The streamlines of both flows are smooth. This changes when the ratio of the volumetric flow rates is increased to 32:1 (Fig. 4.19). Then a recirculation zone forms in the stream at the outside channel wall a short distance before the layers come together, where there are smaller velocities. This can be explained by the high volumetric flow rate forced by the faster stream. The recirculation zone causes a reduction of the effective cross section of the lower channel in such a way that, despite the high velocities, the condition of constant volume is met. Experiments with model fluids have shown that this effect indeed can be observed under real conditions [11,23]. In order to eliminate long residence times and the resulting degradation of the polymeric melt, either different operating conditions or a different die geometry must be chosen.

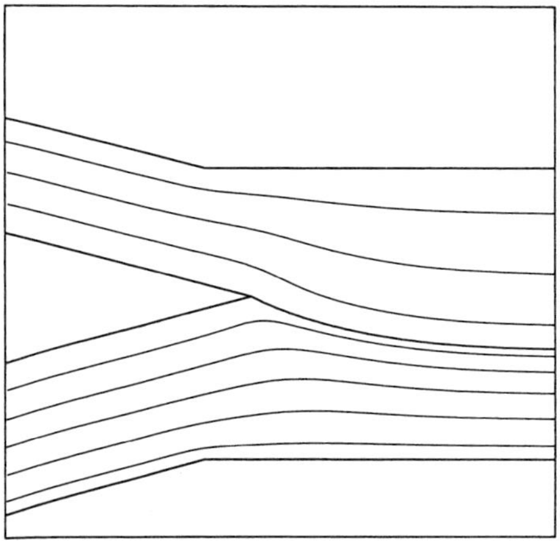

**Figure 4.18** Planar two-layer flow with a flow rate ratio of 2.5:1

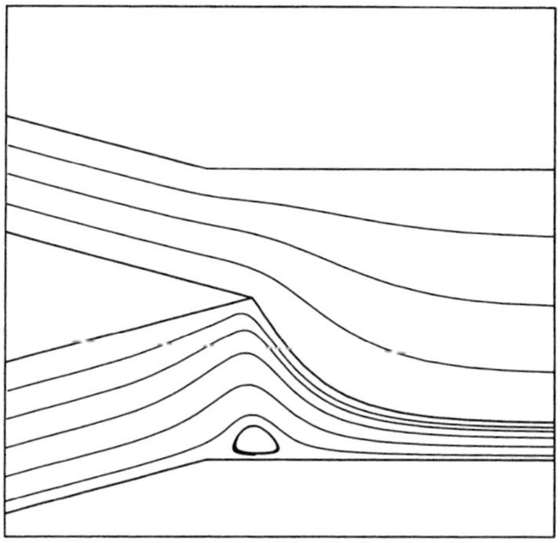

**Figure 4.19** Planar two-layer flow with a flow rate ratio of 32:1

### *Transition from a Rotating Screw Tip to a Mandrel* [23]

An example of the need to use three-dimensional computations even for a relatively simple geometry is depicted in Fig. 4.20. It shows the flow between a rotating screw tip (right) and a stationary tip of a mandrel that is a part of a pelletizer (left). The streamlines of particles that are located in different radial positions on the right side at the inlet into the observed section of the flow channel are depicted. It can be clearly seen that the particles that flow in the proximity of the rotating screw tip rotate strongly and form a helical path. Conversely, the particles near the outside wall of the die are barely affected by the rotation. A short distance behind the screw tip the flow is again two-dimensional and axially symmetrical: because of high friction forces in the stream, the rotation propagates only over short distances.

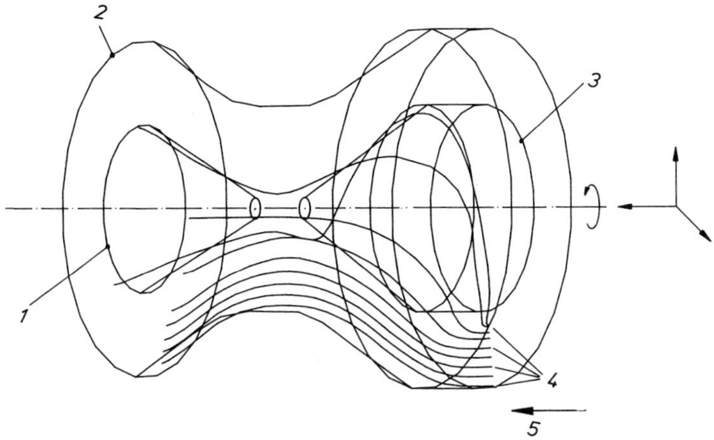


**Figure 4.20** Flow in the space between a rotating screw tip and the mandrel tip of a pelletizing head: (1) fixed mandrel, (2) housing, (3) rotating screw tip, (4) particle streamline, (5) direction of flow

The result confirms the necessity of a three-dimensional flow simulation in spite of the axially symmetrical geometry: because of the rotation of the screw tip, the particles of the melt have velocity components and velocity gradients in all three directions in the space.

Consequently, both the die designer and the processor have the tools to evaluate the necessary changes of the die or process improvements in many cases without costly experiments. Moreover, such a simulation promotes and builds a foundation for the still-indispensable feel for the design of extrusion dies. This is particularly true even when the results of computations are not accurate enough as a result of the simplifying assumptions.

### *Simulation of the Predistribution in a Spiral Mandrel Die [85]*

An example for a three-dimensional, nonisothermal computation, in which not only the flow channel but also the surrounding die material is modeled, is given in [85]. Considered is the flow through the predistributor of a flat spiral mandrel die. The predistributor divides the melt delivered by the extruder into eight substreams (Fig. 4.21), which are then fed into the spiral channels of the main distributor. The main target in the design of the predistributor is to ensure that at each outlet of the predistributor the same mass throughput is given. An inhomogeneous predistribution would lead to an uneven flow in the spiral mandrel die and hence to an uneven wall thickness distribution in the extruded tubular film. This thickness distribution must be expensively compensated for by downstream equipment, such as by using a segmented cooling ring, otherwise it can lead to unnecessary material consumption in thicker film sections or even to the production of waste if individual layers in the film are outside the tolerances and therefore can no longer fulfill their specification. One possible problem in this context is the influence of the temperature distribution on the melt flow in the predistributor. Both the external tempering of the die and the dissipative shear heating can lead to an uneven temperature distribution in the melt. Since the viscosity of the melt is dependent on the temperature, an inhomogeneous temperature distribution would result in an uneven throughput distribution at the predistributor outlets.

**Figure 4.21** Predistributor (left) and main distributor (right)

In order to consider these effects in the design of the predistributor, numerical computations are necessary. Figure 4.22 depicts the computed temperature and throughput distribution in the predistributor for a total throughput of 300 kg/h, a melt temperature at the die inlet of $T_M = 200\,°C$, and an outer die wall temperature of $T_W = 220\,°C$. The material is a PE-HD whose viscosity is modeled according to the Carreau law and the WLF equation. It can be seen that the external tempering of the die leads to an inhomogeneous temperature distribution in the die, which results in higher melt temperatures in the subchannels that lead to outlets 2 and 3. This effect is further reinforced by the dissipative shear heating. The combination of high dissipation near the flow channel wall and the branching of the subchannels also leads to an increase of the melt temperature at outlets 2 and 3. As a result, outlet 3 has the highest melt temperature and outlet 1 and 4 the lowest. As shown on the right-hand side of Fig. 4.22, the temperature distribution is also reflected in the flow rate distribution at the outlets. Outlet 3 has the highest temperatures and therefore the lowest flow resistance, because of which the highest output is also at outlet 3. At outlets 1 and 4, on the other hand, which have the lowest melt temperatures, are the lowest flow rates. The overall result is a difference between the maximum and minimum throughput of approximately 8.4 % of the average throughput.

It should be noted that the inhomogeneities are dependent on the process parameters. For comparison, Fig. 4.23 depicts the computation results for an outer die wall temperature of $T_W = 200\,°C$. All other parameters are identical to the ones in Fig. 4.22. Here, it can be seen that, by lowering the die temperature $T_W$, the throughput distribution gets more homogeneous. In this case, the total throughput difference amounts to 3.6 %. The reason for this is that by setting $T_W$ identical to $T_M$, the influence of the external tempering on the melt temperature is minimized. Here, the melt temperature is only influenced by the dissipation.

In conclusion, the thermally induced throughput inhomogeneities in the predistribution can be partially reduced by adaptation of the process parameters. However, the process parameters cannot be changed freely as they influence several more process and product attributes. As a consequence, it is not possible to completely eliminate the inhomogeneities solely by optimizing the process parameters. Instead, it is necessary to change and adapt the die design, for example by integrating thermal design measures such as heater cartridges into the die. At this point, the benefits of numerical computations are particularly highlighted since they enable the evaluation of a great number of different die designs without having to produce each variation.

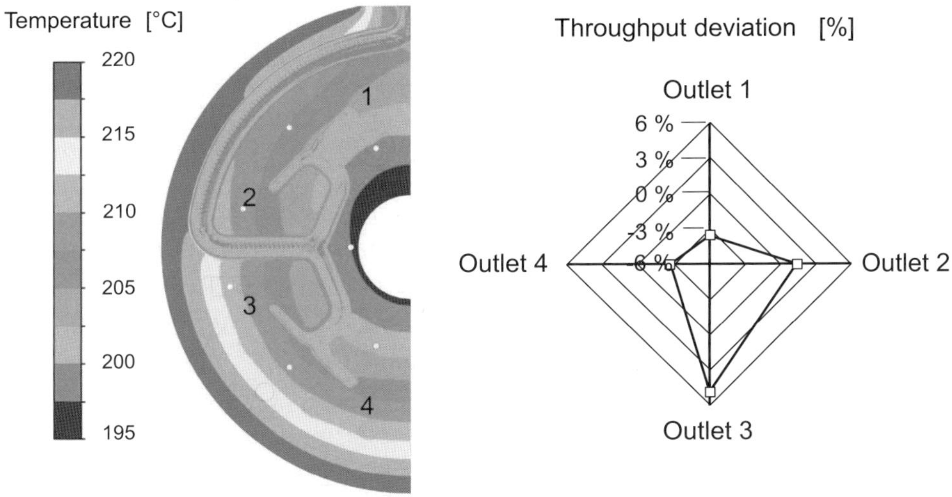


**Figure 4.22** Temperature and throughput distribution in the predistributor (300 kg/h, $T_M = 200\,°C$, $T_W = 220\,°C$)

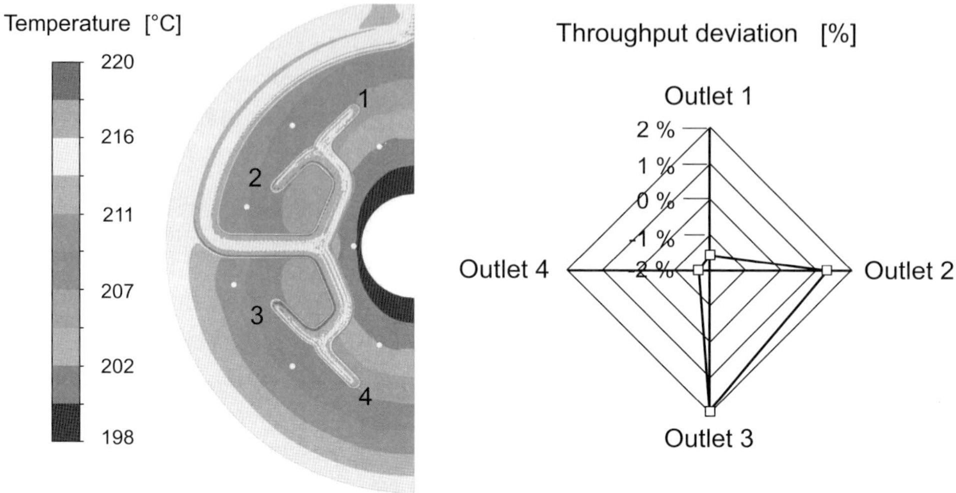


**Figure 4.23** Temperature and throughput distribution in the predistributor (300 kg/h, $T_M = 200\,°C$, $T_W = 200\,°C$)

## ■ 4.5 Consideration of the Viscoelastic Behavior of the Material

When designing extrusion dies, the elastic properties of the melt, that is, the ability of the melt to store an imposed deformation elastically (memory fluid; see Section 2.13), are generally not considered.

The elastic properties bring about effects, however, that are important for the design of the die. These are, for example:

- inlet pressure drop at channel constrictions
- entrance vortices
- extrudate swelling after its exit from the die (the profile and the die cross sections are not the same)
- change in diameter and reduction in length as well as the increase of wall thickness of a tube emerging from an extrusion blow head
- interfacial instabilities between layers or at surfaces due to a viscoelastic turbulence [3]

Neglecting the elasticity of a melt can lead to relatively large errors in the computation of the pressure loss. Reference [12] demonstrates this with an example of flow of a high molecular and hence highly elastic PE-HD through a blow-molding head.

Figure 4.24 is a graphic representation of the pressure loss computed by the explicit difference method versus the measured pressure consumption (at different throughputs). It turns out that the computed values of the pressure consumption are always under the measured ones and that the error increases with the increased pressure consumption (which corresponds to an increased throughput). This error can be probably explained by the fact that the viscoelasticity was neglected in the computations.

The effect of viscoelasticity on the results is even more pronounced in the computation of the extrudate swelling. As shown in numerous publications (e.g., [15,32,33]), the swelling factor *Sw*, defined for circular dies as

$$Sw = D / D_0 \tag{4.70}$$

where

$D_0$ Diameter of the die orifice

$D$ Diameter of the extrudate

depends to a high degree on the elastic properties. For a Newtonian fluid, generally the value of *Sw* is approximately 1.13 [34–36], whereas elastic fluids exhibit considerably higher values; these depend on their relaxation times and the shear rates during the extrusion.

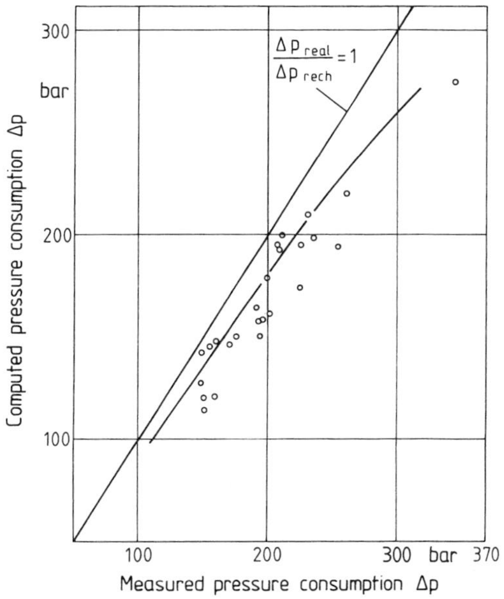


**Figure 4.24** Comparison of computed and measured pressure loss in a blow-molding die

There are essentially two ways to consider elastic properties:

1. Figure 4.25: The velocity field is first computed with the assumption of a purely viscous behavior. Subsequently, the elastically stored deformations are computed from a suitable material model combined with the results from the previous procedure. From that the swelling factor is obtained. The effect of the viscoelasticity on the velocity field is here completely neglected.

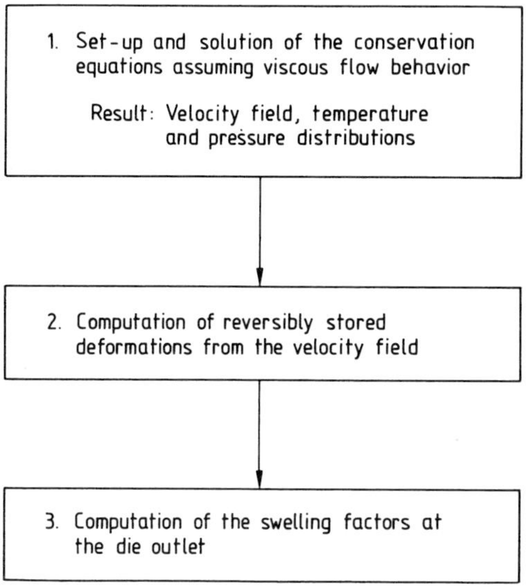


**Figure 4.25** Diagram of computation without the introduction of a viscoelastic material law into the conservation equations

2. Figure 4.26: The velocity field is first computed with the assumption of a viscous behavior. Subsequently, based on the result, the computation proceeds including a suitable material model to obtain a stress field. This stress field is then used as input into the momentum equation, and from that a new velocity field is computed. From that a new stress distribution results, and so on. The iteration is discontinued when either the difference of results from two steps fall under a set limit or if the maximum of the permissible steps is reached. In this procedure the effect of viscoelasticity on the velocity field and the pressure distribution is taken into consideration.

It is obvious that Procedure 1 is considerably simpler than Procedure 2. Procedure 1 was, for example, applied in combination with the explicit difference method [12,16,17] or with FEM [37,38]. Today such computations can be easily implemented by the use of personal computers [12]. Procedure 2 is considerably more demanding because

- a great number of iterations must be performed and
- the computing procedure often does not converge, particularly if there is a strong influence of the elastic behavior of the melt [15,39–41].

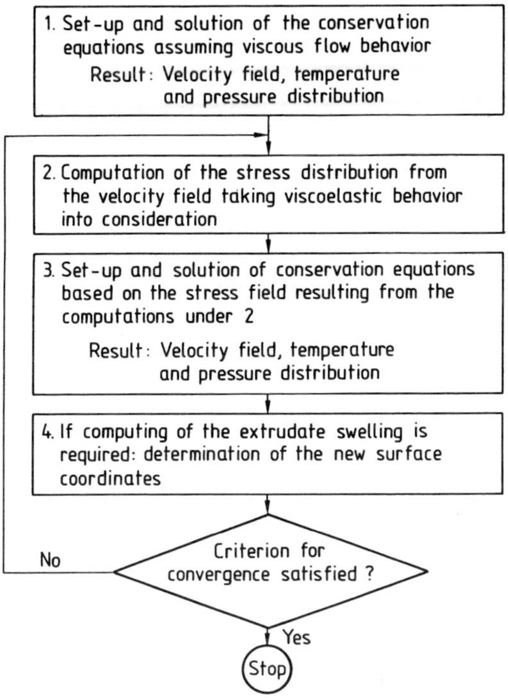


**Figure 4.26** Diagram of the computation with the introduction of the viscoelastic material law into the conservation equations

However, Procedure 2 allows, for example, the following computations:

- vortices at the inlet or a constriction
- shifting of the boundary of a layer due to elastic effects
- velocity distribution with the consideration of elasticity
- the shape of the extrudate emerging from a die

Procedure 2, also described in [15,39–41], is today the subject of intensive research. The goal of these efforts is to improve the convergence and the stability of the computations.

Another problem is the absence of a universal rheological material law that would be valid for any deformation and all polymers, at least for one class of polymers. Although there have already been many material models developed (overviews given for example in [4,47–49]), no model in existence can meet these requirements (see also Section 2.1.3). Therefore, the accuracy of the computations of the extrudate swelling or other effects depending on viscoelasticity is limited.

## 4.6 Computation of the Extrudate Swelling

It has been demonstrated before by Figure 4.25 and Figure 4.26 that there are two possibilities for the computation of the extrudate swelling: either a potential for swelling is determined from the reversibly stored deformations and from that the integral extrudate swelling (increase in cross section) is found [12,16,17], or the unknown coordinates of the emerging surface of the extrudate are considered unknowns in the problem—such as velocities, pressures, and temperatures—and are then computed by using a suitable expansion of the system of differential equations (e. g., [23,39,42]).

One equation suitable for the determination of the free surface of the system determination is valid under the conditions that no mass flows over this surface and that all velocity vectors are tangential to it. Mathematically, it means that the scalar product from the velocity vector $\vec{v}$ and the vector normal to the surface $\vec{n}$ must become zero:

$$\vec{v} \cdot \vec{n} = 0. \tag{4.71}$$

The entire procedure is very demanding, however. Therefore, reference is made to the literature for further details (e. g., [23,39]). A very similar method can be used to determine the true location of the interface between layers [23,43] because the latter is defined by condition of Equation (4.71) the same way as a free surface. An example of such an analysis is shown in Fig. 4.27 [23,30]. It involves the result of the FEM computation of a planar, symmetrical three-layer flow, where the top and bottom layers are the same material. Therefore, only one-half of the die orifice is considered. The FE network is continued beyond the outlet of the die so that the flow can be described even at that point. Newtonian behavior of both melts is assumed here.

At the beginning of the computation, the location of both the layer interface and the free surface is unknown; therefore, both have to be computed by the iterative process using the condition of Equation (4.71) [23].

In the above case, all melt layers have the same viscosity. This results in only a slight swelling of the extrudate. A slight contraction occurs in such a case, where the viscosity of the outer layers (at the same boundary conditions) is only 10 % of the middle layer (Fig. 4.27, center). When the outer layers are made from a highly viscous material (a factor of 10 relative to the middle layer; see bottom of the figure), the extrudate swells the most, because in this case the highest rearrangement of velocities occurs at the die exit. This rearrangement occurs because at the die exit a parabolic velocity profile exists that turns into a plug profile immediately after leaving the die. This leads to an acceleration of the outer layers and a slowdown of the layers in the center region.

### *Combination of the Explicit Difference Method with the Model according to Wortberg and Junk*

As shown above, the explicit difference method produces at least an estimate of swelling behavior relatively easily. It should be kept in mind, though, that it provides only integral factors, not the exact course of the free surface. The integral estimate is very frequently sufficient, at least for profiles with axial symmetry and for those with a high width/height ratio.

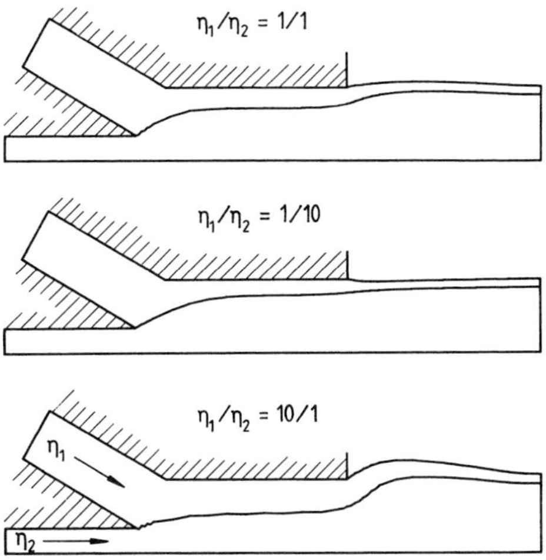


**Figure 4.27** Symmetrical planar three-layer flow with a free surface: comparison for different viscosity ratios between cover layer and middle layer

In the following section, the method by *Wortberg* [16] and *Junk* [17], previously described in Section 2.13, will be applied to the computation of the buildup of reversible elastic deformations in an extrusion die. Here the determination of reversible elastic deformations produced in a volume element is determined directly following the computation of the temperature and velocity fields.

The average, mathematically determined (theoretical) state of deformation of the melts, $\bar{\varepsilon}_R$, existing at the outlet of an extrusion die can then be correlated with the swell factor [16,17], the specific temperature, pressure, and deformation (velocity) history being taken into consideration.

The starting point for the mathematical determination of the reversible deformations $\varepsilon_R$ at the die outlet is the calculated temperature and velocity distributions in the die and the velocity distribution providing information on the local shear and extension rates $\left(\dot{\varepsilon}_S, \dot{\varepsilon}_D\right)$ of the melt in the die (compare Fig. 2.19) and on the deformation buildup shown in Fig. 2.24, taking into consideration the specific relaxation behavior of the processed material.

Since shear and extensional deformations can occur simultaneously in a melt particle, it appears to be appropriate to define a representative rate of deformation $\dot{\varepsilon}_{\mathrm{rep}}$, which results from superimposing the two portions of the deformations [16]:

$$\dot{\varepsilon}_{\mathrm{rep}} = a\dot{\varepsilon}_{\mathrm{s}} + b\dot{\varepsilon}_{\mathrm{D}} \tag{4.72}$$

where

$$\dot{\varepsilon}_{\mathrm{s}} = \dot{\gamma}$$

As shown in [16], the evaluation factors $a$ and $b$ must be determined experimentally for the material processed. Starting with the results of the measurements reported in [44], Wortberg found such a weighting on the basis of a PE-LD:

$$\dot{\varepsilon}_{\mathrm{rep}} = 0.01 \cdot \dot{\varepsilon}_{\mathrm{s}} + \dot{\varepsilon}_{\mathrm{D}}. \tag{4.73}$$

This means that, starting out from the same rates of deformation, the extension deformations have a considerably greater effect on the reversible elastic deformations in this case than do the shear deformations. In order to calculate the reversible deformation from location $z_{\mathrm{k,n}}$ to $z_{\mathrm{k+1,n}}$ (see Fig. 4.2), the representative deformation velocity $\dot{\varepsilon}_{\mathrm{rep}}$ is calculated from the shear rates as their mean value, i.e., $\frac{1}{2}\left(\dot{\gamma}_{\mathrm{k,n}} + \dot{\gamma}_{\mathrm{k+1,n}}\right)$, and the extension rate $(v_{\mathrm{k+1,n}} - v_{\mathrm{k,n}})/\Delta z$. The corresponding temperatures and pressures for determining the actual relaxation times $\tau(\varepsilon_{\mathrm{R}})$ in the computing step under consideration are also obtained as average values from the values at the interpolation points. The numerical relationships given in Fig. 2.24 can therefore be rewritten into computing instructions for determining the change in the reversible deformation in the $n$th grid layer of the flow channel and in the $\Delta t_{\mathrm{k,k+1}}$ time step under consideration [16]:

$$\Delta\varepsilon_{\mathrm{R}} = \frac{2\Delta z}{v_{\mathrm{k,n}} + v_{\mathrm{k+1,n}}}\left[\dot{\varepsilon}_{\mathrm{rep}} - \frac{\varepsilon_{\mathrm{R}_{v,\mathrm{n}}}}{C_1(T)\exp\left(-C_1\varepsilon_{\mathrm{R}_{\mathrm{k,n}}}\right)}\right]. \tag{4.74}$$

This analysis can be carried out for all layers of the FDM grid over the height of the channel. In [16] the initial values of the reversible deformations in the inlet cross section are set equal to the steady-state values for the respective deformation rates.

By this procedure, the distribution of the reversible deformation $\varepsilon_{\mathrm{R}}$ can be calculated numerically over the channel height in each section of the die, as well as its mean value $\bar{\varepsilon}_{\mathrm{R}}$, which in turn can be correlated at the die outlet with the swell factor $Sw$. If the swell factor is defined as the ratio of the dimensions of an extrudate after (index 2) and before (index 1) complete recovery of stored elastic deformation (complete retardation, similar to Equation (4.70)), then the following relationship applies to the extension between the two states:

$$\varepsilon_R = \ln \frac{L_2}{L_1}. \tag{4.75}$$

If the volume is assumed to be constant during the retardation, then for a rectangular profile it follows that

$$B_1 H_1 L_1 = B_2 H_2 L_2. \tag{4.76}$$

A retardation of longitudinal deformations $\bar{\varepsilon}_R$ brings about swell factors in the direction of thickness [16]:

$$Sw_H = \exp\left(K \, \bar{\varepsilon}_R\right) \tag{4.77}$$

and in the direction of width

$$Sw_B = \exp\left(K^* \, \bar{\varepsilon}_R\right) \tag{4.78}$$

with $K + K^* = 1$

Here, $K$ and $K^*$ indicate anisotropy of the values of swelling; they are generally different and depend on the shape of the extrudate cross section. If the cross section is square, then $K = K^*$.

As Equations (4.77) and (4.78) demonstrate, the swell factors can be correlated with the numerically determined mean state of deformation in the melt, $\bar{\varepsilon}_R$.

*Note:* The values of swelling determined by this procedure do not take into consideration the swelling that occurs due to the rearrangement of the velocity profile at the die outlet into a block profile, even for purely viscous fluids. Because of that, especially when the elastic recovery is small, too low a swell factor can result from the computation. Some examples of the results from the computations are presented below.

Figure 4.28 taken from [16] illustrates the effect of the geometry on the buildup and decay of reversible deformations by means of a slit-shaped flow channel with a variable height $H$. The right half of the figure in each case shows the strong effect of convergent regions of the flow channels, where in addition to shear deformation, extensional deformation occurs; the latter must be weighted more heavily with respect to the reversible deformation. This means that the effect of the so-called land zones in extrusion dies is made evident here. Die 1 and die 2 show that deformations, introduced directly at the end of the channel, are more swell-effective than those introduced farther upstream (fading memory of polymer melts).

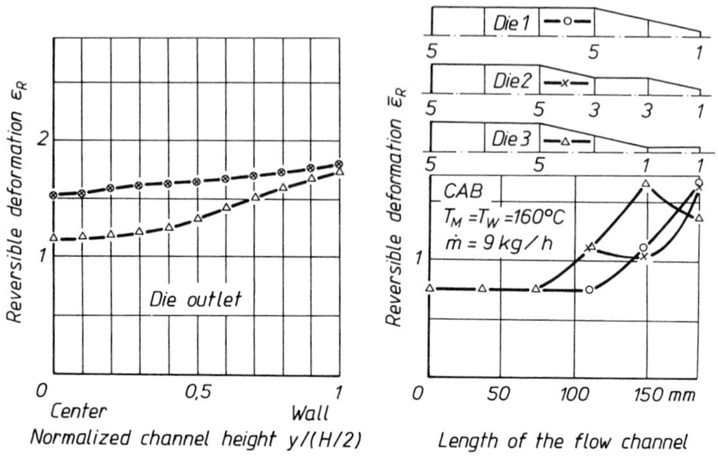


**Figure 4.28** Effect of different channel profiles in the direction of flow on the reversible deformations of CAB (according to [16])

The left side of Fig. 4.28 shows also the effect of the land region on the deformation profile over the cross section of the flow channel. Die 3 shows lower $\varepsilon_R$ values, whose comparatively steep increase near the wall is attributable to the quite high shear rate in the last narrow section of die 3. The land region, therefore, does not have a positive effect on the decay of reversible deformation near the die wall.

In the case of the buildup of reversible, elastic deformations in an axially symmetric channel, deformations in the circumferential direction, in addition to those in the longitudinal direction, must be taken into consideration in convergent/divergent annular slit regions of the channel [12,17].

Figure 4.29 [17] shows the average, mathematically determined longitudinal and circumferential deformation over the length of the flow channel of a crosshead die of an extrusion blow-molding line (compare to Fig. 4.9). The average, reversible, elastic, longitudinal deformation initially increases steeply because of the rapid increase in flow velocity in the inlet region. However, this deformation decays through relaxation processes up to the end of the following channel section with constant pipe diameter. The flow velocity is reduced greatly in the annular cross-sectional region commencing at the mandrel tip, as a result of which negative, longitudinal deformations occur and a long period of time becomes available for relaxation; consequently, the reversible, longitudinal deformations decay strongly. In this region of the channel, however, the average radius of the annulus becomes larger, and the reversible circumferential deformations increase greatly. The increasing flow velocity in the region of the mandrel support subsequently again leads to an increase in longitudinal deformations, while the circumferential deformations can relax.

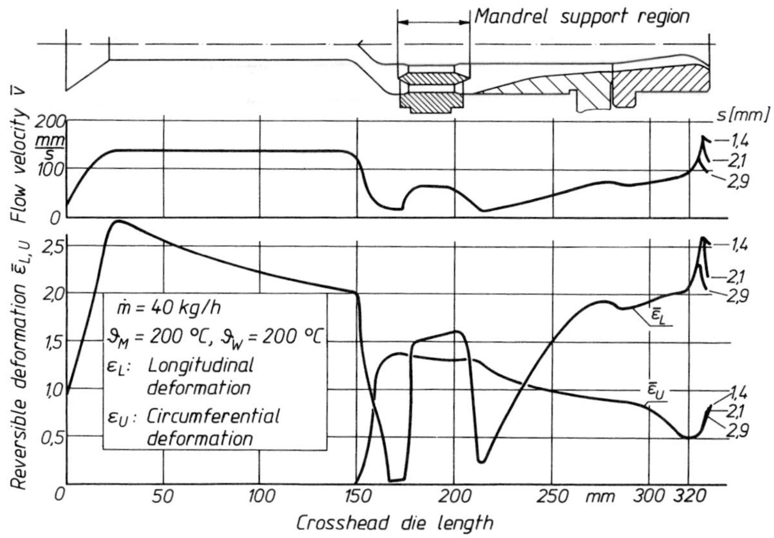


**Figure 4.29** Mean elastic deformations over the length of the channel in a center-fed mandrel support die

The large cross-sectional areas following the mandrel support lead to the decay of longitudinal deformations through compression and relaxation. Subsequently, as the width of the outlet slit and the mean radius of the annulus decrease, longitudinal deformations are built up once again, while circumferential deformations decay. At 320 mm, the location of the smallest mean radius of the annulus, the circumferential deformations start to increase once again while the longitudinal deformations decrease. Figure 4.29 illustrates, moreover, how different widths of the outlet slit affect the reversible deformation and, therefore, the swell.

The effect of the process parameters (throughput and temperature of the material) on the reversible deformation is discussed thoroughly for slit-shaped channels in [16] and for tubular and annular flow channels on an example of a blow-molding head in [12,17].

Figure 4.30 [16] illustrates the correlation between the swell factor $Sw_H$ of an extruded flat ribbon with the calculated mean deformation $\varepsilon_R$ as it is represented by Equation (4.77) for different materials. The maximum deviations from the mean correlation lines are less than 10% for PS and approximately 2% for PE-HD [16].

The procedure introduced here opens up the possibility of including the elastic properties of a material in the design of extrusion dies and enables the designer to analyze the effect of swell behavior qualitatively and to make quantitative predictions of the swelling behavior as process parameters and flow channel geometry are varied, when correlations similar to those in Fig. 4.30 exist.

Thus this analysis shows the strong influence of convergent regions of the channels on swelling, while parallel regions of the channel, such as land regions, lead to a decay of reversible, elastic deformations. It must be kept in mind here, however, that a parallel land at the end of the die once again increases the pressure consumption of the die. An optimized channel geometry, therefore, must be the result of matching pressure loss computations with the computation of reversible deformations.

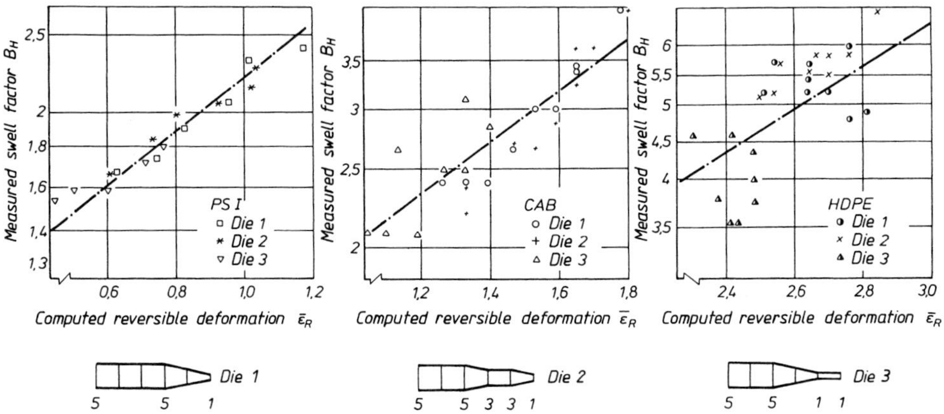


**Figure 4.30** Comparison of measured and computed swell factors for different types of slit dies

## 4.7 Methods for Designing and Optimizing Extrusion Dies

This chapter discusses the topics of designing and optimizing extrusion dies. The design process currently employed in industry is described, and the objectives of die design are presented. In addition, methods are introduced that permit abstraction of the design process in a manner that can be aided by optimization methods, which are also discussed. At the end of the chapter, examples illustrate how such an optimization can be performed, and a brief overview of the current state of the art with regard to automatic optimization of extrusion dies is given.

### 4.7.1 Industrial Practice for the Design of Extrusion Dies

In plastics processing, the primary objective of the rheological design of an extrusion die or distributor is a uniform velocity distribution in the polymer melt at the end of the flow channel. This requirement arises from the desire for a product that changes its dimensions only slightly due to superimposed local velocity profiles after exiting from the die. It is possible to meet this requirement with an appropriate design of the flow channel.

Usually, a distinction is made between standard extrusion dies and profile extrusion dies. Standard extrusion dies involve simple flow geometries (slit dies, mandrel and spiral dies) for extrusion of pipes, solid rods, sheets, and films. The flow channels of these dies are characterized largely by simple pipe and slit geometries with almost constant cross sections. In contrast, profile extrusion dies encompass all dies with which more complex shapes can be extruded. The geometry of the flow channels in profile extrusion dies is often very complex.

The flow conditions in standard extrusion dies and simple profile extrusion dies with flow channels that can be divided into sections having simple geometries can be estimated on the basis of analytical pressure-flow relationships for simple or idealized geometries (circle, slit, annulus (Section 3.4)). To calculate the velocity and pressure distributions, the flow channel is divided into geometrically simple sections. For these sections, simple equations relate the flow $\dot{V}$ to the pressure loss $\Delta p$ via a flow resistance $R$. For a description of the resistance that is independent of the flow, the nonlinear behavior of the melt viscosity with regard to shear rate can be incorporated via the flow exponent $n$ (Equation (4.79)):

$$\Delta p = R \cdot \dot{V}^{n} \tag{4.79}$$

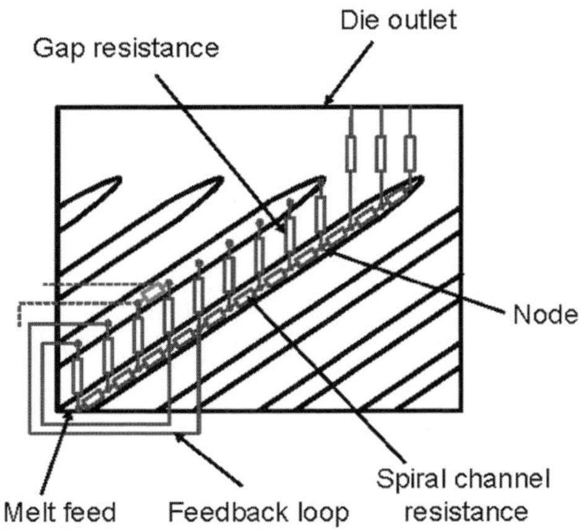


**Figure 4.31** Resistance network

For simple geometries such as convergent pipe and slit and annular flows, the resistance in the flow channels can be calculated using the classical Hagen–Poiseuille equation, in which the non-Newtonian viscosity is taken into account via the exponent [73]. Accordingly, the pressure and velocity distributions in a die comprising several simple geometries can be calculated using these equations. To this end, the resistance of each individual section of the die is determined, and the resistance values are combined into a resistance network [51]. Figure 4.31 shows such a network.

For given boundary conditions (e.g., inlet and outlet pressures), such a network can generally be calculated very quickly using computational rules analogous to Kirchhoff's Laws. Kirchhoff's laws describe conditions at nodes and around loops. The law regarding nodes states that at a node where several resistances are connected to one another, the sum of the incoming currents (flows) must equal the sum of the outgoing currents:

$$\sum \dot{V}_{\mathrm{in,i}} = \sum \dot{V}_{\mathrm{out,j}} \tag{4.80}$$

In addition, the law regarding loops states that in a closed loop the pressure (voltage) drop always equals zero:

$$\sum \Delta p_{\mathrm{k}} = 0 \tag{4.81}$$

Using these rules, a system of equations can be set up for any arbitrary network, and the velocities and pressures in the die can be calculated. If a flow-independent definition is chosen for the pressure–flow relationship, Equation 4.79 must be linearized, and the system of equations describing the network must be solved through iteration. To obtain a linear form, the resistances are converted into a linear and flow-independent form at each iteration:

$$\begin{gathered} R(\dot{V}) = R_{\mathrm{const}}\,\dot{V}_{\mathrm{iterationstep}}^{\mathrm{n}-1} \\ \Delta p = R(\dot{V}) \cdot \dot{V} \end{gathered} \tag{4.81}$$

With the aid of these simple equations, for instance, the geometry of a slit die can be calculated directly via a reverse design process (Section 5.2). This means that the flow channel of a die can be established directly from the desired cross section of the extrudate. Such direct design, however, is possible in only a few special cases. If the network describing the flow channel is more complex, even simple extrusion dies can be designed only via an iterative process in which the geometry is changed incrementally until the desired optimization result is achieved.

The spiral mandrel die will be presented here as a simple example. As with the slit die, it is possible for a given spiral mandrel die to calculate the velocity distribution at the discharge adequately via network theory for a large number of materials. Figure 4.32 shows as an example the calculated velocity distribution at the outlet of a spiral mandrel die. However, direct design of the flow channel is not possible because of feedback effects. Even for these simple geometries, an iterative procedure is necessary to obtain an optimal melt distribution at the discharge.

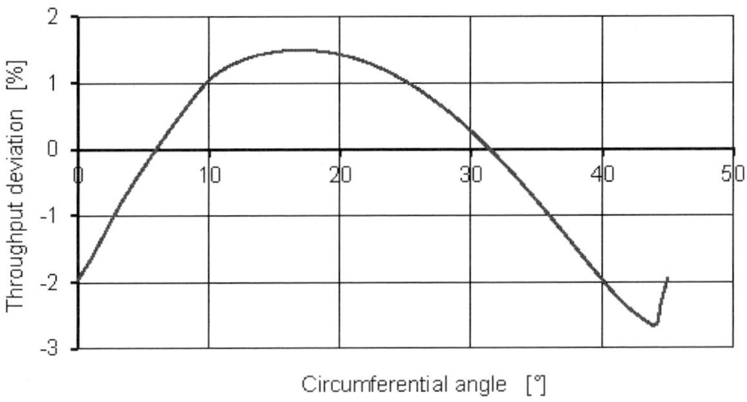


**Figure 4.32** Throughput distribution at the die outlet

For the more complex geometries usually encountered in profile extrusion, a design procedure based on network theory fails because resistances can no longer describe individual channel sections analytically. Moreover, for such dies, there is no method available that permits direct calculation of a suitable flow channel from a desired profile cross section [55,61].

In actual practice, more complex extrusion dies are designed in an iterative process. First, an initial concept for the flow channel is prepared, and then, based on the experience of the designer and practical trials, changes are made over the course of several iterations until the desired velocity distribution is obtained at the exit. For this optimization, the tool maker alters the geometry of the flow channel on the basis of an analysis of the velocity distribution of the melt at the exit from the die. The decision as to "where" and "how much" the flow channel must be modified depends primarily on the experience of the tool maker. According to manufacturers, 10 to 15 iterations are required to optimize complex profile geometries [80].

In recent years, for time and cost reasons, there has been an increasing trend in this area as well to minimize the number of actual trials through the use of computer-aided design. By using simulation programs, it is possible today in many cases to perform computer trials instead of the previous real-world trials that were conducted on actual dies. Here, as in the real world, an iterative design process is employed. For this procedure, the velocity distribution in the flow channel can be calculated by any of several different methods.

Often, the finite element method (FEM) is employed to simulate the flow conditions in complex geometries. This method permits exact calculation of the flow in any three-dimensional flow channel geometries [58,59,65,83]. In addition to the accuracy of the simulation, an advantage of this method is that, after the flow has been calculated via FEM, the velocity and pressure distributions are known at every point in the die, thus permitting a "look inside." This method is, however, very laborious because a finite-element grid must be created for the entire geometry. As a result, use of this method is relatively time-consuming and costly.

The flow conditions in a die and at its exit can, however, also be established via simulation programs that are based on other methods of computation. For instance, the finite difference method (FDM), the boundary element method (BEM) [72], or even a combination of network theory and these methods can be employed.

However, it should be noted that, in spite of advances in simulation methods and the ever-increasing performance of the computers available, it is still not possible today to take into account via a simulation all of the factors that affect the shape of the extruded profile. While mathematical models are available for many physical phenomena, exact measurement of the material parameters used in these models is often lacking. Moreover, incorporation of the usually nonlinear models into simulation programs can lead to numerical instabilities that prevent calculation. The viscoelasticity of polymer melts and the wall slip of several materials are mentioned here as examples that cannot yet be fully described via simulations.

Nevertheless, simulation programs have today become an indispensable tool supporting the design process. The following sections discuss the criteria according to which optimization of an extrusion die can be performed, what boundary conditions must be observed and how this design process can be facilitated through use of a computer.

## 4.7.2 Optimization Parameters

### 4.7.2.1 Practical Optimization Objectives

There are usually several objectives when designing extrusion dies as well as when designing and optimizing flow channels. Depending on the particular boundary conditions, these objectives can be assigned various priorities.

The quality of an extruded product is affected primarily by the flow distribution of the polymer melt at the outlet from the extrusion die. Accordingly, in plastics processing the primary goal of the rheological design of an extrusion die or distributor is usually to achieve a *uniform velocity distribution* in the melt at the exit from the flow channel. Stagnation regions in which the melt can degrade as the result of excessive residence time are to be avoided. Furthermore, *short residence times* of the melt in the flow channel are the objective for materials susceptible to thermal degradation, like PVC. In addition, it is useful in many dies to have the *narrowest possible residence-time spectrum* in order to minimize the lost time and costs associated with purging the die in the event of a material change.

A *low pressure drop* in the die is another criterion. Since the energy that is introduced by pressure forces is converted largely into heat through viscous dissipation, the melt is heated further by this energy input. This further reduces the efficiency of the overall process, which is often limited only by the cooling of the profile. Supplying a high pressure also requires a more powerful and thus more expensive extruder.

There are also other criteria, like maintaining a certain shear stress at the wall. Since any deposits present on the wall cannot be dislodged and carried along when the shear stress is too low, and the extrudate can be degraded when the *shear stress at the wall* is too high, the shear stress at the wall should not be too low or too high. This criterion can also be extended to the *shear rates,* which are closely related to the stresses in the melt.

When designing flow channels, it must also be ensured that the melt is subject to a *continual acceleration* as it moves from the inlet to the die to the outlet. This requirement applies especially to materials that tend to exhibit wall slip, since if the shear stress at the wall increases monotonically along the flow path due to acceleration of the melt a critical shear stress at which a stick-slip effect can appear occurs only once in the die.

Furthermore, with materials that exhibit highly viscoelastic behavior, the die swell should be taken into consideration during the design process in order to promote *uniform die swell* over the entire cross section, since it is possible, for instance, to compensate for this via a higher takeoff speed.

For many dies, an effective design for melt distribution, that is, a *minimal operating point dependence* and a *minimal material dependence,* should also be ensured.

#### 4.7.2.2 Practical Boundary Conditions and Constraints When Designing Flow Channels

When designing dies, the design freedom is limited by a few boundary conditions. For instance, the space available for the flow channel may be limited by the maximum overall dimensions of the die, by the location of other flow channels (in co-extrusion dies), or even by other infeed channels (e. g., in coating dies).

Additional constraints limiting the selection of the flow channel geometry result only from the manufacturability of the desired geometry.

When designing and dimensioning the overall die, it must be considered further that very high pressures are often present in the flow channels, so sufficient mechanical strength must be provided in order to prevent a deformation of the flow channel that could affect the optimization results.

#### 4.7.2.3 Independent Parameters during Die Optimization

As mentioned in the previous section, the freedom of the designer is limited by only a few constraints when designing the flow channel. Subject to the given constraints, the designer has the opportunity to specify the complete flow channel geometry of the die in a manner that takes into consideration the optimization and design goals. There are no design methods today for establishing the starting or basic geometry of the flow channel that represents the basis for the iterative optimization of the flow channel geometry. Thus, the initial design still depends on the experience of the tool maker.

To optimize the starting geometry, the designer specifies geometric parameters that are changed in a subsequent iterative optimization. Selection of the parameters and the amount by which they must be changed in order to achieve the desired optimization goals depend on the skill of the designer.

#### 4.7.2.4 Dependent Parameters during Die Optimization and Their Modeling

The quality of the selected parameters is evaluated by the designer during the optimization. In the course of this evaluation, the quality criteria listed in Section 1.2.1, for instance, may be employed. Estimating the quality of the flow channel with regard to any one of these quality criteria, such as the velocity distribution at the outlet from the die, for instance, is often only subjective. With a mathematical description of a quality criterion, however, an objective evaluation becomes possible. To this end, the quality criteria, subject to the flow conditions in the die, are modeled in such a manner that the quality of the flow channel can be described by a scalar value or index. For instance, the standard deviation of the velocities at the outlet from the die can be used as a quality criterion for the uniform velocity distribution at the outlet of the extrusion die.

Moreover, the introduction of indices is also useful with regard to an automatic optimization of flow channels, since with the aid of such indices the quality of a flow channel can be determined by a numerical algorithm.

If the design process for extrusion dies is considered abstractly, the process is a normal optimization problem with many independent parameters (Fig. 4.33). In this case, the abstract optimization goals discussed in Section 1.2.1 represent the parameters that are affected by the flow channel geometry. The independent parameters of such an optimization problem thus describe either a portion of the geometry or the entire geometry, while the dependent parameters provide a quantitative indication of the degree of success achieved with regard to meeting one or several optimization goals. For a mathematical formulation of such a problem, these relationships must be expressed in the form of objective criteria. For instance, mathematical functions that exhibit an absolute extremum when independent parameters reach an optimum value must be found for the optimization goals. In the terminology of optimization methodology, these mathematical expressions are called quality functions or merit functions.

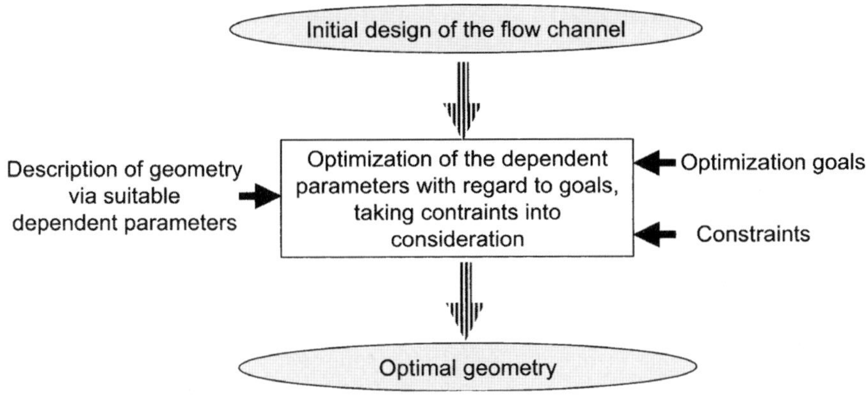


**Figure 4.33** Abstraction of the optimization problem

For instance, as mentioned above, the optimization goal of a uniform velocity distribution at the outlet can be formulated to mean that the standard deviation of calculated velocities at the outlet should be as small as possible. In this case, it is useful to consider the velocities averaged over the height of the individual outlet sections for the various outlet sections, so a possible quality function might appear as follows:

$$f\left(v_1,\ldots,v_n\right)=\sqrt{\sum_n \alpha_i\cdot\left(v_i-\overline{v}\right)^2} \tag{4.83}$$

where $v_i$ represents the mean velocity in one section of the outlet, $\overline{v}$ represents the mean velocity over the entire outlet, and $\alpha_i$ can be used additionally as a weighting factor in order to assign greater importance to sections of the outlet that are especially important for proper operation of the die.

The pressure loss between the die inlet and outlet can, for instance, be employed as a further quality criterion. For this purpose, this value can be normalized with a weighting factor as follows:

$$f\left(\Delta p\right)=\beta\cdot\Delta p \tag{4.84}$$

By means of formulations such as those presented here as examples, quality functions can be described mathematically and employed to evaluate flow channels. It should also be noted, though, that other variables, such as the minimal operating point dependence or the residence-time spectrum, for instance, are more difficult to access. In principle, however, any criteria or combination of criteria can be employed for objective evaluation of an extrusion die.

Usually, optimization with respect to several criteria is desired. For this purpose, a quality function can be defined as the weighted sum of individual criteria:

$$Q = \sum_{i} f_i \tag{4.85}$$

If it is further possible to describe the geometry of the flow channel parametrically in addition to defining the quality function; these geometric parameters can be optimized so that the quality function achieves the desired value. It is possible then to not only create objective evaluation criteria but also to have, within certain limits, a computer automatically perform the iterative optimization in a subsequent step through the use of various optimization strategies. To implement an automatic optimization, a computational method for determining the flow conditions in the extrusion die (e.g., FEM) can be linked with an optimization method (e.g., evolution strategy). With an automatic optimization, the geometric parameters, which define the geometry of the flow channel, can be selected by means of various optimization algorithms until the flow conditions, which are evaluated objectively through quality criteria, achieve an optimal value. In this case, it must be ensured when defining the quality function that not only the objectives of the optimization are expressed in a mathematical model, but also that, when using the optimization algorithms, the mathematical properties of the quality function contribute significantly to convergence of the selected optimization method.

The following section shows how optimization objectives can be achieved with the aid of optimization strategies and how such an optimization can be performed in an abstract sense.

### 4.7.3 Optimization Methods

In research and engineering there are numerous problems that focus on finding an optimum. Viewed mathematically, this involves searching for a maximum or minimum for a specified function in a defined region.

The variety of problems in research and engineering suggests that there are a great number of optimization problems that require special or specific optimization techniques. As a result, several optimization strategies have been developed, and these employ a variety of procedures and algorithms to locate the extremum.

Common to all optimization methods, however, is the search for an extremum (maximum or minimum) in a specified parameter space. Viewed abstractly, every optimization problem with $n$ independent system parameters can thus be considered as the maximization (or minimization) of a merit function [54], so the following holds:

$$F(\underline{x}) = \text{max!} \text{ for } \underline{x} \in R^n \tag{4.86}$$

The merit function $f(\underline{x})$ represents a surface in an $n+1$ dimensional space, where $n$ is the dimensionality of the vector $\underline{x}$, and the function value indicates the quality of the selected parameter set. For an optimization problem with two independent parameters, this surface might appear as shown in Fig. 4.34. In this illustration, the $x$ and $y$ axes define the parameter space. Every point on this $x$-$y$ surface represents a parameter set. If all possible parameter variations are assigned a scalar value that is plotted on the $z$ axis, the result is a surface that describes the quality of the entire parameter space. This surface can exhibit very different shapes for different problems.

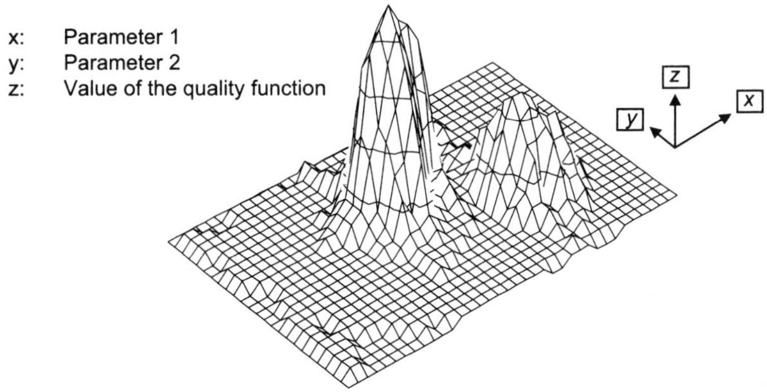


**Figure 4.34** Quality (merit) function

While the individual points that define the surface in its entirety can be calculated, they are not known a priori. The objective of every optimization strategy now is to localize the optimum in the fewest possible calculations.

There are many strategies that an optimization algorithm can follow to select an optimal parameter set. The various optimization strategies can be classified. For optimization problems with multiple independent parameters (geometry points of the channel), the optimization strategies can be classified as deterministic and stochastic methods. For the deterministic methods, a specific principle is employed to proceed from one iteration point to the next, hopefully better point. Given identical starting and boundary conditions, optimization always follows the same course with this type of strategy. With the stochastic methods, a randomness is incorporated into the course of the optimization [78].

The deterministic methods are also called "hill climbing" strategies. These methods can be further classified as to whether or not the gradients of the quality function are incorporated into the strategy. If the information provided by the gradient can be employed, it is often possible to achieve the optimization objective very quickly. In this regard it should be noted, on the one hand, that calculation of the gradient is not always possible and that, on the other, the value of the gradient cannot always be determined reliably or that the calculation is too laborious, thus limiting the use of these methods.

A classification of common optimization algorithms is given in Fig. 4.35. In the next section, various search strategies are presented and the individual methods described briefly. The reader is referred to the references for further reading [77,78,84].

#### 4.7.3.1 Gradient-Free Optimization Methods

##### *Coordinate Strategy*

The coordinate strategy of Gauß and Seidel [54] is an extension of one-dimensional optimization strategies for multidimensional problems. Starting from an initial state with this strategy, only the first parameter is optimized until the quality function reaches a maximum, while the other parameters remain constant. Next, all other parameters are optimized following the same principle. If no improvement in quality results from a modification of the parameters, the optimization sequence is stopped. The one-dimensional strategy for finding the optimum can be freely selected. In the basic form of this method, an exploratory step is taken in both the positive and negative directions.

##### *Rotating Coordinate Strategy*

This strategy developed from the coordinate strategy [75]. In the rotating coordinate strategy, the search direction is not determined by parallels to the individual axes. Instead, a new coordinate system for locating the optimum is introduced. One axis of this coordinate system always points in the direction of the vector formed by connecting the starting point to the end point of the previous iteration step. The coordinate axes for the other dimensions are always orthogonal to this axis and to the other coordinate directions. With this procedure, it is likely to find a favorable search direction, which is usually reflected in a faster rate of convergence.

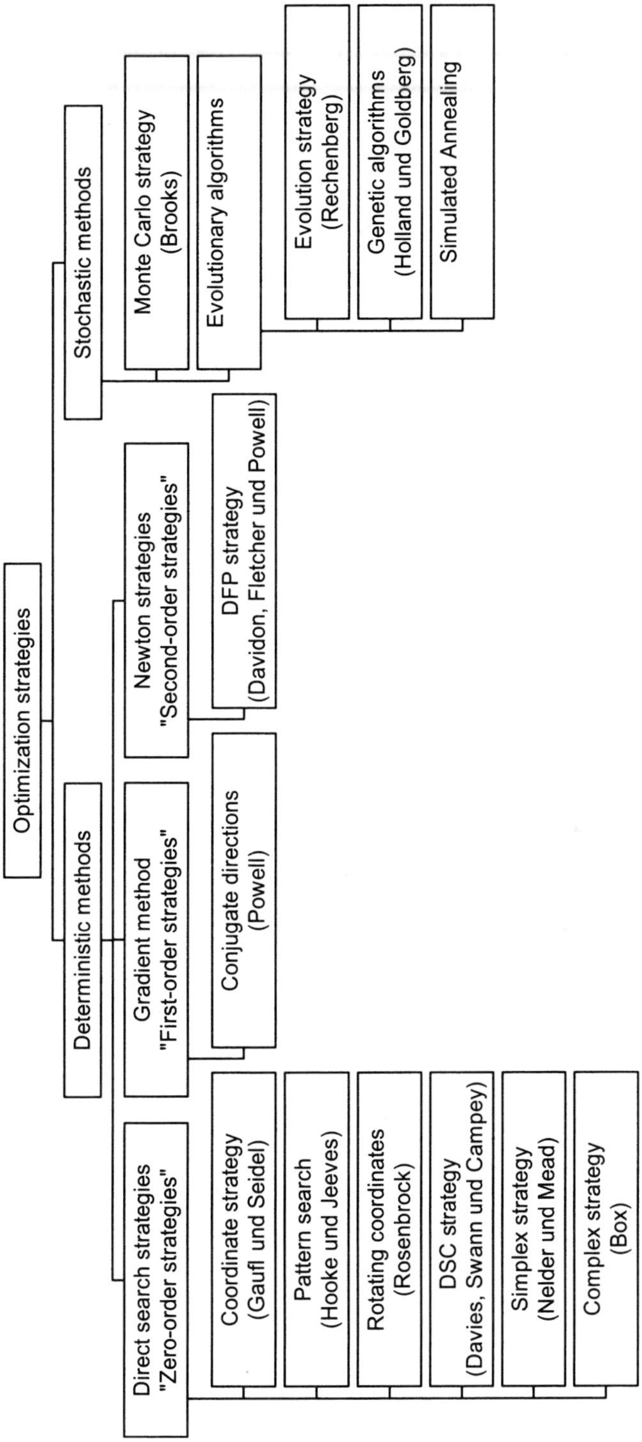


**Figure 4.35** Classification of optimization algorithms

### *Pattern Search (Direct Search)*

The pattern search method of Hooke and Jeeves [62] is an extension of the coordinate strategy. With this optimization method, an exploratory move is first made in a manner similar to the exploratory steps employed with the coordinate strategy. Assuming that the line connecting the starting point to the end point represents an especially promising direction, one iteration step is made in this direction by extrapolation (pattern move). The success of the extrapolation is determined only after evaluating the quality of an additional exploratory step. If the move is not successful, the step is retracted; if successful, the move is continued. By modifying the extent and direction of the extrapolation, it is possible to react to the results from the individual steps so that with only a gradual change in the optimal search direction the length of the extrapolation step increases from time to time, resulting in a shorter search.

### *DSC Strategy*

The DSC strategy, named after Davis, Swann, and Campey, is a combination of the rotating coordinate strategy and a linear search [50]. Starting from an initial location, a linear search for the optimum is conducted in each of the coordinate directions. Starting from the optimal end point determined in this manner, a one-dimensional optimization now follows in the direction of the line connecting the starting point to the end point. Next, the axes of the coordinate system are realigned to form a normalized, rectangular coordinate system, and a new linear search is conducted parallel to these axes. Following this, a one-dimensional optimization is performed in the direction of the line connecting the starting point to the end point.

### *Simplex and Complex Methods*

Another search strategy that differs from those described above follows the simplex method of Neider and Mead [71]. With this method, at least $n + 1$ starting points, representing one parameter set, are specified in an $n$-dimensional search space. The starting points are selected to be equidistant from one another. This generally results in a regular polyhedron called a simplex. Optimization via this strategy proceeds by first evaluating all vertices, each of which describes a parameter set (usually by means of a quality or merit function). In the next step, the vertex with the worst value in relation to the others is identified. This parameter set is discarded and replaced by a new set generated by reflecting the discarded point through the center of the remaining points. If the newly generated vertex produces the worst evaluation in the next iteration, the second-worst vertex is reflected and so forth. If, as a result of this iteration process, the simplex has approached the optimum closely, the simplex is rotated around the vertex with the best evaluation. A closer approach to the optimum can then be achieved by shortening the edge length. Figure 4.36 shows how such a procedure can be performed in a two-dimensional search space. The lines in this figure represent locations having the same value of the quality function. The starting points are identified with $0^1$, $0^2$, and $0^3$. Point 1 results from reflection of point $0^1$ through the center of the line connecting points $0^2$ and $0^3$ and so forth.

A refinement of this method, the complex method, permits incorporation of additional conditions in the form of inequalities [52]. The most significant differences with respect to the simplex method are the greater number of vertices and the increase in size of the polyhedron, called a complex, after each reflection.

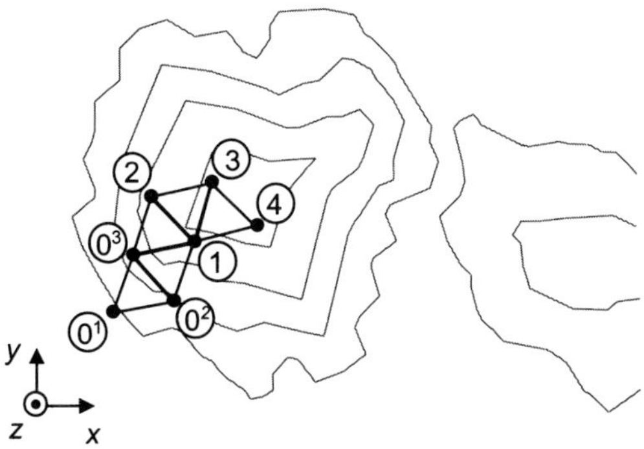


**Figure 4.36** Simplex strategy

### 4.7.3.2 Gradient-Based Optimization Methods

When searching for a minimum or maximum using the "hill climbing" strategies described above, the merit function is evaluated only at discrete locations in the course of locating the optimization objective. The class of gradient-based optimization strategies employs, in addition to the values of the merit function, its first partial derivatives to specify the direction of the next iteration step. Generally, the gradient of the merit function in different directions is first determined at some initial point with this method. The new iteration point is specified in the direction having the steepest gradient. The procedure repeats until the gradient in each investigated direction changes and it can be assumed that an optimum has been located.

Nowadays, gradient methods exist that not only allow steps parallel to the coordinate axes, but also incorporate knowledge gained from the successes and failures of prior iteration steps in order to determine an optimal search direction. With these optimization methods, obtaining the derivatives, which can sometimes be very complex or in many cases cannot be determined at all, can present problems.

If a function is differentiable several times, the information provided by higher derivatives can be used in addition to that from the first derivative. The Nexton strategy, for instance, employs the second derivative in order to represent a function at a given location by means of terms in a Taylor series. In this way, the next best position with respect to the quality function can be determined in a single step without having to conduct a linear search.

The number of optimization methods using partial derivatives is quite large; details can be found in [54].

#### 4.7.3.3 Stochastic Optimization Methods

In contrast to the deterministic methods, the stochastic optimization methods employ a randomness to influence the search. The basic technique within this class of optimization methods is the Monte Carlo strategy, which represents a purely random, or blind, search. With this strategy, parameter values are selected randomly within a specified parameter space, and the resulting values of the merit function are determined. No methodical procedure is employed. When no further improvement in the value of the merit function is observed after evaluating a specified number of parameters, the process stops. Methods that incorporate a methodical procedure in addition to a purely stochastic approach are called evolutionary methods.

#### 4.7.3.4 Evolutionary Methods

The methods in this class find wide application in engineering since they are usually well suited to handling problems with very many independent parameters. In addition, they offer advantages for problems with poorly conditioned quality or merit functions that exhibit undesirable properties such as instabilities, noise, or even undefined values.

The evolutionary methods are optimization methods that perform an optimization in a manner similar to biological evolution. In this case, biological principles are transferred to technical processes (Table 4.2). The evolutionary methods can be divided into two groups. First, there is the *evolution strategy*, which was first employed by Rechenberg [74] for optimization of a flow problem. Then there are the *genetic algorithms*, which were introduced by Holland [66] and Goldberg [60]. There are many variations among these optimization strategies, but only the basic ideas underlying them will be presented here. Additional methods that are comparable to the evolutionary algorithms include *simulated annealing* and *evolutionary programming*, but these will not be considered further here. An overview of all the methods mentioned here can be found in [53,78,84].

**Table 4.2** Transfer of biological principles to the technical optimization process

| Principle | Biology | Engineering |
|---|---|---|
| Replication | Identical duplication of the genetic material | Copying an existing parameter set |
| Mutation | Random, undirected modification of the genotype | Random, undirected modification of the parameter set |
| Birth | Formation of a phenotype | Selecting the new parameter set |
| Life | Survival of the life form in the environment | Evaluating the quality of the selection |
| Selection | Selection of the most well-adapted organism and death of the less well-adapted individuals | Comparing the qualities of the parameter sets and discarding the parameter sets with the lower qualities |

### *Evolution Strategies*

The objective of evolution strategy is to optimize a parameter set through so-called mutation and selection. To this end, new parameter sets are generated from a single parameter set through small, random changes (mutation) to the individual parameters. From the newly generated parameter sets, the one with the best evaluation with respect to the optimization objective is chosen (selection). The procedure followed for optimization by means of an evolution strategy can be divided into the steps below:

1. Specify one or more parameter sets as the starting point.
2. Change all existing parameter sets randomly or mutate the individual parameter sets and thus generate the specified number of "descendants."
3. Select the parameter sets or the "descendants," which are to be used as "parents" for the next generation, and discard the rest.
4. If the optimization objective is not yet reached, return to Step 2.

### *Genetic Algorithms*

In the evolution strategy, the parameters to be optimized are described by real numbers. For instance, the position of a point in space is defined with the aid of real values. Instead of such representations of parameters through real numbers, the genetic algorithms employ series of binary numbers to describe the parameters. Like the chromosomes in biology, these contain the system parameters in coded form.

How such series of numbers appear is shown in Fig. 4.37. In this example, three different parameters are encoded with the aid of series of binary numbers having different lengths. The lengths of these series and the manner used to convert the system parameters into binary code are problem-specific. Once a useful scheme is found for encoding the real numbers into series of binary numbers, the individual parameter sets are changed randomly for the purpose of optimization. When employing genetic algorithms, these random changes are generated through so-called point mutations and crossovers. A mutation changes one digit in a series of binary numbers. The digit that changes is determined randomly. By way of example, Fig. 4.37 shows such a mutation for a parameter encoded by a six-digit binary number. Parameter sets are also changed by crossovers. With the aid of a crossover, two new parameter sets are generated from two existing parameter sets by exchanging values of the binary code starting at a certain digit, which is determined randomly (Fig. 4.37). After this change to the parameter sets, they undergo selection once again. The procedure followed for optimization by means of a genetic algorithm can be divided into the steps below:

1. Specify several parameter sets, which form a starting population.
2. Change all existing parameter sets randomly or perform mutations.
3. Generate new parameter sets through crossover.
4. Select the parameter sets or the "descendants" that are to serve as "parents" of the next generation and discard the rest.
5. If the optimization objective is not yet reached, return to Step 2.

For both of these methods, there are many different ways to generate mutations and select the descendants. For further details, the reader is referred to [77,78].

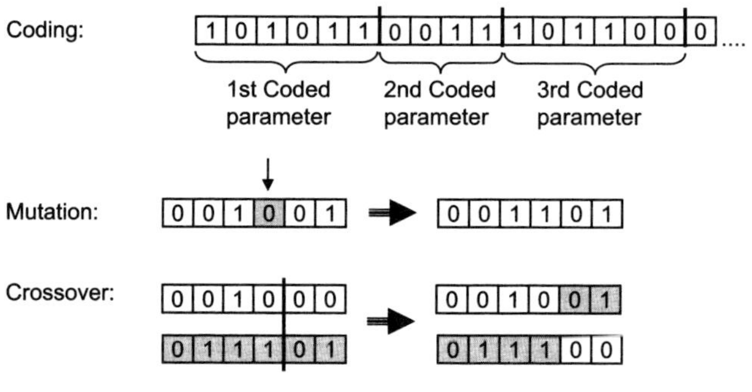


**Figure 4.37** Coding, mutation, and crossover

#### 4.7.3.5 Treatment of Boundary Conditions

Optimization problems in engineering generally involve constraints. This means that not every parameter is freely variable, but rather can change only within specific limits. For instance, when optimizing the throughput from an extruder, the pressure at the inlet to the die is limited to the maximum available from the extruder. If the pressure–throughput relationship of the die is modeled mathematically and, for instance, the die characteristic is approximated by a polynomial for the purpose of optimization, an optimization algorithm to increase the throughput would attempt to increase the pressure endlessly. This can happen because the information regarding the maximum pressure available from the extruder is not contained in the modeled die characteristic. It can be seen in this simple example that the boundary conditions must be incorporated into this and similar problems. To ensure this, the merit function of an optimization problem can be expanded in a manner that incorporates the constraints when evaluating the parameters.

Expressed mathematically, a single optimization problem with constraints is converted into one or more optimization problems without constraints by modifying the merit function. The penalty method and barrier method are two techniques that accomplish this.

### *Penalty Method*

To prevent parameters from passing too far beyond a permissible range and entering an impermissible range during an optimization, penalty terms can be introduced. The value of the penalty terms is added to the value of the original merit function so that a new, modified merit function results. The purpose of this is to modify the value of the merit function in such a way that the quality resulting from parameters that are outside the permissible range is artificially reduced. To achieve this, the penalty functions must be tailored to each problem. The mathematical definition for these penalty terms takes the following form when searching for a maximum:

$$S(\underline{x}) = \begin{cases} = 0 & \underline{x} \in M^n \\ < 0 & \underline{x} \notin M^n \end{cases} \tag{4.87}$$

The effect of the penalty terms is shown graphically in Fig. 4.38 (left). In this illustration, a monotonically increasing merit function is depicted. Since the optimization algorithm can only evaluate the merit function during an optimization (i.e., when searching for a maximum), the value of parameter $s$ was set to be as large as possible in order to achieve the highest possible quality in the model. Boundary conditions were not taken into account in this case. The restrictive constraint or various boundary conditions can, however, be incorporated into the optimization by adding a penalty term with the indicated shape to the values of the merit function. By following this procedure, the values of the merit function are not changed within the permissible range, but the maximum in the function is prevented from occurring far outside the permissible range. With suitable selection of the penalty terms, an optimum that is located near the edge of the permissible range can be found with this method. If, however, the penalty terms are not well matched to the merit function, a parameter set that lies in the restricted range can be selected as the optimum. Preventing this requires the use of so-called barrier functions.

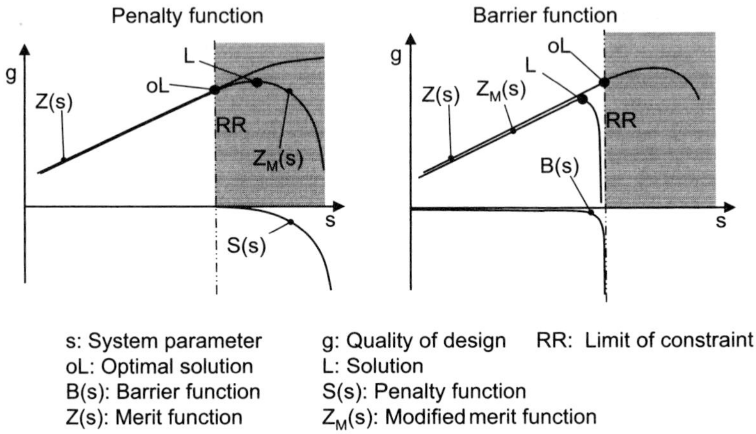


**Figure 4.38** Principles of the barrier and penalty functions for consideration of constraints

***Barrier Methods***

A barrier function is designed to increase indefinitely as the edge of the restricted range is approached (Fig. 4.38, right). This reliably prevents an impermissible parameter set from being selected. However, when using barrier functions, it is not possible to locate an optimum that lies on or near the edge of the restricted range, since, in contrast to the penalty method, values of the merit function are changing also within the permissible range.

## 4.7.4 Practical Applications of Optimization Strategies for the Design of Extrusion Dies

### 4.7.4.1 Optimization of a Convergent Channel Geometry

Finding a useful definition of the quality function that determines the quality of the selected parameters with respect to the quality criteria is the most difficult aspect of successfully applying the optimization methods mentioned above. The following discussion will show how different definitions of the quality function affect the result of an extrusion die optimization when employing optimization strategies.

In the example below, a convergent channel with a reduction ratio of 3:1 was optimized with respect to various criteria by means of the evolution strategy [64]. For the optimization, the flow in the channel was calculated using FEM. Further, the upper limit of the convergent section of the channel was modeled by means of a spline having the points shown in Fig. 4.39 as independent geometric parameters. These points were incorporated into the optimization as parameters, and the finite-element grid was readjusted to the modified channel geometry successively for each iteration step.

If the particle-specific residence time is set as the optimization criterion, the channel assumes the shape shown in Fig. 4.39. Since the flow channel narrows, the melt is accelerated significantly and the mean residence time decreases. As a consequence, however, the pressure drop increases.

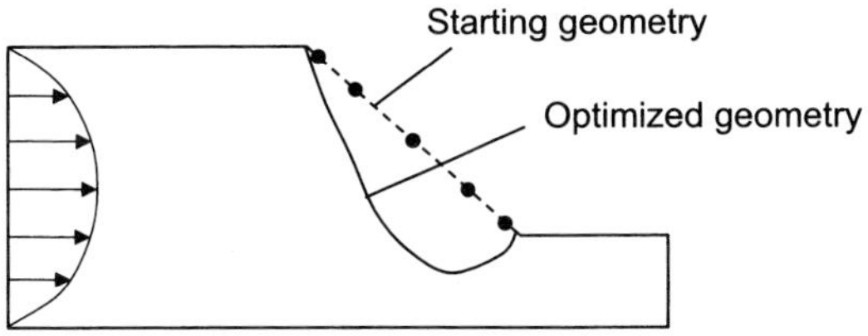


**Figure 4.39** Optimization of a convergent channel geometry

If only the pressure drop in the depicted flow channel is optimized as the quality function, the geometry shown in Fig. 4.40 results. It can be seen that a bulge forms, creating a stagnant area. As a result of this stagnant region, the shear stress at the wall and, thus, the pressure drop in the convergent section are reduced, thus achieving the optimization objective with respect to the selected criterion. From an engineering standpoint, however, this optimization is not acceptable, since stagnant regions lead to degradation of the melt as the result of excessive residence time in the die. This can be avoided, for instance, by combining several optimization criteria. In the present example, the residence times or the shear stresses at the wall could also be taken into account in order to prevent formation of a stagnant region.

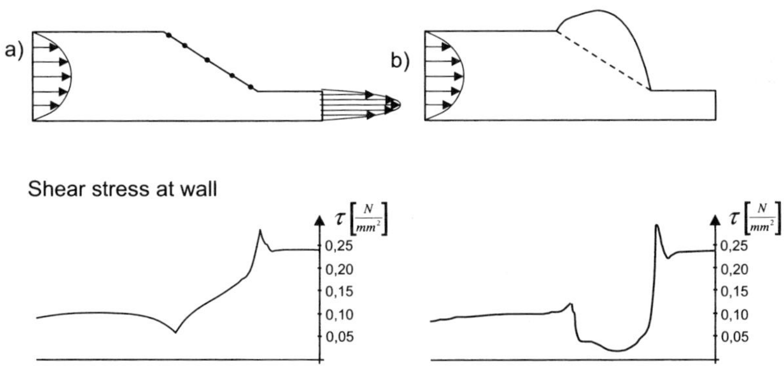


**Figure 4.40** Shear stress at wall before and after optimization

It is clear from these simple examples that various criteria, each of which may be weighted differently, must be incorporated into an optimization.

#### 4.7.4.2 Optimization of Profile Dies

The primary objective when designing an extrusion die is a uniform velocity distribution at the die outlet. Results from automatic optimization of profile extrusion dies with respect to this criterion will be summarized at the end of the chapter. This list makes no claim to completeness, but these research studies show that automatic optimization of complex profiles will be possible in the near future.

In the first example, optimization of a simple profile die is presented. The die is intended to produce a wedge-shaped profile. With this in mind, a starting geometry with a rectangular inlet and wedge-shaped parallel zone is proposed. The starting geometry as well as the velocity distribution calculated by FEM at the outlet of the die are shown in Fig. 4.41.

**Figure 4.41** Velocity distribution before optimization

For optimization of the geometry, five independent parameters describing the height of the flow channel between the inlet and transition zone were selected. The points whose heights are changed during the optimization are marked in Fig. 4.42. When the heights of these points are changed, the associated hexagonal regions are deformed linearly.

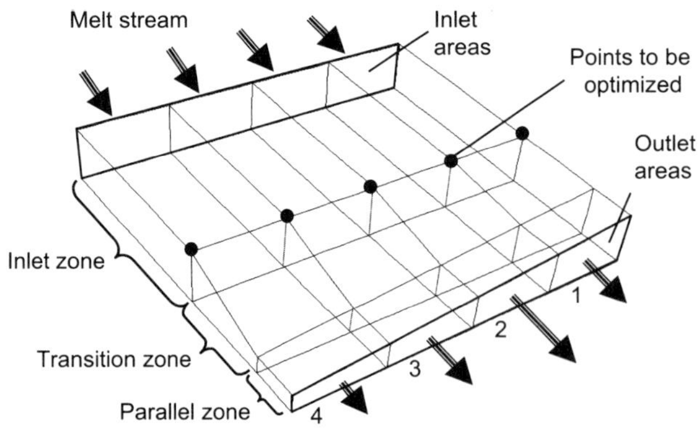


**Figure 4.42** Trial geometry

The simulation method chosen for optimization of the velocity distribution at the die outlet is based on a computational technique that combines network theory and FEM [63,69,82]. This method permits very rapid simulation of the velocity profile within the die, and the changes in melt velocity resulting from changes to the flow channel geometry can be calculated very quickly. Using this technique, it is possible to calculate approximately 300 variations in geometry in one minute on a computer with a Pentium III CPU.

A manual optimization of the depicted starting geometry would attempt to obtain a more uniform velocity distribution at the outlet by increasing the cross sections before the narrower regions of the parallel zone and vice versa. This would force the melt to flow more strongly in the region that initially has less flow.

If an optimization cycle is started with the upper nodes between the inlet and transition zone free to vary during optimization, this is exactly what happens [68]. The three cross sections shown at the right in Fig. 4.42 are narrowed, and the region to the left is expanded by displacing the vertex upward. In this example, a simple evolution strategy that generated five new parameter sets (descendants) from each initial parameter set (parent) was chosen as the optimization method. The best parameter set was selected from the descendants and served as a starting point for the next generation.

The velocity distribution resulting from the optimization is shown in Fig. 4.43. It can be seen clearly that the change in geometry produced an improvement in distribution of the melt. It should be noted that further optimization of the mean velocities was performed for only four outlet segments. By dividing the flow channel into more segments, an even more uniform velocity distribution could have been achieved through additional computational effort.

Optimized geometry

**Figure 4.43** Velocity distribution after optimization

An automatic optimization of a more complex profile die for producing PVC profiles was performed by Szarvasy [81]. FEM was employed as a simulation method in this example to calculate the velocity and pressure distributions. In addition, the viscosity and wall slip behavior of the melt were modeled through incorporation of equations with exponents. The die is designed such that, after entering the die, the entire melt stream is divided into four substreams that are recombined in the die land (Fig. 4.44). The optimization proceeded in two steps. In the first step, the individual volumetric flows necessary to provide a uniform velocity distribution across the segments were determined. In the second step, the geometry of each individual partial channel was optimized to achieve an identical pressure drop in each partial channel at the throughput specified in the first step. The shape of the channels before and after the optimization as well as the simulated velocity distribution at the die outlet are shown in Fig. 4.45. The advantages that such an automatic optimization yields in actual practice become clear if the desired profile is compared to those obtained with various die geometries. Figure 4.46 shows cross sections of the desired profile, together with the unoptimized and the optimized extrudate cross sections after the calibration unit. It can be seen clearly that through optimization the wall thickness distribution of the profile was improved noticeably.

Additional automatic optimizations based on the finite-element method for simulation of the velocity and pressure distributions will be discussed briefly in the following.

Reddy et al. [76] optimized a simple slit die that was divided into two segments having identical widths but different heights. Here, the local lengths of the die lands were automatically designed to achieve a uniform velocity distribution over the width of the outlet.

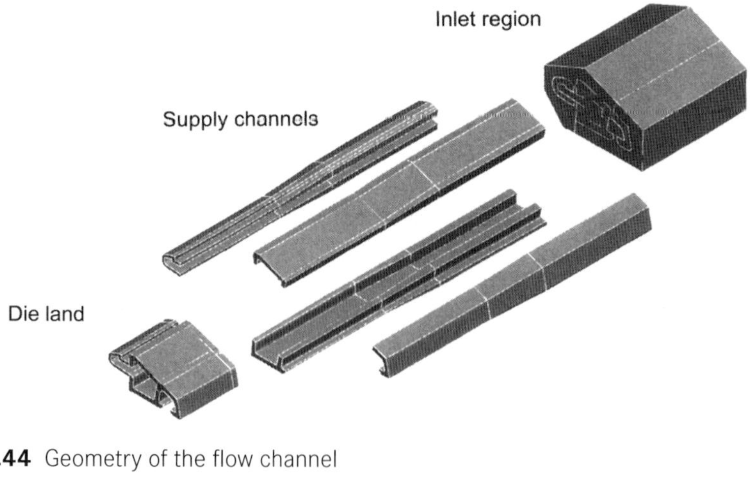


**Figure 4.44** Geometry of the flow channel

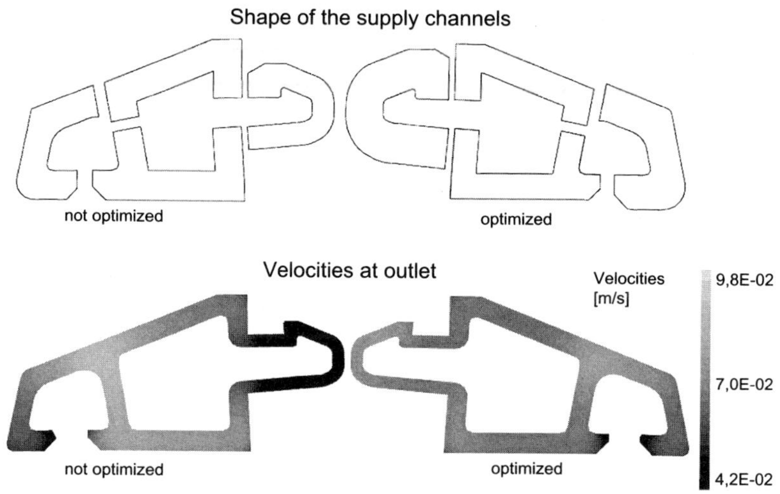


**Figure 4.45** Results of optimization

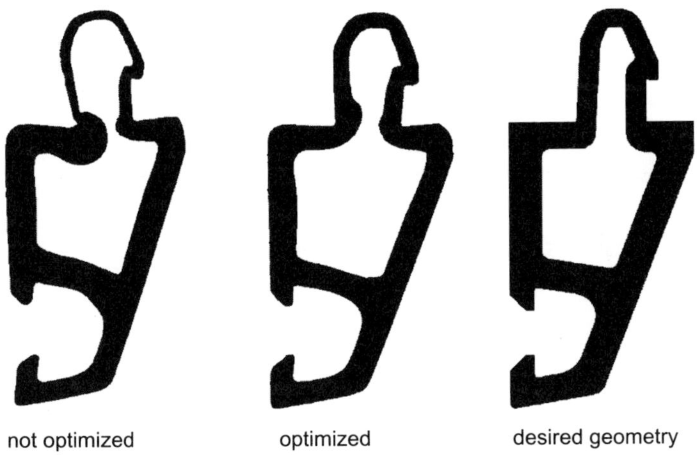


**Figure 4.46** Geometry of the extruded profile

Nóbrega et al. [70] optimized a die with outlet segments having different widths for producing cross-shaped profiles. For this optimization, an optimization zone upstream from the die land was introduced for each segment. Each of these zones has an initial section with the dimensions of the die land and then expands into a wider cross section, which remains constant up to the end of the optimization zone. During the optimization, the ratio of the length of the narrow section to the length of the wide section was optimized to achieve an identical mean melt discharge velocity from all four segments.

In work performed by Sienz and Marchal [67,79], a flow channel for extrusion of a dual-lumen catheter was optimized by means of gradient-based methods. The overall flow channel was divided by ribs into two channels that were modified through optimization to provide a uniform velocity profile at the outlet. To achieve this, three design variables were used to specify the cross-sectional geometry.

It should be noted as this chapter concludes that, although the optimization of profile extrusion dies is still largely based on a "trial and error" approach at present, very promising steps have been made to automatically optimize such dies with the aid of flow simulation programs coupled with optimization algorithms. In the future, it is probable that the design of extrusion dies for producing profiles in materials whose flow behavior can be described well mathematically will be greatly facilitated, if not completely automated.

Nevertheless, for some materials, it will remain difficult in the future to accurately describe all of the factors affecting the melt in the extrusion process with the aid of simulation calculations. More extensive research is probably still required to predict in advance, for instance, the viscoelastic behavior of a polymer melt that causes, among other effects, the die swell experienced by the profile after leaving the outlet from the die. The problem of material shrinkage as well as the phenomenon of wall slip cannot yet be handled completely by means of simulation calculations. The reasons for this are that either the mathematical models are not yet satisfactory or it is not yet possible to obtain reliable measurement results for the parameters in the material models.

## Symbols and Abbreviations

| | |
|---|---|
| $a, b$ | Valuation factors for shear rate and extension rate |
| $A, B, C$ | Carreau parameters |
| $B$ | Slit width |
| $Bi$ | Biot number |
| $c_p$ | Specific heat capacity |
| $D$ | Diameter |
| $D_0$ | Diameter of the die orifice |
| $g$ | Gravitational constant |
| $H$ | Slit height |
| $k$ | Coefficient of heat transfer between the contact layer melt and die wall |
| $K, K^*$ | Factor for the ratio of thickness and width swelling |
| $L$ | Length |
| $n$ | Flow index |
| $\vec{n}$ | Normal vector |
| $N$ | Node number |

| | |
|---|---|
| $NT$ | Number of layers in the difference grid |
| $\Delta p$ | Pressure loss |
| $\vec{\dot{q}}$ | Vector of heat flux |
| $R$ | Flow resistance |
| $R_b$ | Radius of pipe bend |
| $R_p$ | Radius of pipe |
| $s$ | Distance |
| $Sw$ | Swelling factor |
| $Sw_B$ | Swelling factor of an extruded flat band in the direction of its width |
| $Sw_H$ | Swelling factor of an extruded flat band in the direction of its thickness |
| $t$ | Time |
| $T$ | Temperature |
| $T_M$ | Mass temperature |
| $T_W$ | Temperature at flow channel wall |
| $U$ | Internal energy |
| $\overline{v}$ | Average velocity |
| $v_{max}$ | Maximum velocity |
| $v_P$ | Specific volume |
| $v_x, v_y, v_z$ | Velocity components in the direction of coordinates $x, y, z$ |
| $W_i$ | Weight function |
| $x, y, z$ | Coordinates |
| $\dot{\gamma}$ | Shear rate |
| $\varepsilon$ | Residual |
| $\varepsilon_R$ | Reversible deformation |
| $\overline{\dot{\varepsilon}_R}$ | Mean (average) reversible deformation |
| $\varepsilon_{rep}$ | Representative rate of deformation |
| $\dot{\varepsilon}$ | Extension rate |
| $\eta$ | Shear viscosity |
| $\theta$ | Cylindrical coordinates |
| $\lambda$ | Heat conductivity |
| $\rho$ | Density |
| $\tau$ | Extra-stress tensor |
| $\phi$ | Heat source |
| $\phi_i, \phi_N$ | Shape function |
| $\Omega$ | Area (or volume) of an elment |

## 4.8 References of Chapter 4

1. Bird, R.B., Stewart, W.E. and Lightfoot, E.N: Transport Phenomena, 2nd Ed., Wiley, London (2006)
2. Schade, H. and Kunz, E.: Strömungslehre, 4th ed., de Gruyter, Berlin ( 2013)
3. Böhme, G.: Strömungsmechanik nicht-newtonscher Fluide, 2nd Ed., Teubner, Stuttgart (2000)
4. Bird, R.B., Armstrong, R.C. and Hassager, O.: Dynamics of Polymer Liquids. Vol. 1: Fluid Mechanics. Wiley, New York (1987)
5. Jischa, M.: Konvektiver Impuls-, Wärme- und Stoffaustausch. Vieweg, Braunschweig (1982)
6. Winter, H.H.: Temperaturänderung beim Durchströmen von Rohren. In: Praktische Rheologie der Kunststoffe. VDI-Verl., Düsseldorf (1978)
7. Winter, H.H.: Ingenieurmäßige Berechnung von Geschwindigkeits- und Temperaturfeldern in strömenden Kunststoffschmelzen. In: Berechnen von Extrudierwerkzeugen. VDI-Verl., Düsseldorf (1978)
8. Winter, H.H.: Wärmeübertragung und Dissipation in Scherströmungen von Polymerschmelzen. Adv. Heat Transfer 13 (1977) p. 205
9. Sebastian, D.H. and Rakos, R.: Interactive software package for the design and analysis of extrusion profile dies. Adv. Polym. Tech. 5 (1985) 4, pp. 333–339
10. Schmidt, J.: Wärmetechnische Auslegung von Profilkühlstrecken mit Hilfe der Methode der Finiten Elemente. Thesis at the RWTH Aachen (1985)
11. Strauch, Th.: Ein Beitrag zur Theologischen Auslegung von Coextrusionswerkzeugen. Thesis at the RWTH Aachen (1986)
12. Hüsgen, U.: Thermische und rheologische Berechnungen im Bereich Blasformen. Thesis at the RWTH Aachen (1988)
13. Menges, G. et al.: FEM in der Werkzeugkonstruktion. CAD-CAM Report 6 (1987) 5, pp. 66–76
14. Törnig, W., Gipser, M. and Kaspar, B.: Numerische Lösung von partiellen Differentialgleichungen der Technik, 2nd ed. Teubner, Stuttgart (1985)
15. Crochet, M.J., Davies, A.R. and Walters, K.: Numerical Simulation of Non-Newtonian Flow. Elsevier, Amsterdam (1984)
16. Wortberg, J.: Werkzeugauslegung für die Ein- und Mehrschichtextrusion. Thesis at the RWTH Aachen (1978)
17. Junk, P.B.: Betrachtungen zum Schmelzeverhalten beim kontinuierlichen Blasformen. Thesis at the RWTH Aachen (1978
18. Zienkiewicz, O.C.: The Finite Element Method. Maidenhead, Berkshire, New York: McGraw-Hill, 6th ed. (2005 )
19. Chung, T.J.: Finite Element Analysis in Fluid Dynamics. New York: McGraw-Hill (1978)
20. Altenbach, J. and Sacharov, A.S. (Eds.): Die Methode der finiten Elemente in der Festkörpermechanik. Hanser, München (1982)
21. Bathe, K.-J.: Finite Element Procedures in Engineering Analysis. Englewood Cliffs: Prentice Hall (1982)
22. Schwarz, H.R.: Methode der finiten Elemente,3rd ed. Teubner, Stuttgart (1991)
23. Schwenzer, C.: Finite Elemente Methoden zur Berechnung von Mono- und Coextrusionsströmungen. Thesis at the RWTH Aachen (1988)
24. Hinton, E. and Campbell, J.S.: Local and global smoothing of discontinuous finite element functions using a least squares method. Int. J. Numer. Methods Eng. 8 (1974) pp. 461–480
25. Lee, R.L. et al.: Smoothing techniques for certain primitive variable solutions of the Navier-Stokes equations. Int. J. Numer. Methods Eng. 14 (1979) pp. 1785–1804

26. Hövelmanns, N. et al.: Rechnergestützte Planung und Auslegung von Extrusionsanlagen. In:Proc. 14th IKV Colloquium, Aachen (1988)
27. Cushman, J.H.: Difference schemes or element schemes? Int. J. Numer. Methods Eng. 14 (1979) pp. 1643–1651
28. Vlachopoulos, J.: Should you use finite difference or finite element methods for polymer flow problems? In: Proc. 35th SPE ANTEC, Montreal, Canada (1977)
29. Menges, G. et al.: Wärmeausgleichsrechnung in der Kunststoffverarbeitung mit der FEM. Kunststoffe 77 (1987) 8, pp. 797–802
30. Schwenzer, C.: Simulation von Kunststoffströmungen mit der FEM. In: Proc. CAT '88, Stuttgart. Konradin-Verlag, Leinfelden-Echterdingen (1988)
31. Menges, G. and Schwenzer, C.: Entwicklung von Konstruktionshilfen für Profilwerkzeuge. AIF research report, Aachen (1985)
32. Leonov, A.I. and Prokunin, A.N.: On the stretching and swelling of an elastic liquid extruded from a capillary die. Rheol. Acta 23 (1984) pp. 62–69
33. Orbey, N. and Dealy, J.M.: Isothermal swell of extrudate from annular dies; Effects of die geometry flow rate and resin characteristics. Polym. Eng. Sci. 24 (1984) 7, pp. 511–518
34. Nickell, R.E. et al.: The solution of viscous incompressible jet and free-surface flows using finite element methods. J. Fluid Mech. 65 (1974) 1, pp. 189–206
35. Allan, W.: Newtonian die swell evaluation for axisymmetric tube exits using a finite element method. Int. J. Numer. Methods Eng. 11 (1977) pp. 1621–1632
36. Batchelor, J. et al.: Die swell in elastic and viscous fluids. Polymer 14 (1973) 7, pp. 297–299
37. Masberg, U.: Einsatz der Methode der finiten Elemente zur Auslegung von Extrusionswerkzeugen. Thesis at the RWTH Aachen (1981)
38. Menges, G. et al.: Numerical simulation of three-dimensional non-Newtonian flow in thermoplastic extrusion dies with finite element methods. In: Numerical Analysis of Forming Processes. Pittman, J.F.T. and Zienkiewicz, O.C. et al. (Eds.) et al. (Eds.) Wiley, New York (1984)
39. Gesenhues, B.: Rechnerunterstützte Auslegung von Fließkanälen für Polymerschmelzen. Thesis at the RWTH Aachen (1984)
40. Mitsoulis, E. et al.: Numerical simulation of entry and exit flows in slit dies. Polym. Eng. Sci. 24 (1984) 9, pp. 707–715
41. Keunings, R. and Crochet, M.J.: Numerical simulation of the flow of a fluid through an abrupt contraction. J. Non.-Newt. Fluid Mech. 14 (1984) pp. 279–299
42. Luo, X.L. and Tanner, R.I.: Finite element simulation of long and short circular die extrusion experiments using integral models. Int. J. Numer. Methods Eng. 25 (1988) pp. 9–22
43. Dheur, J. and Crochet, M.J.: Newtonian stratified flow through an abrupt expansion. Rheol. Acta 26 (1987) pp. 401–413
44. Münstedt, H.: Rheologische Eigenschaften einiger Kunststoffschmelzen. In: Praktische Rheologie der Kunststoffe. VDI-Verl., Düsseldorf (1978)
45. Richtmyer, R.D. and Morton, K.W.: Difference Methods for Initial-Value Problems, 2nd ed. Interscience Publ., New York (1994)
46. Poloshi, G.N.: Mathematisches Praktikum. Teubner, Leipzig (1963)
47. Pearson, J.R.A.: Mechanics of Polymer Processing. Elsevier, London (1985)
48. Carreau, P.J. and De Kee, D.: Review of some useful rheological equations. Can. J. Chem. Eng. 57 (1979) pp. 3–15
49. Astarita, G. and Marrocci, G.: Principles of Non-Newtonian Fluid Mechanics. McGraw Hill, London (1974)

50. Box, M.J., Davies, D. and Swann, W.H.: Non-linear Optimization, Oliver Boyd, Edinburgh, Scotland (1969)
51. Booy, M.L.: A network flow analysis of extrusion dies and other flow systems. Polym. Eng. Sci. 22 (1982) 7, pp. 432–437
52. Box, M.J.: A new method of constrained optimization and a comparison with other methods. Computer Journal 8 (1965) pp. 42–52
53. Bäck, T. and Schwefel, H.P.: An overview of evolutionary algorithms for parameter optimization, Evolutionary Computation 1 (1993) 1, pp. 1–23
54. Bronstein, I.N. and Semendjaev, K.A.: Taschenbuch der Mathematik, 7th Ed. Harry Deutsch Verlag, Frankfurt A.M. (2008)
55. Cohen, A. and Isayen, A.I.: The concept of inverse formulation and its application to die flow simulation, In: Modeling of Polymer Processing (Recent Developments) Hanser Verlag, München (1991)
56. Chmiel, H. and Schümmer, P.: Eine neue Methode zur Auswertung von Rohrrheometer- daten. Chem. Eng. Sci. 43 (1971) 23, pp. 1257–1259
57. Czyborra, L.: Offene Fragen bei der rechnerischen Auslegung von Werkzeugen für die Profilextrusion. Ph.D. presentation, IKV, Aachen (1992)
58. Fang, S.: Auslegung und Optimierung von Profilwerkzeugen, komplexer Gemotrien mittels FEM. In: Extrusionswerkzeuge – Schwerpunkt Profilwerkzeuge, VDI-Verlag GmbH, Düsseldorf (1996)
59. Fang, S.: Konzipierung und Dimensionierung komplexer Extrusionswerkzeuge mittels der Finiten Elemente Methode. Thesis at University Stuttgart (1999)
60. Goldberg, D.E.: Genetic Algorithms in Search, Optimization and Machine Learning. Addison Wesley, Reading, MA (1989)
61. Haberstroh, E., Hoffmann, K. and Kropp, D.: Rechenmöglichkeiten bei der Extrusion von Elastomeren. Kautsch. Gummi Kunstst. 50 (1997) 4, pp. 304–311
62. Hooke, R. and Jeeves, T.A.: Direct search solution of numerical and statistical problems. J. Assoc. Comput. Mach. 8 (1961) 2, pp. 212–229
63. Huneault, M.A., Lafleur, P.G. and Carreau, P.J.: Evaluation of the FAN technique for profile die design. Int. Polym. Process. 11 (1996) p. 1
64. Hoffmann, K.: Fließkanalberechnung an Profilwerkzeugen mit Finite-Elemente-Methoden, In: VDI-Kunststofftechnik Tagung „Extrusionswerkzeuge“ (1993) pp. 30–49
65. Hoffmann, K.: Grundlegendes Auslegen und anschließende Überprüfung der Ergebnisse von Extrusionswerkzeugen mittels FEM, In: Extrusionswerkzeuge – Schwerpunkt Profilwerkzeuge, VDI-Verlag GmbH, Düsseldorf (1996)
66. Holland, J.H.: Adaptation in Natural and Artificial Systems. The University of Michigan Press, Ann Arbor, MI (1975)
67. Marchai, J.M. and Goublomme, A.: Parametric optimization of extrusion dies through numerical simulation, In: 3rd ESAFORM Conference on Material Forming (2000)
68. Michaeli, W. and Kaul, S.: Computer-aided optimization of extrusion dies, In: ANTEC, SPE Annual Technical Conference, May 6–10 (2002) San Francisco
69. Michaeli, W., Kaul, S. and Wolff, T.: Computer-aided optimization of extrusion dies. J. Polym. Process. (2001) 21 pp. 225–237
70. Nóbrega, J.M., Carneiro, O.S., Pinho, F.T. and Oliveira, P.J.: Flow balancing of profile extrusion dies, In: ANTEC, SPE Annual Technical Conference, Paper presented May 6–10 (2001) Dallas, TX
71. Neider, J.A. and Mead, R.: A simplex method for function minimization. Computer Journal 7 (1965) pp. 308–313
72. Oswald, T.: Boundary Elements XVII. Computational Mechanics Publications (1995)

73. Plajer, O.: Angewandte Rheologie, Plastverarbeiter 29 (1978) pp. 4–10
74. Rechenberg, I.: Evolutionsstrategie: Optimierung technischer Systeme nach Prinzipien der biologischen Evolution. Frommann-Holzboog, Stuttgart (1973)
75. Rosenbrock, H. H.: An automatic method for finding the greatest or least value of a function. Computer Journal 3 (1960) pp. 212–229
76. Reddy, M. P., Schaub, E. G., Reischneider, L. G. and Thomas, H. L.: Design and optimization of three-dimensional extrusion dies using adaptive finite element method, In: ANTEC, SPE Annual Technical Conference, Paper presented (1999) New York
77. Schwefel, H. P.: Numerische Optimierung von Computer-Modellen mittels der Evolutionsstrategie. Band 26 der Reihe Interdisciplinary systems research, Birkhäuser Verlag, Basel (1977)
78. Schwefel, H. P.: Evolution and Optimum Seeking. John Wiley & Sons, New York (1995)
79. Sienz, J., Marchai, J.-M. and Pittman, J.: Profile extrusion die design using optimisation techniques and an expert system, In: 3rd ESAFORM Conference on Material Forming (2000)
80. Szarvasy, I., Sienz, J., Pittman, J. and Hinton, E.: Computer-aided optimisation of profile extrusion dies: Definition and assessment of the objective function. Int. Polym. Process. 1 (2000) p. 15
81. Szarvasy, I.: PVC Extrusion die design: Parallel finite element based optimisation of an upstand bead profile die, In: 3rd ESAFORM Conference on Material Forming (2000)
82. Wolff, T.: Rechnergestützte Optimierung von Fließkanälen für Profilextrusionswerk- zeuge. Thesis at RWTH Aachen (2000)
83. Wolff, T.: Rechnergestützte Auslegung von Extrusionswerkzeugen, In: Tagungsumdruck der Fachtagung, Neuigkeiten in der Extrusion, Süddeutsches Kunststoff-Zentrum, Würzburg (1996)
84. Yao, X.: An overview of evolutionary computation. Chinese J. Adv. Software Res. 3 (1996) p. 1
85. Hopmann, C. and Yesildag, N.: Numerical investigation of the temperature influence on the melt predistribution in a spiral mandrel die with different polyolefins. J. Polym. Eng. 35 (2015) p. 10

# 5 Monoextrusion Dies for Thermoplastics

After having developed the fundamental theoretical considerations for the design of flow channels of extrusion dies in the preceding chapters, now in Chapter 5 the focus will be on types of design, applications, and the layout of dies for the discharge of a single melt. These will be referred to as monoextrusion dies, in contrast to coextrusion dies (see Chapter 6). In discussing this subject, the extrusion dies will be divided into groups according to the geometry of their exit cross section since as a rule dies with similar exit cross sections are practically equal in the way they are designed, regardless of the extruded product.

## 5.1 Dies with Circular Exit Cross Section

Dies with a circular exit cross section are used for the production of pellets and the extrusion of fibers, strands, and solid rods. The latter two shapes are extruded generally through axially fed dies, while the melt flow for fibers is frequently deflected.

### 5.1.1 Designs and Applications

#### *Pelletizer Die Plates*

Pelletizer die plates form plastic strands that are subsequently cut into pellets. In principle, there are two pelletizing methods, which differ in the sequence of the individual steps involved [1]:

- Hot pelletizing: Strand extrusion → cutting → cooling
- Cold pelletizing: Strand extrusion → cooling → cutting

In both methods, the pelletizer die takes over the task of extruding the strands; in hot pelletizing, it acts, in addition to that, as a cutting plate (counterknife). There are essentially two types of knife arrangements: centric and excentric (Fig. 5.1).

The advantage of the excentric arrangement is that there is no need to deflect the melt in the die. This is particularly important for heat-sensitive melts, because there is no stagnation caused by their deflection in the flow channel. When designing the die plate for the hot pelletizing process, it is necessary to take into consideration the cooling of the pellets, which follows the cutting step. Different methods of cooling as well as the typical process conditions are shown in Table 5.1 [1].

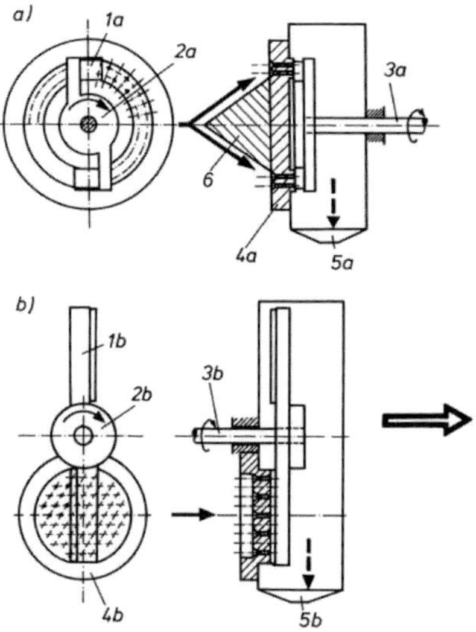


**Figure 5.1** Hot pelletizers [1]: (a) pelletizer with a knife shaft in the center; (b) pelletizer with an excentric knife shaft:1a and 1b, knife;2a and 2b, knife holder; 3a and 3b, knife shaft; 4a and 4b, pelletizer die; 5a and 5b, pellet collector housing; 6, torpedo

The boreholes in the pelletizer die plate (Fig. 5.2) typically have a conical entry with the entry angle kept at less than 30° to eliminate secondary flows. In addition, the holes generally have a relatively small length-to-diameter ratio (*L/D* less than 10) [2].

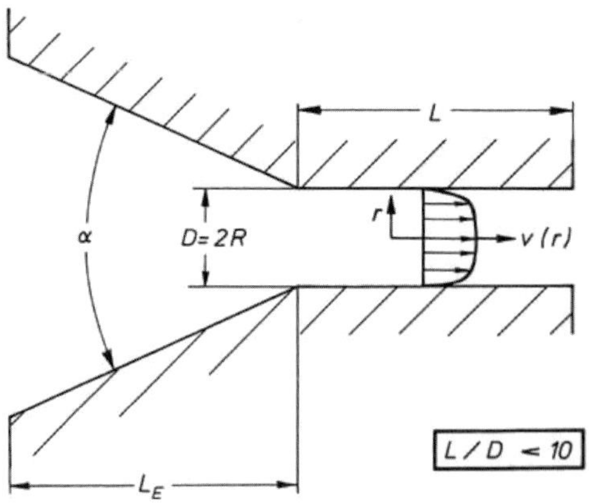


**Figure 5.2** Pelletizer die (geometry)

**Table 5.1** Hot Pelletizing Process [1]

| Process | Type of product | Throughput | Pelletizer die | | | Melt pressure | Swelling[a] of pellets | Knife | | | | Cooling water | |
|---|---|---|---|---|---|---|---|---|---|---|---|---|---|
| | | | Heating | Number of holes | Throughput per hole | | | Number | Circumference velocity | Power input | Working life | Throughput | Temperature |
| Cutting in/Cooling in | – | kg/h | – | – | kg/h | bar | – | – | m/s | kW | h | $m^3/h$ | °C |
| Air/Air | PVC rigid/soft | 600–1800 | Electric | 1100 | 0.55–1.65 | – | 0.6 | 2 | 5–8 | 2 | 50–500 | – | – |
| Air/Water | PE-HD | 1700–2800 | Steam | 286 | 6–10 | 20–40 | 0.8 | 2 | 22–27 | 9 | 400–500 | 60 | 50–70 |
| Water/Water | PP | 1900–2800 | Steam | 192 | 10–12 | 60–100 | 0.7 | 6 | 20 | 44 | 250–600 | 60 | 30–50 |

a) Borehole diameter/pellet diameter.

The arrangement of the holes in the die plate is either linear or circular. To avoid solidification of the melt in the holes during the underwater pelletizing and damage resulting from the melt pressure, the boreholes are partially insulated by the use of special inserts (Fig. 5.3). This arrangement allows a higher temperature melt to be brought to the cooled zone at the die outlet [3]. The pelletizing plate is heated either electrically or by high-pressure steam.

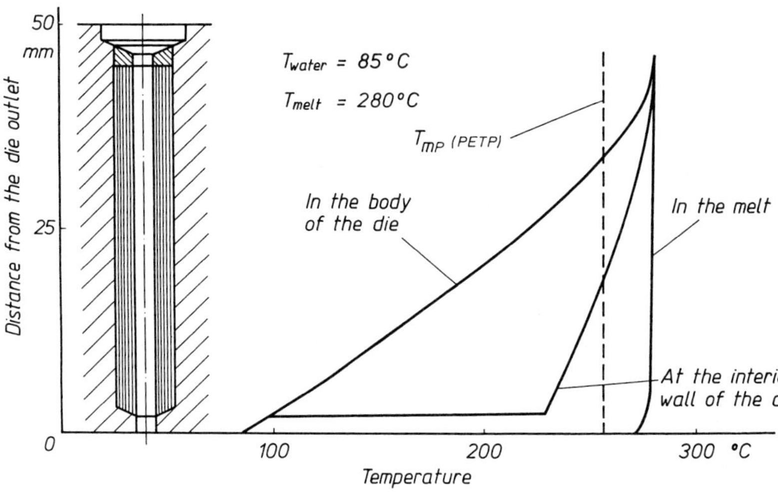


**Figure 5.3** Underwater pelletizer die [3]

During startup, a pressure peak of 500 to 700 bar can occur in front of the plate; during the steady-state operation, the pressure loss still can be up to 250 bar [3]. At the same time, the mass flow through a single hole can be up to 15 kg per hour [4]. A comprehensive description of the pelletizing technique is given in [2,5].

### *Screens*

In order to eliminate dirt particles from a plastics melt or remove particles of the same material with a different (mostly higher) degree of polymerization, various screening or filtering devices are used between the extruder and the die [6]. The following filters are used successfully [7,8]:

- Filters from sand grit
- Filters from sintered metal (prepared by powder metallurgy)
- Filters from wire mesh with different weaves
- Filters from nonwoven wires

The filtration of a plastic melt requires that the pressure losses are kept at a permissible level and that certain requirements for flow uniformity and serviceability and so on are met [6]. Mesh gaps are between 5 μm and 2–3 mm, defining different screening qualities [8]. To keep the residence time and the distribution of the melts in the filtering device to a minimum, only such filtering arrangements are practical that have large filter areas in a housing of small volume [6]. These requirements are best met by a concentric arrangement of filtering inserts in a candle form or as disks. The filtering devices are stretched over or between perforated supports that allow the filtered melt to proceed to the exit from the housing [6].

The effectiveness of a filter is determined by the amount of mass throughput, filtering area, and pore size and by the flow rates and pressure losses resulting therefrom, as well as by the physical properties of the filter [6,9]. Mass throughputs over 1000 kg per hour are possible in the production of fibers with filtering areas of 0.5 to 7 $m^2$ [6].

Automatic melt filtration systems are an interesting development in this area. They offer advantages with regard to compactness, undisturbed flow, and maintenance [10,11]. An example is shown in Fig. 5.4. A screen pack is placed between two perforated disks. A so-called feeler is attached on the inlet side, which is capable of sweeping over the entire cross section of the disk by rotating. This rotation starts whenever a pressure limit is exceeded due to increasing clogging of the screens. As soon as the feeler leaves its parking position, the contaminated melt flows in the opposite direction through the screen pack and perforated disks due to the pressure buildup. This reverse flow presses the filtered dirt particles through the holes that were covered by the feeler. This way the dirt particles removed from the screen pack pass along the hollow drive shaft of the feeler to the outside. Depending on the polymer and the degree and kind of contamination, the system shown here allows throughputs up to 500 kg per hour [10].

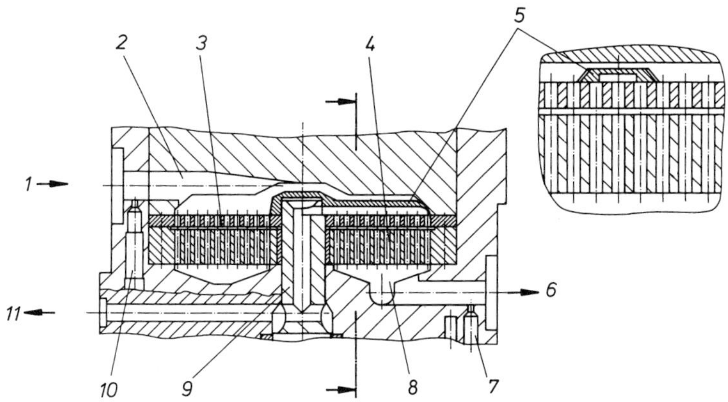


**Figure 5.4** Melt filter [10]: 1: contaminated melt; 2: feed (inlet) chamber; 3: screen pack; 4: breaker plate pack; 5: cleaning fingers; 6: clean melt; 7, 10: pressure transducer well; 8: outlet chamber; 9: drive shaft; 11: concentrated contaminant

Besides a gradual rotation of a rotating screen disk, another possibility of continuous filtration is shown in Fig. 5.5. The melt is split up and flows through two or more screen packs. As soon as the pressure drop of one screen pack is too high, this pack is moved out of the flow path so that it can be cleaned or replaced. In the meantime, the screw speed is increased at constant throughput through the remaining screen pack until the renewed pack is back in position. Throughputs of up to 15,000 kg/h are possible [12].

**Figure 5.5** System for a continuous filtration via splitting up the melt through different screen packs [Nordson Kreyenborg]

Continuous systems are usually beneficial for large lot sizes and few material changes because no stopping of production for filter cleaning is necessary. Especially for continuous systems, degassing, such as by bleed grooves, is required.

A very good description containing many useful hints for the construction and computation (involving pressure losses) of screen packs and the supporting perforated disks is given in [6,13].

#### *Spinnerets*

For spinning plastics melts of, for example, polypropylene, polyamide, or polyester into filaments or yarns, so-called spinnerets are used. These contain a very high number of boreholes (spinning nozzles). These spinnerets are frequently placed in a horizontal position so that the filaments can be pulled away downward [14,15]. If a melt with a low viscosity is processed, a gear pump is placed between the extruder and the spinneret that ensures a uniform melt stream [16].

The spinnerets can be of circular shape with diameters from 40 to 80 mm, with the plate thickness between 10 and 45 mm and containing 10 to 1000 spinning holes, or they can be rectangular. Typical measurements of the latter can be 60 mm × 60 mm to 150 mm × 450 mm, with a plate thickness between 20 and 30 mm. The number of holes can be up to 10,000 [15]. The individual boreholes in the plate can be arranged in a linear or circular fashion; the distances between them are typically 6 to 10 mm [15]. The individual holes have, as a rule, an exit diameter between 0.2 and 0.6 mm with *L/D* ratios from 1 to 4. The diameter of the inlet hole is between 2 and 3 mm with a transition angle between 60° and 90° into the outlet region [15]. Smaller transition angles may produce fibers with a better surface. The best results for polypropylene monofilaments were achieved according to [17] with transition angles less than 20°. To produce filaments of good quality, the surface of the borehole has to be extremely smooth and the edge of the outlet sharp and without burrs [14,15].

Besides round holes, other geometries, such as triangular, Y-shaped, and T-shaped holes are used [14,15].

Under production conditions, the spinning plates are subjected not only to pressures up to 300 bar and to high mechanical stresses but also to high temperatures (up to 300 °C), which can cause strong corrosive attack. The latter is made worse by cleaning the spinning plates in salt baths. Hence, cleaning in salt baths is unusual, in contrast to, for example, vacuum pyrolysis or hydrolysis. Therefore, all of the above stresses have to be taken into consideration when selecting the material for this purpose. Useful data can be found in [14,15,18].

### *Solid Rods*

Solid rods from plastics with diameters up to 500 mm and also flat bars with cross-sectional dimensions up to 250 mm × 100 mm [19] are produced by dies that differ from others by the particular arrangement in which the calibrating section is attached to the melt shaping die by a flange (see Fig. 5.6 [20]). Since the calibration section has to be cooled very strongly, there must be good thermal insulation between it and the die [19–21].

The flow channel in the die is shaped in such a way that the melt exits from a relatively small diameter, approximately 8 to 10 mm [20], and enters the attached guiding zone of the calibration section at a relatively steep angle. A solid skin is formed in contact with the cooled wall, and a wedge- or cone-shaped melt zone, defined by a parabolic cooling curve, is formed in the rod (Figs. 5.6 and 5.7). The melt pressure acts thereby on the inclined solidified surface, and this may lead to increased friction forces in the calibrator and thus to jamming of the profile. For that reason, it is important to build the calibration line just so long so that by its end the solidified layers are so thick that they can resist not only the pressure from the melted core but also the takeoff tension.

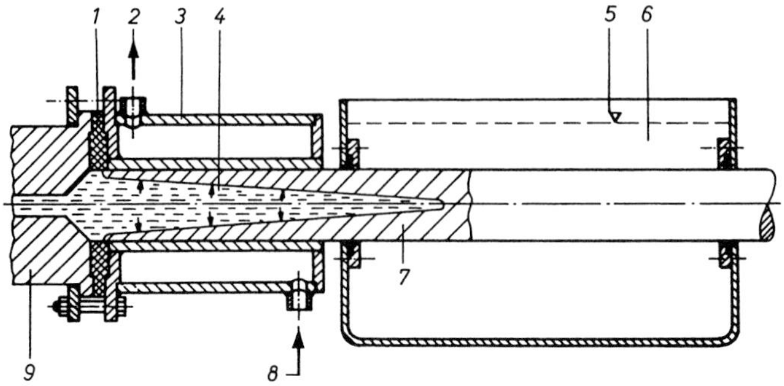


**Figure 5.6** Solid bar extrusion: 1: heat separation; 2: cooling water outlet; 3: calibrating section; 4: melt; 5: water level; 6: water bath; 7: solidified layer; 8: cooling water inlet; 9: die

**Figure 5.7** Real course of the freeze line for a material change trial [Ensinger]

The volume shrinkage due to the cooling can produce voids in the rod. To avoid this, the extruder has to operate at a sufficiently high pressure. This required pressure as well as the value of the pressure at which the jamming in the calibrator occurs determine the operating conditions for the given extrusion line.

The pressure limits can be altered under some circumstances by coating the surface of the calibrator with PTFE, for example, or by feeding a lubricant between the surfaces of the extrudate and the calibrator [20,21]. However, it may happen that the pressure required to eliminate a void can no longer be applied. Then, profile brakes, which slow down the exit speed of the profile from the calibrator [20], have to be installed. For products with larger cross sections and resulting slow cooling rates, the typical line speeds are, for example, 2.5 m/h for a 60 mm and 0.5 m/h for a 200 mm polyamide round rod [19]. Often, tempering has to be done on rods with thick walls to eliminate internal (residual) stresses that are due to process conditions. When materials with a low melt viscosity are processed, it is often necessary to install a valve between the extruder and the die that increases the melt pressure and thereby its homogeneity [19–21].

## 5.1.2 Design

When designing perforated plates, there are a variety of requirements that have to be met. They differ with the application: the plate may be used for pelletizing or as a support for screens.

For a pelletizer plate, the following requirements are typical:

- Low pressure loss
- Uniform geometry of the extrudate over time (stable flow of melt, elimination of melt break) and across the plate (equal flow through all holes)
- Forming strands to the required diameter (i. e., consideration of the extrudate swelling)
- Elimination of zones of stagnation in front of the plate

Typical requirements for a breaker plate are as follows:

- Low pressure loss
- Support for the screen packs
- Elimination of stagnation zones in front of and behind the breaker plate

A correct design is achieved only when its rheological and mechanical aspects are in tune (for mechanical design, see Section 9.1).

The pressure loss of a breaker plate limits the maximum possible throughput and hence the productivity of the equipment. This is particularly important when only a low discharge pressure is available or if there is another pressure-consuming part located after the screen pack, such as a blown film die. The permissible pressure loss is limited by the mechanical strength of the plate; the boreholes reduce the mechanical strength. This has to be taken into consideration [22]. Careful mechanical design of the perforated plates for hot pelletizing is extremely important because if the plate bends, which can happen quite easily, it does so in the direction of the knife, which in turn cannot fit well, resulting in a poor cut (Fig. 5.1). The pressure loss of a plate consists of the following components: inlet pressure loss, capillary pressure loss, and outlet pressure loss. The inlet and outlet pressure losses are caused by the rearrangement of the velocity profile from approximately plug flow in front of the orifice to a parabolic profile in the orifice and back to plug flow behind it (Fig. 5.8).

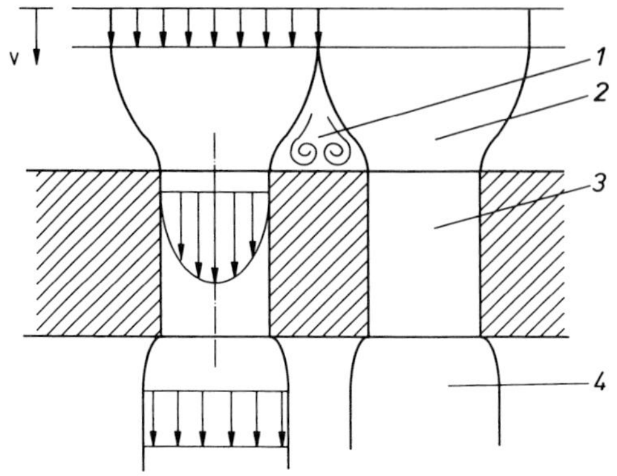


**Figure 5.8** Velocity profiles during the flow through a breaker plate: 1: stagnation zone; 2: inlet region; 3: orifice; 4: melt strand

Perforated plates are often made with a conical borehole in the inlet zone (Fig. 5.9) [23,24] to eliminate stagnation zones, secondary flows, and unstable whirls at the inlet. The *L*/*D* ratio of the cylindrical part of the orifice runs usually under 10 [2] and the opening angle of the inlet between 30° and 90°.

A simple calculation of the pressure loss for short holes ($L/D < 10$) is not possible since the pressure losses at the inlet are at least of the same order as the pressure losses in the cylindrical part of the nozzle or even higher [2,25–28]. Rather, the following procedure is recommended:

1. Calculate the viscous pressure loss in the cylindrical part of the hole, for example by the method of the representative viscosity:

$$\Delta p_{\text{pipe}} = \frac{8 \cdot \eta_{\text{rep,pipe}} \cdot \dot{V}}{\pi \cdot R^4} \cdot L \tag{5.1}$$

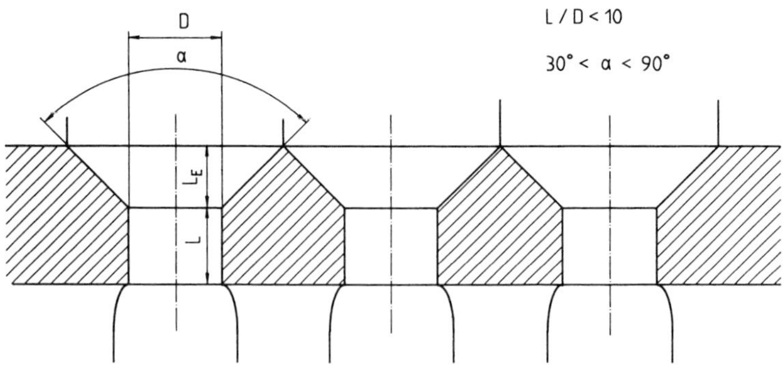


**Figure 5.9** Breaker plate with conical boreholes

with $\eta_{\text{rep,pipe}} = f\left(\dot{\gamma}_{\text{rep,pipe}}\right)$ (determined, e.g., by the *Carreau* or *Ostwald-de Waele* constitutive equation)

$$\text{and} \quad \dot{\gamma}_{\text{rep,pipe}} = \frac{4 \cdot \dot{V}}{\pi \cdot R^3} \cdot e_\circ \tag{5.2}$$

where $e_\circ = 0.815$.

2. When determining the inlet pressure loss, the fact that there is a good correlation between the inlet pressure loss $p_{\text{entry}}$ and the shear stress at the wall $\tau_W$ in the established flow can be utilized. This correlation, tabulated for many polymers, is independent of temperature (Fig. 5.10) [29]. For such a situation where the area for the flow passage in the capillary is large relative to its length, the above correlation is practically independent of the geometry (flow from an infinitely large container). The relationship between the inlet pressure loss and the shear stress at the wall can be determined from results from a capillary rheometer. For these, the Bagley correction is applied or the measurements are arranged in such a way that they are performed using a capillary with a slit and at least one pressure gauge in front of the capillary and two along the capillary.

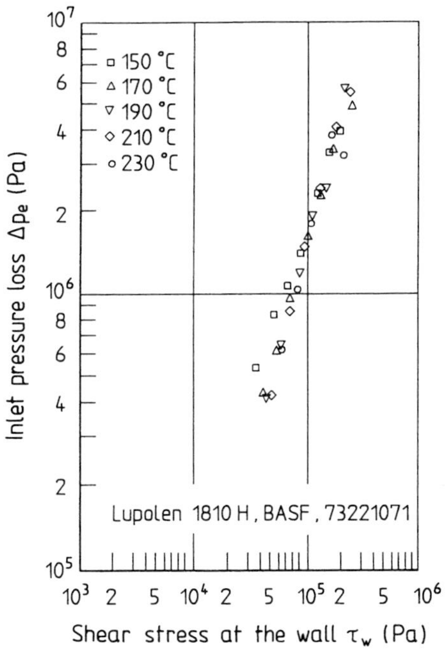


**Figure 5.10** Relationship between the inlet pressure loss and the shear stress at the wall

To evaluate the correlation for the calculation of the total pressure loss, the shear stress at the wall is determined in the second step from the calculated pressure loss using the momentum balance:

$$\tau_W \cdot A_W = \Delta p_{pipe} \cdot A_{cross\_section}, \tag{5.3}$$

$$\tau_W = \frac{\Delta p_{pipe} \cdot R}{2L}. \tag{5.4}$$

Thus, the corresponding inlet pressure loss can be found from the shear stress calculated above:

$$\Delta p_{entry} = f\left(\tau_W\right) \tag{5.5}$$

3. The total pressure loss of the orifice is obtained from the addition of the values of the viscous pressure loss in the cylindrical part of the nozzle and the inlet pressure loss:

$$\Delta p_{total} = \Delta p_{pipe} + \Delta p_{entry}. \tag{5.6}$$

This procedure also gives good results for channels with a conical inlet zone provided the opening angle is greater than 60°. Also, for these nozzles, the shear stress at the wall in the cylindrical part of the channel must be calculated.

When more accurate data are required, it is reasonable to determine the pressure loss in the capillary rheometer for the material in question and use the capillary having the same inlet shape and length as the hole that will be used in the intended application. The mass flow for a single orifice can be up to 15 kg per hour [4]. This can be used for an approximate layout of the nozzle.

In the described procedure, the interaction of the streams through each hole was not considered. Besides experimental study, only a demanding three-dimensional flow computation using FEM can be applied to analyze this problem.

The condition for a uniform extrudate geometry over a certain period of time is that no flow instabilities occur. These have several sources: depending on the material and operating conditions, an unstable flow in the inlet zone, which leads to a winding corkscrew-like extrudate, or a melt fracture, which manifests itself in a rough nonuniform surface of the extrudate. The melt fracture results predominantly from the partial loss of the wall adhesion in the channel (Fig. 5.11; see also Section 3.5.2) [30–33]. Generally, the occurrence of flow instabilities depends on the level of the rate of deformation, which means the productivity of the equipment is limited by these effects. It depends on the material being processed whether instabilities in the inlet or melt fracture will occur and which of these two effects will occur in which range of deformation.

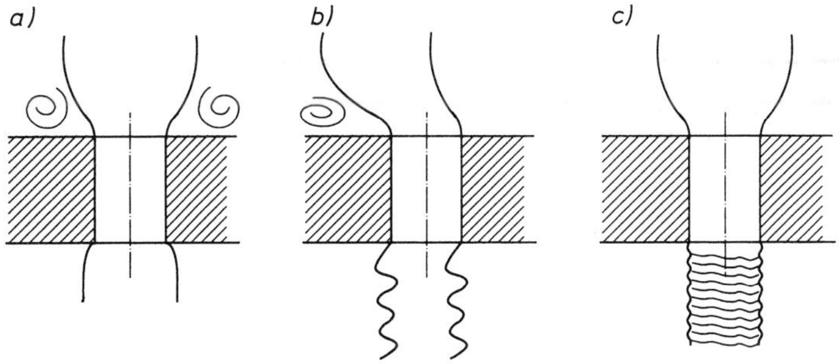


**Figure 5.11** Flow instabilities: (a) stable flow, (b) unstable inlet region, (c) melt fracture

PE-HD is the known exception here: in the range of high shear rates, a condition of slipping occurs that is due to a complete loss of wall adhesion, which produces extrudates with good surfaces [33].

The design of the channel can influence the occurrence of instabilities at the inlet and the resulting corkscrew-shaped extrudate. A conical inlet zone and a long, cylindrical land region reduce the occurrence of the problem.

A uniform volume flow through all holes of a plate can be achieved by such an arrangement, where the flow path from the melt entrance (extruder adapter) is equal to each hole. As an alternative, a suitable predistribution must be ensured such that all boreholes are supplied with the melt uniformly. In general, it is true that a higher pressure loss of the individual holes improves the uniformity of the individual streams.

When deciding on the diameter of the holes of the perforated plate, consider that the strand will be larger than the hole because of extrudate swelling. This is particularly important when designing plates for underwater pelletizing because the size of the extrudate is not affected by an additional stretching.

An exact computation of the extrudate swelling in advance is at this time possible only in a few cases and only after an exact experimental examination of the material and the respective channel geometry. A generally valid relationship is provided by the phenomenological correlation between the extrudate swelling and the shear stress at the wall, similar to the correlation between the inlet pressure loss and the shear stress at the wall (for the former, see Chapter 7).

Another consideration in design of the perforated plates is that the size of the holes is not too large, causing a screen rupture. When considering that, the arrangement and sizing of the screen pack as well as its total pressure loss have to be taken into account. For the purpose of the calculation of the pressure loss, the screen pack can be regarded as an accumulation of perforated disks [34,35]. When using the perforated plates, it is also important that there are no stagnation zones behind them that would increase the residence time of the material and can lead to degradation. For that purpose, perforated plates are often made with a conical outlet (Fig. 5.12).

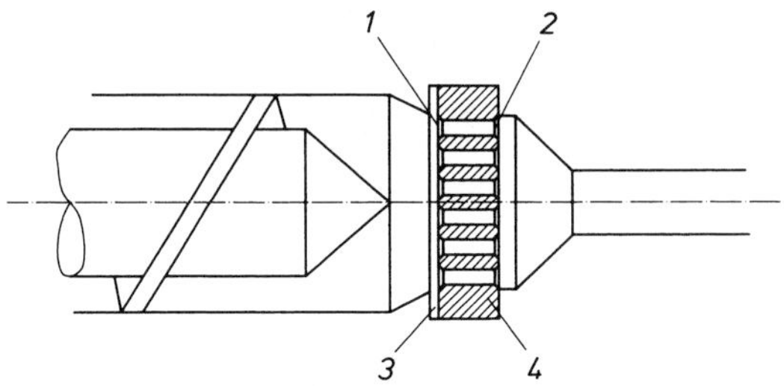


**Figure 5.12** Example of a breaker plate assembly: 1: inlet; 2: outlet; 3: screen pack; 4: breaker plate

## 5.2 Dies with Slit Exit Cross Section

Dies with flat slit-shaped exit cross sections are used for the production of extrudates with large width-to-thickness ratios, such as films and sheets, as well as for the coating of substrate webs. The distinction between the flat extrudates film and sheet is difficult to make since both the thickness and the ability to be wound are brought in for the decision about the category [36–40]. (The ability to be wound means that there is no permanent deformation or damage of the product.) Considering the fact that both notions run smoothly into each other, an arbitrary borderline can be set at the thickness of 0.5 to 0.7 mm.

All dies with a slit-shaped opening show a great similarity in the way they are made up. The substantial differences between them are based on the geometry of the manifold (melt distribution channel) used. The variety of manifolds will be discussed in the following section.

### 5.2.1 Designs and Applications

An extrusion die for sheets or films shapes a round melt strand into a flat rectangular sheet. The distribution of the melt is accomplished by a manifold followed by a flow resistance region—also referred to as the land.

As shown in Figs. 5.13 and 5.14, the melt flow enters the die in the center and is spread symmetrically on both sides to form an even flat extrudate. Experiments to evaluate the feeding of the melt to one side only, that is, asymmetrically, and distributing it were discontinued because of technical difficulties.

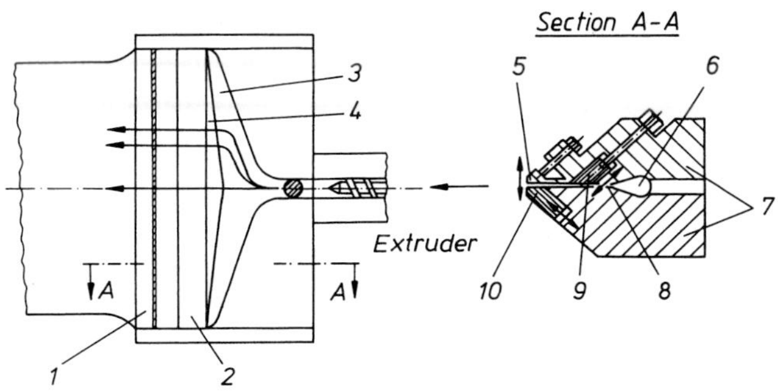


**Figure 5.13** Flat slit die for film extrusion: 1: lip (adjustable); 2: choker bar (locally adjustable); 3: manifold; 4: land; 5: flex lip; 6: body of the die

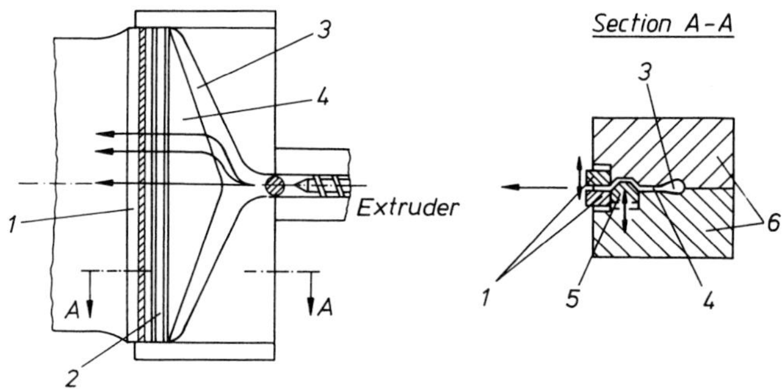


**Figure 5.14** Flat slit die for the extrusion of sheets: 1: lip (adjustable); 2: choker bar (locally adjustable); 3: manifold; 4: land; 5: body of the die

In order to obtain as uniform a melt flow across the entire width of the die as possible, the flow channel geometry has to be designed with the consideration of the material properties, operating conditions, and rheology. This procedure is discussed in Section 5.2.2.

Differences in the flow across the die during its operation almost always occur and can be regulated by several methods. In flat film extrusion, local differences in the flow rate are usually evened out by local changes in the gap width of the die lips. This is done, for example, by a slight bending of the lip by pressing screws (flex lip); the material is weakened and thus provides the required point of bending (Fig. 5.13). The limited capacity of steel for deformation allows deflections of up to about 2 mm [41]. The lip recovers because of the elasticity of the steel and the melt pressure.

In exceptional cases, flat film dies have an additional device, called a choker bar or restrictor bar (Fig. 5.13). This is a locally adjustable (bendable) profile strip in the interior of the die, which is capable of making coarse adjustments only because of its massive size.

For sheet extrusion, the choker bar is standard equipment to ensure a uniform melt distribution. But still, flex lips are also used here for fine adjustments (Fig. 5.14). Moreover, there is also the possibility to control the melt distribution and thus the gauge of the extrudate by changing the temperature profile across the die. This method is, however, used only in special cases.

Extrusion dies for coating of flat substrates are built the same way as dies for flat sheets without a choker bar; the flow regulation, as needed, can be accomplished by adjustable lips. In order to be able to extrude films and sheets by a wide slit die with different average thickness, very massive lips are used, which can generally be adjusted or changed only when the unit is at a standstill.

Typically, the length of the lips in wide slit dies is between 30 and 90 mm [37], although it can be much higher in some cases.

Depending on the swelling or stretching behavior of the processed material, the lips, which run parallel across the entire width of the die, can be adjusted by several percent closer or more open to achieve the required thickness of the extrudate [42]. However, this is only feasible when the melt is not purposely stretched after it leaves the orifice.

In thick sheet extrusion, the melt usually leaves the die in the horizontal direction and also proceeds this way to the calender stack. However, the melt for the fabrication of cast film or coated webs can be discharged at an angle or straight down, either tangentially or perpendicularly, onto a cooling roll (casting roll) or onto a carrier web running on a guide roll. Therefore these two types of dies are called casting dies. The melt has to be rerouted in these cases by means of an elbow between the extruder and the die.

The exit gaps for the extrusion of flat film and for extrusion coating are between 0.25 and 0.7 mm [43]. Usual die widths are between 1500 to approximately 4000 mm [44].

Sheet can be extruded with a thickness of up to 40 mm, whereby the widths can be up to 4000 mm [43,45–47]. The operating width of a flat slit die can be reduced with the help of deckling systems. The simplest form, which can be applied to all flat slit dies, is the external deckling. Here, the outlet cross section is deckled by a metal plate. However, the external deckling leads to dead spots, which can cause problems when thermally sensitive materials are processed [48,49]. Therefore, alternatively, internal deckling systems are used in which displacement parts are inserted into the manifold and land so that not only the outlet cross section but the whole flow channel is changed. Such internal deckling systems can only be used in T-manifolds because of the required straight shape of the insert. In flat slit dies with other manifolds, internal deckling systems can be used only in the die land. In this case, dead spots are not avoided; instead their volume is reduced compared to external deckling systems. Internal deckling systems can usually only be inserted at a standstill of the extrusion line. In comparison, external deckling systems can be adjusted during the running process.

As was mentioned at the beginning of this chapter, the main concern in the rheological configuration of the flow channels of flat slit dies is to obtain as uniform a melt flow across the width of the lip opening as possible. Now, in order to transform the circular flow as it comes from the extruder into a flat shape, a special flow channel system is necessary. There are several types, and the most important ones are shown in Fig. 5.15. These are referred to as T-manifold, fishtail manifold, and coathanger manifold. There are additional types under development [50,51], but none of them has found widespread use so far. The dies with the coathanger manifold are the most frequently used today. The reason for this is that, if properly designed, they provide a good distribution that can to a high degree be independent of the operating conditions.

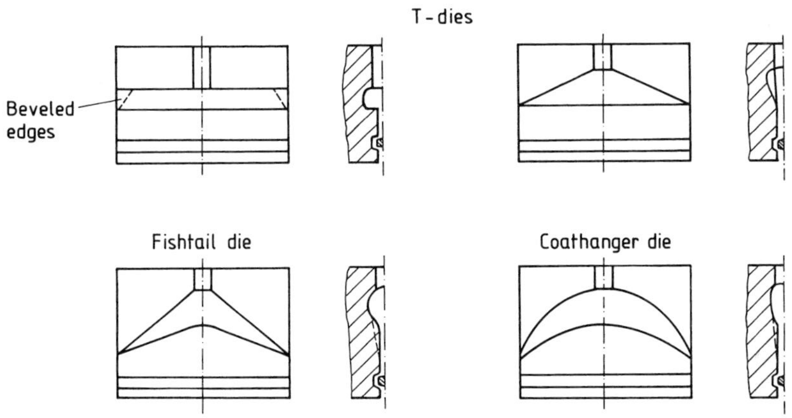


**Figure 5.15** Flat slit dies: manifold shapes

A disadvantage of this system is the high manufacturing costs that are due to the complicated flow channel geometry involved (Fig. 5.16).

Flat slit dies (or sheet dies) with a T-manifold are very simple and therefore much less expensive to make. They are most frequently used for extrusion coating lines. This type of die is not suitable for the processing of temperature-sensitive materials. The flow distribution can be optimized or influenced by geometric modifications of the manifold as well as of the choker bar. Theoretical studies for the design of such dies are given in [52–54].

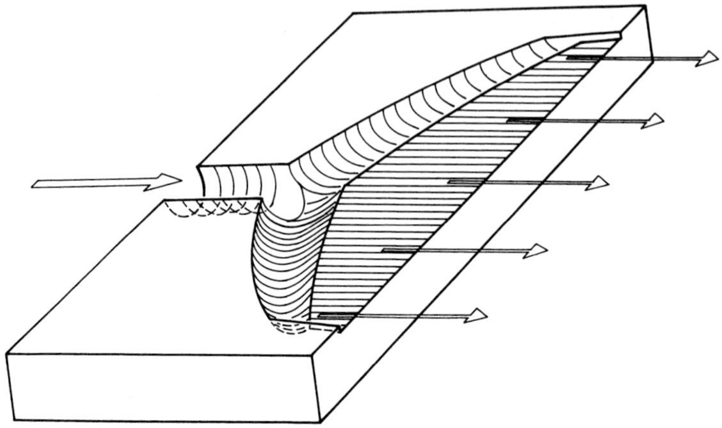

**Figure 5.16** Principle of the coathanger manifold

With regard to distribution quality and production costs, the fishtail manifold lies in the middle between the two previously mentioned systems. Studies determining its rheological configuration are presented in [55–57]. As is the case with the T-manifold, the fishtail manifold by itself does not bring about a uniform-enough melt distribution, so corrective elements, such as a choker bar, have to be added.

As mentioned before, the increasingly complicated geometry of flow channels represents higher technological demands and thus higher costs. Because of that, often in practical applications compromises are made between the designed and actually used melt distribution systems [48]. The costs of the flat extrusion die include another factor, namely the area in the die that is subjected to pressure. As this area increases in size, more mechanical effort must be expended to prevent the pressure from opening up the die. This phenomenon is referred to as clam shelling. If this happens, it has a negative effect on the thickness distribution across the width of the product [48].

All flat slit dies depicted in Fig. 5.17 [41] have a so-called symmetrical central division: the flow channel is incorporated into both halves of the die (top and bottom). An asymmetrical central division is a design where the flow channel is incorporated in one-half of the die and the other half remains planar. Such a die is, of course, less expensive to make, but it has certain flow-related drawbacks compared to the die with a symmetrical division.

Flat slit dies operate with pressure losses up to 200 bar [36,43]. Coating dies operate at much lower pressure losses mainly because the materials for coating have a lower melt viscosity at the processing temperature [48].

When processing high-viscosity melts, the pressure at the die entry can reach 400 bar [43]. In such cases, precautionary measures must be taken to endure such large forces. This can be done by a locking mechanism with toggles or hydraulic pistons [16] or by the use of so-called single block dies, which consist of a single, mechanically extremely stable, U-shaped die block. In the U-shaped recess of this die, there are strips containing the distribution channel as well as the die lips [43]. Moreover, dies with a flow channel rerouting (Fig. 5.18) have been proposed and actually used [36,43,58]. The rerouting of the melts leads to less stress on the screws holding the die together (a smaller area is subjected to pressure). Furthermore, the lips are, in fact, not forced apart, but only slightly displaced relative to each other under the existing loads [48].

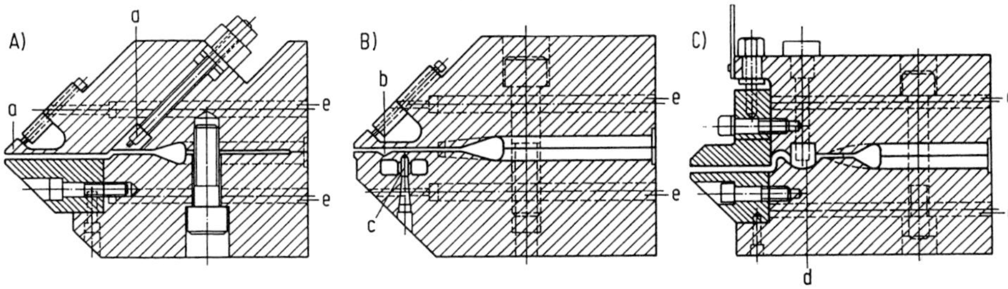


**Figure 5.17** Flat slit dies (Johnson [41]) (A) for films and thin sheets, (B) for films, and (C) for thick sheets (slabs): (a) flexible lip (flex lip), (b) transverse channel for flow equalization, (c) locally deflectable channel wall, (d) choker bar, (e) cartridge heater

The most commonly used shapes of the choker bar are shown in Fig. 5.17. Here also belongs the choker bar mounted at an angle under 45 degrees (in the die on the left), which is very advantageous from a flow point of view. However, with this choker bar, as well as with the one shown on the right, dead spots occur. In these dead spots, local degradation of the material as well as deposition of additives, fillers, pigments, and so on and so-called plate-out occur frequently. For that reason there are, if possible, no choker bars used in dies for extruding PVC.

The local adjustment of the choker bar, that is its bending, is done by tractor and pressing screws, which are alternately arranged over the width of the die at 40 to 60 mm intervals. In some cases, differential push/pull screws and push/pull screws are used which, when turned clockwise and counterclockwise, move the choker bar up and down, respectively. Flexible (locally bendable) lips are commonly used today for flat film extrusion and coating dies because they allow a relatively fine regulation of the thickness distribution without any appreciable disturbance of the flow. Moreover, this is only possible during the operation of the machine, something that is critical also when sliding lips are used. In that case, only a coarse adjustment can be made because of the friction between the lips and the body of the die and because of possible play in the setting of the adjustment screws [48].

Another interesting possibility to influence the melt distribution is shown in the middle of Fig. 5.17. In that case, a region of the bottom half of the die is weakened mechanically so that a membrane, as it were, is formed across the entire width. It can be very slightly deformed by pressure screws, which provide limited additional regulation of the flow.

The membrane section is followed by a groove-shaped, transverse channel, which also runs across the entire width of the die and is connected to the above regulating element. Its purpose is to equalize the pressure and, therefore, the flow through the die [59,60].

The regulation of the flow over the choker bars and by adjusting the lips is difficult particularly for wide dies because of the large number of adjustment screws; it requires a great deal of experience and feel on the part of the operator.

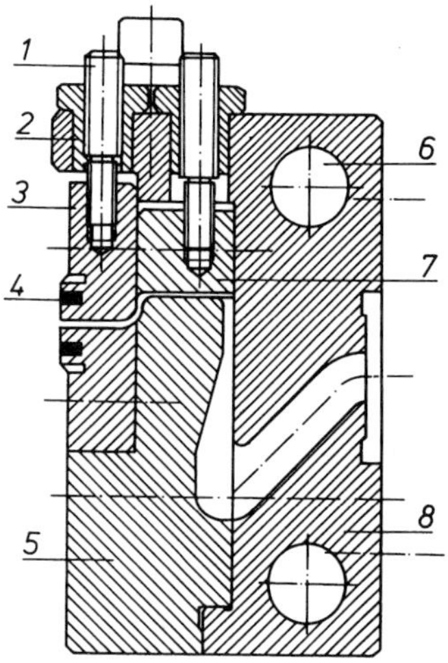


**Figure 5.18** Flat slit die with angled flow channel [36]: 1: differential screw; 2: threaded bushing; 3: adjustable die lip; 4: lip heating; 5: block with manifold; 6: borehole for cartridge heater; 7: choker bar; 8: die block

Because of that, most of the wide flat slit dies are equipped with adjustment elements controlled electrically that cause the required bending of the lips and choker bars. A system that contains a hydraulically operated screw-driving system on a rail alongside the die, which can be moved up against each adjusting screw of the choker bar and adjust it, is described in [47,61,62]. A wide flat slit die is described in [40] that contains a choker bar adjustable locally up and down by means of small electric motors with highly stepped-down gears. Several systems for automatic lip adjustment were developed. Here also electric motors are utilized as control elements for the adjusting screws of the lips [44]. The setup shown in Fig. 5.19 [63,64] is the most widely used and well established on the market. The flexible lip in this system is adjusted by a heat expansion bolt, the expansion of which is regulated by electric heating. When the heating required for the expansion of the bolt is switched off, the lips can be reset quickly by an additional cooling of the bolt with air. This system is used successfully to automatically control thickness in flat sheet extrusion [65,66].

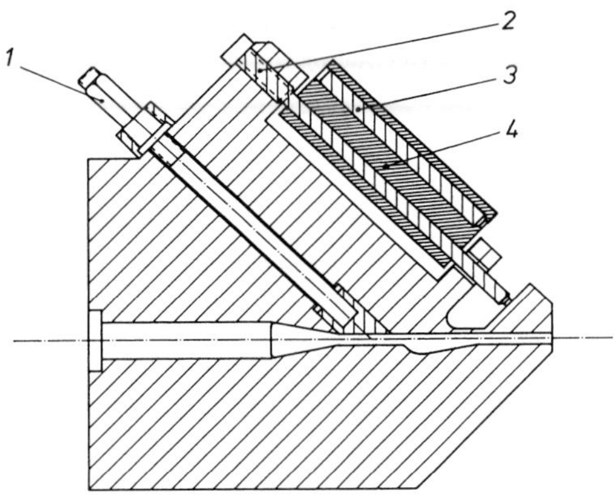


**Figure 5.19** Flat slit die with heated die bolts for gap adjustment (Welex) [64]: 1: bolt for the choker bar adjustment; 2: bolt for the lip adjustment; 3: bolt heating; 4: heating block

Alternatively, an extrusion die was demonstrated that uses so-called translators operating on the piezoelectric principle for the control elements for the lip adjustment in [67]. Such a piezotranslator can vary its length from fractions of millimeters up to several millimeters depending on the applied voltage. Its reaction time is only a few milliseconds, so it provides a very fast gauge control [67].

The wide flat slit dies are generally heated by electricity, either by cartridge heaters located in the body of the die or by heating bands placed against the outside of the die. Inductive heating systems are in an experimental stage [16]. In order to attain as uniform a die temperature across its entire width as possible, heat pipes, which are very efficient in heat transfer, are sometimes used [68].

If the lips are massive and the plastic melt flows with difficulty, rod-shaped lip heaters placed lengthwise are sometimes used in addition to the usual heating systems. Selectively controllable heating systems for the regulation of flow as suggested in [69] and tried on the die described in [40] have had many problems and therefore have not found widespread practical use.

### 5.2.2 Design

The manifold of the wide flat slit die serves the purpose of distributing the melt across the width of the slit in such a manner that there is the desired volume flow distribution across the die orifice. Generally, a uniform gauge of the extrudate is required. However, there are applications, such as sheets or films, that are oriented uniaxially or biaxially in the subsequent procedure. In that case, the sheet leaving the die has to have a certain gauge profile [69,70]. The second requirement is that the melt has a relatively uniform history of deformation, temperature, and residence time. Finally, it is expected that the above requirements are met over a wide range of operating conditions and for a variety of materials.

In the following sections, first, the analytical formulas for the design of manifolds for wide flat dies that provide a uniform melt distribution will be discussed, and then a numerical method for the design of a regular manifold and a wide flat die manifold with equal flow length will be described.

When designing wide slit dies with a manifold analytically, it is advantageous to assume the desired flow distribution and then consider the pressure losses resulting from the corresponding volume flow and geometry.

If equal mean flow rates of the melt (i. e., uniform flow distribution) exist at the exit cross section of a wide flat die, the flow resistance of all pathways of the flow through the die must be equal (Fig. 5.20); that is, the different flow pathways of the incoming melt must be neutral.

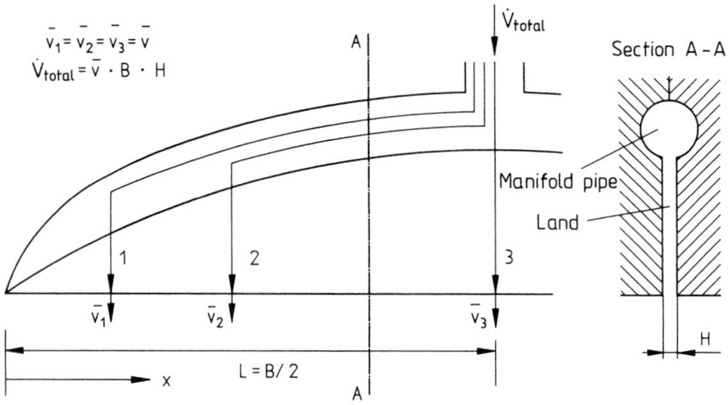


**Figure 5.20** Pathways of the flow in a flat slit die with a manifold

This means, at the same time, that the volumetric flow rate in the pipe of the manifold in front of the point of entry (influx) up to the end of the pipe (i. e., across the entire width of the die) falls off to zero in a linear fashion. These relationships can be formulated in the following way using the coordinate system shown in Fig. 5.21 (because of symmetry, it is sufficient to consider only half of the die):

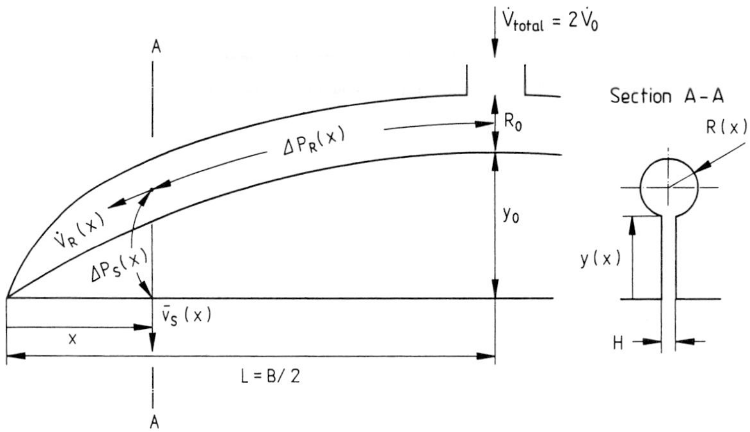


**Figure 5.21** System of coordinates in a flat slit die

$$\bar{v}_{slit}(x) = \text{const.} = \frac{\dot{V}_{total}}{B \cdot H} \tag{5.7}$$

$$\dot{V}_{pipe}(x) = \frac{x}{L} \cdot \dot{V}_0 \tag{5.8}$$

$$\dot{V}_0 = \frac{\dot{V}_{total}}{2}, \tag{5.9}$$

$$\Delta p(x) = \Delta p_{pipe}(x) + \Delta p_{slit}(x) = \text{const.} \tag{5.10}$$

The mathematical formulation of the requirement for equal flow resistance on all flow paths across the entire width of the die results from Equation (5.10):

$$\frac{\partial \Delta p(x)}{\partial x} = 0. \tag{5.11}$$

These basic relationships are valid for all design work pertaining to manifolds. Furthermore, the following restrictions apply:

- Isothermal, steady-state, laminar, and fully developed flow
- Incompressibility of the melt (i.e., $\varrho = \text{const}$)
- Ignoring the inlet and outlet effects
- Wall adhesion

The following simplifications are applied to the design formulas shown in the sections below:

- Consideration of the manifold as a pipe or a flat slit
- Constant slit height in the land area

#### 5.2.2.1 T-Manifold

A rheologically optimal design of the flow channel of a T-manifold is generally not possible. The reason is that, in the case of a constant height of the die land, the flow resistance of the flow path in the middle of the channel is always lower than the resistance of all other flow paths. This is why there is always a throughput maximum in the middle of a T-manifold and a throughput minimum at the side boundaries. Because of this, the criterion in the design of the T-manifold should be to keep the flow resistance in the manifold transverse to the extrusion direction as low as possible compared to the flow resistance of the land.

If a manifold with a circular cross section (radius $R$) is considered, in which the volume flow rate $\dot{V}$ is to be distributed evenly over the width $B$, the concept of the representative viscosity leads to the following pressure loss in the manifold:

$$\Delta p_{\text{pipe}} = \int_{x=0}^{x=B/2} \frac{8 \cdot \eta_{\text{rep,pipe}} \cdot \dot{V}(x)}{\pi R^4} \cdot e_{\bigcirc} \, dx = \int_{x=0}^{x=B/2} \frac{8 \cdot \eta_{\text{rep,pipe}}}{\pi R^4} \cdot e_{\bigcirc} \frac{\dot{V}_{\text{total}} \cdot x}{B} dx \tag{5.12}$$

For the pressure loss in the die land with the length $L$ and height $H$ it follows:

$$\Delta p_{\text{slit}} = \frac{12 \cdot \eta_{\text{rep,slit}} \cdot \dot{V}_{\text{total}} \cdot L}{BH^3} \cdot e_{\square} = \int_{x=0}^{x=B/2} \frac{8 \cdot \eta_{\text{rep,pipe}}}{\pi R^4} \cdot e_{\bigcirc} \frac{\dot{V}_{\text{total}} \cdot x}{B} dx \tag{5.13}$$

$$\begin{aligned} \frac{\Delta p_{\text{pipe}}}{\Delta p_{\text{slit}}} &= \frac{\dfrac{8 \cdot \dot{V}_{\text{total}}}{\pi R^4 \cdot B} \cdot e_{\bigcirc} \cdot \int_{x=0}^{x=B/2} \eta_{\text{rep,pipe}}(x) \cdot x dx}{\dfrac{12 \cdot \eta_{\text{rep,slit}} \cdot \dot{V}_{\text{total}} \cdot L}{B \cdot H^3} \cdot e_{\square}} \\ &= \frac{2}{3} \cdot \frac{1}{\eta_{\text{rep,slit}}} \cdot \frac{H^3}{\pi R^4 \cdot L} \cdot \frac{e_{\bigcirc}}{e_{\square}} \cdot \int_{x=0}^{x=B/2} \eta_{\text{rep,pipe}}(x) \cdot x dx \end{aligned} \tag{5.14}$$

If an upper boundary is given for this relation, the minimum radius of the manifold can be determined iteratively for a given geometry of the die land.

#### 5.2.2.2 Fishtail Manifold

For the fishtail manifold (Fig. 5.22), the length of the land across the entire width is assumed to be linear:

$$y(x) = y_0 \cdot \frac{x}{L}. \tag{5.15}$$

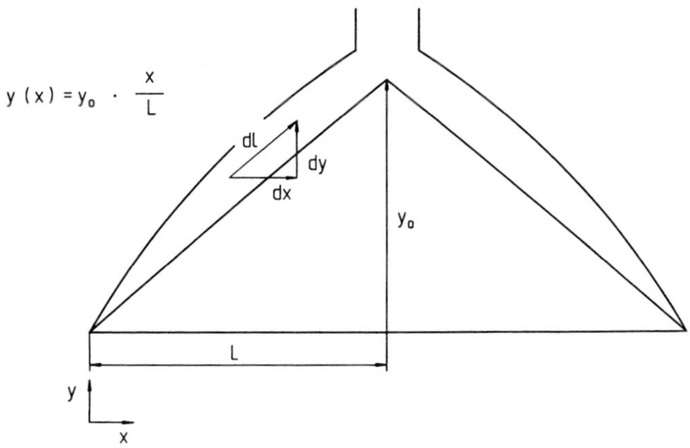


**Figure 5.22** Fishtail manifold

The pressure loss in the manifold is obtained by the application of the method of representative data:

$$\Delta p_{\text{pipe}}(x) = \int_x^L \frac{8\eta_{\text{rep,pipe}}(x)\dot{V}_0 x}{\pi R^4(x) L} \mathrm{d}l. \tag{5.16}$$

The pressure loss in the slit is

$$\Delta p_{\text{slit}}(x) = \frac{12\eta_{\text{rep,slit}}(x)\cdot \dot{V}_0 \cdot y_0}{L^2 \cdot H^3} \cdot x. \tag{5.17}$$

If considering also the relationship between the length of the manifold and the width of the die in Equation (5.16)

$$\mathrm{d}l = \frac{\mathrm{d}x}{\cos\left(\arctan\left(\frac{y_0}{L}\right)\right)}, \tag{5.18}$$

we obtain the course of the manifold radius across the entire die width using Equation (5.11):

$$R(x) = \left[\frac{1}{3\pi}\cdot\frac{\eta_{\text{rep,pipe}}(x)}{\eta_{\text{rep,slit}}(x)}\cdot\frac{B\cdot H^3 x}{y_0 \cdot \cos\left(\arctan\left(\frac{2\cdot y_0}{B}\right)\right)}\right]^{\frac{1}{4}}. \tag{5.19}$$

The advantages of the fishtail manifold are as follows:

- Relatively simple techniques are used in its manufacture because of the linear course of the distribution channel.
- When computing the pressure loss in the manifold, Equation (5.16), the real length of the distribution channel is taken into account, Equation (5.18). This is particularly important for manifolds with very long lands.

The main disadvantage of the fishtail manifold is the great change in the shear rate across the length of the distribution channel. This is equivalent to a variable loading of a material across variable flow pathways.

#### 5.2.2.3 Coathanger Manifold

Unlike in the fishtail manifold, in the coathanger manifold the course of the length of the land across the width is not assumed, but rather developed with the goal of keeping the stress on the melt as uniform and as low as possible.

One possible way of doing this is the study of the mean residence time along the streamline. The mean residence time is calculated from the mean velocity in the flow cross section and the length of the flow path:

$$\bar{t}(x) = \int_x^L \frac{1}{\bar{v}_{\text{pipe}}(x)} \mathrm{d}x + \frac{y(x)}{\bar{v}_{\text{slit}}(x)} \tag{5.20}$$

where

$$\text{the mean velocity in the manifold } \bar{v}_{\text{pipe}}(x) = \frac{\dot{V}_{\text{pipe}}(x)}{\pi R^2(x)}, \tag{5.21}$$

$$\text{the mean velocity in the land } \bar{v}_{\text{slit}}(x) = \text{const} = \frac{\dot{V}_{\text{total}}}{B \cdot H}. \tag{5.22}$$

A formula for the design of a flat wide slit die with a manifold that includes an optimized residence time distribution is such that the cross section of the distribution channel of the manifold is shaped so as to minimize the residence time along the flow path that goes through the entire manifold [71]. This means, that for

$$\bar{t}(x=0) = \int_0^L \frac{1}{\bar{v}_{\text{pipe}}(x)} \mathrm{d}x \tag{5.23}$$

a minimum has to be found.

The search for the minimum value from Equation (5.23) yields as a solution the following shape of the cross section of the distribution channel:

$$R(x) = R_0 \left(\frac{x}{L}\right)^{\frac{1}{3}}. \tag{5.24}$$

Furthermore, it can be shown that the condition of the minimum of the residence time in the distribution channel is equivalent to the condition of constant shear stress at the wall in the distribution channel [71].

The shape for the land follows by using Equation (5.24):

$$y(x) = y_0 \left(\frac{x}{L}\right)^{\frac{2}{3}}. \tag{5.25}$$

The geometric shapes of functions (5.24) and (5.25) are independent of material behavior and the operating conditions. However, $R_0$ and $y_0$ are affected by the flow behavior of the melt. For the sinh law by *Prandtl-Eyring,* the following relationships result:

$$\frac{4\dot{V}_0}{\pi R_0^3 \cdot C} = \frac{1}{3}\frac{\Delta p}{A}\cdot\frac{R_0}{L}\cdot\phi\cdot\left(\frac{1}{3}\cdot\frac{\Delta p}{A}\cdot\frac{R_0}{L}\right) \tag{5.26}$$

for the maximum radius of the distribution channel and

$$\frac{6\cdot V_0}{H^2\cdot C\cdot L} = \frac{1}{2}\frac{H}{y_0}\cdot\frac{\Delta p}{A}\cdot\psi\cdot\left(\frac{1}{2}\cdot\frac{H}{y_0}\cdot\frac{\Delta p}{A}\right) \tag{5.27}$$

for the maximum length of the land. The functions $\varphi(u)$ and $\psi(r)$ are summations of the hyperbolic functions of the form

$$\phi(u) = \frac{8}{u^2}\left[\frac{1}{2}\cosh u - \frac{1}{u}\sinh u + \frac{1}{u^2}(\cosh u - 1)\right] \tag{5.28}$$

$$\text{with } u = \frac{1}{3}\frac{\Delta p}{A}\frac{R_0}{L}, \tag{5.29}$$

for the flow in the channel and

$$\psi(r) = \frac{3}{r^2}\cdot\left(\cosh r - \frac{1}{r}\sinh r\right) \tag{5.30}$$

$$\text{with } r = \frac{1}{2}\frac{H}{y_0}\frac{\Delta p}{A} \tag{5.31}$$

for the flow through the slit.

Note that $R_0$ and $y_0$ cannot be determined explicitly from Equations (5.26) and (5.27), respectively. To determine these values, an iterative computational procedure or nomograms have to be used [72] (Fig. 5.23).

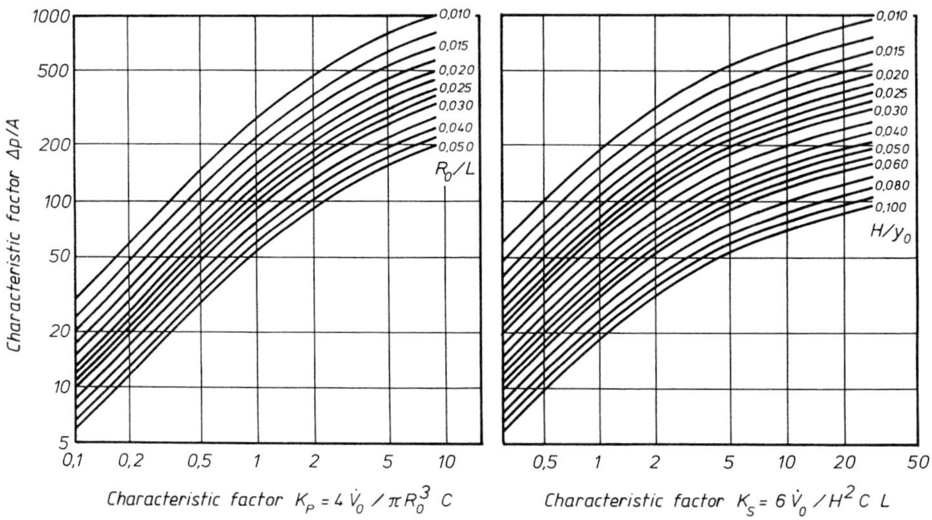


**Figure 5.23** Nomograms for the design of flat slit dies utilizing the sinh law [71]

The following procedure is recommended when using nomograms:

1. The choice of $R_0$ is based on empirical data, and $\Delta p$ is determined from Fig. 5.23 (left). Repeat until $\Delta p$ lies within the required range.
2. The assumption of $H$ is based on empirical data, and $y_0$ is determined from Fig. 5.23 (right) with $\Delta p$ from Step 1. Repeat until $y_0$ lies within the desired range.

The procedure as presented previously permits the design of wide flat slit dies with consideration of the residence times and the shear stress at the wall in the distribution channel. However, a die can be designed by using this method for only one material (flow properties of which can be represented by the sinh law) and for only one specific operating condition (mass throughput and mass temperature). Deviation from the operating conditions from the flow behavior leads to a change in the function of the distribution channel.

The manufacture of a die with a nonlinear course of the land length (see Equation (5.25)) is rather difficult and is only possible with the use of numerically controlled milling machines.

Another procedure for the design of wide flat slit dies is based on the method of representative data that allows the computation of the pressure loss by the fundamental die equations valid for Newtonian liquids [73].

The advantage of this method is that any material law can be used for the description of the relationship between shear rate and temperature.

An additional condition for the determination of the contour of the distribution channel required here [73] is that the representative shear rate in the distribution channel (manifold) remains constant, which means

$$\dot{\gamma}_{\text{rep,pipe}} = \frac{4\dot{V}_{\text{pipe}}(x)}{\pi R^3(x)} e_{\circ} = \text{const.} \tag{5.32}$$

From the relationship between the volumetric flow rate in the channel and the path coordinate (Equation (5.8)) and from Equation (5.32) it follows that

$$\dot{\gamma}_{\text{rep,pipe}}(x) = \frac{4\left(\dot{V}_0 \frac{x}{L}\right)}{\pi R^3(x)} e_{\circ} = \frac{4\dot{V}_0}{\pi R_0^3} e_{\circ} = \dot{\gamma}_{\text{rep,pipe}}(L) \tag{5.33}$$

Resolving for *R(x)*, the following is obtained:

$$R(x) = R_0 \left(\frac{x}{L}\right)^{\frac{1}{3}}. \tag{5.34}$$

where $R(L) = R_0$.

This contour is identical to that found by using the conditions of minimum residence time in the distribution channel and constant shear rate at the wall in the distribution channel (Equation (5.24)).

The total pressure loss along the flow path consists of the pressure loss in the distribution channel and the pressure loss across the land. It is obtained from the following relation (similar to Equations (5.16) and (5.17)):

$$\Delta p(x) = \Delta p_{\text{pipe}}(x) + \Delta p_{\text{slit}}(x) = \int_x^L \frac{8\eta_{\text{rep,pipe}}(x)\cdot \dot{V}_0 x}{\pi R^4(x) L} \mathrm{d}l + \frac{12\eta_{\text{rep,slit}}(x)\dot{V}_0}{LH^3} y(x). \tag{5.35}$$

The relationship between d*l* and d*x* is not known here, so it is simplified by setting

$$\mathrm{d}l = \mathrm{d}x \tag{5.36}$$

From the assumptions of a constant shear rate in the distribution channel (Equation (5.31)) and constant flow velocity at the exit from the land gap (Equation (5.7)) with a constant slit height in the land region, it follows that

$$\eta_{\text{rep,pipe}} = \text{const} \neq f(x), \tag{5.37}$$

$$\eta_{\text{rep,slit}} = \text{const} \neq f(x). \tag{5.38}$$

By the differentiation of Equation (5.35), setting the result to zero, and using Equation (5.34), the following differential equation for the course of the length of the land results:

$$\frac{dy(x)}{dx} = \frac{\eta_{rep,pipe}}{\eta_{rep,slit}} \cdot \frac{2H^3 x}{3\pi R_0^4 \cdot \left(\frac{x}{L}\right)^{\frac{4}{3}}}. \tag{5.39}$$

After integration of Equation (5.39), the following relationship results:

$$y(x) = \frac{\eta_{rep,pipe}}{\eta_{rep,slit}} \cdot \frac{H^3 \, L^{4/3} x^{2/3}}{\pi R_0^4} = \frac{\eta_{rep,pipe}}{\eta_{rep,slit}} \cdot \frac{H^3 \, L^2}{\pi R_0^4} \cdot \left(\frac{x}{L}\right)^{2/3}. \tag{5.40}$$

With $y(L) = y_0$, the result is

$$y(x) = y_0 \left(\frac{x}{L}\right)^{2/3} \tag{5.41}$$

as well as

$$y_0 = y(L) = \frac{\eta_{rep,pipe}}{\eta_{rep,slit}} \frac{H^3 L^2}{\pi R_0^4}. \tag{5.42}$$

The maximum radius of the distribution channel $R_0$ and the maximum length of the land $y_0$ are mutually related through Equation (5.42) and through the viscosities (and thus by shear rates) in the channel and the slit:

$$\eta_{rep,pipe} = f\left(\dot{\gamma}_{rep,pipe}\right) = f\left(\frac{4\dot{V}_0}{\pi R_0^3} e_{\bigcirc}\right), \tag{5.43}$$

$$\eta_{rep,slit} = f\left(\dot{\gamma}_{rep,slit}\right) = f\left(\frac{6\dot{V}_0}{LH^2} e_{\square}\right). \tag{5.44}$$

As quantities defining the geometry, $B$, $\dot{V}_{total}$, $H$, and $\Delta p$ can be assumed, and $R_0$ and $y_0$ can be calculated. For this purpose, the following procedure can be used:

1. Calculation of the representative shear rate and viscosity in the die land:

$$\dot{\gamma}_{rep,slit} = \frac{6 \cdot \dot{V}_{total}}{BH^3} \cdot e_{\square} \tag{5.45}$$

$$\eta_{rep,slit} = \eta(\dot{\gamma}_{rep,slit}) \tag{5.46}$$

2. Calculation of the die land length $y_0$:

$$\Delta p_{\text{slit}} = \frac{12 \cdot \eta_{\text{rep,slit}} \cdot \dot{V}_{\text{total}}}{BH^3} y_0 \tag{5.47}$$

$$\text{with} \quad y = y_0 \left( \frac{x}{L} \right)^{2/3} \tag{5.48}$$

3. Iterative calculation of $R_0$ with Equation (5.49). This equation can only be solved iteratively because $\bar{\eta}_{\text{rep,pipe}}$ is also dependent on $R_0$.

$$R_0 = \left( \frac{\eta_{\text{rep,pipe}}}{\eta_{\text{rep,slit}}} \cdot \frac{H^3 \cdot L^2}{\pi y_0} \right)^{1/4} \tag{5.49}$$

$$\text{with} \quad R(x) = R_0 \cdot \left( \frac{x}{L} \right)^{1/3} \tag{5.50}$$

For all design formulas presented so far, it holds true that (theoretically) an absolutely uniform distribution is achieved only with the specific design material at the design-specific point of operation. Therefore, these design procedures are referred to as depending on operating conditions. When expanding this design using the method of representative data by adding the condition of equal shear rates in land and manifold, a procedure that is independent of the operating conditions results. This means that

$$\dot{\gamma}_{\text{rep,slit}} = \dot{\gamma}_{\text{rep,pipe}}. \tag{5.51}$$

From this it follows that

$$\eta_{\text{rep,slit}} = \eta_{\text{rep,pipe}}. \tag{5.52}$$

From Equation (5.51) it follows that

$$\frac{4\dot{V}_0}{\pi R_0^3} e_{\bigcirc} = \frac{6\dot{V}_0}{LH^2} e_{\square}. \tag{5.53}$$

Solving for $R_0$, the result is

$$R_0 = \left( \frac{2}{3\pi} \frac{e_{\bigcirc}}{e_{\square}} LH^2 \right)^{1/3}. \tag{5.54}$$

Putting $R_0$ into Equation (5.42), $y_0$ results:

$$y_0 = \left( \frac{3}{2} \frac{e_{\square}}{e_{\bigcirc}} \right)^{4/3} \left( \pi H L^2 \right)^{1/3}. \tag{5.55}$$

As can be seen from Equations (5.53) and (5.55), after introducing the condition of equal shear rate in the flow channel and the land gap, the operating conditions and the material do not have any influence on the result of the design.

The design is independent of the operating conditions, as required, and therefore it always produces a distribution channel that theoretically distributes the melt uniformly.

Another advantage of this formula is that it produces equal shear rates and equal melt stress on every flow pathway.

The disadvantage of this design independent of the operating conditions is that the calculated maximum lengths of lands are relatively large. The large land areas combined with high pressure drops may lead to very high forces causing opening of the die, which often cannot be prevented mechanically. Therefore, the design of wide slit dies by the method independent of processing conditions is feasible only in exceptional cases, such as only for narrow dies. On the other hand, the design of the flow channel by this method is practical for side-fed mandrel dies (see Sections 5.3.3, 9.2, and 9.3).

The pressure loss of the manifold system can be computed by using Equation (5.35). Since all flow pathways are equivalent with regard to the pressure drop based on the chosen conditions, it is sufficient to consider only the pressure drop that results from the flow across the maximum land width in the middle of the die ($x = L$); from that the total pressure loss for the die can be determined:

$$\Delta p = \frac{12 \cdot \eta_{\text{rep,slit}} \cdot \dot{V}_0 \cdot y_0}{L \cdot H^3}. \tag{5.56}$$

The computation of the residence time along a streamline can be calculated with the help of Equation (5.20). If the relationship between the flow through the manifold, its radius, and the current time coordinate is included in that equation, the following is the result:

$$\bar{t}\left(x\right) = \frac{\pi R_0^2 L^{1/3}}{\dot{V}_0} \int_x^L x^{-1/3}\, \mathrm{d}x + \frac{H y_0 L^{1/3}}{\dot{V}_0} x^{2/3}. \tag{5.57}$$

The residence time in the die can be calculated from the mean velocity at the exit of the die and the maximum length of the land in the center of the die:

$$\bar{t} = \frac{L H y_0}{\dot{V}_0}. \tag{5.58}$$

1. Definition of the maximum permissible pressure drop and of the height of the land gap from the empirical data, determination of the maximum length of the land in the center of the die (Equation (5.56)).

2. Determination of the maximum radius of the distribution channel from Equation (5.54). Since this value enters into the determination of the representative viscosity in the channel, an iterative procedure is meaningful.
3. An estimation of the dependence of the die on the processing conditions can be done by taking $R_0$ found in Step 2 and determining the new value $y_0$ for a different operating point or another material and then comparing it with the value obtained in Step 1. The smaller the difference between the two values of $y_0$, the less pronounced is the dependence on the operating conditions.

A disadvantage of design methods discussed above is, when applied to the coathanger manifold, that the length of the distribution channel projected onto the slit at the outlet and not the exact length is used for the calculations. This leads to an insufficient thickness of the extrudate at the outside regions of the die, particularly if the maximum length of the land is great.

Starting with the condition consideration of the exact length of the distribution channel, an analytical procedure for the design of the coathanger manifold was developed [74]. The power law is chosen to represent the flow, and an additional condition set for the design is the requirement of an equal mean residence time on all flow pathways. This requirement is equivalent to the condition of equal local residence time at any location in the die:

$$\mathrm{d}\bar{t}_{\text{slit}}(x) = \frac{\mathrm{d}y}{\bar{v}_{\text{slit}}(x)} = \mathrm{d}\bar{t}_{\text{pipe}}(l) = \frac{\mathrm{d}l}{\bar{v}_{\text{pipe}}(l)} = \text{const.} \tag{5.59}$$

(For the definition of $\mathrm{d}l$ see Fig. 5.24.)

The above leads to the following relationship of the radius of the distribution channel and the current time coordinate:

$$R(x) = R_0 \left(\frac{x}{L}\right)^{1/3} \tag{5.60}$$

with

$$R_0 = \left[\frac{(3n+1)}{2(2n+1)}\right]^{\frac{n}{3(n+1)}} \left(\frac{H^2 L}{\pi}\right)^{1/3}. \tag{5.61}$$

The functional correlation between the radius of the flow channel of the manifold and the current time coordinate (Equation (5.60)) corresponds to the relationships upheld by conditions of minimum residence time in the distribution channel (Equation (5.24)) and constant representative shear rate in the distribution channel (Equation (5.34)).

For the slope of the land length, the following relationship applies:

$$\frac{dy}{dx} = \left[ \frac{\left(\pi^2 R^4(x)/H^2\right)}{x^2 - \left(\pi^2 R^4(x)/H^2\right)} \right]^{1/2} . \tag{5.62}$$

After integration, the course of the length of the land along the flow coordinate is

$$y(x) = \frac{3}{2} k^{1/2} \left[ \left[ x^{1/3} \sqrt{x^{2/3} - k} \right. \right. \\ \left. \left. + k \cdot \ln \left( \frac{x^{1/3} + \sqrt{x^{2/3} - k}}{\sqrt{k}} \right) \right] \right] \tag{5.63}$$

with

$$k = \left( \frac{3n+1}{4n+2} \right)^{\frac{4n}{3n+3}} \cdot (\pi H)^{\frac{2}{3}} . \tag{5.64}$$

An interesting variation of the design method described above is represented by the layout of a manifold for a wide slit die with a slit shape distribution channel [75].

The distribution channel is not shaped as a pipe in this case, but as a plane slit (Fig. 5.24). This shape of the channel is advantageous from the point of view of the rheological design and of the manufacturing technology. Dead spots in the distribution channel will not occur because the flow is a combination of flow in the direction of the channel and the direction of the slit (Fig. 5.25).

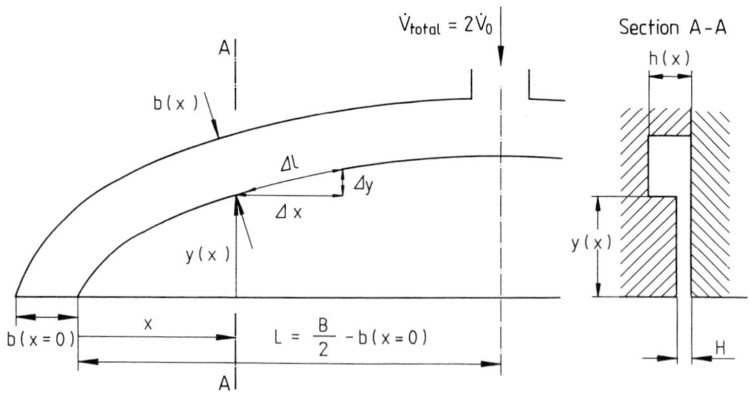


**Figure 5.24** Manifold for a flat slit die with a slit-shaped channel

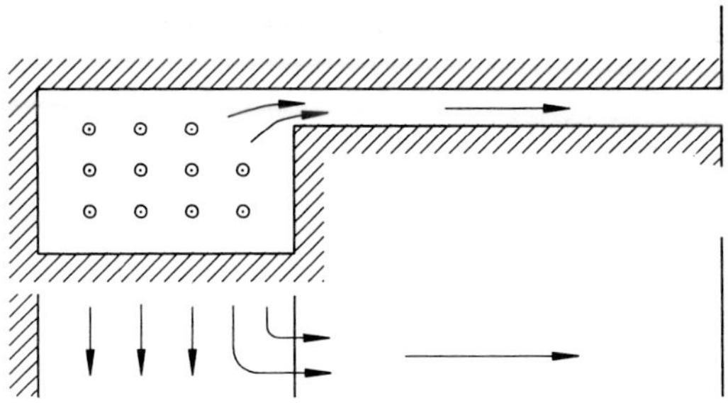


**Figure 5.25** Flow in the slit-shaped distribution channel (schematic)

What is striking about the die with the constant distribution channel width $b(x)$ presented above is the fact that the cross section of the distribution channel does not become zero at the edges of the die; instead, it supplies melt to the outer region. In reference [75] the design of this type of die is described using a power law and requiring an equal residence time on each flow path for constant width of the distribution channel $b(x)$ and for a constant ratio of height/width $(h(x)/b(x))$ in that channel. This design is particularly simple when the method of representative data is applied. The basis here is the correlation between the flow coordinate and the length of the distribution channel:

$$dl = \sqrt{dx^2 + dy^2} \tag{5.65}$$

and the correlation between the pressure loss across the land slit and the pressure loss in the distribution channel:

$$\frac{dp}{dl} = \frac{dp}{dy}\frac{dy}{dl}. \tag{5.66}$$

A rearrangement of Equation (5.65) yields

$$\frac{dy}{dx} = \frac{1}{\sqrt{\left(\frac{dl}{dy}\right)^2 - 1}} \tag{5.67}$$

When combining Equations (5.66) and (5.67), a relationship for the determination of the course of the land length is obtained:

$$\frac{dy}{dx} = \frac{1}{\sqrt{\left(\frac{dp}{dy} / \frac{dp}{dl}\right)^2 - 1}}. \tag{5.68}$$

The volumetric flow rate in the distribution channel of this special type of die cannot be described by Equation (5.8), rather, the following relationship is valid here:

$$\dot{V}_D(x) = \frac{\dot{V}_0}{B/2}(x+b). \tag{5.69}$$

with the pressure gradient in the land slit

$$\frac{dp}{dy} = \frac{12\eta_{rep,slit}\ \dot{V}_0}{\frac{B}{2}\cdot H^3} \tag{5.70}$$

and in the distribution channel

$$\frac{dp}{dl} = \frac{12\eta_{rep,D}\ \dot{V}_0(x+b)}{bh^3(x)\frac{B}{2}} \tag{5.71}$$

This inserted into Equation (5.68) yields

$$\frac{dy}{dx} = \frac{1}{\sqrt{\left(\frac{\eta_{rep,slit}}{\eta_{rep,D}}\left(\frac{h(x)}{H}\right)^3 \frac{b}{b+x}\right)^2 - 1}}. \tag{5.72}$$

As an additional condition for the determination of the contour of the distribution channel, the requirement for a constant shear rate in it is introduced:

$$\dot{\gamma}_{rep,D} = \text{const.} = \frac{6\dot{V}_D(x)}{bh^2(x)} e_\square \tag{5.73}$$

and refers to the end of the channel:

$$\frac{6\frac{\dot{V}_0}{B/2}(x+b)}{bh^2(x)} e_\square = \frac{6\frac{\dot{V}_0}{B/2}b}{bh_E^2} e_\square. \tag{5.74}$$

From that

$$h(x) = h_E\sqrt{1+x/b} \tag{5.75}$$

with the slit height at the end of the distribution channel $h_E$:

$$h(x=0) = h_E. \tag{5.76}$$

Setting the above into Equation (5.72) and integrating the function for the land length we obtain

$$y\left(x\right)=\frac{2b}{k}\sqrt{k\left(1+\frac{x}{b}\right)-1} \tag{5.77}$$

with

$$k=\left[\frac{\eta_{\text{rep,slit}}}{\eta_{\text{rep,D}}}\left(\frac{h_{\text{E}}}{H}\right)^{3}\right]^{2}. \tag{5.78}$$

For $h_{\text{E}} = H$ it follows that $\dot{\gamma}_{\text{rep,D}} = \dot{\gamma}_{\text{rep,slit}}$ and hence $\eta_{\text{rep,slit}} = \eta_{\text{rep,D}}$ (design independent of operating conditions), $k$ becomes equal to one, and the following simple relationship for the course of the slit length is gained:

$$y\left(x\right)=2\sqrt{x\cdot b}. \tag{5.79}$$

#### 5.2.2.4 Numerical Procedures

As an alternative to the analytical methods of design presented above, which allow the determination of the remaining geometrical parameter after selection of the shape of the land length or the dimensions of the distribution channel, there is a possibility to do the design numerically.

To accomplish this, the geometry of the die is divided into segments; the volumetric flow rates and pressure balances are done in each of them (Fig. 5.26) [76]. For each segment $n$, the following relationship between incoming and outgoing volumetric flow rate in the distribution channel applies:

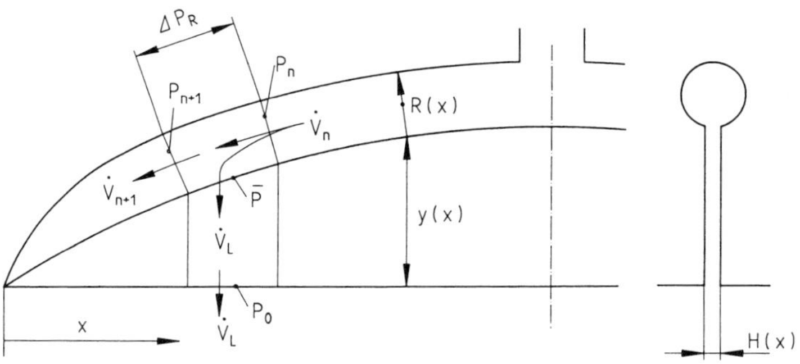


**Figure 5.26** Segmentation of a flat slit die for the numerical calculation

$$\dot{V}_{\text{n}}=\dot{V}_{\text{n+1}}+\dot{V}_{\text{L}}. \tag{5.80}$$

The volumetric flow rate $\dot{V}_L$ from the manifold is a function of the pressure in the center of the segment and the slit geometry:

$$\dot{V}_L = f\left[\left(\bar{p} - p_0\right), H(x), y(x)\right]. \tag{5.81}$$

The pressure in the center of the segment is calculated as the mean value from the pressures at the segmental joints:

$$\bar{p} = \frac{p_n + p_{n+1}}{2}. \tag{5.82}$$

The pressure loss in the distribution channel from one segmental joint to the other is ultimately determined from the volumetric flow rate throughout the channel and its geometry:

$$p_n = p_{n+1} + \Delta p_{pipe}, \tag{5.83}$$

$$\Delta p_{pipe} = f\left(\dot{V}_n, \dot{V}_{n+1}, R(x)\right). \tag{5.84}$$

Starting with known values for $\dot{V}_n$ and $p_n$ at the boundary of the segment, the $\dot{V}_L$ can be determined by iteration applying Equations (5.80) and (5.84). Given the distribution of the outgoing volumetric flow rate, that is, $\dot{V}_L$ for each segment and two of the three geometric parameters $R(x)$, $y(x)$, and $H(x)$, it is possible to find the course of the third geometric parameter by iteration. The process starts with the iterative determination of the geometry at the edge of the die, since the values for the pressure in the distribution channel (surrounding pressure) and the volumetric flow rates in the distribution channel (generally equal to zero) are known as initial values. The advantages of the numerical design of the flow channel are as follows:

- The design of a die with any specified geometry, such as a wide flat slit die with a slit height that is not constant, in the land region is possible.
- Any distribution of the exiting volume flow can be specified. This enables the design of dies that purposely have a nonuniform volume flow distribution, for example.
- Additionally, the inlet pressure losses during the overflow from the distribution channel into the land slit can be considered. During the overflow, the melt is subjected to a stretch (Fig. 5.27). Consequently, the leakage flow has to overcome not only the pressure loss in the slit $\Delta p_S$ but also the inlet pressure loss from the channel to the slit $\Delta p_e$. The magnitude of this inlet pressure loss is dependent on the geometry and the velocity of the passing flow, that is, on the extent to which the melt is stretched and on the extensional viscosity. The latter is the measure

of the resistance of the melt to elongational deformation. The inlet pressure loss can be taken into account by applying Equation (5.81) and subtracting this pressure loss from the driving pressure $\bar{p}$, thus:

$$\dot{V}_L = f\left[\left(\bar{p} - \Delta p_E - p_0\right), H(x), y(x)\right]. \tag{5.85}$$

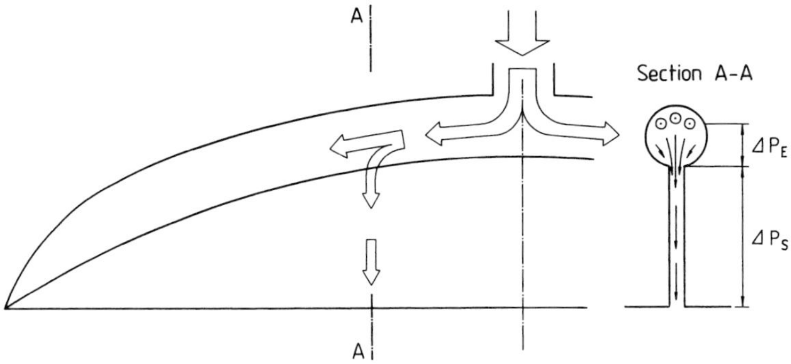


**Figure 5.27** Inlet pressure drop due to the flow from the distribution channel into the land gap

The fact that the inlet pressure loss $\Delta p_{entry}$ depends on the volumetric flow rate through the slit and the geometric conditions has to be considered:

$$\Delta p_{entry} = f\left[\dot{V}_L, H(x), R(x)\right]. \tag{5.86}$$

The disadvantages of the numerical design of the flow channel are as follows: a great deal of programming is necessary, the design requires long computing times, and mathematical stability problems frequently occur in many applications.

#### 5.2.2.5 Considerations for Clam Shelling

So-called clam shelling (opening up) due to the existing internal pressures is one of the problems in the design of wide slit dies. The pressure distribution in a wide slit die is depicted schematically in Fig. 5.28. Figure 5.29 shows the resulting deformation of the die plates (see also Section 9.3 on this). The clam shelling increases the melt flow in the center region of the die.

The effect of this widening can be taken into account during the die design only when the magnitude of the deflection as a function of the pressure distribution can be determined. To compensate for the widening, the die can be made in such a way that if it widens during the operation, it has the computed flow channel geometry. This means that the die is made with a much smaller land slit in the middle region.

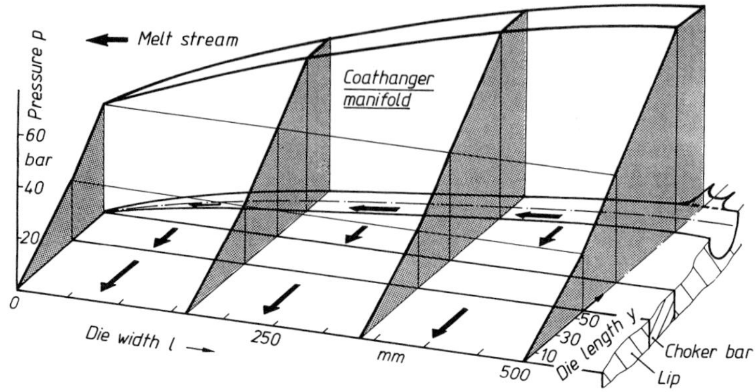


**Figure 5.28** Typical pressure distribution in a flat slit die

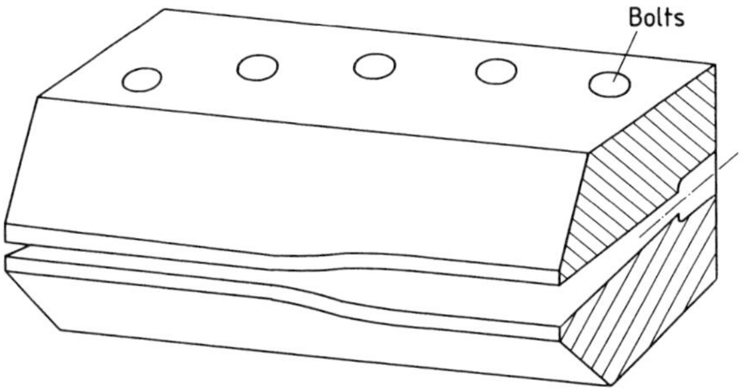


**Figure 5.29** Clam shelling due to internal pressure in a flat slit die

Another possibility is to take the widening into account during the rheological design of the land length or of the contour of the distribution channel. This means that the manifold has to be designed with the height of the land slit changing across its width [70].

The possibilities of predicting the die widening by computation are discussed in Section 9.3 [77].

### 5.2.2.6 Unconventional Manifolds

Manifold systems that accomplish the melt distribution across a distribution channel and a land region are designed in such a way that for a uniform emerging flow there is an equal resistance to flow on each flow path. A situation where each flow path has equal flow history as to the residence time and shear action cannot be achieved by these systems (fishtail, coathanger). Therefore, alternatives have been recommended that are designed in such a way that each flow path is made the same length [78]. When observing the flow paths in a wide slit die (Fig. 5.30) that is made as a divergent slit channel with a constant slit height, it is obvious that the flow paths in the die center are shorter than those at its edges. Under the assumption of a point-like melt feed, the following relationship for minimum and maximum length of flow are valid:

$$l_p(0) = y_0, \tag{5.87}$$

$$l_p(L) = \sqrt{y_0^2 + L^2}. \tag{5.88}$$

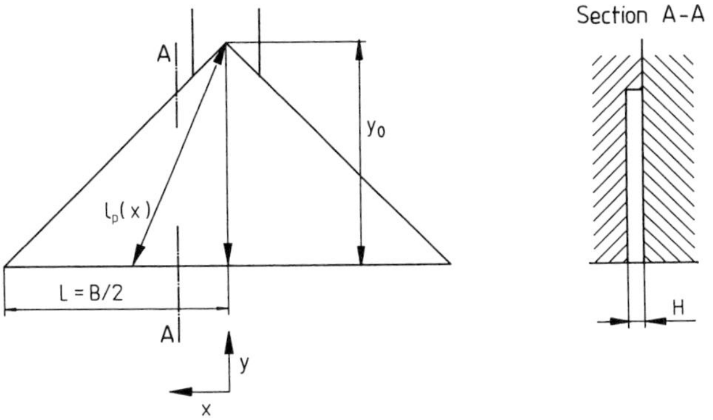


**Figure 5.30** Streamlines in a divergent flow channel

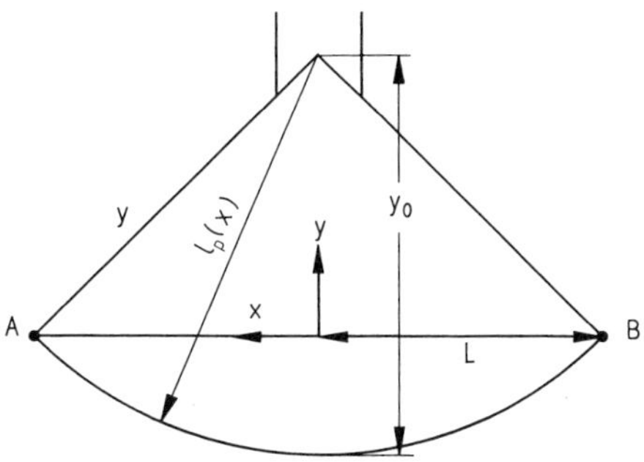


**Figure 5.31** Streamlines in a centrally injected circular section

A flow channel geometry with equal lengths on each flow path can be derived from a study of a centrally injected circular sector (Fig. 5.31). In this case, all flow paths are equally long; however, the melt discharge is not located on the desired plane. This can be corrected by curving the circular sector in the third dimension ($z$ direction) so that all flow paths end on the line A–B.

Such a manifold has been used successfully in injection molding as a predistributor for a film sprue [79].

In Fig. 5.32, there is a simple suggestion for the solution of the design of such a manifold. Each flow path is raised in the middle to such an extent ($h(x)$) that the length of the flow path ends on the edge of the die. The amount of elevation depends on the point of exit $x$. With the projected length of the flow path

$$l_p(x) = \sqrt{y_0^2 + x^2}, \tag{5.89}$$

the length of flow at the edge of the die

$$l_p(L) = \sqrt{y_0^2 + L^2} = l(x) \tag{5.90}$$

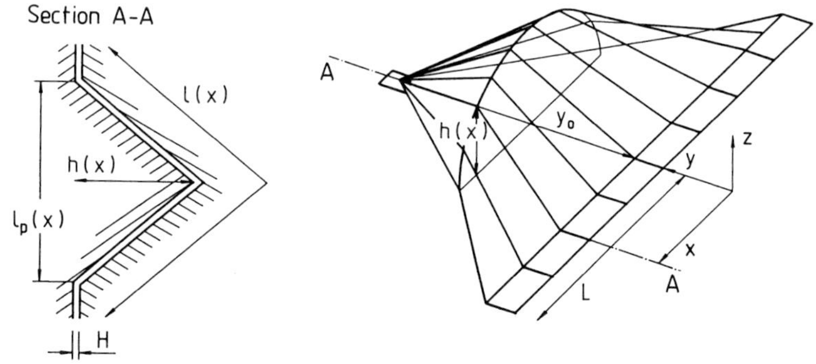


**Figure 5.32** Flat slit manifold with streamlines of equal length

and the resulting length of the flow path

$$l(x) = 2\sqrt{(l_p(x)/2)^2 + h^2(x)} \tag{5.91}$$

the $h(x)$ for each flow path follows:

$$h(x) = \frac{1}{2}\sqrt{L^2 - x^2}. \tag{5.92}$$

It is important to keep in mind that all the quantities refer to the point of exit $x$. The above simple suggestion for a solution is not the optimum from the rheological point of view because of the sharp break in the flow path, but it still was chosen for the illustration of the process for the determination of the geometry for a wide slit die with flow paths of equal length.

The advantages are

- independence of the material being used,
- independence of operation conditions,
- uniform distribution, and
- identical flow history on each flow path.

Disadvantages of these dies are as follows:

- The die when built is rather high. For the type of adjustment of the flow path described above, the maximum height in the center of the die can be a quarter of the width of the exit slit. Therefore, the application of this principle is feasible for narrow dies only.
- Both halves of the die can be made only by CNC milling machines. Moreover, the machining times are long and the material consumption is very high because a great deal of it has to be machined away to produce a deep channel.
- The setting up of the control program for the CNC machines is very time consuming, considerably more so than that for conventional wide slit dies with a manifold.

#### 5.2.2.7 Operating Performance of Wide Slit Dies

As explained earlier, it is possible to design extrusion dies with wide slit openings that are independent of the operating conditions. When the width of the die is more than 500 mm, the long lands resulting from the design procedure give rise to high pressure losses, which lead to high separating forces. These can be held back only by extremely thick die plates. Because of this limitation, the wide slit dies are in most cases designed as dependent on operating conditions.

Since these dies are operating under a variety of operating conditions and with a variety of materials, their process behavior, that is, the effect of changes in material and operating conditions on the melt distribution, is an important design criterion. A simple method to determine if the die is dependent on the process conditions is to do the design for a spectrum of operating conditions and then compare the resulting geometries (Fig. 5.33) [80].

The larger the change in the resulting geometry at various operating conditions, the more process-dependent the die is.

The procedure described above provides a reference point for the design, but an evaluation of the resulting melt distribution is not possible.

This evaluation is only possible after the computation of the effect of the operating point on the geometry of coathanger manifolds, designed to be independent of operating conditions [80] and die flow simulation [76,81–84].

For the above computation, the distribution channel is divided into segments, including the corresponding slit segments (Fig. 5.26). A flow and pressure balance is carried out successively for each segment. The procedure here is similar to the numerical design of the flow channel, the only difference being that in this case the geometry is fully specified and the exiting flow is calculated.

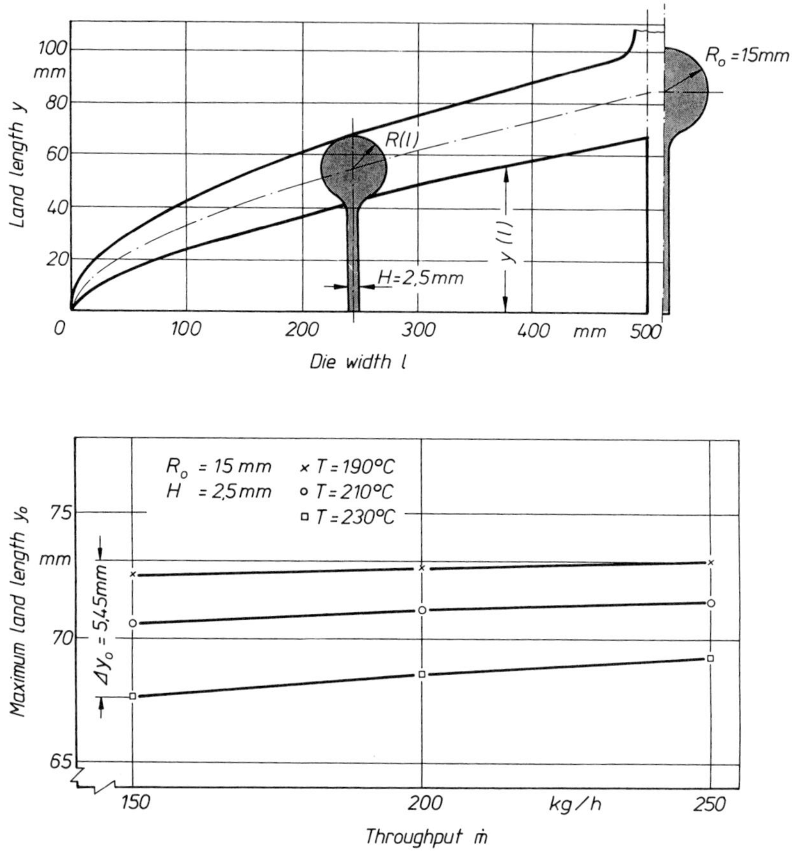


**Figure 5.33** Effect of operating conditions on the geometry of a coathanger manifold designed to be independent of the operating conditions [80]

The check of the distribution is started at the entry point in the middle of the die, because here the incoming flow rate in the distribution channel (equal to half of the total flow rate) is known.

In addition to that, the value of the unknown inlet pressure has to be specified, and then the more accurate value is determined by a series of iterations.

The calculated distributions for a wide slit die for one design point, for a different mass flow rate, for a different mass temperature, and after a change of material are shown in Figs. 5.34 and 5.35.

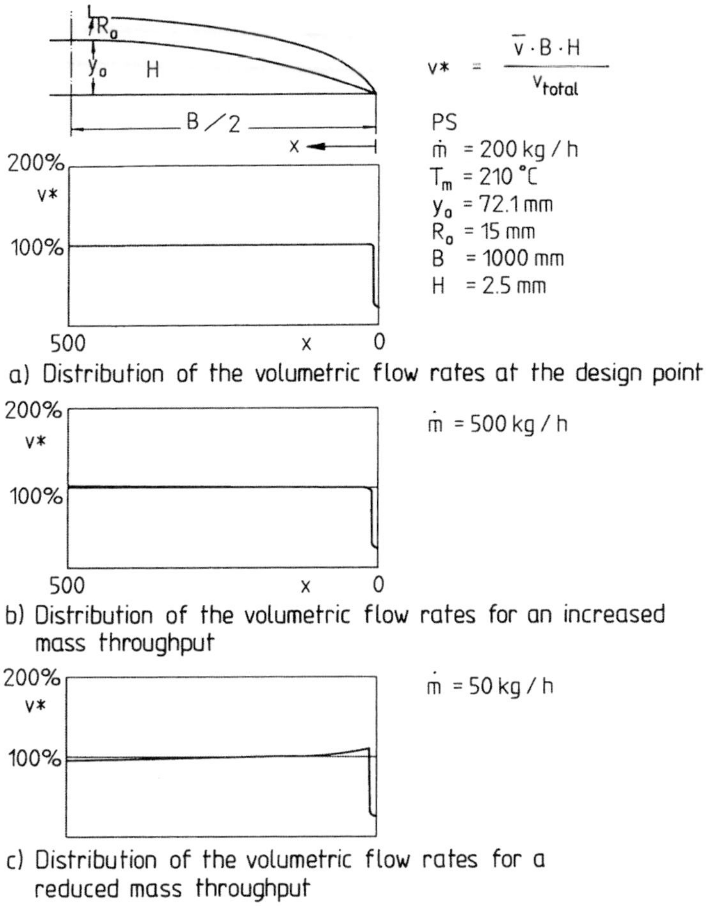


**Figure 5.34** Results of computations for volumetric flow rate distributions, I

It can be seen that the increase in the flow rate has practically no effect on the distribution, and a decrease leads to a slight melt excess at the edge. The raised temperature has a similar effect. A considerable worsening of the distribution effect is shown by a die designed for PS and then recalculated for PE-LD.

By means of such a flow simulation, as shown, it is possible to estimate the size of an operating window of a sheet die.

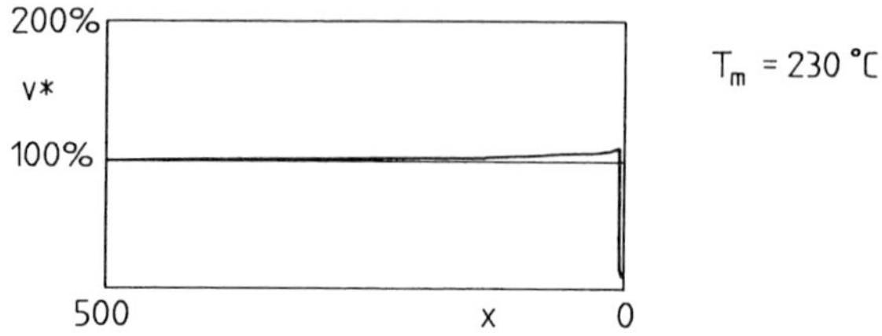


a) Distribution of the volume flow rates for an increased melt temperature

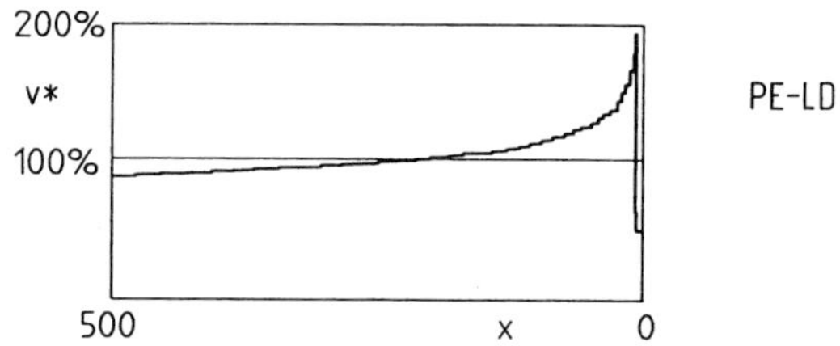


b) Distribution of the volume flow rates for a different polymer

**Figure 5.35** Results of computations for the volumetric flow rate distributions, II

## 5.3 Dies with Annular Exit Cross Section

Dies for the extrusion of melt strands with an annular cross section are used for the extrusion of pipes, hoses, tubular film, and tubular parisons, for extrusion blow molding as well as for cable and pipe jacketing.

Extrusion of pipes and nonreinforced hoses is done mainly with straight dies. For the extrusion of tubular films, reinforced hoses, and parisons as well as for jacketing, the extruder is at an angle—most often 90°—to the direction of the extrudate.

To produce the above-mentioned annular products, mainly the die types shown in Fig. 5.36 are employed:

- Mandrel support die (also referred to as spider-supported or support ring die)
- Screen pack die
- Side-fed mandrel die
- Die with spiral distributor (also called spiral mandrel die)

All these dies have a common feature, namely a parallel land zone located at the die exit, at the so-called outer die ring. The land zone allows the melt to relax and thus reduce at least part of the deformation energy imposed on it during the process. This zone is not present in blow-molding dies with an adjustable width of the exit gap.

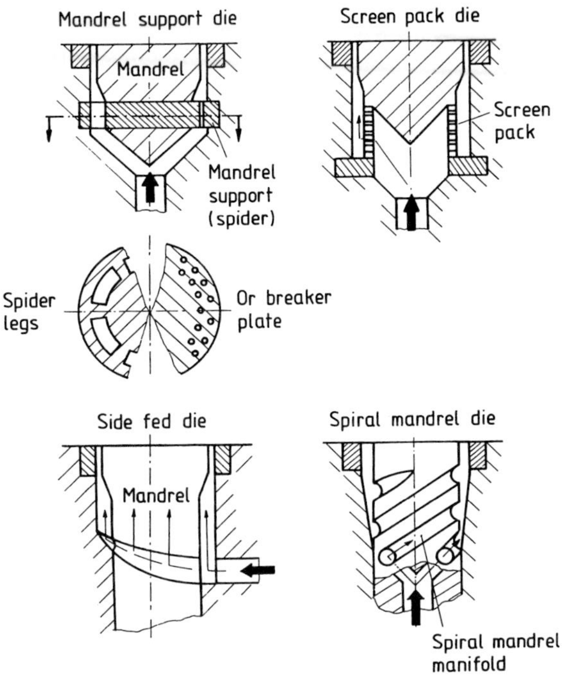


**Figure 5.36** Dies with an annular exit cross section (principle)

A separate temperature control of this die exit zone allows an effective control of the surface quality of the extrudate.

Moreover, the position of the outer die ring can be adjusted, that is, centered relative to the inner mandrel, thus modifying the circumferential flow at the exit.

## 5.3.1 Types

### 5.3.1.1 Center-Fed Mandrel Support Dies

The melt stream, supplied by the extruder in a circular flow channel, is transformed in such a die by the mandrel into an annular stream. The melt is divided in the mandrel support zone into several separate melt streams and flows around the spider legs. The connecting, converging mandrel region, in which pipe dies become narrower at an angle of 10–15° [85] and in which the separate streams come together once again, is followed by a die ring with parallel channel walls. The ratio of the length of the parallel land region to the height of the exit gap for the pipe dies varies between 10:1 and 30:1 [36,45,85]. The ratio of the mandrel support diameter to the exit diameter for the pipe dies is usually 1.4 to 1.6 for PVC and 2 for PE; however, these values also depend on the diameter of the die [85]. The essential parts of the die, such as mandrel housing and outer die ring, are generally exchangeable, so one center-fed mandrel die can be used for several extrudate geometries. These parts are not hatched in Fig. 5.37. In order to achieve a uniform melt flow through the annulus, the outer die ring can be displaced radially by the centering screws mounted around the circumference. A clamping ring pretensioned with disk springs is often used in this design [86]. The disk springs provide an adequate contact pressure so that a good seal is achieved but yet permit the outer ring to move.

From all the dies shown in Fig. 5.36 the center-fed mandrel dies were the ones most widely used in the past because, being fed into the center, they generally provide a good melt distribution and distribute the melt independently of the operating conditions. Because of these advantages, this design is used in particular for technical and residence-time-sensitive materials, especially PVC [87].

However, they have a disadvantage in that spider legs cause flow marks or weld lines. These are not always visible as local thin sections and streaks, but they are always present structurally and can lead to mechanically and optically poor sections [88].

The high degree of melt orientation in the vicinity of the spider legs is among other things the reason for the flow marks. The orientation is produced by the high velocity gradients in this zone when the melt adheres to the walls and particularly due to the large extension of the melt particles near the end of the spider legs, as shown in [88]. Differences in density due to differences in temperatures of the melt and the spider legs can also play a role [89].

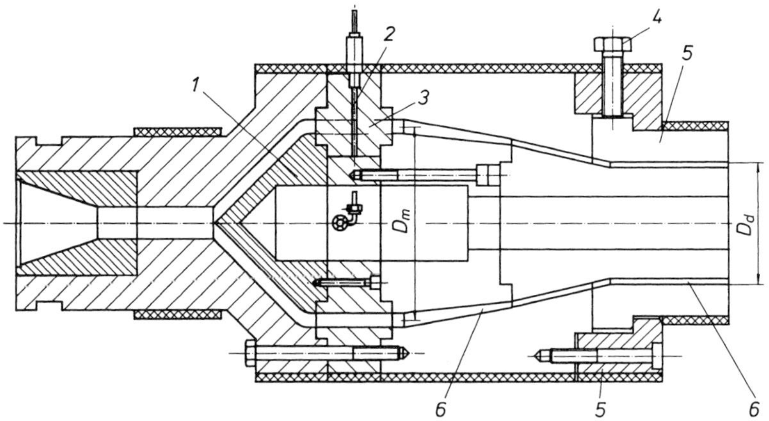


**Figure 5.37** Mandrel support die (principle): 1: mandrel tip; 2: mandrel support; 3: spider leg; 4: centering screw; 5: die ring; 6: relaxation zone

In order to reduce the flow marks, which are unavoidable with the use of center-fed mandrel support dies, there are essentially three possibilities [88,90]:

- An increase in the mass temperature or the residence time in the die
- Uniform distribution along the circumference using a smearing device, for example

- Producing a uniform structure over the entire circumference by orienting the molecules longitudinally, and between the spider legs

To accomplish that, the following concepts were proposed in [88]:

- The use of elements, rotating about the die axis, in order to eliminate the flow marks [91]. The disadvantage of this system is that the rotating elements require an additional drive and can cause sealing problems.
- Coating the spider legs with a nonwetting material, such as PTFE. The disadvantage here is that such a coating wears off rapidly in practical use [92].
- Lengthening the flow path and hence increasing the residence time. This is possible, but the total pressure loss of the die must not exceed the maximum permissible value (Fig. 5.38).
- Heating the spider legs. This often is difficult because of the relative small size of the legs, but gradual improvements are achievable [89] (Fig. 5.38).
- Forcing the stream to follow a strongly convergent flow after the spider legs. This is almost always done in practice as shown in Fig. 5.37.
- Attaching restrictor beads or breaker plates. Both are used in practice for polyolefins [15,85].
- Using smear devices (Fig. 5.38). Multiple helical flutes are cut in the mandrel and the housing with opposite pitch. This induces melt flow components in the tangential direction and, thus, leads to blurring, which is a broader spreading of the flow markings on the circumference of the extrudate. Multiple-flighted smear devices mounted on only one side can also be used successfully.

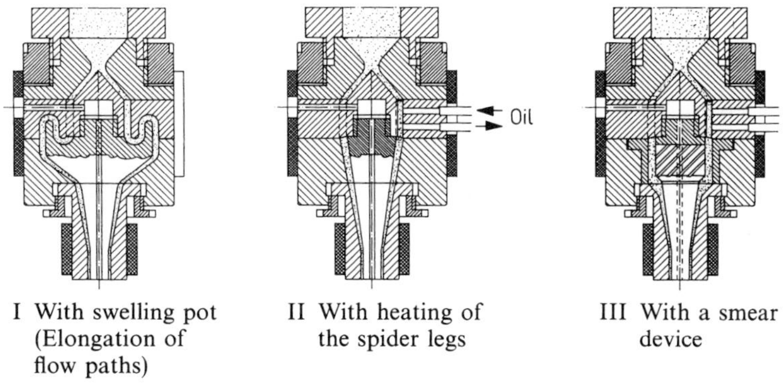


**Figure 5.38** Center-fed mandrel support dies and different methods to reduce weld lines [88]

Some of the mandrel support systems that are designed so as to reduce the formation of weld lines are shown in Fig. 5.39. It should be noted here that a perforated plate (screen plate) should not be used as the mandrel support element because it has unsatisfactory strength [85]. Moreover, for exactly the same reason as the tangential mandrel supports, it does not meet the requirements for the spreading of the defects over a larger area of the circumference completely [93]. The mandrel supports with a supporting ring and offset spider legs used especially in parison dies of the extrusion blow-molding machines are considerably better. The way they work is that the flow marks (defects) do not extend all the way through the entire wall for the extrudate, which is advantageous. However, the flow resistances of the passage areas must be matched accurately in order to achieve equal mean passage velocities [94].

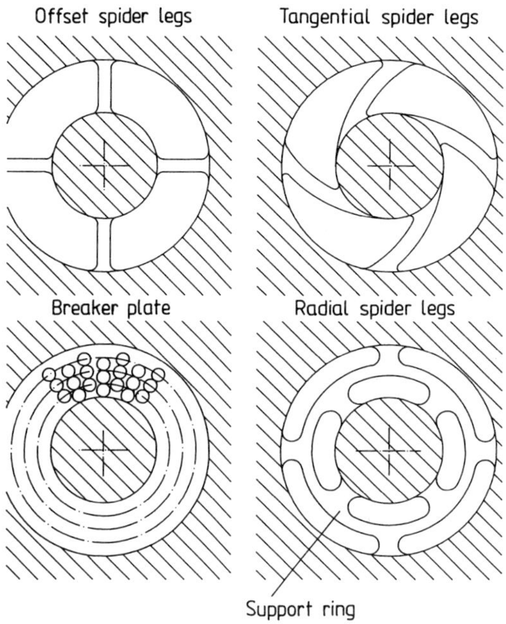


**Figure 5.39** Mandrel support system

The spider legs must be designed with the flow in mind and should not be longer than 30 to 80 mm and not wider than 9 to 12 mm [85] and should run out with a sharp angle of approximately 8°. The passage gap heights are in practice 10 to 25 mm, and the number of spider legs increases approximately with the square of the diameter [85].

The corners at the root of the spider legs should be designed so that no regions of stagnation arise, which could cause a degradation of heat-sensitive materials.

The center-fed mandrel support dies can be subjected to pressures up to 600 bar [85], so the spider legs must be dimensioned such that the forces resulting from the high pressures can be withstood safely.

From the point of view of mechanical strength, the outside diameters of the center-fed mandrel support dies do not exceed 700 mm [48].

### 5.3.1.2 Screen Pack Dies

This type of die is mainly used for the extrusion of large-diameter pipes from polyolefins [85,95]. After the melt enters such a die, it comes up against a cone-shaped mandrel (torpedo) by which it is diverted from the main axial flow to the radial (outside) direction (Fig. 5.40). The melt then passes through a tubular body with numerous boreholes (diameter 1 to 2.5 mm), the so-called screen pack. To this screen pack the adjacent mandrel is attached. After the melt is diverted once again into the axial direction, it flows through a parallel die land with an adjustable die ring in a similar way as in the center-fed mandrel support die. The section of the die following the screen pack is interchangeable to a degree to accommodate different extrudate dimensions.

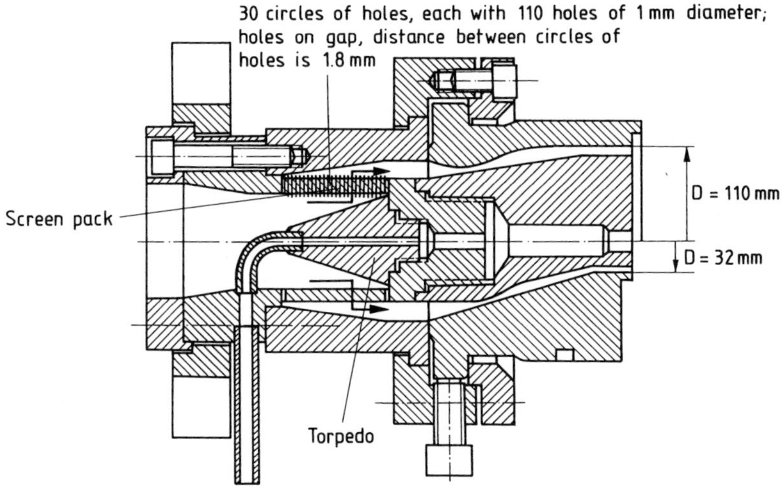


**Figure 5.40** Screen pack extruder head (for pipes with diameters 32 to 110 mm) [Hoechst]

In a way, the screen pack die can be regarded as a special form of the center-fed mandrel support die, in which there are only very small total pressure losses (approx 70 to 120 bar) due to the large passage area of the screen pack [85]. This has a direct effect on the cost efficiency of the process (higher throughputs, lower energy consumption) as well as on the quality of the extrudate and often lower mass temperatures. For safety reasons, especially during the startup, the screen pack dies today are designed for pressures up to approximately 300 bar [85].

A great advantage of these dies, when compared to the center-fed mandrel support dies, is that they are built more compactly: a ratio of mandrel exit diameter to screen pack diameter of 1.4 is possible. As a result, they are considerably lighter (50 to 65% of the weight of comparable center-fed mandrel support dies) and can therefore be handled much easier [85]. Since a lot of weld lines are created after the flow is recombined, screen pack dies are only suitable for melts characterized by a very good welding behavior, but not for optical products. These dies are therefore in particular used for polyolefin pipes due to robustness, easy handling, and versatile usability [87].

#### 5.3.1.3 Side-Fed Mandrel Dies

In a side-fed mandrel die, the melt is brought in at an angle, typically 90°. This is necessary especially if something must be passed through the mandrel, such as supporting or cooling air or a semifinished product to be jacketed.

The entering melt is passed around the mandrel by way of a manifold or even a manifold system, which may be mounted on the mandrel or the outer body of the die. This way, the melt stream is gradually deflected from the radial entry flow direction to the axial exit flow direction. A typical mandrel with manifold is shown in Fig. 5.41.

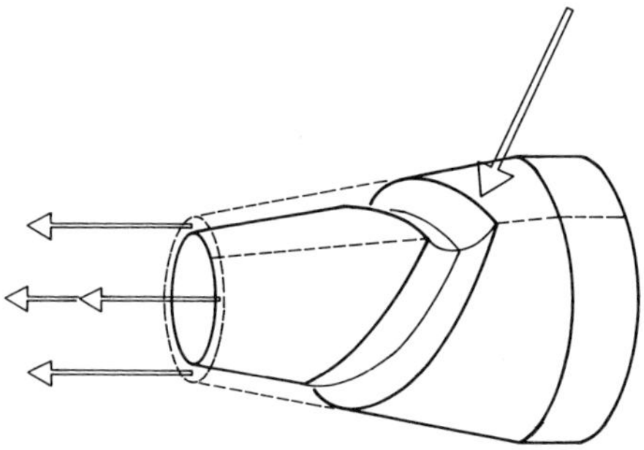

**Figure 5.41** Mandrel of a side-fed die with a coathanger manifold (principle)

The development of the flow geometry in this die can be visualized by imagining the wide slit manifold shown in Fig. 5.16 to be wrapped around a truncated cone. The difficulty with these dies lies in designing the manifold in such a way that the material exits uniformly through the annular opening around the entire circumference.

When using side-fed mandrel dies, a similar problem occurs as described previously, namely weld lines; the methods to deal with them are also the same. Another design form of the mandrel, a so-called heart-shaped curve, is shown in Fig. 5.42. In this system, the melt stream is divided into two separate melt streams; each of them is guided on flow paths around one-half of the mandrel (the lengths of the pathways should be equal if possible) until they are combined once again. Two weld lines are formed, which is one more than in the design shown in Fig. 5.41. The name heart-shaped curve is derived from the fact that the torpedo on the mandrel has a heart-shaped contour. The characteristic difference from the mandrel support discussed previously (Fig. 5.41) is that here the melt stream flows around the torpedo and not over it, as it does in the flat slit manifold (over a die land) or in a mandrel with a manifold. This heart-shaped form is used frequently, particularly when attempting long and equal flow pathways for all melt particles.

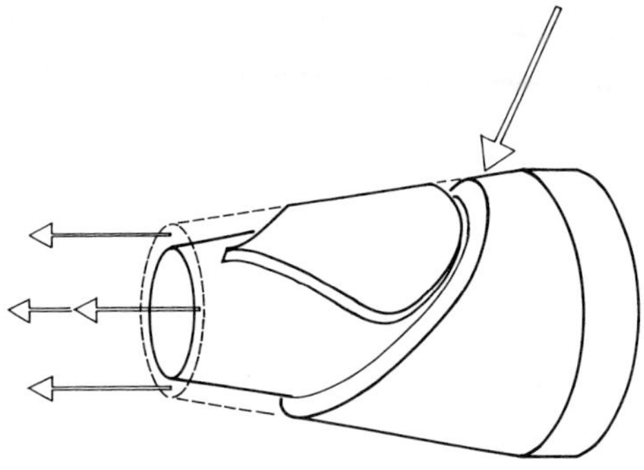

**Figure 5.42** Mandrel of a side-fed die with a heart-shaped manifold (principle)

As in all other types of extrusion dies, the side-fed mandrel support die has a parallel land region at the exit with a die ring that can be centered.

#### 5.3.1.4 Spiral Mandrel Dies

In the classic axial spiral mandrel die (Fig. 5.43), the melt stream supplied by the extruder is initially divided into several separate streams. Star-shaped and ring-shaped melt-distributing systems can be used for this purpose, the latter type being preferred when large bores or many film layers are required in the mandrels [96]. These primary distributors discharge into spiraling channels that are machined into the mandrel in the form of multiple threads. In recent years, this approach to supply each spiral channel individually has sometimes been replaced by coathanger distributors extending around the circumference, followed by spiral channels in the form of a smear thread [97,98]. The advantage of this design is the smaller primary distribution system, above all when working with a large number of spiral channels and melt layers.

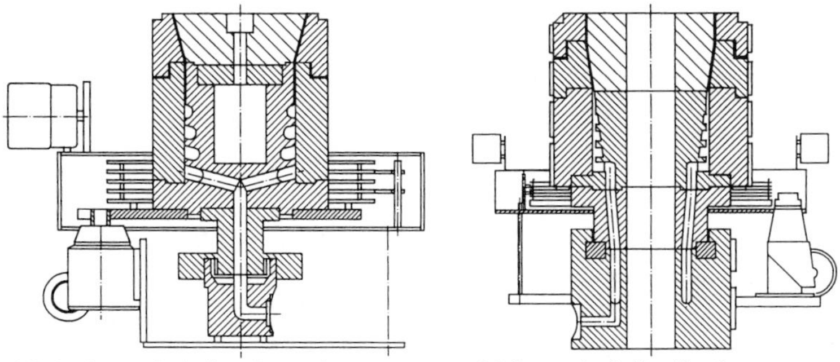

*1* Spiral mandrel die: Central feed – Star shaped melt distribution Radial boreholes – Spiral channels

*2* Ring spiral distribution: Central borehole for internal air exchange – Ring-shaped melt distribution – Axial boreholes

**Figure 5.43** Spiral mandrel dies for blown film extrusion [Reifenhäuser]

In the spiral section, the channel depth decreases steadily, and the gap between the mandrel and the outer part of the die (die housing) usually increases steadily in the direction of extrusion. In this way, the melt stream flowing through one spiral divides itself into two streams. The first stream flows axially over a land formed between two spirals; the other continues to follow the course of the spiral. As a result, every point around the annular discharge gap receives melt formed by the superposition of the tangential melt streams from all channels of the spiral distribution system. Since there are no weld lines with this arrangement, good thermal homogeneity (i. e., a uniform melt temperature) is achieved in addition to the desired physical homogeneity. The complete elimination of weld lines and flow marks is one of the greatest benefits of this melt distribution system [99].

In contrast, in radial spiral distributors (Fig. 5.44), the channels do not wind around the mandrel axially in the conventional way, but rather lie in a plane in a radial arrangement resembling the grooves in a phonograph record (Fig. 5.45). Now the melt can be introduced from the outside instead of centrally from below. The channel depth decreases toward the center so that the melt, which initially flows only in the channel, is gradually forced over the land. The melt is finally redirected toward the die outlet and, if desired, can be combined with additional melt layers.

Specific advantages of the radial arrangement over the axial arrangement include greater flexibility as a result of the modular design, good thermal separation of the individual modules, smaller differences in residence times, smaller amounts of melt volume within flow channels when there are many layers, and the ability to retrofit easily to existing systems. The long common flow paths can result in melt instabilities and thus create problems when there are many melt layers. In addition, radial distributors are more susceptible to deformation of die plates as a result of the melt pressure and have more sealing surfaces [100].

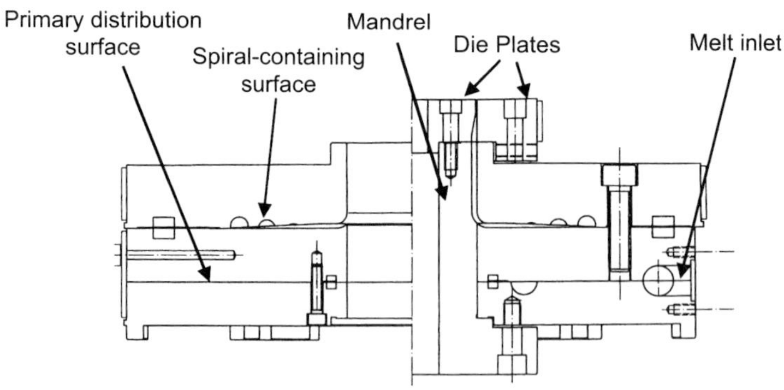


**Figure 5.44** Radial spiral distribution module [IKV, Kuhne]

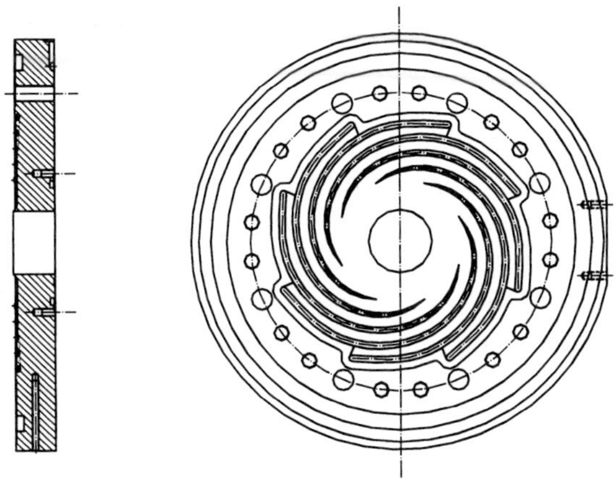

**Figure 5.45** Spiral arrangement

A combination of an axial and a radial spiral mandrel die is represented by the conical spiral mandrel die (Fig. 5.46). The spirals are positioned along truncated cones instead of within a flat plane. This variation offers advantages of both systems in a multilayer setup (Section 6.1.3): as always, two layers are within one plate, and the conical spiral mandrel die is more compact than the classical stack die. Furthermore, adding further layers does not imply a layerwise increasing diameter like for cylindrical spiral mandrel dies [101].

The outlet section of spiral distribution systems is also usually interchangeable, can be centered, and has a die land, and its temperature can be controlled separately.

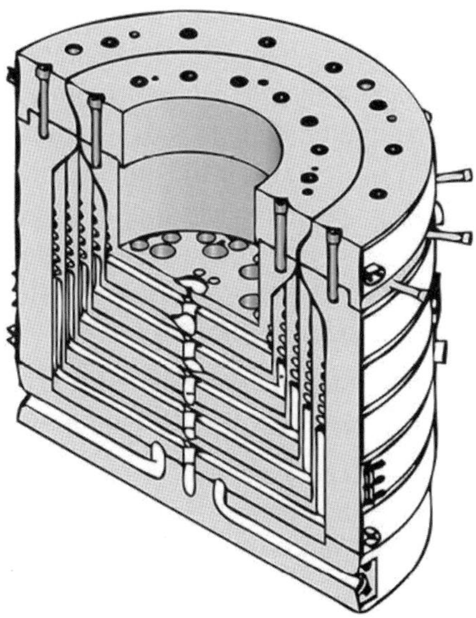

**Figure 5.46** Conical mandrel die

## 5.3.2 Applications

### 5.3.2.1 Pipe Dies

Pipes are extruded predominantly by longitudinally fed (in-line) dies. So-called crosshead or offset dies are used only in combination with internal cooling or internal calibration. For this reason, the use of side-fed dies is limited [16,36,45,85], but in coextrusion they are sometimes applied to add an outer layer. The usual diameters of extruded pipes range between several millimeters up to 2.5 meters [87] with wall gauges up to 60 mm [95].

The construction of the dies used for this purpose depends on the type of material to be processed and the desired dimensions of the finished pipe. Thus, pipes made from rigid PVC are generally extruded with in-line dies, the largest outside diameters being approximately 630 mm [85,95]. Large polyolefin pipes are predominantly produced by screen pack dies as well as by spiral mandrel dies [87,95, 96,99,102,103]. The latter usually works with higher pressure losses than the former, which usually run typically at 70 to 120 bar.

As with other dies, with pipe dies the mandrel and die ring can be exchanged so that pipes with different outside diameters and wall gauges can be produced [85]. In [104] a die is shown that allows the extrusion of pipes with different wall thicknesses by using a slightly conical (approx. 2°) outer ring region and a mandrel that can be adjusted mechanically or hydraulically in the axial direction even during operation.

Dies for pipes with a diameter greater than 630 mm can no longer be centered while in operation and are therefore centered with a template before the run.

Small and medium-sized rigid PVC pipes are produced using center-fed dies with the ratio of mandrel support to mandrel diameters between 1.4 and 1.6 to 1. In the design of these dies, careful attention is paid to avoiding areas of melt stagnation (dead spots) during operation. These dies are designed for pressures of approximately 250 to 600 bar [85]. Polyethylene pipes with smaller diameters are extruded with the mandrel support/mandrel diameter ratio of 2:1; this ratio can become even 1.3:1 for larger diameters [36,85]. If this ratio becomes too small, the danger exists that the melt, having passed the spider legs, will not be able to flow together again, which can cause flow lines. When this happens, for example breaker plates (perforated disks) are inserted behind the spider legs to homogenize the melt.

In pipe dies with diameters greater than 140 mm, the mandrel has a separate temperature control in addition to the usual temperature control of the die ring [85].

The properties of a large polyolefin pipe, especially the creep rupture strength, are strongly affected by the internal stresses introduced by the subsequent cooling and calibrating and possibly by crosslinking or degradation of the material during the processing [96,99,105]. Therefore, the temperature of the melt should be kept as low as possible, as should the mechanical stresses on the melt. If this is achieved by the use of spiral mandrel dies, then the main pressure drop occurs in the spiral mandrel section due to its large exit gap and not in the die land region. This is in contrast to blown film dies [96,99].

To improve handling, the dies for large pipes are mounted on a mobile carriage. This is not surprising considering that, for example, the weight of a 630 mm center-fed pipe die is approximately 2.8 metric tons [85,95].

Smaller, thin-walled pipes can be produced using multiexit dies, by means of which high extruder outputs can be utilized better and with less stress on the material due to the lower melt velocities in the individual sections of the die, which can manifest itself in lower shrinkage. Furthermore, the surface appearance is better at lower flow velocities, and it is simpler to maintain the required tolerances [104,106].

Figure 5.47 shows a multiple-pipe die with an individual flow regulation with the purpose of attaining equal mean velocities of the outflowing melt.

A summary of patent literature pertaining to extrusion dies for the production of pipes and hoses is given in [107].

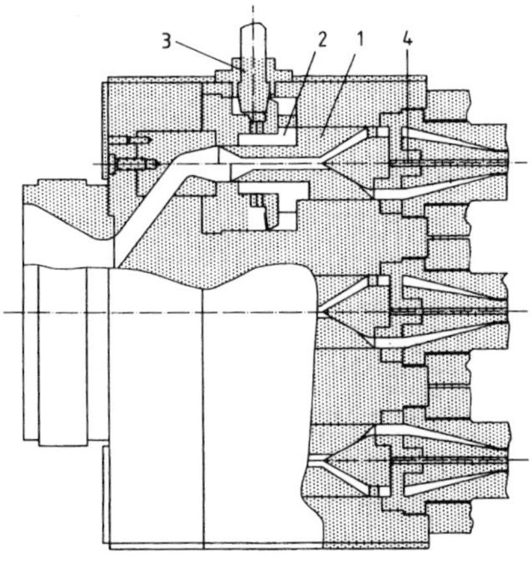


**Figure 5.47** Triple die for pipe extrusion with flow control [104]: 1: sliding sleeve; 2: threaded sleeve with a bevel gear; 3: drive shaft with a bevel gear; 4: mandrel

#### 5.3.2.2 Blown Film Dies

In the extrusion of blown film today, spiral mandrel dies are most frequently used. The almost always necessary change in the direction of melt flow from horizontal to vertical is accomplished by the use of an elbow placed in front of the die [16,48,96,108,109]. Center-fed dies with spiders and flow resistance zones or similar designs used in the past have not been overly successful [86,110].

The width of the gap at the outlet of the blown film dies is usually between 0.6 mm (thin film) and 1.6 mm (construction film); the outside diameter at that location can be several centimeters up to 2.5 m [48]. If the level of longitudinal orientation in the film needs to be raised by a higher take-up speed, the width of the gap can be increased to 2.8 mm, which also leads to more relaxation in the orifice area. The parallel walls in the land region affect the dimensional tolerances as well as the quality of the film; defects in the flow channel surface have a particularly negative effect in this region [111]. The longer this region, the better the optical properties of the film. The actual length (as ultimately manufactured) of the parallel die land is usually a compromise between the length required for a satisfactory melt relaxation and the length still having an acceptable pressure drop.

The shrinkage of a blown film, an important quality attribute, is affected by the deformation of the melt in the shaping die as well as by the blow-up and take-off conditions after it emerges from the die. The effect of the design of the flow channel of the die on the melt shrinkage was proven for the first time in [110]. Another factor to be considered in this context, the subsequent blow-up of the film (the formation of the so-called bubble), also introduces reversible deformations in the material. These can be much greater than those imposed by the die, particularly if the blow-up ratio is high.

Figure 5.48 illustrates the percentage pressure consumption in blown film dies with a spiral mandrel with star-shaped and ring-type distributors. The high pressure consumption in the die land and the relatively low percentage pressure drop in the spiral mandrel melt distributor are clearly demonstrated. The total pressure drop in both dies is between approximately 200 and 230 bar. In blown film dies for the processing of polyolefins, these values can be as high as 350 bar, while the dies are designed for pressures up to 500 bar [48]. Nowadays, the dies are designed for pressures up to 600 bar, while processing is usually done up to 500 bar due to more viscous materials and cooler operating temperatures.

When compared to pipe dies and extrusion blow molding dies, blown film dies are provided with special turning devices. Their purpose is to spread bands with thicker gauges around the circumference so that an accumulation of the material in one place, which leads to a heavier band on the wind-up roll, is eliminated [16,112]. A extrusion die-based solution can be used with center-fed as well as side-fed dies. Much effort is required to ensure a good seal, to devise ways of cleaning, and for support of the revolving parts by suitable bearings [48]. DC motors are used for this purpose, and the rotating parts of the dies move slowly, usually at 5 to 20 minutes per revolution. An example of such an arrangement is given in Fig. 5.49 [48]. This concept is still in the market and in that case mostly used for monolayer blown films. However, reversing take-up units are predominant and state of the art to homogenize the distribution of potential thick and thin places along the circumference, resulting in cylindrical wind-up rolls [101].

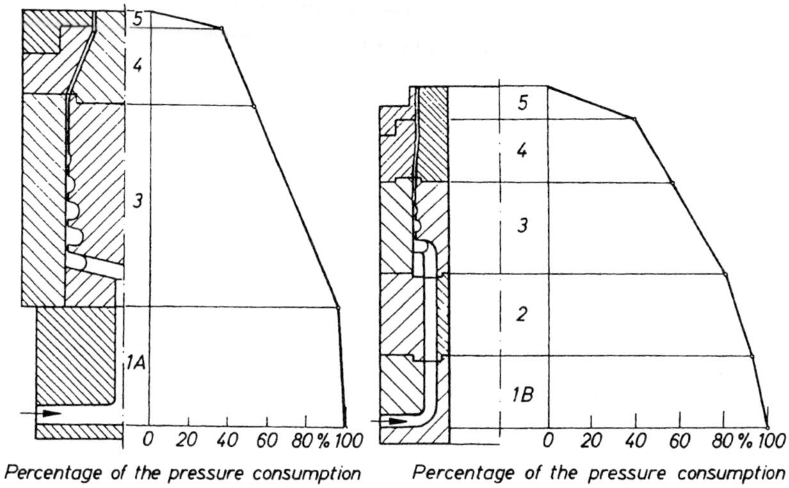


**Figure 5.48** Pressure loss in a blown film die [96]: 1A: inlet (star-shaped distributor); 1B: inlet (ring-type distributor); 2: rotational part; 3: spiral mandrel distributor; 4: transition; 5: parallel die land

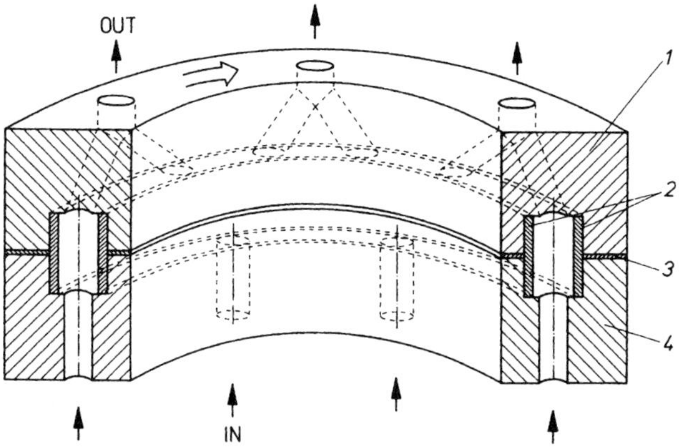


**Figure 5.49** Distribution channel of a rotating blown film die [110]: 1: rotating part; 2: sealing sleeve; 3: butting ring; 4: stationary part

### 5.3.2.3 Dies for the Extrusion of Parisons for Blow Molding

In extrusion blow molding, the tubular parison is always ejected perpendicularly downward from the die. Since the extruder is usually arranged horizontally in this process, the term "crosshead die" is often used for a parison die.

Parison dies are divided into those in which the melt is shaped continuously and those with melt accumulators, in which the extrudate, called a parison, is pushed out rapidly after the necessary volume is plasticized and accumulated in the storage chamber.

In center-fed parison dies, elbows (bent pipes) are usually used between the extruder and the die that serve the purpose of changing the flow direction from horizontal to vertical. When designing such an elbow, it is important to select a proper ratio of the internal radius to the channel radius of the elbow to achieve favorable hydrodynamic properties; this ratio should be, according to [113], between 6 and 8 in order to prevent secondary flows (see also Section 4.4.5). Moreover, when designing the channel, it is recommended in [113,114] to design convergent, circular channels with taper angles no larger than 30° and convergent annular channels with taper angles less than 25° and differences between outer and inner angle up to 15°. In-line mandrel dies and side-fed dies are used as melt-distributing systems in parison extrusion. For coextrusion blow-molding equipment, spiral mandrels are also often selected for that purpose [115].

When considering side-fed mandrel dies as existing systems producing either single or multiple streams, which subsequently overlap [75], it can be concluded that variations frequently used in the past with a circumferential ring-shaped groove of a large diameter followed by a flow channel with a high flow resistance (principle of overflowing dam) do not meet today's quality requirements [75].

The situation is different when one looks at the side-fed mandrel dies with heart-shaped manifolds. In this case, the melt is divided into two or four partial streams, which are then lead by means of an empirically determined distribution curve around the mandrel and then partially overlap (offset heart-shaped manifold) [93,116,117]. The overlapping has the effect that the flow marks, which are formed automatically in this system, do not penetrate through the entire thickness of the wall. This weak spot can be further reduced when two manifolds shifted 180° apart produce an asymmetrical wall thickness that is thinnest at the point where the bonding seams form. The overlapping of both melt streams must be arranged in such a way that the resulting wall thickness is uniform along the entire circumference.

Classical parison dies include mandrels with coathanger manifolds and a land (Fig. 5.41), allowing the formation of a parison that exits the die with a uniform wall thickness and velocity around the circumference. The negative effect of the flow marks resulting in this process can be offset by the method described above, namely by the use of a concentric arrangement of two manifold systems [75]. Modern parison dies have spiral mandrels to avoid weld lines.

Figure 5.50 depicts an interesting variation, namely a parison head that has two coathanger manifolds attached to a mandrel (Fig. 5.51). Each manifold sweeps across 180 degrees. The parison produced by this method has two flow marks, which are deflected tangentially by slanted flights attached to the mandrel on its circumference. This way a continuous weak spot in the axial direction is eliminated [75].

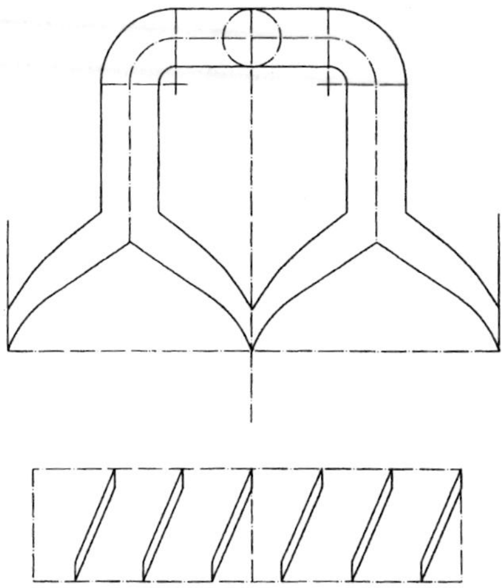

**Figure 5.50** Blow-molding head, BKSV model [Bekum]

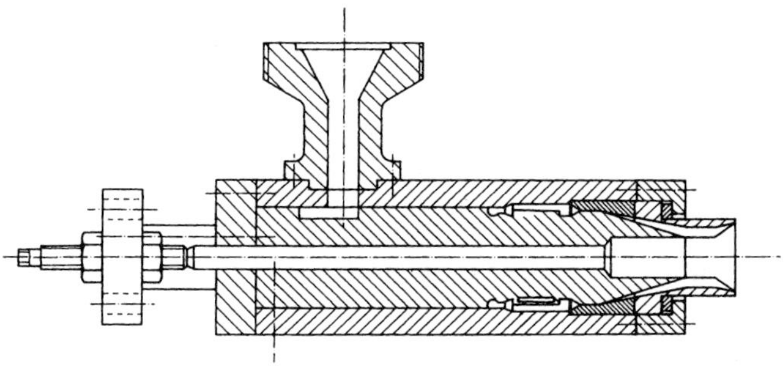

**Figure 5.51** Melt distributor system: double coathanger with attached guiding lands [Bekum]

Mandrel support dies for blow molding are most commonly set up with a support ring and offset spider legs (Fig. 5.52). These eliminate defects extending throughout the wall, and, being center-fed, they present no problems with the melt distribution [94,116]. When designing the above dies, it is important to match the separate streams accurately in order to avoid partial advance of the melt, which can lead to fluctuations in wall thickness and to surface defects [94].

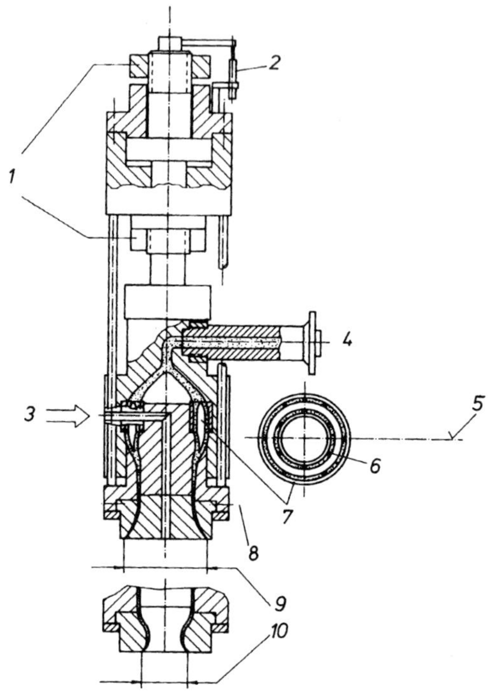


**Figure 5.52** Parison die with a support ring with offset spider and offset spider legs [Battenfeld-Fischer]: 1: stroke limiter; 2: motion pickup; 3: supporting air; 4: extruder connection; 5: die dividing plane; 6: spider leg; 7: mandrel support (spider); 8: nozzle centering; 9: maximum nozzle diameter; 10: minimum nozzle diameter

The advantages and disadvantages of these mandrel support dies with offset spider legs are discussed in [116]; disadvantages are the high pressure drop and the related high shear deformation. When processing thermally sensitive materials like PVC, it is recommended to use parison dies with a mandrel support having two spider legs (torpedo), as shown in Fig. 5.53 [118]. By this design the zones with detrimental flow can be eliminated to the highest degree. Moreover, it is possible to position both spider legs in the farther-located split plane of the mold located downstream. Thus, the flow lines are hardly noticeable in the finished product [118].

The pressure losses occurring in the mandrel support dies are examined in [119].

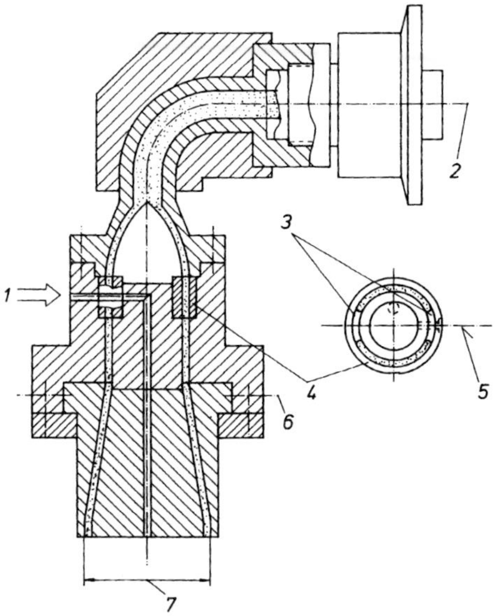


**Figure 5.53** Parison die with two spider leg mandrel support [Battenfeld-Fischer]:
1: supporting air; 2: connection extruder;
3: spider legs; 4: mandrel support (spider);
5: die dividing plane; 6: nozzle centering;
7: maximum nozzle diameter

### *Parison dies for the continuous forming of preforms*

Both previously described side-fed mandrel dies and in-line mandrel dies in widely varying geometries are used in this application. When blow-molding articles with a volume of 5 to 10 liters, multiple heads are used [120] that produce up to eight parisons simultaneously fed by a single extruder. The major problem here is the distribution of the melt into the individual dies with often different local material and die temperatures as well as flow resistances [121], which result in different exit velocities of the individual parisons. In addition, cycle times are increased by wide forming dies and, hence, longer traverse paths. In multiple heads, side-fed dies are frequently used as melt distributors; because of their compact design, they allow for smaller axial distances than in-line mandrel dies [75].

### *Parison heads for discontinuous preparation of preforms*

These parison heads, also referred to as accumulator heads, are used in cases where a large amount of melt is needed for sizeable parisons and when the melt strength of the material used is so low that the parisons would draw down and neck or even tear off if the continuous process is used. Another factor is the danger of an excessive cooling of the parison during the continuous process.

Modern accumulator heads have volumes from 1 to 400 liters [75]; for example, in the processing of polyethylene, accumulator heads are used starting with the melt consumption of 2.5 liters per shot. There are many accumulator systems of different designs in use [113,116,117,122]. The ram accumulator heads are considered to be state of the art [75]. They contain a ram serving the purpose of ejecting the melt (Fig. 5.54).

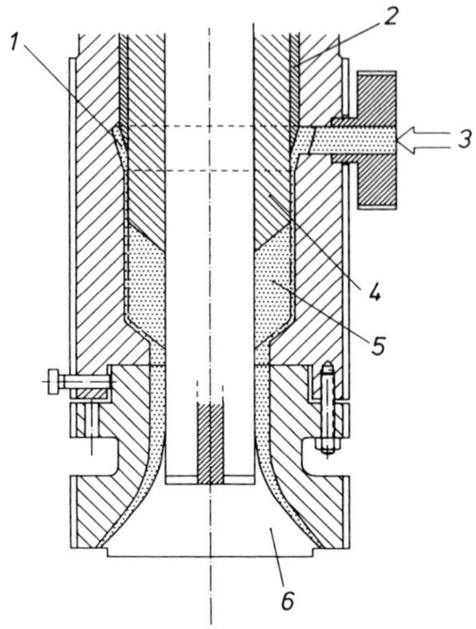


**Figure 5.54** Blow-molding head with an annular ram accumulator: 1: annular groove; 2: sleeve; 3: melt feed; 4: annular ram; 5: accumulator; 6: adjustable mandrel

Moreover, all modern designs guarantee the FIFO principle (first in, first out), which means that the melt entering the accumulator first is also ejected first. This way the residence time is kept to a minimum. The ejector piston is moved upward by the pressure of the melt during the filling phase.

The differences between the various designs are in the melt injection and distribution, located above the accumulator.

Thus, Fig. 5.55 (left) shows a parison die with a mandrel fed from two sides as well as a sleeve pulled down far guiding the ram [120]. On the right side of Fig. 5.55 there is a manifold system with two offset heart-shaped curves that bring about the overlapping of the melt streams. The advantage of this design, too, is that the surface of the ram is constantly flushed by the melt.

Figure 5.56 shows the so-called two-tier heart-shaped curve system [123], which represents the further development of the bolt-type mandrel support described in [116]. Here two melt tubes are formed in two separately fed planes with heart-shaped distributors and subsequent ring channels. These two tubular melt streams combine downstream, while the flow marks caused by the heart-shaped curves are opposite one another. Sections through the two distributor planes with the heart-shaped inner torpedo are shown in [123].

In all types of parison dies, the flow channel can have widely varying geometries: cylindrical, divergent, or convergent. In convergent and divergent orifices, the widths of the exit gap can be altered by a relative movement of the mandrel and the die in the axial direction while the parison is emerging. In crosshead parison dies, the last mandrel section is moved, while in in-line parison dies, the last die section is moved.

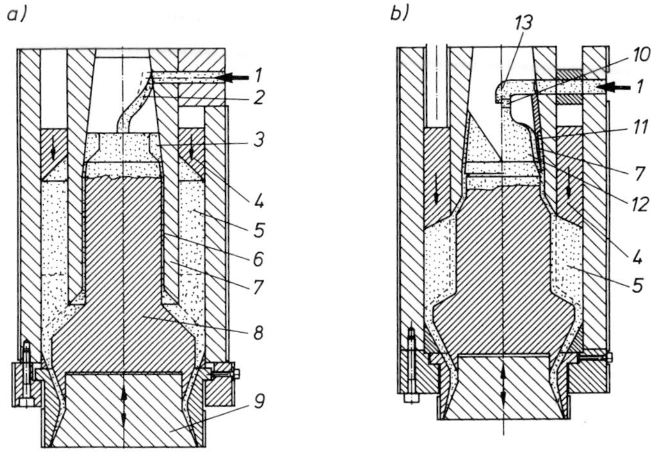


**Figure 5.55** Parison dies with tubular ram accumulator and different melt distribution systems [119]: (a) with an annular groove distributor, (b) with a double offset heart-shaped curve: 1: melt feed; 2: distributing channels; 3: ring channel (annular groove); 4: annular ram; 5: accumulator; 6: gap; 7: sleeve; 8: torpedo; 9: mandrel (adjustable); 10: aperture; 11: external melt stream; 12: internal melt stream; 13: end point of the forked section

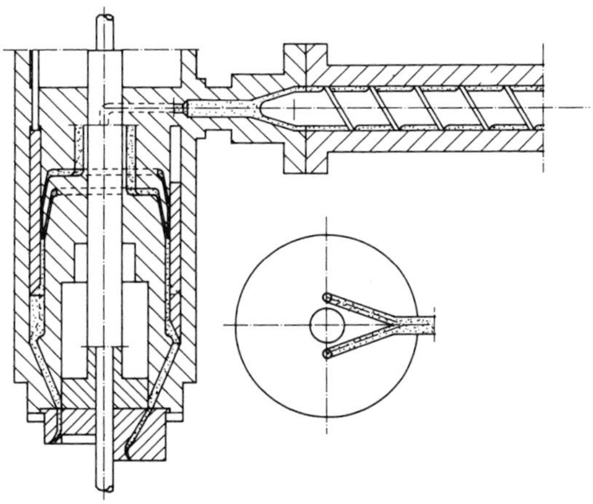

**Figure 5.56** Two-level heart-shaped distributor [Battenfeld-Fischer]

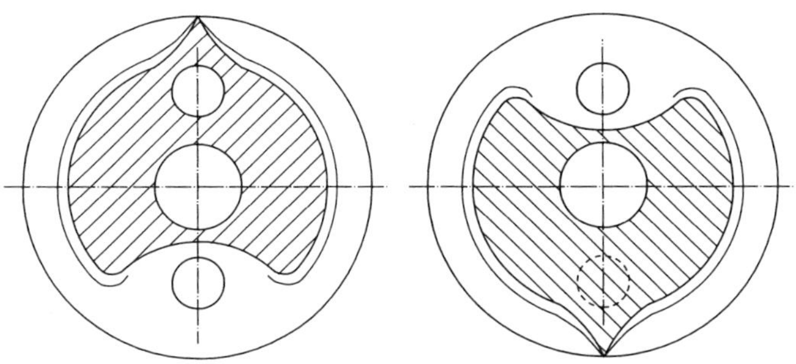

**Figure 5.57** Two-level heart-shaped distributor (schematic sectional view) [Battenfeld-Fischer]

As a rule, the adjustment is made hydraulically, while programming instrumentation is employed to predetermine the desired profiles of parison wall thickness [124]. For the production of hollow bodies with a complex geometry, so-called PWTC (partial wall thickness control) systems have found a wide use. These allow a control of the wall thickness along its circumference within certain limits [125,126]. If the use of this type of gauge control is not possible, the outside die rings are adjusted in their geometry, such as shaped radially [113,127].

The outside die rings usually have their own temperature control independent of the rest of the head and can be centered. The mandrels, as a rule, are not heated. The dimensions of the orifice are chosen with regard to the swelling behavior of the material to be processed [121,128]. The relationships used for the selection and computation of the geometry of the orifice with regard to the geometry and dimensions of the final product are discussed in [113,129].

#### 5.3.2.4 Coating Dies

Coating dies belong to the group of extrusion dies with an annular exit cross section. Their characteristic is that they apply a layer of melt to a semifinished product that is led through a core tube designed for this purpose. Typical applications for this process are coating of electrical conductors (cable coating), conductors of light waves (fiber optics), and pipes. The purpose of the coating is electrical insulation, surface improvement, an improvement in protection against corrosion, and so on [121]. The extruder in coating lines is usually positioned at an angle of 30° to 90° to the direction of the movement of the semifinished product, the perpendicular direction being predominant.

The following statements relate to the wire-coating dies because they represent the most natural application for the process. Figure 5.58 shows an example of the general features of a wire-coating die. In principle there are two distinct die designs for this coating process [16,130]: the pressure coating die and the tube coating die.

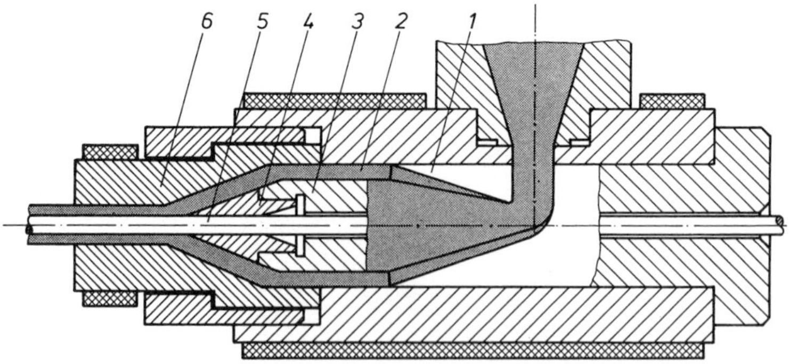


**Figure 5.58** Basic design of a wire-coating die: 1: manifold (here a heart-shaped curve); 2: annular slit; 3: torpedo (hollow mandrel); 4: mandrel orifice (torpedo tip guiding the wire); 5: wire; 6: die ring

#### *Pressure Coating Die*

For pressure coating, the conductor is enveloped by the melt under pressure within the die (Fig. 5.59, top). This procedure is preferred for the manufacture of cable coatings (primary coating), especially when coating stranded conductors where the melt should also penetrate between the individual conductor wires to avoid air inclusion. To further reinforce this, a vacuum may be applied.

Pressure coating has an additional benefit, namely an improved adhesion of the melt to the conductor [16].

#### *Tube Coating Die*

When using the tube coating die, a melt tube similar to that from a tubing die is extruded and applied to the conductor outside the flow channel by stretching due to the take-off speed and sometimes aided by the use of vacuum (Fig. 5.59, bottom). This method is often used for coating already-insulated cores, that is, for applying additional coatings (secondary coatings), and for coating metal pipes with polyethylene.

The advantage of this method is that the individual layers are truly coaxial [130].

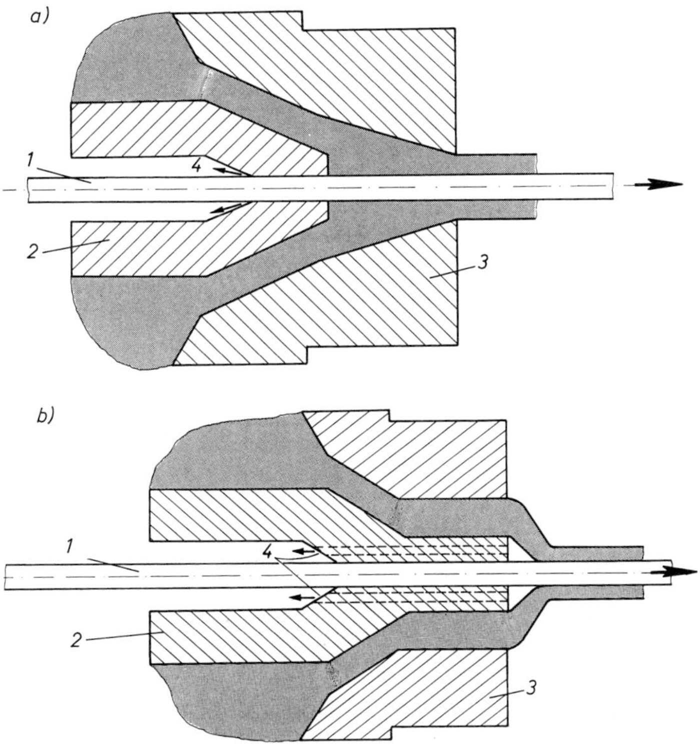


**Figure 5.59** Wire-coating dies: (a) for pressure coating (top), (b) for tube coating (bottom): 1: wire; 2: torpedo; 3: die body; 4: vacuum (if necessary)

With the applications previously discussed, in coating it is also important to achieve a uniform exit of the melt along the circumference by the use of an appropriate melt-distribution system.

To accomplish this, side-fed mandrel dies with heart-shaped curves and coathanger manifolds are predominantly used [131,132], where the latter, because of their good distribution characteristics and appropriate design, are preferred. Spiral mandrel dies have not been used, as far as it is known, but the use of spiral-shaped manifolds is described in the literature [92,114,133].

After the manifold system, the mandrel and the die housing taper toward the conductor (Fig. 5.58) in order to gradually increase the flow rate and match it with the speed of the wire.

The guide tip, as well as the die that can be centered and separately heated, can be exchanged to accommodate different conductors and provide a variety of coatings with different gauges. In order to prevent the melt from leaking into the mandrel core because of the sometimes considerable pressures that occur during operation of coating dies, the clearance between the conductor and the tip must be kept at a minimum. For the pressure coating, the clearance can be approximately 0.05 mm; for the tube coating between 0.2 and 0.3 mm [16].

Because of this small clearance and the high wire speeds, the guide tip is subjected to substantial wear. Therefore, it is often made with inserts made of diamond, aluminum silicate, or hard metal alloys [36,130]. The centering of the die ring is done by bolts. Only very small adjustments of the melt flow are necessary when the dies are properly designed because the throughput changes exponentially with the gap width by a power of up to three.

Figure 5.60 shows a precentered die, for which, as the name indicates, no additional centering is required [130]. It is claimed that a further advantage of this die is that the conductor is already surrounded by a thin layer of melt as it passes through the inner conductor guide. This reduces the wear of the guide and improves the adhesion between the conductor and the polymer.

The geometry of the orifice of the outer die ring affects the permissible maximum coating speeds as well as the surface quality of the extrudate. The latter is positively affected by less taper and longer die lands [134]. According to [114], die lands of 0.2 to 2 *D* are suitable for plasticized PVC (PVC-P) and 2 to 5 *D* for polyethylene. The product quality is influenced by the size of the die relative to the dimensions of the product to be coated. If the die is too large, the residence times are too long; if it is too small, the extrudate dimensional control can be adversely affected because the die provides an insufficient damping action between the screw and the die exit. In some wire-coating dies, the mandrel or the mandrel tip can be shifted in order to change the resistance to eliminate flow markings caused by the distributor system. As with pipe dies and blown film dies, only mandrels of large dies can have a separate temperature control.

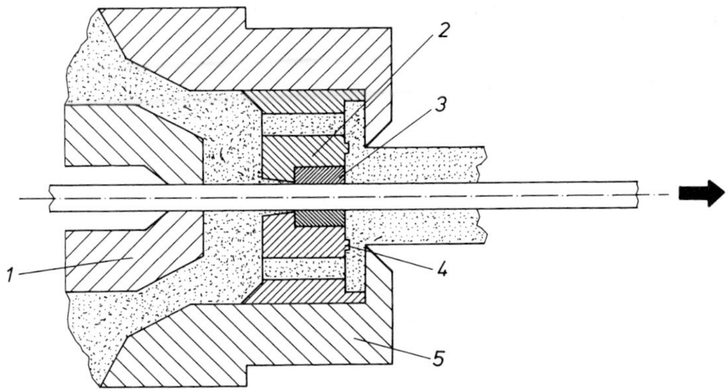


**Figure 5.60** Precentered pressure coating die [130]: 1: wire guide; 2: internal wire; 3: wear-resistant insert; 4: restrictor ring; 5: die ring

## 5.3.3 Design

When contemplating a rheological design of dies with annular cross sections, distinctions must be made between different types of dies.

In mandrel dies and screen pack dies, the melt distribution along the circumference is good as a result of the basic geometry. Important here are the total pressure loss and the sectional forces for the mechanical design; minimum and maximum shear stresses at the wall for the occurrence of stagnation zones; melt fracture, shear, and temperature history for the evaluation of the load on the melt; and, finally, the configuration of the mandrel support zone for the elimination of stagnation zones and flow instabilities.

The design of the mixing grooves is very similar to the design of spiral mandrel distributors and therefore will be discussed in Section 5.3.3.3.

The foremost objective in the rheological design of side-fed mandrel dies, of heart-shaped curve manifold dies, and of spiral mandrel dies is a good melt distribution at the die outlet, as is the case with slit dies.

For the jacketing dies for pressure coating especially, a rheological consideration pertaining to the coating zone is necessary.

### 5.3.3.1 Center-Fed Mandrel Dies and Screen Pack Dies

For center-fed mandrel dies and screen pack dies, generally the same rules apply. The only major exception is the calculation of the pressure loss for the screen pack; for that purpose, the procedure for the calculation of the pressure drop for breaker plates can be used.

In mandrel support dies, four different zones have to be considered (Fig. 5.37) [135]:

- The diverting zone, in which the melt stream emerging from the extruder is diverted 90° and is distributed by the mandrel tip into an annular cross section.
- The mandrel support zone, in which the mandrel is held by either spider legs or by a perforated plate.
- The relaxation zone, where the flow marks, caused by the spider legs, can be distributed and where the melt flow becomes uniform along the circumference.
- The land region, where the extrudate is shaped to the proper dimensions (diameter and wall thickness).

When designing the diverting zone, it is important to keep in mind that the length of the mandrel tip is in the same order of magnitude as the nominal diameter of the die, $D_m$, and that the increase in the cross-sectional area should be continuous along the flow path [135,136] in order to eliminate stagnation zones. For the design of the mandrel support, both rheological and mechanical factors are important. The goal in the design of this zone is to make the spider legs in such a way that they can resist the forces due to pressure and shear acting on the mandrel and that they have minimum effect on the flow. Therefore, the rheological and mechanical design factors are closely intertwined in this zone. The mechanical design of a mandrel die will be discussed in more detail in Section 9.2.

The process of the rheological design depends on the mandrel support system used. Examples of some usual types are shown in Fig. 5.39. The rules derived in Section 5.1.2 can be readily applied to the computation of flow in the area of the breaker plate mandrel support.

The computation of the flow in the spider-leg zone can be done by applying the principles for the computation of an annular gap or a flat slit. However, it is necessary to take into account the reduction of the actual cross section of flow by the spider legs. This can be done most advantageously by determining the flow in the spider region by multiplying the volumetric flow rate with the quotient from the cross section without the spider legs and the remaining area of the cross section (Fig. 5.61).

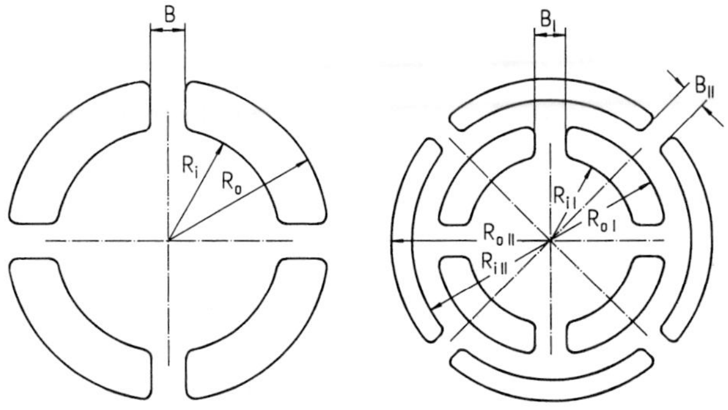


a) Spider leg mandrel support

b) Mandrel support with a supporting ring

**Figure 5.61** Spider geometries

The pressure drop in the mandrel support zone is

$$\Delta p_{\text{leg}} = f\left[\dot{V}^*, R_{\text{i}}, R_{\text{o}}, \eta(\dot{\gamma})\right], \tag{5.93}$$

$$K = \frac{\pi\left(R_{\text{o}}^2 - R_{\text{i}}^2\right)}{\pi\left(R_{\text{o}}^2 - R_{\text{i}}^2\right) - N_{\text{leg}} \cdot B \cdot \left(R_{\text{o}} - R_{\text{i}}\right)}, \tag{5.94}$$

$$\dot{V}* = \dot{V} \cdot K. \tag{5.95}$$

The flow situation is somewhat more complicated in the mandrel support zone. The inside and outside annular gaps have to be considered as two connected parallel flow resistances; it is not known beforehand how the volume stream will be divided between the two gaps. A computation of the flow distribution is possible by utilizing the condition that the pressure drop across the inside gap is equal to that of the outside gap:

$$\Delta p_{\text{I}} = \Delta p_{\text{II}}. \tag{5.96}$$

Also the following is true:

$$\dot{V}_{\text{I}} = \dot{V}_{\text{II}} = \dot{V}_{\text{total}}. \tag{5.97}$$

By combining Equations (5.94), (5.95), and (5.97) with (5.96), the flow rates are determined by iteration.

Applying the method of representative data for the computation of pressure loss in the annular cross section, the following relation is obtained:

$$\frac{\Delta p}{L} = \frac{8\eta_{\text{repI}} \cdot \dot{V}_{\text{I}} \cdot K_{\text{I}}}{\pi\left(R_{\text{o I}}^2 - R_{\text{i I}}^2\right) \cdot \overline{R}_{\text{I}}^2} = \frac{8\eta_{\text{repII}} \cdot \left(\dot{V}_{\text{total}} - \dot{V}_{\text{I}}\right) \cdot K_{\text{II}}}{\pi\left(R_{\text{o II}}^2 - R_{\text{i II}}^2\right) \cdot \overline{R}_{\text{II}}^2} \tag{5.98}$$

with

$$\bar{R} = R_o \cdot \left(1 + k^2 + \frac{1 - k^2}{\ln k}\right)^{\frac{1}{2}} \quad \text{and} \quad k = \frac{R_i}{R_o} \tag{5.99}$$

and with

$$\frac{\eta_{rep} K}{\left(R_o^2 - R_i^2\right) \bar{R}^2} \Big|_{I,II} = a_{I,II} \tag{5.100}$$

It follows:

$$\dot{V}_I = \dot{V}_{total} \cdot \frac{a_{II}}{a_I + a_{II}}. \tag{5.101}$$

Equation (5.101) can only be solved by iteration since the volumetric flow rates enter into $a_I$ and $a_{II}$ through the shear-rate-dependent viscosity.

This allows the computation of pressure losses, shear stresses at the wall, volumetric flow rates, and representative shear rates in the mandrel support zone and the evaluation of conditions in actual dies. When designing the relations of the slit heights of the inside and outside annular channels, it is appropriate to strive for a uniform distribution of the flow. The optimum location of the support ring, given its height, can be calculated from Equations (5.96) and (5.98). After solution of the problem, the following relationship is obtained:

$$R_{o\,I} = \sqrt{R_{i\,I}^2 + \frac{\eta_I \cdot K_I}{\eta_{II} \cdot K_{II}} \cdot \left(\frac{\bar{R}_{II}}{\bar{R}_I}\right)^2 \cdot \left(R_{o\,II}^2 - R_{i\,II}^2\right)} \tag{5.102}$$

Since $R_{o\,I}$ and $R_{i\,II}$ are related through the height of the support ring and, further, since $\eta_{rep,I}$, $K_I$, and $\bar{R}_I$ depend on $R_{o\,I}$ and $R_{i\,I}$, Equation (5.102), too, can be solved only by iteration.

For the length of the spider legs, a value equal to 0.5 to 0.7 $D_m$ (based on strength studies) is recommended [135].

The melt deformations occurring in the mandrel support zone can be relaxed in the following relaxation zone. In this zone of continuous compression, with reduced flow cross sections, the flow disturbances caused by the walls of the spider legs can also relax. The usual ratios of channel cross-sectional area (immediately after the mandrel support) to the area of the die exit cross section are 5 to 7 [88] with the length of the relaxation zone of 2.5 $D_m$ [135].

From a design point of view, it is possible to attain the ratio of the maximum diameter of the outer ring orifice $D_d$ to its nominal diameter $D_m$ at 1.25 [88]; for screen pack dies the maximum value of $D_d/D_m$ is 1.4 [95].

In blow-molding dies, the melt elasticity often has a negative effect when the $D_d/D_m$ ratios are high. The melt tube shrinks considerably in dimension when emerging from the die.

The maximum values for $D_d/D_m$ depend on the type of the polymer used and its molecular weight (1 for PE-HD, 0.8 for high molecular weight PE-HD). When designing the geometry of the flow channel for the entire head, it is important that a stable melt flow is ensured and that no instabilities, such as those due to the loss of wall adhesion, occur. This is particularly critical for PE-HD; in this case the representative shear rate must not be higher than 30 $s^{-1}$ [88].

#### 5.3.3.2 Side-Fed Mandrel Dies

In side-fed mandrel dies, the melt is always supplied by the extruder to the die at an angle, mostly at 90°. This means that the melt must pass in an appropriate way around the mandrel in order to attain a uniform velocity at the exit of the die.

A simple solution to this requirement is to add a ring-shaped groove with a large cross section that runs around the mandrel and to attach a zone with an annular slit to it. This subsequent zone has to have a high resistance to flow (Fig. 5.62). This type of die can be seen as a T-manifold wrapped around a cylinder [137,138]. However, it provides a very poor melt distribution with a wide spectrum of residence times and is therefore not used in industrial applications.

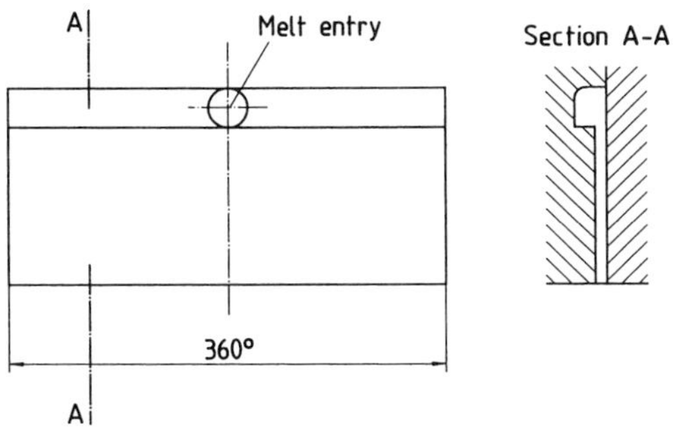


**Figure 5.62** Development of a flow splitter with an annular groove and a parallel die land

Another simple solution is to insert an annular torpedo around where the melt has to flow (Fig. 5.63). This ensures that the flow path of the melt entrance to the melt exit on the side of injection is equal to the flow path on the opposite side.

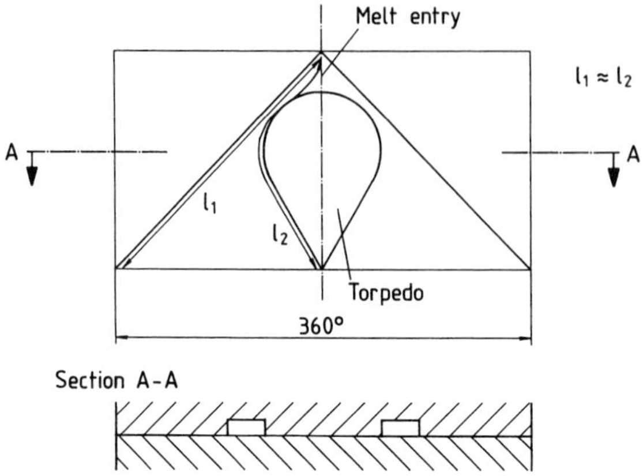


**Figure 5.63** Development of a flow splitter with a spreading body (heart-shaped manifold)

The distribution in this type of die with a heart-shaped manifold is poor; a good distribution can only be obtained by a subsequent region with parallel die lands with a high flow resistance [139]. Another disadvantage of this die is that two characteristic weld lines are formed because of the splitting of the flow.

The design of side-fed mandrel dies with a theoretically fully uniform melt distribution is possible if the design principles developed for fishtail and coathanger manifolds are applied. This is permissible as long as the mandrel diameter is substantially greater than the height of the gap in the land zone and the diameter of the distributor channel (Fig. 5.64).

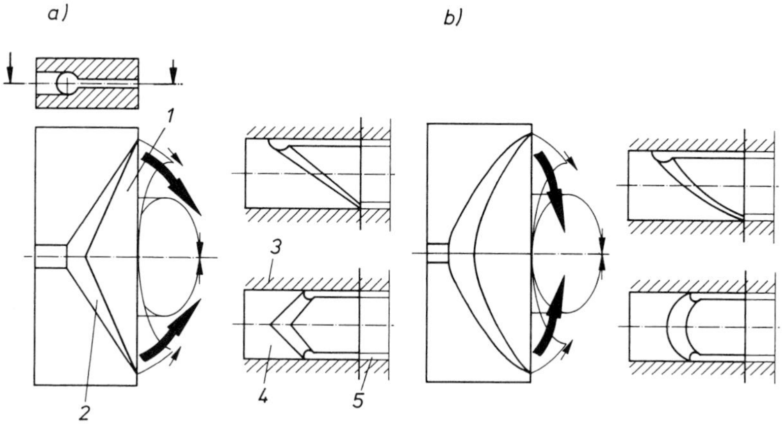


**Figure 5.64** Development of a (a) fishtail manifold and of a (b) coathanger manifold: 1: die land; 2: distribution channel; 3: housing; 4: mandrel; 5: parallel die land

In contrast to the wide flat slit manifold, the high pressure drop and resulting internal pressures can be handled more easily in crosshead dies, because the pipe-shaped housing only has to be designed for the internal pressure. This generally allows a manifold design independent of the operating conditions.

However, because of the resulting great maximum length of the lands resulting from the above design procedure, it is prudent to choose a design strategy that takes into account the exact length of the distribution channel. Generally, an exit section is incorporated at the end of a side-fed mandrel die where the melt stream is shaped to the desired diameter. A particularly space-saving arrangement with a low pressure consumption is a so-called conical flow splitter, which has the distribution channel wrapped around a truncated cone and not around a cylinder (Fig. 5.65).

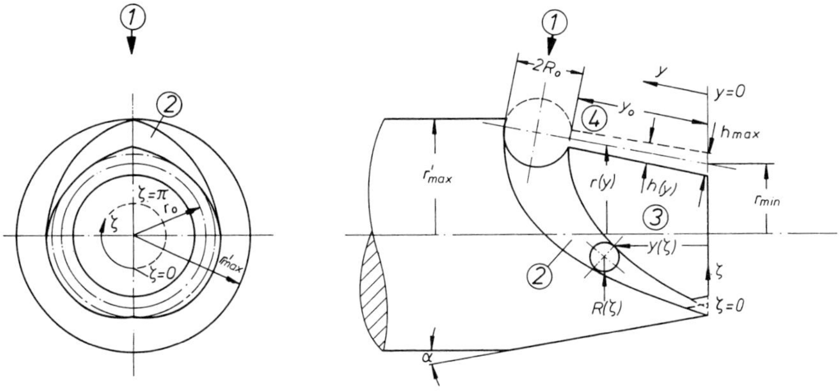


**Figure 5.65** Tapered mandrel with a coathanger manifold: 1: melt inlet; 2: distribution channel; 3: die land region; 4: slit

Since the mandrel is tapered, the distribution channel cannot be made with a constant slit height in the land region [140]. Rather, it is necessary to increase the land slit height in the direction of flow in such a way that the shear rate in the slit remains constant. From the relationship for the representative shear rate

$$\dot{\gamma}_{\text{rep,slit}} = \frac{6 \cdot \dot{V}_0}{\pi r(y) \cdot h^2(y)} \cdot e_\square \tag{5.103}$$

it follows with $r(y)$ and $r_{\min}$ that

$$h(y) = \sqrt{\frac{r_{\min}}{r(y)}} \cdot h_{\max} \tag{5.104}$$

$$r(y) = r_{\min} + y \cdot \sin \alpha. \tag{5.105}$$

For the design independent of the operating conditions, the maximum radius of the distribution channel $R_0$ at the point is

$$R_0 = 0.889 \cdot \left(r_{min} \cdot h_{max}^2\right)^{1/3}. \tag{5.106}$$

The maximum land length $y_0 = y(\zeta = \pi)$ on the surface line is

$$y_0 = h_{max}^{1/3} \cdot r_{min}^{1/3} \cdot \left(6.29 \cdot h_{max}^{1/3} \cdot \sin\alpha + 5.016 \cdot r_{min}^{1/3}\right). \tag{5.107}$$

The maximum radius of the mandrel $r'_{max}$ can be calculated from the above, thus:

$$r'_{max} = r_{min} + \left(2R_0 + y_0\right) \cdot \sin\alpha + \frac{h_{max}}{2} \cdot \cos\alpha. \tag{5.108}$$

This value is important for the estimation of the space required for the head.

The radius of the distribution channel as a function of distance is obtained from the requirement for the constant shear rate in the distribution channel:

$$R(\xi) = R_0 \cdot \left(\frac{\xi}{\pi}\right)^{1/3}. \tag{5.109}$$

For the course of the land length, the following relationship results:

$$y(\xi) = y_0 \cdot \left(\frac{\xi}{\pi}\right)^{2/3} \cdot \left(\frac{\pi^{1/3} \cdot \xi^{2/3} \cdot h_{max}^3 \cdot r_{min} \cdot \sin\alpha + 4R_0^4}{\pi h_{max}^3 \cdot r_{min} \cdot \sin\alpha + 4R_0^4}\right) \tag{5.110}$$

The pressure loss in the manifold and the mean residence time $\bar{t}$ can be computed from the following equations:

$$\Delta p_{total} = \frac{8 \cdot V_0 \cdot \eta_{rep,slit}}{\sin\alpha\pi \cdot h_{max}^3 \cdot r_{min}^{3/2}} \left[\left(r_{min} + y_0 \cdot \sin\alpha\right)^{3/2} - r_{min}^{3/2}\right] \tag{5.111}$$

and

$$\bar{t} \approx \frac{2\pi \cdot r_{min}^{1/2} \cdot h_{max}}{3\sin\alpha \cdot \dot{V}_0} \left[\left(r_{min} + y_0 \cdot \sin\alpha\right)^{3/2} - r_{min}^{3/2}\right]. \tag{5.112}$$

#### 5.3.3.3 Spiral Mandrel Dies

In contrast to the wide flat slit dies or side-fed mandrel dies with coathanger or fishtail manifolds, in the design of spiral mandrel dies it is not possible to find a procedure producing a suitable geometry from a specified flow distribution at the exit. This is further complicated by the fact that in a spiral mandrel die the spectrum of the residence times and the leakage volume flow along the spiral have to be brought in as additional design criteria.

In this case, the determination of the geometry by computation is not possible; therefore, the process has to be reversed: a flow simulation is performed for a given geometry, which provides the flow distribution along the spiral and the exit gap [81].

For the computation a number of assumptions have to be made:

- The flows in the spiral and the annular slit are developed and do not affect each other.
- The leakage flow, leaving the spiral, has no effect on the flow in it.
- The effect of curvatures in the mandrel is neglected (same consideration as for the problem above). This assumption is permissible because the diameter of the mandrel is considerably greater than the diameters of the channel and the annular slit.
- The spiral is divided into segments for the purpose of the computation. From the assumption that the isobars in the die run parallel to the exit gap, it follows (Fig. 5.66):

$$\Delta p_{\text{slit}} = \Delta p_{\text{spiral}}$$

With pressure gradients along the annular slit (land) and along the spiral, respectively, the following relationship results:

$$\frac{\mathrm{d}p}{\mathrm{d}z}\bigg|_{\text{slit}} = \frac{\mathrm{d}p}{\mathrm{d}l}\bigg|_{\text{spiral}} \frac{1}{\sin\phi}. \tag{5.113}$$

A good description of the geometry of the spiral distributor is crucial for the accuracy of the computation. A reasonable way to do that is the description of the geometry through the base points for the slope of the spiral, through the dimensions of the spiral channel and the height of the annular gap.

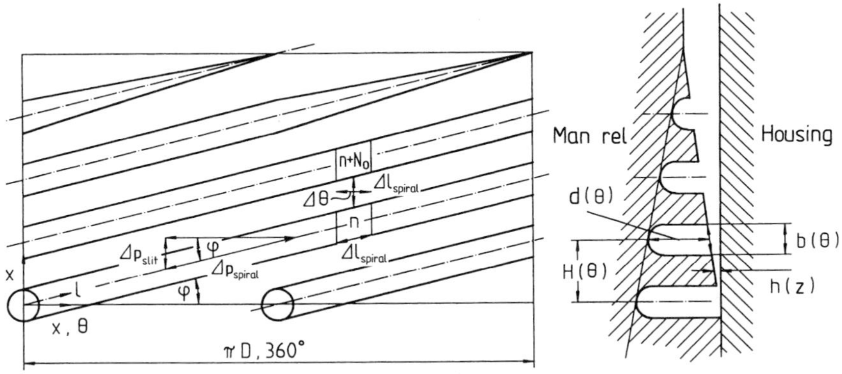


**Figure 5.66** Development geometry of a spiral distributor

In order to facilitate the establishment of the relationship between individual segments, it is recommended to express the position ($H(\theta)$), the width ($b(\theta)$), and the depth ($d(\theta)$) of the spiral in terms of the circumferential angle and for the height of the annular gap in terms of the mandrel height ($h(z)$). During the computation, a volumetric flow rate balance for each segment is taken (see Fig. 5.67):

$$\dot{V}_{\text{spiral}}(n)+\dot{V}_{\text{slit}}(n)=\dot{V}_{\text{spiral}}(n+1)+\dot{V}_{\text{slit}}(n+N_0) \tag{5.114}$$

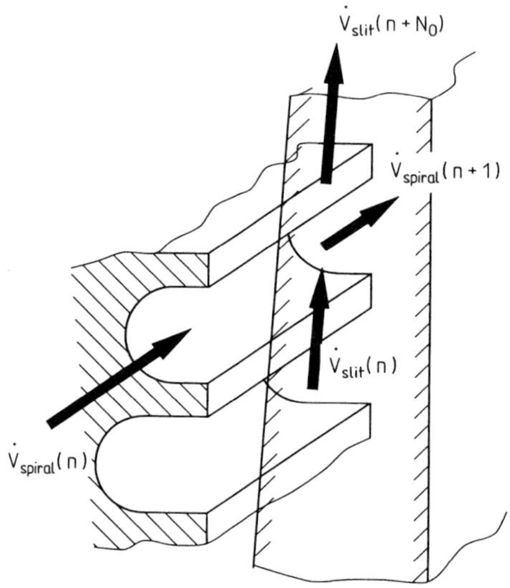


**Figure 5.67** Volumetric flow rate balance at a segment of a spiral distributor

with

$$N_0=\frac{360^{\circ}}{N_{\text{spiral}}}\cdot\frac{1}{\Delta\theta}. \tag{5.115}$$

The leakage volumetric flow rate $\dot{V}_L$ leaving the spiral corresponds to the difference between the outgoing and incoming volumetric flow rates through the slit:

$$\dot{V}_L=\dot{V}_{\text{slit}}(n+N_0)-\dot{V}_{\text{slit}}(n). \tag{5.116}$$

In the next step, the pressure loss in the spiral channel is calculated:

$$\frac{dp}{dl}\bigg|_{\text{spiral}}=f\left[\dot{V}_{\text{spiral}},b,d,\eta(\dot{\gamma})\right] \tag{5.117}$$

with

$$\dot{V}_{\text{spiral}}=\frac{\dot{V}_{\text{spiral}}(n)+\dot{V}_{\text{spiral}}(n+1)}{2}. \tag{5.118}$$

From this and Equation (5.113), the pressure loss across the slit area is obtained. From this pressure loss, the volumetric flow rate through the slit can be determined:

$$\dot{V}_{\text{slit}}\left(n+N_0\right)=f\left[\frac{\mathrm{d}p}{\mathrm{d}z}\Big|_{\text{slit}},h,\eta\left(\dot{\gamma}\right)\right]. \tag{5.119}$$

Because Equation (5.117) as well as Equation (5.119) cannot be solved analytically, the iterative procedure must be applied.

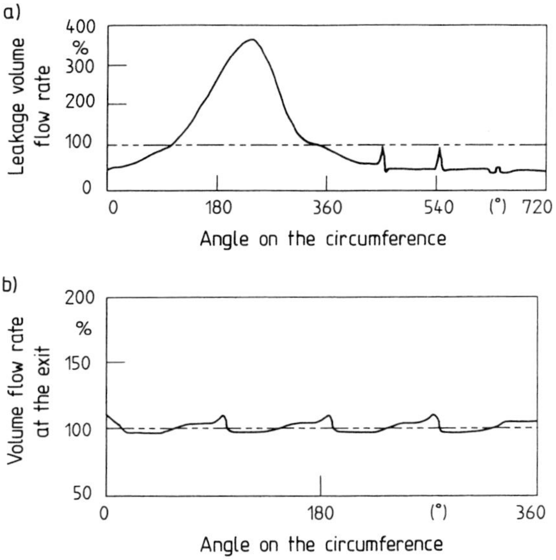


**Figure 5.68** Results from the computation of the flow simulation for a spiral distributor: (a) leakage volume flow rate, (b) volume flow rate at the exit (four spirals)

The computation starts at the beginning of the spiral, where the flow rate through the spiral (volume flow coming in from the extruder divided by the selected number of spiral turns) and the incoming flow rate through the slit are known. The latter is equal to zero up to the first overlap, that is, up to the beginning of the next spiral turn:

$$\dot{V}_{\text{slit}}\left(n\right)=0 \quad \text{for} \quad 1\leq n\leq N_0. \tag{5.120}$$

With this, the flow rate leaving the slit up to the first overlap $\dot{V}_{\text{slit}}\left(n\right)$ for $N_{0+1} \leq n \leq (N_0 + N_0)$ can be calculated.

These computed leakage flow rates exiting from the slit are used in the next step of the computation, the incoming slit flow.

This procedure is repeated up to the last overlap; here the exiting leakage flow equals the flow from the die. The typical results from the computation of the leakage flow and of the outgoing flow for a spiral distributor designed with four spirals are shown in Fig. 5.68.

The distribution of the volume flow at the die exit is very uniform; nevertheless there are still four maximums caused by the four spirals. The maximum deviations from the uniform exit volumetric flow rate for the case in question are +8.3 % and −4.2 %. Significant amounts of material in the mandrels at the end of the mandrel zone lead to pulsation marks and an inhomogeneous film thickness along the circumference.

For the design not only a uniform exit flow is important. Another factor of importance is the requirement that the leakage flow, as shown in Fig. 5.68 (top), increases steadily from one-half of the steady leakage flow (steady leakage flow equals the flow entering the spiral divided by the length of the spiral) until it reaches a maximum value of three to four times the steady leakage volume flow at approximately one-third of the spiral length; it then decreases steadily until it becomes zero at approximately two-thirds of the length of the spiral and then stays at this value until the end of the spiral [141]. The highest value of leakage flow is equivalent to maximum mixing.

In spite of the possibilities offered by the flow simulations for the design of spiral distributors, the experience of the designer is of considerable importance because of the many geometric variables (number of turns, spiral slope, spiral depth, slit height).

#### 5.3.3.4 Coating Dies

For coating dies, mainly side-fed mandrel dies with heart-shaped, fishtail, and coathanger distributors are used. Not only a good melt distribution is important, but also, particularly in pressure coating, a proper configuration of the die orifice, which is matched to the thickness of the coating and the speed of the wire or pipe being coated.

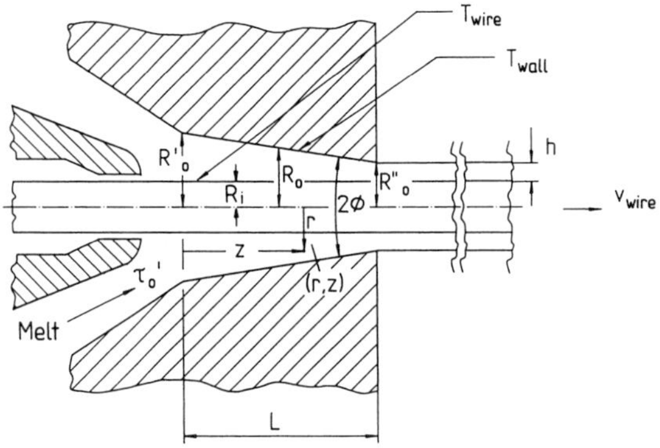


**Figure 5.69** Longitudinal section through a pressure coating die [142]

For the study of the flow and temperature conditions within a pressure coating die, the flow in an annular slit can be assumed. In the zone where the melt meets the wire until the exit from the die, the boundary condition at the inside diameter must be changed. Under the assumption of a good adhesion of the melt to the wire, a drag flow is induced by the moving wire and the viscosity of the melt. A longitudinal section through a pressure coating die for a high-speed coating is shown in Fig. 5.69, for which the die was subjected to computations of flow and temperature fields using a finite difference program [142]. The computed velocity and temperature profiles at different points in the die for a PE-LD are shown in Fig. 5.70.

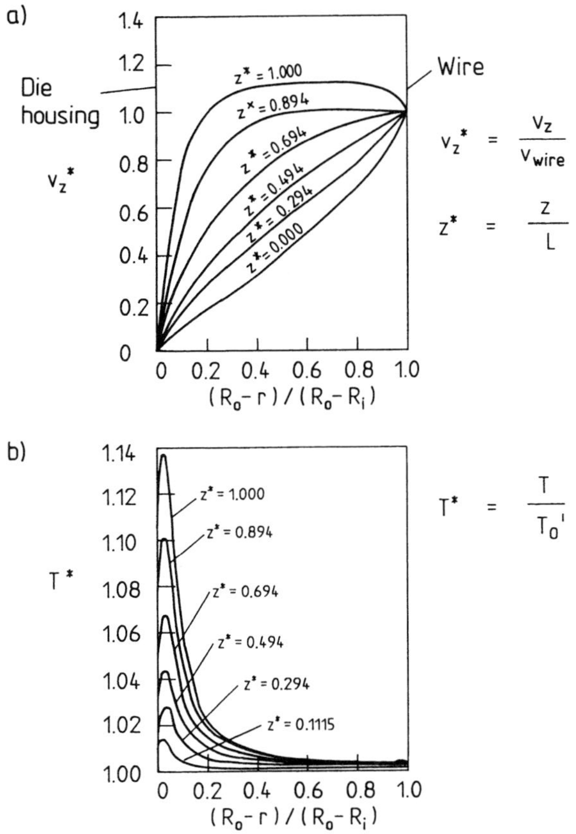


**Figure 5.70** Results from flow and temperature computations for a pressure coating die: (a) velocity profile, (b) temperature profile

The strong influence of the drag flow is clearly seen in the region where the melt meets the wire ($z^* = 0$). Toward the die exit this effect decreases when compared to the pressure flow resulting from the decreasing channel cross section. At the die exit ($z^* = 1$), the flow velocity has its maximum within the flow stream.

The strong increase in temperature at the die wall is caused by the high shear rate at the wall.

An FDM computation of the flow conditions for very high speeds of the wire showed that the drag flow caused by the wire can have repercussions on the velocity profile before the point where the melt meets the wire [143]. The geometry of the die under study is shown in Fig. 5.71.

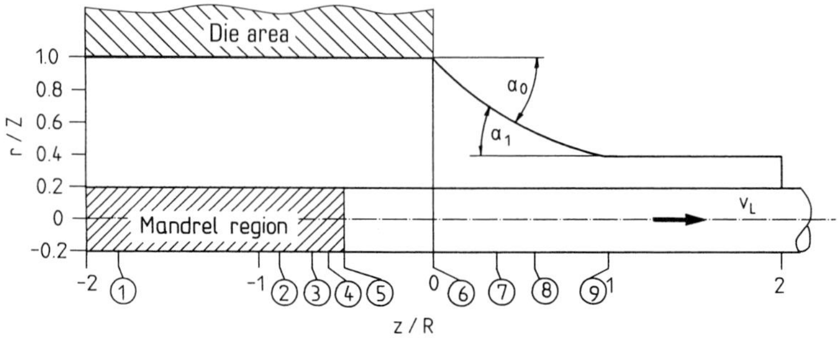


**Figure 5.71** The geometry of the wire-coating die under study [183]

The investigation was done on a melt with constant viscosity. It was assumed that the takeoff speed of the wire is 15 times greater than the mean velocity of the melt at the entrance to the die, $\bar{v}_z$.

The computed velocity profiles for different sections along the coating die are depicted in Fig. 5.72.

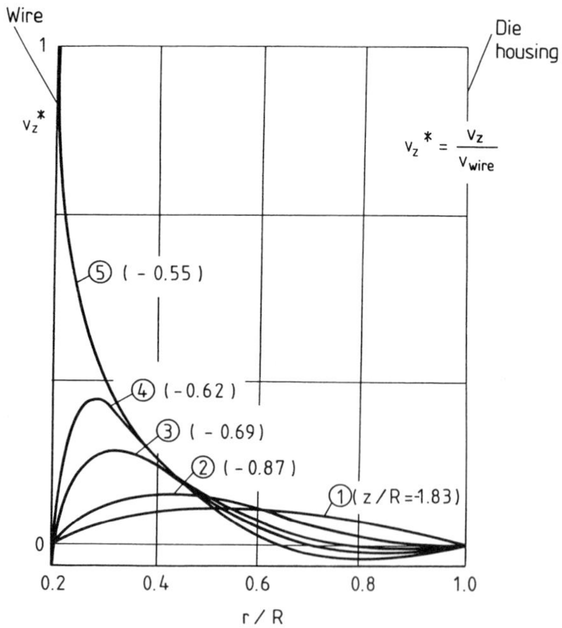


**Figure 5.72** Velocity profiles in the wire-coating die (section cuts 1 to 5, see Fig. 5.71)

It is obvious that already in the mandrel zone ($z/R < -0.55$) there is an effect of the drag flow. The strong effect of the drag flow causes negative components of flow in the outside region of the annular slit.

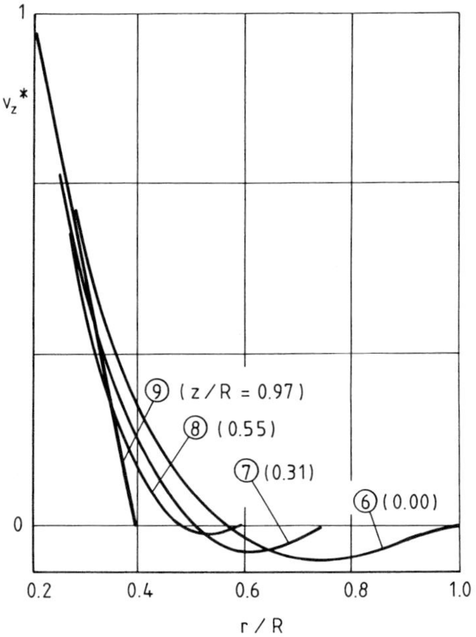


**Figure 5.73** Velocity profiles in the wire-coating die (section cuts 6 to 9, see Fig. 5.71) [183]

This yields the shape of the streamlines in the die shown in Fig. 5.74. The counterflow leads to the formation of a vortex in the flow channel. An additional result is the dimensionless pressure profile in the die as a function of the drag number $S$ (Fig. 5.75). The drag number is equal to the ratio of the haul-off speed of the wire $v_{wire}$ and the mean velocity of the melt at the die entrance $\overline{v}_z$.

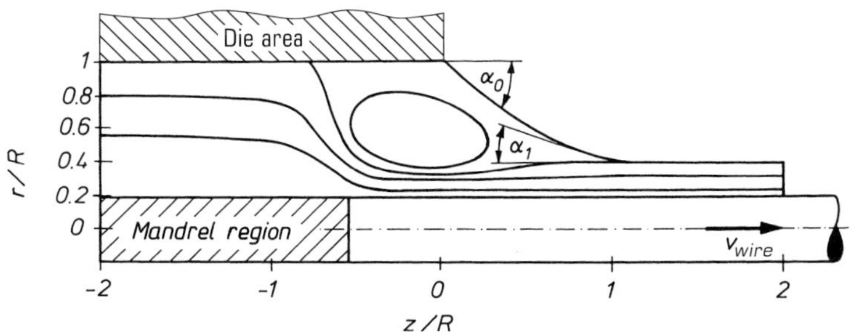


**Figure 5.74** Streamlines in a wire-coating die

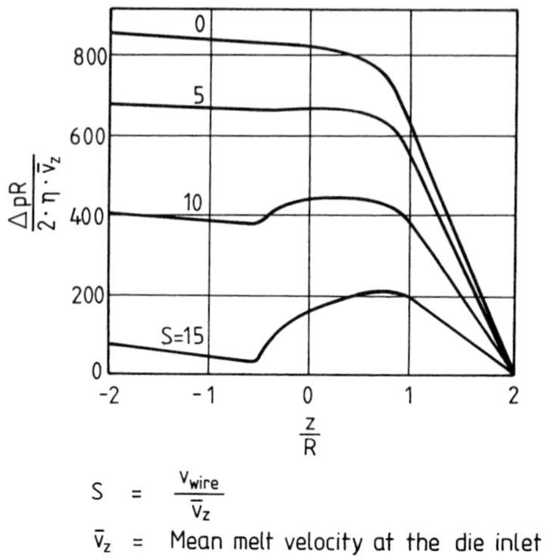


**Figure 5.75** Dimensionless pressure profile in a wire-coating die as a function of the drag number *S*

For high drag numbers, a pressure gradient against the direction of flow can be observed in the region where the melt meets the conductor. The computation of flow can provide important information, such as the temperature distribution and the resulting load on the melt, the pressure drop of the die, and the portion of the wire haul-off force that is caused by the viscous stresses.

## 5.4 Formulas for the Computation of the Pressure Loss in Flow Channel Geometries other than Pipe or Slit

The formulas for the design of dies introduced in Chapter 5 are applicable in principle to ideal one-dimensional flow channel geometries (pipe, flat infinite slit, annular slit). The above flow channel cross sections are not, in many cases, suitable for manufacturing purposes, however. Therefore, in the following sections, several formulas for the computation of flow in other flow channels will be introduced.

Another problem is that often such flow channels are needed where the area of the cross section changes in the direction of flow. Such a case was already presented in Chapter 3, where the procedure of dividing convergent or divergent flow channels into segments of constant cross-sectional area for the purpose of computing the corresponding pressure loss was applied.

The principal shapes of distribution channels are slits with a small width-to-height ratio, rectangular with one semicircular side, teardrop shape, and half teardrop (Fig. 5.76).

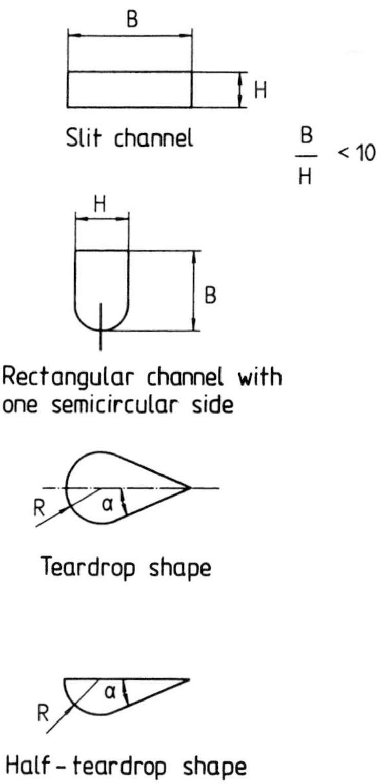


**Figure 5.76** Shapes of flow channels suitable for milling

There are a variety of formulas available for the computation of the pressure loss for the above geometries:

1. Formulas covering simple flow channel geometries with equal area.
2. Formulas covering simple flow channel geometries with constant volume/area or circumference/cross section ratios (equal hydraulic diameter):

$$d_{\text{hyd.}} = \frac{4 \cdot A}{U}. \tag{5.121}$$

3. Formulas dividing nonideal cross sections into ideal cross sections that can be subjected to computation.

   This method can be used for the computation of pressure loss of slit channels with an edge effect or rectangular channels with one semicircular side. The method of dividing a nonideal cross section is demonstrated in Fig. 5.77. The foundation of this procedure is in the study of isovels (lines of equal velocity). In order to take the edge effect into account, the flow channel borders are replaced with semicircles with diameters equal to the height of the flow channel.

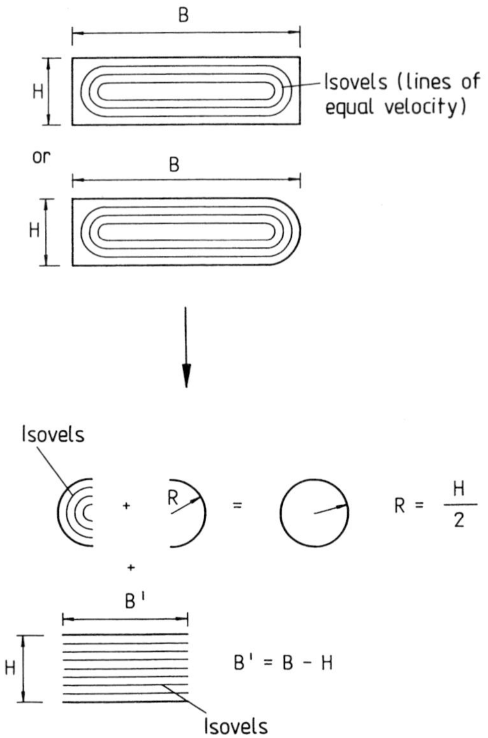


**Figure 5.77** Analysis of a nonideal cross section for the computation of the pressure loss

In order to explain the procedure, the derivation of the pressure loss using the method of representative data will be briefly demonstrated.

The pressure loss is equal in both substitution cross sections:

$$\frac{dp}{dx} = \frac{8\eta_{rep,pipe} \cdot \dot{V}_{pipe}}{\pi \cdot R^4} = \frac{12\eta_{rep,slit} \cdot \dot{V}_{slit}}{B' \cdot H^3}. \tag{5.122}$$

Furthermore, the following is true:

$$\dot{V}_{pipe} + \dot{V}_{slit} = \dot{V}_{total}. \tag{5.123}$$

Now the relationship between the total volumetric flow rate $\dot{V}_{total}$ and the pressure drop is required. After combining Equations (5.122) and (5.123), and then rearranging, the following equation results:

$$\frac{dp}{dx} = \frac{\dot{V}_{total}}{\left(\frac{\pi \cdot R^4}{8\eta_{rep,pipe}}\right) + \left(\frac{B' \cdot R^3}{12\eta_{rep,slit}}\right)}. \tag{5.124}$$

There are several ways to determine the viscosities $\eta_{pipe}$ and $\eta_{slit}$.

The viscosities can be determined separately for the two substitute geometries.

For reasons of continuity, the velocity and viscosity profiles on the boundary between the semicircle and slit substitute geometries must be equal.

Therefore, it is permissible to start with the same viscosity in both substitute cross sections. The solution of Equation (5.122) can be obtained by iteration.

4. Computation of the pressure losses of the nonideal flow channel geometry by means of the finite difference method (FDM) or finite element method (FEM). These procedures are very accurate.

5. The use of correction factors, called flow coefficients, for the computation of the pressure loss. These coefficients are useful for the correction of results per methods 1 or 2 and must be first determined as functions of the flow geometry, the melt properties (flow exponent), and the process conditions, for example per (3) or (4) or by experiments.

   The application of the flow coefficients has the advantage that similarities between cross sections, for which the pressure loss is being determined, can be utilized.

   The determination of the flow coefficient function based on the results from (3) will be shown below.

   When the result from (3) is related to the pressure loss of a one-dimensional flow through a slit having a constant height/width ratio, the flow coefficient $f_p$ results:

$$f_{pipe} = \frac{\frac{dp}{dx}\Big|_{slit}}{\frac{dp}{dx}\Big|_{substitute\,geometry}}. \tag{5.125}$$

When setting

$$\bar{\eta} = \eta_{rep,slit} = \eta_{rep,pipe} \tag{5.126}$$

and combining Equations (5.124) and (5.126), the result is

$$f_{pipe} = 1 + \frac{H}{B} \cdot \left(\frac{3\pi}{32} - 1\right) \quad \text{for} \quad H/B \leq 1 \tag{5.127}$$

and

$$f_{pipe} = 1 + \frac{B}{H} \cdot \left(\frac{3\pi}{32} - 1\right) \quad \text{for} \quad H/B > 1. \tag{5.128}$$

The flow coefficient is now dependent only on the height/width ratio of the cross section under consideration. For very small or very large *H/B* ratios, the flow coefficient approaches unity. In such cases the edge effect may be neglected (Fig. 5.78).

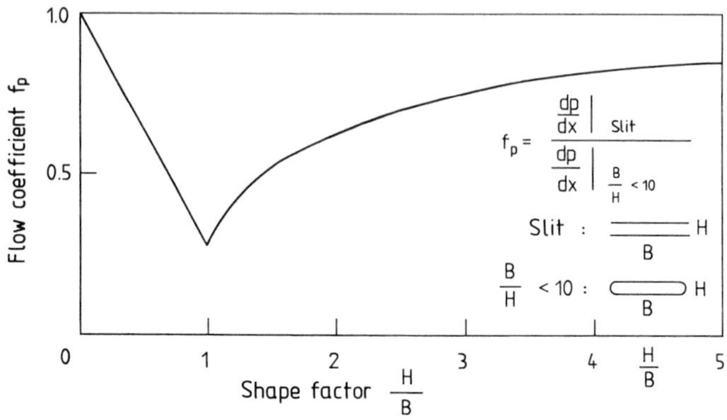


**Figure 5.78** Flow coefficient function for the computation of pressure loss in slits with the consideration of the edge effect (width/height < 10)

A flow coefficient function determined by the application of FEM is depicted in Fig. 5.79 [144]. Here the pressure loss in a rectangular channel with one semicircular side was set into a relationship with a pressure loss of a pipe with an equal hydraulic diameter.

The studies involving computations with FEM show that the flow coefficients are independent of the material, the operating conditions, and the absolute value of the rectangular cross section with one semicircular side. Considering the above, it appears that determination of the flow coefficient by means of FEM computations is the most reasonable way.

Performing the computations requires a great deal of work at first; however, the actual computation of pressure losses from the thus-determined flow coefficient function is fast and accurate.

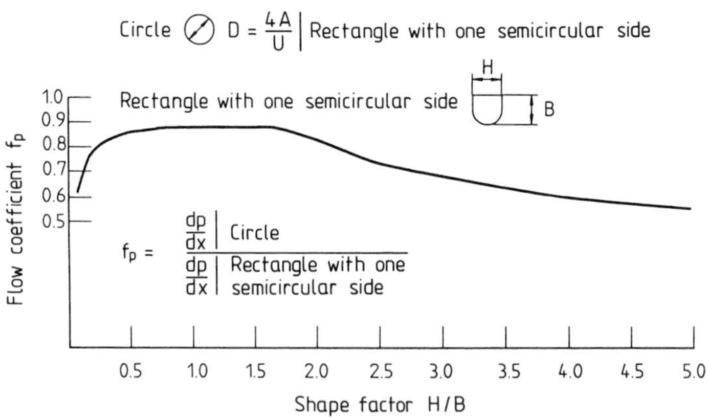


**Figure 5.79** Flow coefficient function for the computation of pressure loss for rectangular flow channels with one semicircular side

# 5.5 Dies with Irregular Outlet Geometry (Profile Dies)

The term "profiles" has become well established in the rubber and plastics processing industry for extruded semifinished products with an irregular cross section, although generally this term also includes simple extrudates, such as pipes, slabs, and solid bars. Nevertheless, this practical definition of profiles will be adopted and will cover all products extruded from dies with other than annular or rectangular shape.

The production of profiles, which have exact measurements and shapes and a relatively low cost, is the most difficult task in extrusion technology [20,145–147]. The reason for that is the seemingly unlimited variety of profiles with different shapes and sizes and the difficulty, sometimes even impossibility, of designing the required dies using relatively simple and proven theoretical computations and studies. It should be stated then that the design of dies of many profiles still depends to a high degree on empirical procedures and requires a great deal of experience on the part of the die designer.

## 5.5.1 Designs and Applications

In view of the large number of profile geometries, a rough division can be made between open and hollow profiles.

Hollow profiles, as their name already indicates, contain closed hollows, as is the case with window frame profiles. Conversely, a U-profile is an open profile [20,145,148].

An additional, more detailed classification is given in Fig. 5.80 [149,150]:

- Tubular profiles with relatively uniform wall thickness and rounded corners
- Hollow profiles with an outer wall that encloses hollow chambers, while the walls can have different gauges, and also sharp corners and fins can be present.
- Profiles with hollow chambers and legs that can point to the outside; different wall thicknesses are possible
- Solid profiles without chambers; arbitrarily shaped cross sections are possible
- Composite profiles in all the above forms, while the individual sections are made from the same material with different color or properties (e.g., hardness) or from a different material altogether.
- Core profiles as hollow profiles with a core of a different material (e.g., steel, wood).

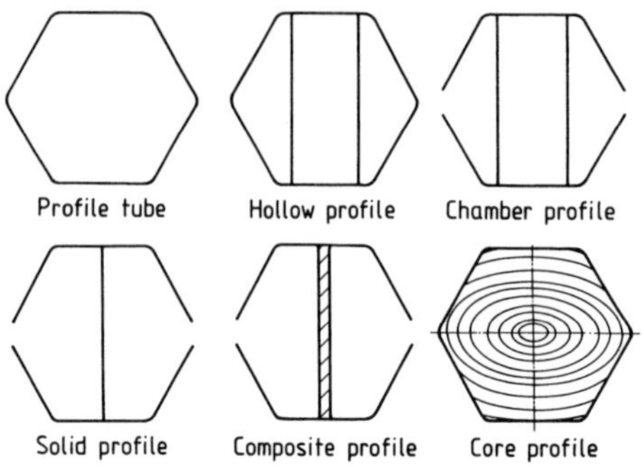


**Figure 5.80** Profile types [149,150]

Almost all profile dies are fed axially, and the profile, still in the molten stage, exits with an approximate contour of the semifinished product. In the design of the flow channel and of the cross section of the die orifice, swelling (recovery of reversible deformations stored in the material) and shrinkage (volume contraction of the material during subsequent cooling), the rearrangement of the velocity profile in the die orifice to the block-shaped profile on the outside, and the so-called draw-down must be taken into consideration [143,145,151,152].

Generally, profile dies can be divided into three groups [20,145,148,153]:

- Plate dies
- Stepped dies
- Profile dies with a gradual change in cross section

### *Plate Dies*

Plate dies (Fig. 5.81) consist of a die base with a die plate, which can be exchanged easily and quickly. These dies are mostly used for small profiles. There are very sharp changes in the flow channel geometry in this type of die. This can lead to dead spots that can—especially in the processing of rigid PVC (PVC-U)—result in degradation of the polymer.

Furthermore, it is not possible to attain high extrusion speeds [20,148] or to achieve a high dimensional accuracy because of the precipitous narrowing of the cross section. Therefore, even though they are easily manufactured at low cost, these dies are seldom used in the processing of plastics. Their use is restricted primarily to the extrusion of plasticized PVC (PVC-P) or the simplest profiles from rigid PVC (PVC-U) when only short runs are required [20,148]. However, these dies are still widely used for the processing of rubber.

From the point of view of flow, the die is designed strictly empirically; it is important to make sure that the plate is thick enough (approx. 5 to 20 mm [154]) in order to allow local flow adjustments to be made, done by changing the land length (length of the flow resistance zone) [20,148].

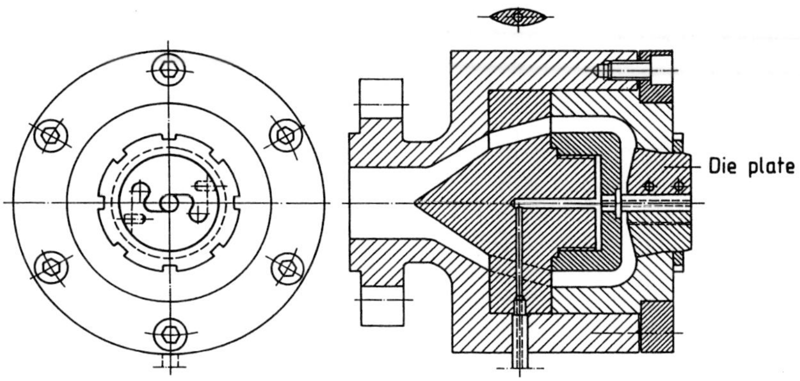


**Figure 5.81** Plate die (profile die) [145]

To avoid deformation by the pressure of the melt, the die plate is sometimes reinforced by welding on bridges (Fig. 5.82) [154].

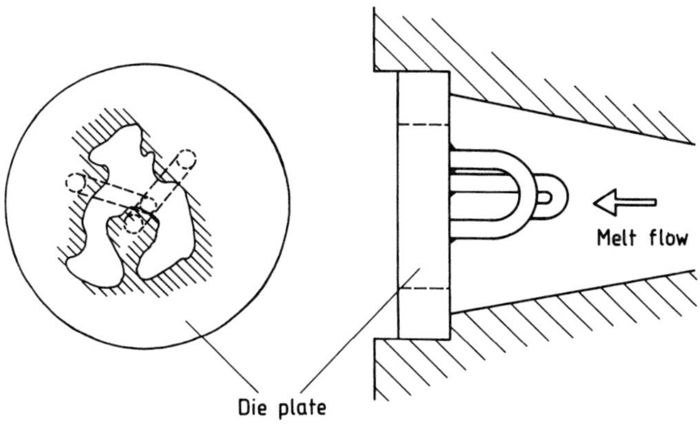


**Figure 5.82** Reinforcing bridges on a die plate [154]

Because the overall concept of plate dies is rather poor from the point of view of flow, close attention must be paid to accurate temperature control in the die [154].

The insert dies, described in [154] and depicted in Fig. 5.83, are derived from plate dies. Exchangeable inserts, which are rheologically considerably more advantageous, are used in the insert die.

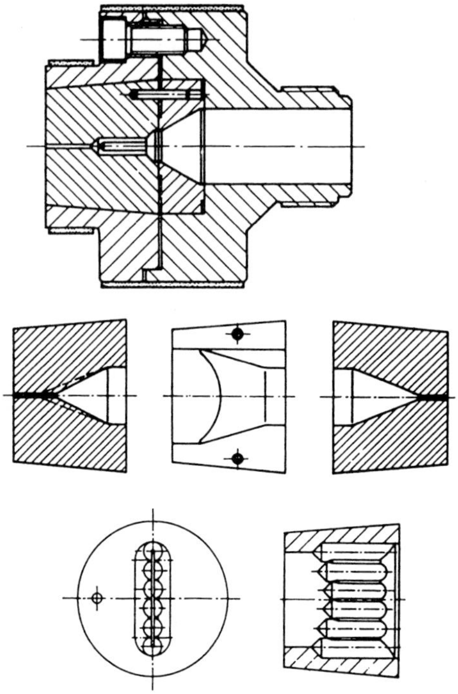

**Figure 5.83** Modular die with die inserts for profile extrusion [155]

### *Stepped Dies*

Stepped dies (Fig. 5.84) show a stepwise change in the flow channel. They are formed by connecting several die plates of a plate die in series. The contour is worked out in each disk; only at the inlet of each disk are the edges beveled.

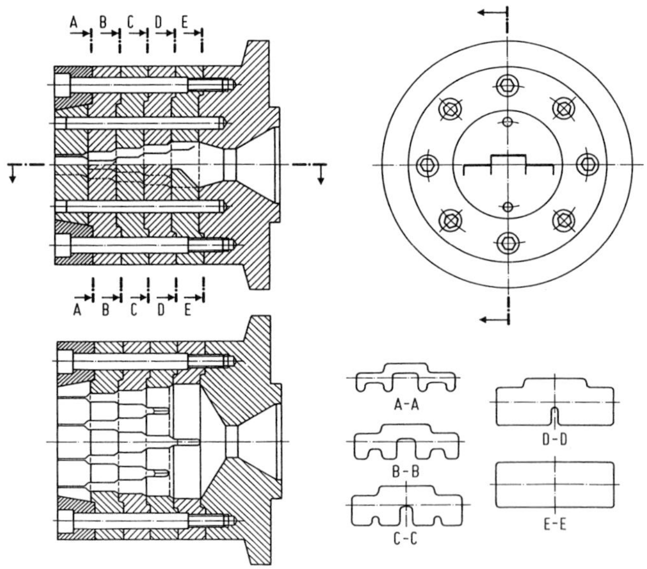


**Figure 5.84** Multistage die [145]

These transitions, too, are critical in processing of rigid PVC (PVC-U); thus, the comments made about the plate dies are applicable to the process behavior and the use of multistage dies as well. Therefore, these dies are used for simple profiles only [20,148].

#### *Dies with a Gradual Change in Cross Section*

These dies are always required whenever profiles with a high dimensional accuracy are produced at high extrusion speeds (Fig. 5.85). Therefore, the following have to be taken into consideration in the determination of the flow channel geometry [145]:

- The flow channel must not have any dead spots (stagnation zones)
- From the time it enters the die until it leaves, the melt must be accelerated as steadily as possible until it reaches the desired exit velocity in the die land, just before the exit. Decelerations, or increases in cross section, should be avoided, if possible, unless caused by the mandrel (core) support.
- The design of the die should be kept simple, and it should be possible to take it apart easily either for cleaning or modifications of the flow channel (compare Fig. 5.86) if required.

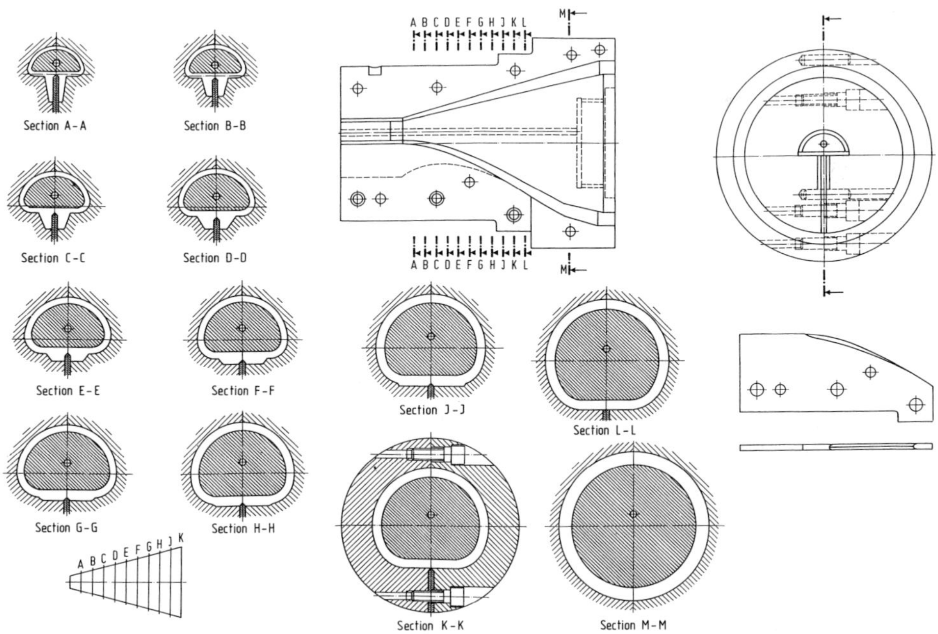


**Figure 5.85** Profile die with a continuous change of the flow channel cross section [148]

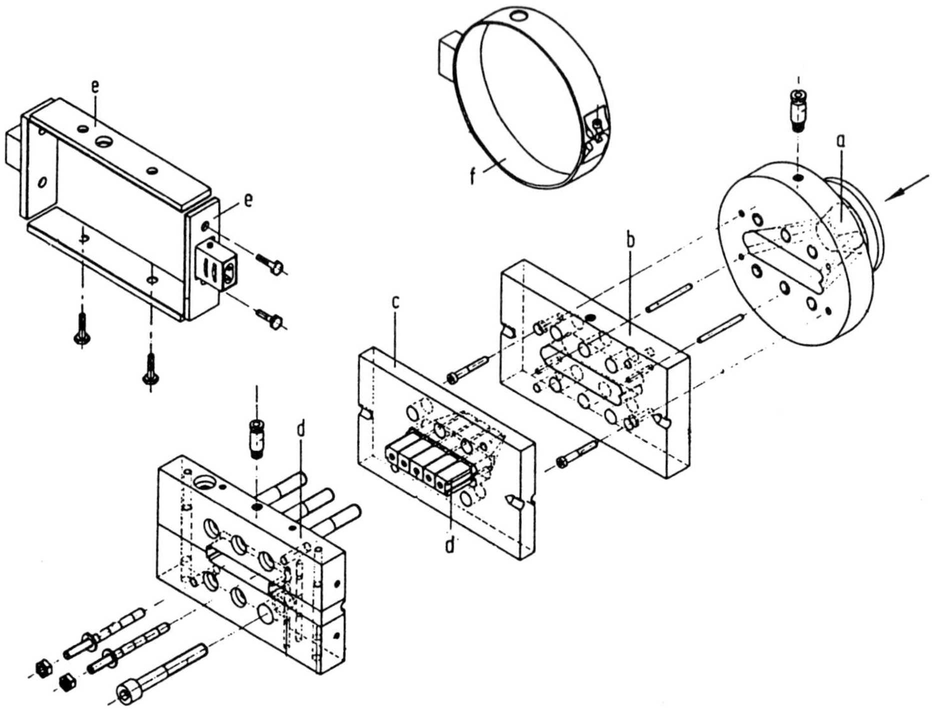


**Figure 5.86** Profile die [149]: (a) heating plate, (b) heater band, (c) parallel die land, (d) transition part (mounting plate), (e) entrance part

From this it follows that profile dies with channels of gradually changing cross section consist of three essential sections [20]:

- Feed section (connecting part)
- Transition part (serving partly as a supporting plate)
- Parallel die land (orifice)

These can be seen in Figs. 5.86 and 5.87. However, this separation is not always so obvious, and the individual sections often merge into one another (Figs. 5.88 and 5.89). The contour of the parallel die land corresponds approximately to that of the profile.

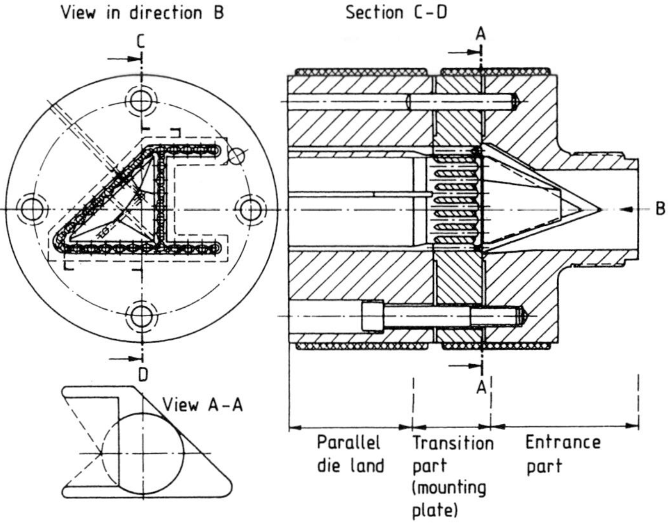


**Figure 5.87** Profile die for an end profile [155]

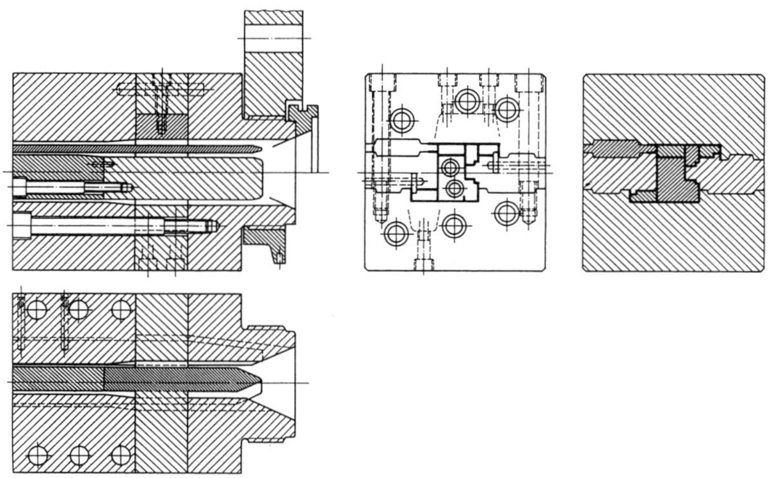

**Figure 5.88** Window profile die [145]

Figures 5.85 and 5.88, as well as Fig. 5.90, show elaborate profile dies with very complicated torpedoes (cores, mandrels) that, in contrast to pipe dies, are rigidly connected to the outer die ring. The centering of profile dies, therefore, cannot be adjusted. For the reason of geometry, the spider legs in such dies are in some cases designed to be mechanically weak so that the core floats in the melt stream and centers itself. However, this is successful only if the design of the flow channel is correct. In contrast to the system of mandrel support, mandrel, and mandrel support tip in pipe extrusion (Fig. 5.37), the core in the profile extrusion is usually made from one piece together with the support plate on which it is mounted. Such a part, although extremely complicated, is entirely appropriate, since it would be very difficult to reassemble the individual core segments after the necessary corrections have been made in starting up the die [143,145]. As in the case of center-fed dies, the spider legs are streamlined so they do not interfere with the flow, and all edges between spider legs and housing or core are rounded off to avoid stagnation zones.

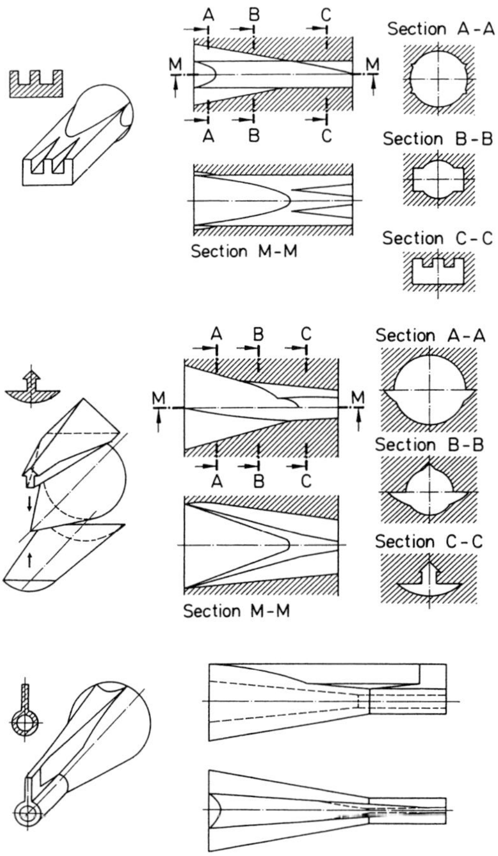


**Figure 5.89** Transitions in profile dies [156]

Figure 5.90 shows boat-shaped depressions behind the supporting fins in the flow direction. These depressions are intended to bring about the advance of the melt and to ensure that the material fins of the extruded profile can be formed with a sufficient amount of melt [143].

In addition, Fig. 5.90 shows that air is supplied to the hollow chambers of the profile to prevent collapse of the profile at the orifice. This is particularly important when starting up the profile extrusion line [143]. The air supply should always be available for the hollow profiles.

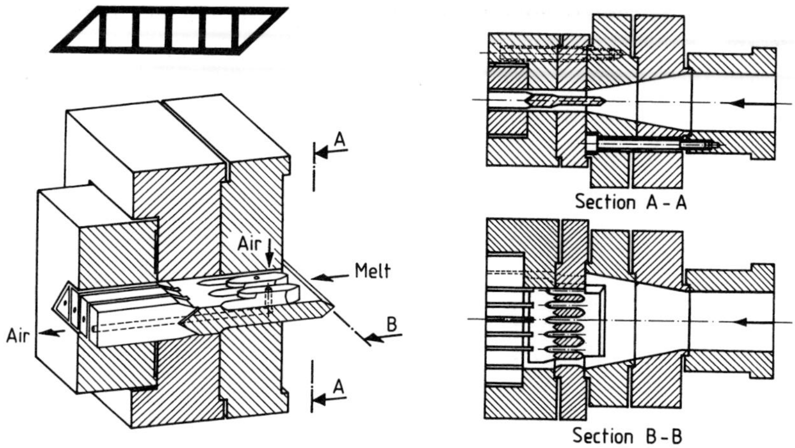


**Figure 5.90** Profile die [143]

As already mentioned, mass accumulations should be avoided in profiles. If, however, two different melt streams must flow next to each other in the die, they can be kept separate by a thin dividing fin (Fig. 5.91). The two melt streams do remain connected, however; the fin does not go through all the way, so a continuous weld line (and therefore a defect) is avoided in the extruded product [143,155]. The dividing fin should end a short distance (approx. 3 mm [157]) before the die exit, at which point the two separate melt streams can once again merge completely.

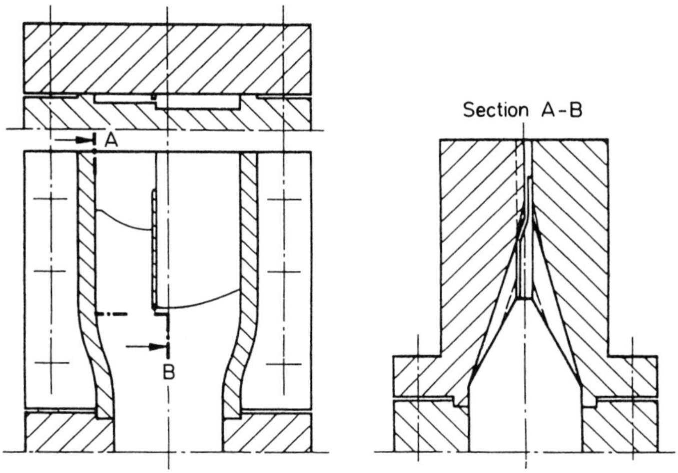


**Figure 5.91** Profile die with a dividing fin [155]

The necessary flow restriction zones of different length in the wide and narrow regions of the channels are clearly shown in Fig. 5.91. The flow restriction zone is shorter near the walls, which restrict the channel at the sides, so that the additional frictional effect of the walls can be taken into consideration in the distribution of the flow.

The temperature of the torpedo in profile dies is generally not controlled. Besides the heating system for the die, there is an additional heating system for the die land.

### 5.5.2 Design

There are essentially five factors that affect the quality of an extruded profile [149]:

- Accuracy of dimensions
- Shape accuracy in cross section as well as along the length
- Accuracy in function
- Surface appearance
- Special features

In order to meet these requirements, the design has to start only when it is clear that the profile is suitable for the material and process. While doing this, the following design rules should be observed [20,143,145,148,155,158]:

- The profile cross section should be kept as simple as possible. Internal walls should be avoided since they cannot be cooled directly and therefore can cause sink marks in the profile (Fig. 5.92).
- The wall thickness of the internal walls should be 20 to 30 % smaller than the thickness of the external wall. The edges should be rounded (their radius should be 0.25 to 0.50 times the wall thickness) [158].
- The profile should be designed in such a way that it keeps its shape for a short time after it exits the die but is still soft.
- Mass accumulations and sudden changes in the wall thickness have to be avoided since they cause difficulties in the melt distribution in the die, problems in cooling (sink marks due to differential shrinkage), and profile distortion.
- Hollow chambers should not be too small, otherwise the torpedoes (cores) can become too small and will not operate satisfactorily.
- Flat strips on the profile should be as short as possible because they are cooled faster (sometimes on both sides), causing a distortion of the profile.
- A symmetrical profile or a profile with a rotational symmetry causes the least distortion because the stresses during the cooling balance themselves.
- The axis of the profile, which is determined from the center of gravity, should coincide with the axis of the screw in order to reduce the differences in the flow paths.

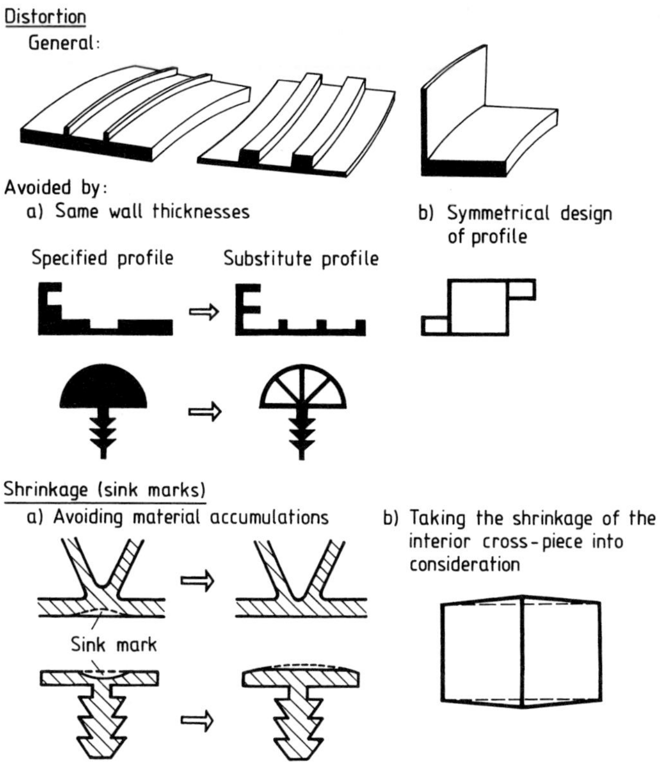


**Figure 5.92** Recommendations for designing profiles [143,158]

After the cross section of the profile is established, the layout of the flow channel follows.

The goal of the layout of the flow channel is a uniform average melt velocity at the die exit. Also, the swelling of the extrudate and shrinkage (upon cooling) must be taken into account so that the profile has the proper contour after it leaves the die. Besides that, as is generally the case in the design of the flow channel, attention has to be paid to eliminating stagnation zones and ensuring that the spectrum of the residence times in the die is rather narrow. When defining the dimensions of the exit cross section of profile dies, the following factors must be considered: proceeding from the desired extrudate cross section, compensations (positive or negative) must be made for swelling of the extrudate after emerging from the die for shrinkage on cooling and drawdown, that is, the difference between the mean velocity of the melt emerging from the die and the take-up speed.

### *Swelling*

The swelling of the material at the exit from the die is caused, on the one hand, by the rearrangement of the velocity profile at the die orifice from a parabolic shape in the flow channel, due to wall adhesion, to a plug flow. This results in local extensions and compressions in the emerging extrudate and, therefore, leads to distortion of its cross section. On the other hand, the reversible deformations stored in the material will relax (viscoelasticity, see Section 2.1.3) (Fig. 5.93). These stored deformations are caused by the stretching in the transitions of the flow channel cross section as well as shear actions.

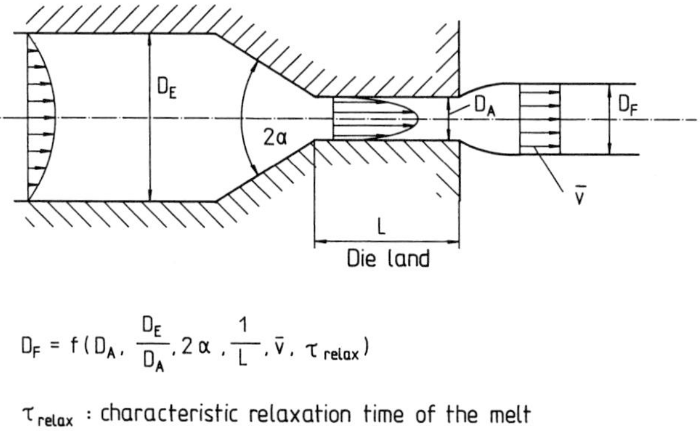


**Figure 5.93** Extrudate swelling and its causes

A study of the effect of geometry and operating conditions on the swelling behavior of an extrudate of a rectangular cross section [159] could not yield any generally valid rules for the prediction of swelling by calculation, but still a number of correlations have been found. The investigation was done on a PVC-type compound and involved the correlation between the relaxation zone, inlet angle, and apparent shear rate and the swelling based on the area and the swelling in the height and width of a rectangular profile. It was found that the swelling increases when the relaxation zone was shortened and the shear rate increased (Fig. 5.94). From that a function has been derived of the following form:

$$S = a + b \cdot e^{(-t_v/c)} \tag{5.129}$$

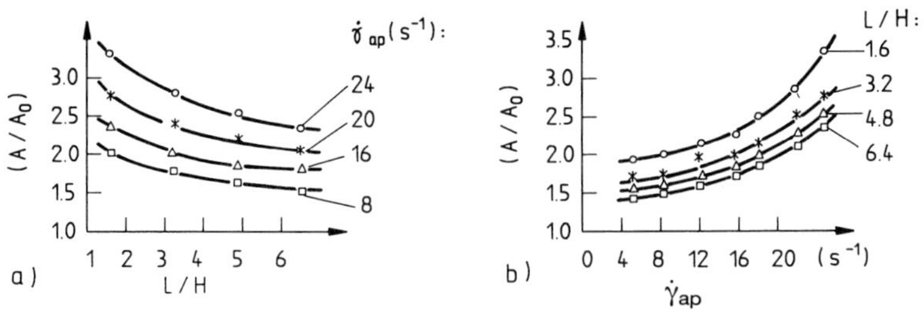


**Figure 5.94** Measurements of the swelling ratio: (a) swelling ratio as a function of the length of the die land, (b) swelling ratio as a function of the apparent shear rate

Equation (5.129) represents the relationship of the swelling potential based on shearing and rearrangement of the velocity profile (a) with the deformations occurring in the inlet zone (b) and the ratio of the residence time in the relaxation zone to the characteristic relaxation time of the melt ($t_v/c$). The values of swelling are plotted against the ratio for the length of the relaxation zone and apparent shear rate (which is proportional to the residence time) in Fig. 5.95. It can be seen that the values of swelling for longer residence times follow Equation (5.129); the swelling potential due to shear and rearrangement of the velocity profile is $a = 1.5$. For low residence times, the curves for different relaxation zones deviate from each other. The reason for this is that the deformations imposed in the inlet zone can already relax during the entry. The result of this is that $b$ (state of deformation of the melt at the entry into the relaxation zone) depends on the operating conditions. This cannot be represented by a simple equation such as Equation (5.128).

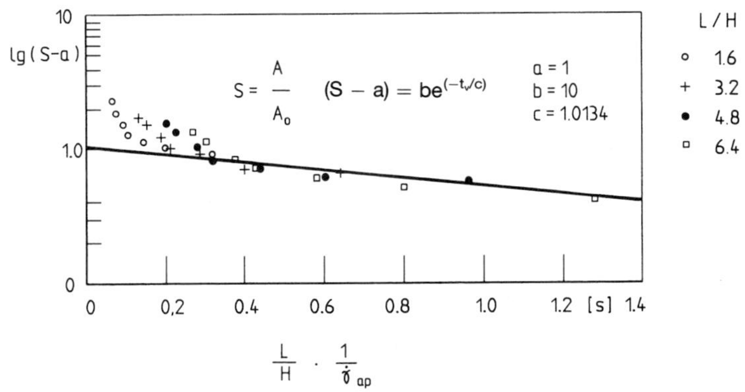


**Figure 5.95** Relationship between the swelling ratio and the residence time in the die land

Another important point is the difference between swelling in the thickness and the width. It is well known that the swelling is the most pronounced in the direction of the greatest velocity gradient, that is, in the smallest dimension (Fig. 5.96). A direct relationship between the ratio of height to width and the distribution of swelling cannot be given because the direction and the amount of deformation imposed in the inlet are also important factors.

A prediction of swelling by computation using FEM is possible, although with limitations. For this purpose, the flow behavior of the melt has to be expressed by a material law that considers its elasticity. The material data required for the material laws of this type exist only in few cases. The necessary three-dimensional computation of flow requires a long computation time.

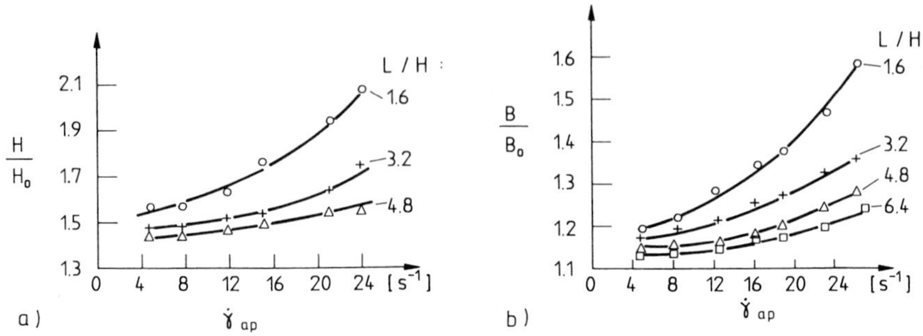


**Figure 5.96** Results from measurements of the swelling ratio: (a) swelling in the direction of height as a function of the apparent shear rate, (b) swelling in the direction of width as a function of the apparent shear rate

The difficulties in the computation of swelling have led to the situation that in practice the values resulting from experience are used for the corrections of the cross section of the die orifice. These values are, of course, dependent on material, geometry, and operating conditions and can be used only as a starting point (Table 5.2).

**Table 5.2** Reduction of the Exit Cross Section to Accommodate Swelling

| Material | Reduction | References |
|---|---|---|
| Rigid PVC: | 10% (wall thickness 1–2 mm) | [145,154] |
| | 3–6% (wall thickness 3–4 mm) | [154] |
| HI-PVC: | 10–20% | [145,154] |

### *Shrinkage*

When cooling an extruded profile from the melt temperature to ambient temperature, a volume contraction occurs that is referred to as shrinkage. The amount of shrinkage can be determined from the *p-v-T* diagram of the polymer. The longitudinal shrinkage can be calculated from the volume shrinkage under the assumption of an isotropic behavior.

### *Drawdown*

The profile is pulled from the die into the calibration unit by using a light stretching in order to put it into contact with the cold wall of the calibrator quickly. To compensate for this so-called drawdown, the cross section of the orifice must be increased. The values for this, obtained from experience, are summarized in Table 5.3.

**Table 5.3** Increase in Orifice Cross Section to Accommodate Drawdown

| Material | Increase in orifice cross section | References |
|---|---|---|
| Rigid PVC | 8–10% (small profiles) | [145,155] |
| | 5–10% | [16,157] |
| | 3–5% (large profiles) | [145,155] |
| Plasticized PVC | 12–15% | [16,157] |
| PE | 15–20% | [16,157] |
| PS | 8–10% | [16] |
| PA | 20% | [16] |

The estimation of the required cross section of the orifice can be based on the data from swelling, shrinkage, and drawdown. If there are no data of this kind, the cross section can be made 10–15% smaller than the cross section of the profile. This way a correction is possible later, if necessary.

After the design of the orifice cross section, the flow distribution in the die must be considered. A uniform exit velocity is achieved by matching the pressure losses in all regions of the die. To do this, the orifice cross section is first divided into cross sections that can be subject to simple computations.

Since these partial cross sections generally have different flow resistances, provided the flow velocity of the melt is uniform, the lengths of the flow channel sections (lengths of the flow resistance zones) will be adjusted in such a way that there is the same pressure loss for each flow pathway.

The process is explained briefly using the profile depicted in Fig. 5.97 (compare to Section 7.4.1). The ratio of the lengths of the flow resistance zones is $L_{pipe}/L_{slit}$.

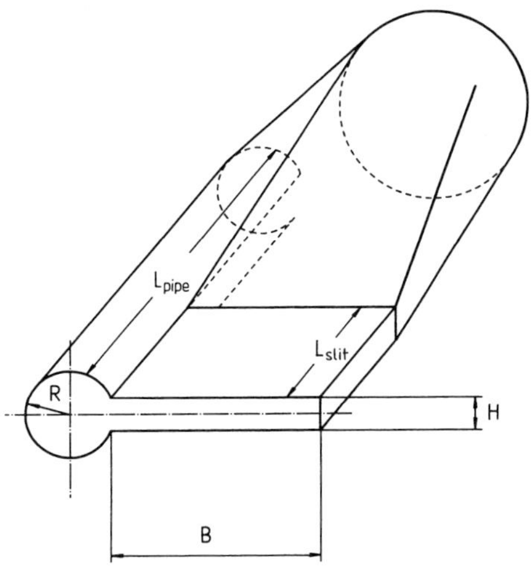


**Figure 5.97** Profile die with adjusted lengths of the flow resistance zones

The exit cross section is divided into a pipe and a flat slit. With the condition of equal mean flow rate, the expressions for the volumetric flow rates are obtained:

$$\dot{V}_{\text{pipe}} = \overline{v} \cdot \pi R^2, \tag{5.130}$$

$$\dot{V}_{\text{slit}} = \overline{v} \cdot B \cdot H. \tag{5.131}$$

The condition of equal pressure loss on each flow pathway leads to

$$\Delta p = \frac{8\eta_{\text{rep,pipe}} \cdot \overline{v} \cdot \pi \cdot R^2}{\pi \cdot R^4} \cdot L_{\text{pipe}} = \frac{12\eta_{\text{rep,slit}} \cdot \overline{v} \cdot B \cdot H}{B \cdot H^3} \cdot L_{\text{slit}} \cdot \tag{5.132}$$

The ratio of the lengths of the flow resistance zones is then

$$\frac{L_{\text{pipe}}}{L_{\text{slit}}} = \frac{3}{2} \frac{\eta_{\text{rep,slit}}}{\eta_{\text{rep,pipe}}} \left( \frac{R}{H} \right)^2 . \tag{5.133}$$

The properties of the material and the operating conditions enter into the ratio of the viscosities. The viscosities can be determined from the representative shear rates and a constitutive equation. The representative shear rate as a function of the average melt velocity can be determined utilizing Equations (5.130) and (5.131) and

$$\dot{\gamma}_{\text{rep,pipe}} = \frac{4\overline{v}}{R} e_{\bigcirc} \tag{5.134}$$

$$\dot{\gamma}_{\text{rep,slit}} = \frac{6\overline{v}}{H} e_{\square} \tag{5.135}$$

Of course, this computation has built-in errors because the following factors are neglected:

- The mutual influence of the partial streams
- The cross-flow
- The effects of obstructing sidewalls

In spite of this, the computation is acceptable in many cases, especially when profiles have a constant wall thickness that is smaller than the width of the fins (see Fig. 5.98).

The computation of the flow by the use of FEM offers help when designing the flow resistance zones of profiles with complicated shapes. When applying this method, the above mentioned errors can be eliminated at the cost of developing the FEM structure and computer-time-intensive, complex three-dimensional flow computation [e.g., 160].

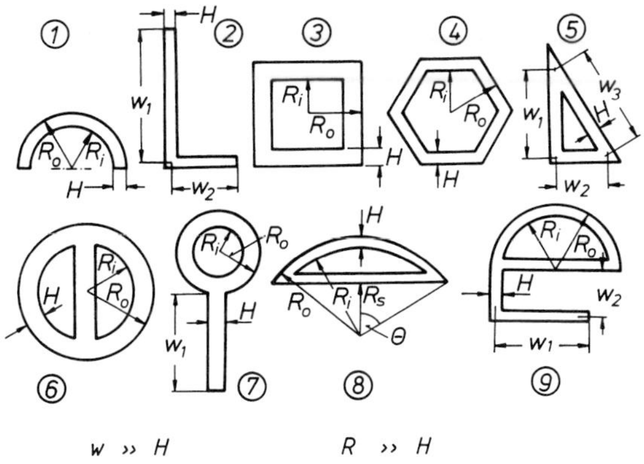


**Figure 5.98** Profiles of cross sections that can be calculated easily

Figure 5.99 shows the result of a 3D-FEM computation of a profile cross section in the core region shown in the same figure in the lower right. Because of symmetry, only half of the profile is considered. The FE mesh reproduces the region of the flow channel in which the melt flows toward the core and where the profile is formed; the individual pathways are drawn in.

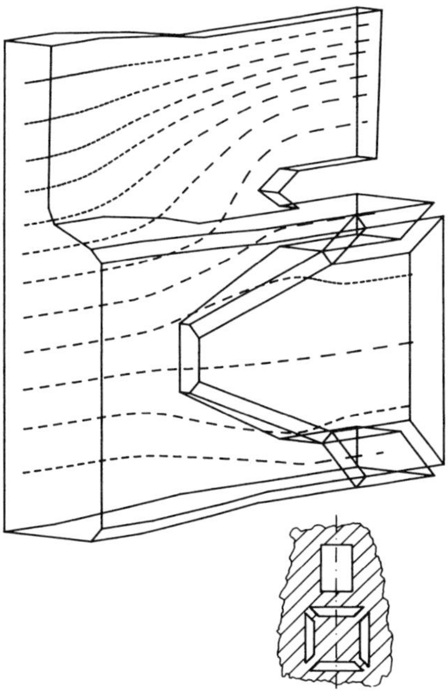

**Figure 5.99** Streamlines for a flow through a profile die (3D-FEM computation)

When designing flow channels, the following general rules should be taken into account [145]:

- The cross section should be reduced steadily in the direction of the flow starting at the extruder exit at an angle of approximately 12° until it reaches the exit cross section. The flow resistance of the parallel die lands should be about 10 to 12 times greater than that of the preceding channel. The maximum length of the parallel die land is 90 mm. From this limiting value and the computed ratios of the flow resistance zones in the individual partial cross sections, the individual lengths of the parallel die lands are obtained. Furthermore, it is important to keep the $L/H$ ratio between 20 and 50 [154,155].
- The edges at the spider legs are rounded off ($R = 0.2H$) and run out at an angle of about 8°.

After the die is finished, it is tested with the material to be used at the required throughput. If the flow at the exit is not uniform, the die lands are shortened in steps.

Frequently, the die is hardened before the last correction in order to allow changes that may be necessary if the die is distorted during the hardening process.

The costs for the preliminary tests and corrections of a profile die may be as much as 10 to 50% of the costs of the profile die [20].

## 5.6 Dies for Foamed Semifinished Products

After the previous sections dealt thoroughly with dies for the production of extrudates from solid plastic materials, the following section will describe dies as they are used for the extrusion of foamed products. Here too, profiles with and without hollow chambers as well as coextrudates of solid and foamed materials are produced.

The prerequisites for a production without interruptions and a high-quality product are proper screw geometry and compound formulation, correct design of the die and calibration, and appropriate processing conditions [161]. One can distinguish between chemical and physical foaming. In chemical foaming, a blowing agent is added to the hopper. At high temperatures, the blowing agent decomposes and a gas is released that initiates the foaming. In physical foaming, a blowing agent (e.g., a supercritical gas) is directly injected via special dosing technology into the melt, and foaming is initiated behind the die. Depending on the desired density reduction, the kind of blowing agent is selected: for densities below 400 kg/dm$^3$, physical blowing agents are used.

There are essentially two methods to extrude foamed semifinished products:

- Expansion of the melt containing the foaming agent immediately after it emerges from the die
- Expansion of the melt containing the foaming agent at some later point

In the first method, the expansion is done either free to the outside and subsequently the extrudate moves through the calibration and cooling line, or the expansion occurs in a device that is attached by a flange to the die. This is called the expansion to the inside. These procedures are mainly used for the production of foamed sheets, which are stretched over a "cooling bulb" [162] after being extruded, mainly from an annular die, and expanded freely.

In addition to the above, there are methods where two parts of one melt with and without blowing agent are fed into a coextrusion die, producing, for example, a profile consisting of a foamed core and a solid inner and outer skin [163].

The second method mentioned is used mainly for processing polyethylene, which contains not only a foaming agent but also a crosslinking agent. The conventionally extruded, not yet crosslinked and not yet expanded (mostly) flat extrudate, also called the matrix, is subsequently first crosslinked and then expanded in an oven [164]. To produce the melt extrudate, the usual wide slit dies are used.

The process of direct expansion after the exit from the die orifice requires special design considerations that will be discussed below.

It is necessary to ensure that the gas, for example splitting off from the blowing agent, remains dissolved in the melt in the extruder and in the die: there must be no premature foaming in the extruder or in the die. Such premature foaming would result in gas bubbles being torn open by the flow of the melt along the wall of the die and would lead to a product with a poor and rough surface.

Furthermore, it has to be kept in mind that the gas-loaded melt has a low viscosity and, therefore, flows easily through the die. Therefore, it is necessary to build up a sufficient pressure in both the extruder and the die by a proper design of the flow channel. This is done by cooling the melt using melt coolers and by cooling the extruder and die. By doing so, the viscosity is increased and the pressure level stays high. A high pressure gradient at the die outlet is recommended to keep the blowing agent dissolved and to improve the foam nucleation right after the die exit. For example, for the extrusion of rigid foam from PVC, pressures of 100 to 300 bar are considered sufficient [161]. Rheological studies dealing with foam extrusion are reported in [165].

### 5.6.1 Dies for Foamed Films

In order to build up the necessary counterpressure, the annular exit gap is narrowed toward the orifice. If the film is relatively thick, the die lips may be cooled in order to increase the flow resistance [166]. This type of die can be designed with the slit at an angle to the outside (mostly 45 degrees) in order to eliminate the formation of wrinkles during the expansion.

Spider legs are quite a problem here because the flow marks are even more pronounced in the foamed sheet than they are in the solid one [167]. Therefore, the number of spider legs is kept to a minimum in such dies. One or two spider legs are commonplace, and they are arranged in such a fashion that, when the sheet is later slit sideways, the knit lines (flow marks) lie in the cutting plane [166].

The methods of minimizing the effect of flow marks are discussed in [156] and Section 5.3.1.1. For example, by arranging flow resistance beads in the mandrel zone behind the passage through the mandrel support, the quality of an extruded foamed sheet from PE was improved.

### 5.6.2 Dies for Foamed Profiles

Many types of dies for this purpose will be discussed briefly in the following section. A more detailed discussion of this subject is in [161].

***Longitudinally Fed, Solid Profile Dies***

These dies exhibit a high flow resistance and are employed for the production of thin-walled profiles with a small cross-sectional area (Fig. 5.100). By comparison with solid profile dies, it can be seen that a short (1–2 mm long) and narrow (approx. 10% of the channel height) flow restriction zone is arranged in the exit region of the die. The function of this flow restriction zone is to greatly reduce the pressure near the die exit and in doing so initiate the foaming process.

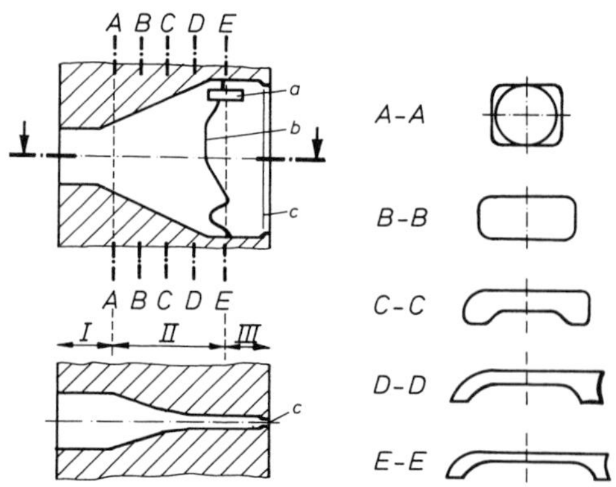


**Figure 5.100** Die for foam profiles (possibilities for flow corrections [161]): I: feed zone, II: transition zone, III: die land; (a) separating fin, (b) different length of the die land, (c) flow restriction zone

### *Plate Dies with Torpedos*

These dies are similar to those generally used in profile extrusion, the torpedo ensuring adequate buildup of pressure in the extruder (Fig. 5.81). The construction with the easily exchangeable die plates permits several geometrically similar profiles to be produced with the same die. The flow can be corrected easily at the die plate.

### *Dies with Restrictor Grids and (If Necessary) Torpedos*

Thin-walled restrictor grids [168] are used to build up a high flow resistance while reducing the flow cross section as little as possible. The flow can be regulated by the local length of the restrictor grid (Fig. 5.101). This design is suitable when foamed profiles with a large cross section, large wall thicknesses, and varying wall thickness distribution are to be produced. The individual strands formed in the restrictor grid are brought together and fused in the die exit region.

Torpedos can be introduced in such a grid. They can affect the flow distribution, greatly change the volumetric weight of the foamed profile in certain selected regions, and even form cavities [169].

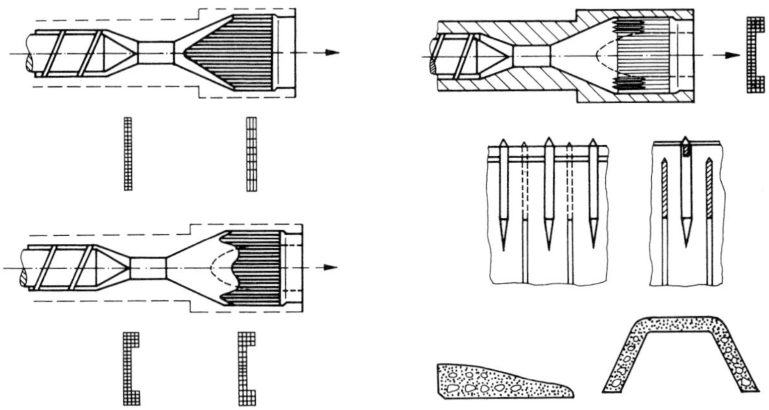

**Figure 5.101** Dies with restrictor grid [168] or restrictor grid and torpedo body [169] for foam profiles

### *Dies for Hollow Profiles*

These correspond in their configuration to those of dies for melts without a blowing agent. In detail, however, the following should be realized: a flow restriction zone and a short parallel die land (length approx. 5 to 10 times the channel height) should be installed in the die exit region; and the compression ratio between the core/mandrel support zone and the die orifice should be 10:1 to 15:1.

The design of a profile die for a rigid foam from PVC is even more difficult than that for a die for solid profiles from rigid PVC because of the expansion of the material on leaving the die orifice in addition to the swell and the rearrangement of the velocity profile [161]. Since the expansion process also depends on various factors, such as type of blowing agent, formulation, temperature, and mixing, it is prudent—in the absence of practical experience—to be guided by the values for the dimensions of the die orifice given in Table 5.4 [161].

**Table 5.4** Dimensions of Die Orifice for the Required Profile Dimensions for PVC-U

| Final size of profile (calibrated) | | Dimensions of the die orifice | |
|---|---|---|---|
| | mm | mm | % of the final size of profile |
| Profile width | 10 to 30 | 6 to 18 | 60 |
| | 30 to 60 | 21 to 42 | 70 |
| | 60 to 100 | 48 to 80 | 80 |
| | 100 to 150 | 90 to 135 | 90 |
| | over 150 | 150 | 100 |
| Profile thickness | up to 3 | up to 1.5 | 50 |
| | 3 to 8 | 1.35 to 3.6 | |
| | 8 to 12 | 4 to 6 | 50 |

These values are kept purposely small in order to make reworking possible. They also depend on the ratio of the width to thickness and according to [161] are valid for ratios of 3 to 15. Above this ratio, the width is no longer corrected; below it, the width is adjusted in the same way as the thickness [161]. The design of a die for foamed profiles should take into account that the flow channel must become more narrow continuously toward the die orifice. Enlargements in the flow channel are to be avoided, as are potential areas of stagnation [161,170].

## 5.7 Special Dies

### 5.7.1 Dies for Coating of Profiles of Arbitrary Cross Section

To this group belong naturally dies for the coating of profiles of any shape, whether for the improvement of the surface or protection against corrosion [20]. The profile moves through a (hollow) mandrel, the same way as in the standard coating procedure; the mandrel has a properly designed fit and a contour with a tight tolerance. Also, in this case the application of the material onto the profile to be coated is assisted by a vacuum.

Figure 5.102 shows, as an example, a die for the coating of aluminum foils of a thickness of approximately 50–200 μm with CAB for edgings. The strips of aluminum run flat into the die, are coated in it, and are subsequently shaped and cooled in the calibrator [171].

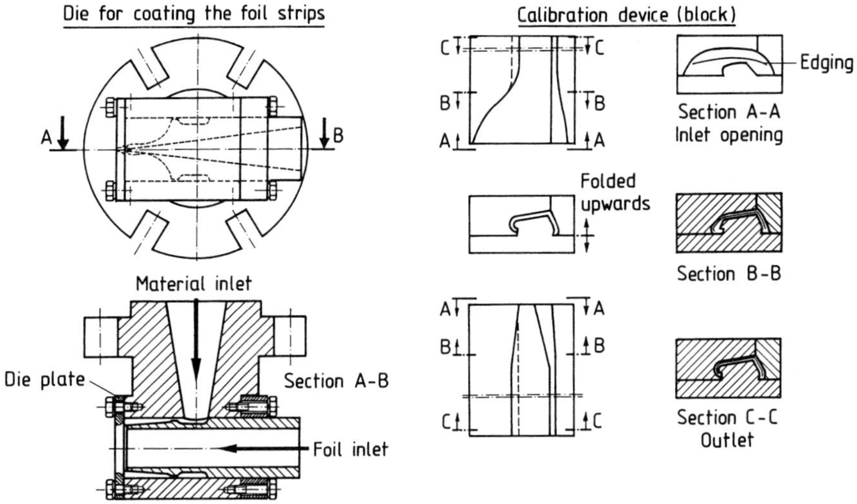


**Figure 5.102** Die and calibration device for the extrusion of edging (CAB-coated aluminum foil) [171]

To avoid scratching of the aluminum foil, the foil entry channel is highly polished. The melt is applied onto the foil almost tangentially. An axial adjustment of the mandrel allows an optimum setting of the orifice gap.

The calibration line should be no longer than 15 to 25 cm in this case so that profile is not torn off by the increasing friction in the calibration and cooling sections [171].

## 5.7.2 Dies for the Production of Profiles with Reinforcing Inserts

There is a wide variety of manufacturing methods for placing reinforcing inserts into extruded plastic and rubber profiles. In [172] there is a detailed treatment of this subject. According to that review, there are basically three possibilities for producing reinforced semifinished products:

- Supply of an insert, prefabricated independent of the extrusion process, into the jacketing die
- First extrusion of the inside layer, which is subsequently reinforced, then application of the outside coat

- Simultaneous production of the reinforcing insert with the extrusion of the profile. This method is used, for example, to coat a wire that is spirally wound in the die and simultaneously coated with the melt.

An example for the second method is given in [173]. The production of a reinforced hose is described, where, in the first step, the so-called raw hose is extruded from plasticized PVC. After cooling, this hose is wrapped with yarns in a winding machine. In order to attain a good adhesion, the hose is subsequently heated on the surface, and the tightly woven yarn easily melts into the plastic. The last step is then jacketing with an outside coat in an appropriate jacketing die [173].

### 5.7.3 Dies for the Production of Nets

Nets from plastics can be extruded by using center-fed or side-fed dies that have the exit region as depicted in Fig. 5.103 [36,174,175]. The outer die ring can be turned continuously or in steps. In a side-fed die, the mandrel can also be turned, in addition to the rotation of the outside die ring. The tooth system can be in the gap-on-gap or the tooth-on-gap position. Half as many strands are extruded in the first position as in the second, that is, the points of intersection of a net are formed; in the second position the net yarns are formed, linking the points of intersection of the net. Nets of different structures can be produced by varying the programmed motion of the rotary parts [174,175].

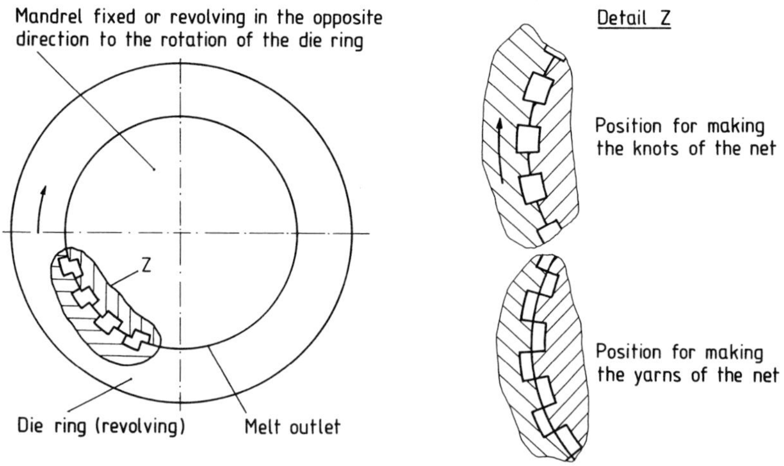


**Figure 5.103** Exit gap of a die producing extruded nets

### 5.7.4 Slit Die with Driven Screw for the Production of Slabs

This special design of a wide slit die is used mainly in Japan for the extrusion of wide and thick slabs from rigid PVC. A separately driven screw in the die body distributes the melt supplied by the extruder (Fig. 5.104). The melt can continuously escape laterally along the length of the die over a flow resistance zone and into a connected lip region. This way a flat sheet or slab is formed. The local flow control can be done by a choker bar or by an adjustable lip (Flexlip) [176].

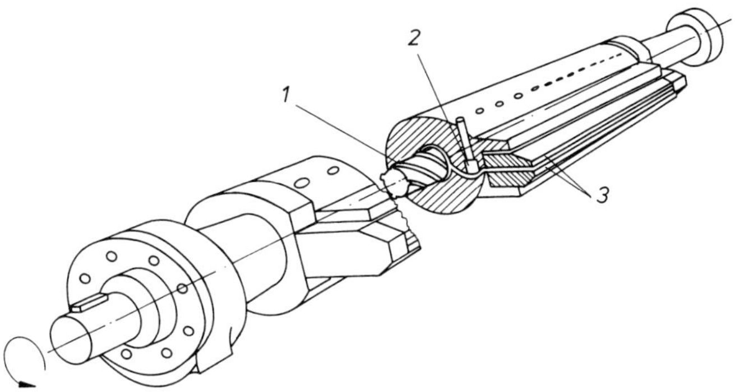


**Figure 5.104** Screw die for sheet extrusion [177]: 1: screw; 2: choker bar; 3: adjustable lips

One advantage of this technically complex and quite expensive system is that no melt stagnations occur and, therefore, the danger of degradation of the rigid PVC is reduced. Hence, the amount of stabilizer can be reduced, resulting in cost savings [177]. A theoretical analysis of the flow of the melt in the screw and in the connected flow resistance zone and lip region is given in [178].

#### Symbols and Abbreviations

| | |
|---|---|
| $A$ | Constant (sinh law of Prandtl and Eyring) |
| $A_Q$ | $A$ of a cross section |
| $A_W$ | Wall surface |
| $a$ | Swell potential due to shearing and rearrangement of velocity profiles |
| $B'$ | Share of the channel width of the substitute cross section |
| $B$ | Width |
| $b$ | Width of the slit-shaped manifold |
| $b$ | Deformations at the inlet |
| $C$ | Constant (sinh law of Prandtl and Eyring) |
| $c$ | Characteristic relaxation time of the melt |
| $D$ | Diameter |
| $D_d$ | Nozzle diameter |

| | |
|---|---|
| $D_m$ | Nominal die diameter |
| $d_{hyd}$ | Hydraulic diameter |
| $e_{\bigcirc}$ | Ratio of representative radius to outer radius of a circular channel |
| $e_{\square}$ | Ratio of representative height to total height of a slit channel |
| $f_p$ | Flow correction coefficient |
| $H$ | Height |
| $h$ | Height |
| $h_E$ | Height of the slit at the end of the distribution channel |
| $L$ | Length |
| $L_E$ | Inlet length |
| $l$ | Coordinate in direction of the manifold |
| $\dot{m}$ | Mass flow rate |
| $N_{slit}$ | Number of slits |
| $N_u$ | Number of wrap angle increments up to the first overlap of spirals |
| $N_{spiral}$ | Number of spirals |
| $n$ | Flow exponent (power law) |
| $n$ | Index number |
| $p$ | Pressure |
| $\overline{p}$ | Average pressure in the center of a calculation segment |
| $p_0$ | Ambient pressure |
| $\Delta p$ | Pressure loss |
| $\Delta p_{entry}$ | Inlet pressure loss |
| $\Delta p_{pipe}$ | Pressure loss in a circular pipe |
| $\Delta p_{slit}$ | Pressure loss in a slit channel |
| $\Delta p_{total}$ | Total pressure loss |
| $R$ | Radius |
| $R_a$ | Outer radius |
| $R_i$ | Inner radius |
| $r$ | Radius |
| $S$ | Swell |
| $T$ | Temperature |
| $T_m$ | Mass temperature |
| $T_{MP}$ | Melting temperature |
| $T_0$ | Reference temperature |
| $t_V$ | Residence time in the die land |
| $\overline{t}$ | Mean residence time |
| $\overline{t}_{V\,total}$ | Mean residence time of the whole channel width |
| $U$ | Circumference |

| | |
|---|---|
| $v$ | Velocity |
| $\overline{v}$ | Mean velocity |
| $\overline{v}_z$ | Mean melt velocity in extrusion direction |
| $\dot{V}$ | Volume flow rate |
| $\dot{V}_L$ | Leakage flow rate of the manifold |
| $x$ | Coordinate in width direction |
| $y$ | Coordinate in extrusion direction |
| $y_0$ | Maximum length of die land |
| $\alpha$ | Angle |
| $\dot{\gamma}_{rep}$ | Representative shear rate |
| $\dot{\gamma}_{rep,pipe}$ | Representative shear rate in a circular pipe |
| $\dot{\gamma}_{rep,slit}$ | Representative shear rate in a slit channel |
| $\zeta$ | Angle (cylinder coordinates) |
| $\eta_{rep}$ | Representative viscosity |
| $\eta_{rep,pipe}$ | Representative viscosity in a circular pipe |
| $\eta_{rep,slit}$ | Representative viscosity in a slit channel |
| $\theta$ | Circumferential angle (cylinder coordinates) |
| $\rho$ | Density |
| $\tau_W$ | Shear stress at the wall |
| $\phi(u)$ | Auxliary function, see Equation (5.28) |
| $\psi(r)$ | Auxiliary function, see Equation (5.30) |

### Abbreviations

| | |
|---|---|
| CAB | Cellulose acetobutyrate |
| FDM | Finite difference method |
| FEM | Finite element method |
| PA | Polyamide |
| PE-HD | High-density polyethylene |
| PE-LD | Low-density polyethylene |
| PTFE | Polytetrafluoroethylene |

## 5.8 References of Chapter 5


1. Wichardt, G.: Heißgranuliervorrichtungen. In: Granulieren von thermoplastischen Kunststoffen. VDI-Verl., Düsseldorf (1974)
2. Knappe, W: Rheologisches Verhalten von Kunststoffschmelzen in Granulierlochplatten. In: Granulieren von thermoplastischen Kunststoffen. VDI-Verl., Düsseldorf (1974)
3. Siemetzki, H.: Unterwassergranuliervorrichtungen, Bauart Barmag. In: Granulieren von thermoplastischen Kunststoffen. VDI-Verl., Düsseldorf (1974)
4. Martin, G.: Unterwassergranuliervorrichtungen, Bauart Berstorff. In: Granulieren von thermoplastischen Kunststoffen. VDI-Verl., Düsseldorf (1974)
5. N. N.: Granulieren von Thermoplasten: Systeme im Vergleich. Gesellschaft Kunststofftechnik, VDI-Verlag., Düsseldorf (1999)
6. Hensen, F.: Einsatz von Großflächenfiltern und dynamischen Mischern zur Qualitätsverbesserung bei der Kunststoffverarbeitung bzw. Spinnextrusion. Plastverarbeiter 33 (1982) 12, pp. 1447-1454
7. N. N.: Neue Entwicklungen bei der Filtration von Polymerschmelzen. Chemiefasern Text. Ind. 34 (1984) 6, pp. 409-411
8. Schmitz, T. et al.: Schutz für empfindliche Oberflächen, Kunststoffe 100 (2010) 4, pp. 54-57
9. Morland, C. D. and Williams, B.: Selecting polymer filtration media. Fiber Prod. (1980) pp. 32-44; 65
10. N. N.: Auch bei starker Rohstoffverschmutzung störungsfreie Produktion. Plastverarbeiter 35 (1984) 4, pp. 162-163
11. Bessemer, C.: Continuous extruder filtration enhances product quality and throughput. Plast. Eng. 39 (1983) pp. 29-31
12. Wöstmann, S.: Peripherie für Extrusionsanlagen - hinter dem Extruder. VDI-Seminar Extrusion thermoplastischer Kunststoffe, Frankfurt (2013)
13. Carley, J. F. and Smith, W. C.: Design and operation of screen packs. Polym. Eng. Sci. 18 (1978) 5, pp. 408-415
14. Schwab, H.: Düsen für das Spinnen von Chemiefasern: Formen, Materialien und Herstellungsverfahren. Chemiefasern. Text. Ind. 27 (1977) 9, pp. 767-775
15. Funk, W. and Schumm, R.: Spinndüsen - Bauteile für die Chemiefaserindustrie. Chemiefasern Text. Ind. 22 (1972) 6, pp. 518-522
16. Mink, W: Extruder-Werkzeuge. Plastverarbeiter 15 (1964) 10, pp. 583-590 and 15 (1964) 11, pp. 652-657
17. Poller, D. and Reedy, O. L.: Designing dies for monofilaments. Mod. Plast. 41 (1964) 3, pp. 133-138
18. Fourné, F.: Synthetic Fibers, Chapter 4.6: Melt Spinning Plants. Hanser Verlag, Munich (1999)
19. Hinz, E.: Erfahrungen beim Herstellen von Vollprofilen aus thermoplastischen Kunststoffen. In: Extrudieren von Profilen und Rohren. VDI-Verl., Düsseldorf (1974)
20. Schiedrum, H. O.: Profilwerkzeuge und -anlagen für das Extrudieren von thermoplastischen Kunststoffen. Plastverarbeiter 19 (1968) 6, pp. 417-423 and 19 (1968) 8, pp. 635-640 and 19 (1968), pp. 728-730
21. Voigt, J. L.: Kontinuierliches Herstellen von Polyamid-Stäben. Kunststoffe 51 (1961) 8, pp. 450-452
22. Franz, P.: Fließverhalten von PVC in Granulierlochplatten. In: Granulieren von thermoplastischen Kunststoffen. VDI-Verl., Düsseldorf 1974
23. Giesekus, H.: Sekundärströmungen und Strömungsinstabilitäten. In: Praktische Rheologie. VDI-Verl., Düsseldorf 1978, pp. 162-171

24. Kleinecke, K. D.: Zum Einfluß von Füllstoffen auf das rheologische Verhalten von hochmolekularen Polyethylenschmelzen. II. Untersuchungen in der Einlaufströmung. Rheol. Acta 27 (1988)
25. White, J. L.: Critique on flow patterns in polymer fluids at the entrance of a die and instabilities leading to extrudate distortion. Appl. Polym. Symp. 9 (1973) 20, pp. 155–174
26. Han, C. D.: Influence of the die entry angle on the entrance pressure drop, recoverable elastic energy and onset of flow instability in a polymer melt flow. J. Appl. Polym. Sci. 17 (1973) pp. 1403–1413
27. Ryder, L. B.: End correction implication in die design and polymer processing. SPE J. 17 (1961) 12, pp. 1305–1309
28. Han, C. D. et al.: Measurement of the axial pressure distribution of molten polymers in flow through a circular tube. Trans. Soc. Rheol. 13 (1969) 4, pp. 455–466
29. N. N.: Kenndaten für die Verarbeitung thermoplastischer Kunststoffe. Ed.: VDMA. Rheologie. Hanser, München (1986)
30. Bartos, O.: Fracture of polymer melts at high shear stress. J. Appl. Phys. 68 (1961) p. 6
31. Ertong, S.: Charakteristische Strömungseffekte bei viskoelastischen Flüssigkeiten: Schmelzebruch, Strahlaufweitung. Rheologieseminar, Institut für Verfahrenstechnik, Aachen 1988, p. 5
32. Vinogradov, C. V. and Insarova, N. I.: Critical regimes of shear in linear polymers. Polym. Eng. Sci. 12 (1972) p. 5
33. Li, Ch.H. et al.: Two separate ranges for shear flow instabilities with pressure oscillations in capillary extrusion of HDPE and LLDPE. Polym. Bull. 9 (1986) pp. 15
34. Masberg, U.: Auslegen und Gestalten von Lochplatten. In: Filtrieren von Kunststoffschmelzen. VDI-Verl., Düsseldorf (1981)
35. Ehrmann, G. et al.: Berechnen und experimentelles Ermitteln der Druckverluste verschiedener Siebgewebe. In: Filtrieren von Schmelzen. VDI-Verl., Düsseldorf 1981
36. Domininghaus, H.: Einführung in die Technologie der Kunststoffe, 2nd, company paper, Hoechst AG, Frankfurt
37. Oertel, H.: Herstellen von tafelförmigem Halbzeug aus Polyolefinen. Plastverarbeiter 17 (1966) 1, pp. 17–23
38. Hensen, F.: Anlagenbau in der Kunststofftechnik, postdoctoral thesis at the RWTH Aachen (1974)
39. Kaehler, M.: Der Einschneckenextruder und seine Anwendung bei der Platten- und Folienherstellung. Kunstst. Plast. 20 (1973) 3, pp. 13–16
40. Michaeli, W.: Zur Analyse des Flachfolien- und Tafelextrusionsprozesses. Thesis at the RWTH Aachen (1976)
41. N. N.: Company paper, Johnson-Leesona Company
42. Fischer, P. and Ortner, A.: ABS-Tafeln problemlos extrudieren. Kunststoff-Berater (1971) 5, pp. 380–385
43. Fischer, P. et al.: Maschinen- und verarbeitungstechnische Fortschritte beim Herstellen von Kunststoff-Tafeln und -Folien. Kunststoffe 61 (1971) 5, pp. 342–355
44. Schumacher, F.: Extrudieren von Breitschlitzfolien und Platten. Kunststoffe 77 (1987) 1, pp. 96–99
45. Heimlich, S.: Extruder-Werkzeuge. Kunststoff-Berater (1973) 12, pp. 883–885
46. Donovan, J. S.: Wide sheet lines in the US. Br. Plast. 43 (1970) 5, pp. 121–125
47. Fischer, P. and Ortner, K. A.: Neue Dimensionen in der Extrusion von Tafeln. Plastverarbeiter 22 (1971) 10, pp. 741–742
48. Predöhl, W.: Herstellung und Eigenschaften extrudierter Kunststoffolien. Postdoctoral thesis at the RWTH Aachen (1977)
49. Gregory, R. B.: Advances in extrusion die design. Mod. Plast. 44 (1967) 3, pp. 132–135

50. Schenkel, G.: Vom Kreis zum Rechteck – Probleme der Extrusion von Flach- und Schlauchfolien. IKT-Colloquium, Stuttgart (1979)
51. Röthemeyer, F.: Formgebung eines Plattenspritzkopfes nach Theologischen Gesichtspunkten. Kunststoffe 56 (1966) 8, pp. 561–564
52. Weeks, D. J.: Berechnungsgrundlagen für den Entwurf von Breitschlitz- und Ringdüsen. Br. Plast. 31 (1958) 4, pp. 156–160; Br. Plast. 31 (1958) 5, pp. 201–205 and Kunststoffe 49 (1959) 9, pp. 463–467
53. McKelvey, J. M. and Ito, K.: Uniformity of flow from sheeting dies. Polym. Eng. Sci. 11, (1971) 3, pp. 258–263
54. Pearson, J. R.A.: Non-Newtonian flow and die design. Plast. Inst., Trans. J. 32 (1964) 99, pp. 239–244
55. Ito, K.: Designing fish-tail die. Jpn. Plast. (1970) 7, pp. 27–30
56. Ito, K.: Designing coat-hanger die. Jpn. Plast. 2 (1968) 1, pp. 35–37 and 3 (1969) 1, pp. 32–34
57. Chejfec, M. B.: Profilierung der Extrusionsdüsen mit flachem Austrittswinkel. Plast. Massy (1973) 12, pp. 31–33
58. N. N.: DE-PS 11 80 510 Reifenhäuser KG, Troisdorf (German patent)
59. Barney, J.: Trends in extrusion die design. Plast. Des. Process. 14 (1974) p. 2
60. Barney, J.: Design requirements for PVC film and sheet dies. Mod. Plast. 46 (1969) 12, pp. 116–122
61. Leggewie, E. and Wurl, G.: Extrudieren von Folien und Platten über die Breitschlitzdüse. Kunststoffe 63 (1973) 10, pp. 675–676
62. N. N.: DE-OS 1779546 Reifenhäuser KG, Troisdorf (German patent)
63. N. N.: US Patent 3,940,221 Welex, Blue Bell, PA
64. N. N.: Welex/LFE Autoflex: Automatic Sheet Profile Control System. Company paper. Welex, Blue Bell, PA
65. Rudd, N. E.: Automatic Profile Control: Technology and Application. SPE Tech. Pap. (1978) pp. 565–567
66. N. N.: Now extrusion control is affortable by everybody. Mod. Plast. Int. 57 (1979) pp. 36–39
67. N. N.: Automatische Flachfoliendüse mit Piezotranslatoren. Kunststoffe 78 (1988) 7, p. 595
68. Friedrich, E.: Extrudieren von Methacrylaten. Kunststoffe 47 (1957) 4, pp. 218–223
69. Pfeiffer, H. et al.:DE-OS 35 03 721 AlBreitschlitzdüse zum Extrudieren eines thermoplastischen Kunststoffes (1986) (German patent)
70. Pfeiffer, Herbert: DE-OS 3534 407 Al Hoechst AG (1987) (German patent)
71. Görmar, E. H.: Beitrag zur verarbeitungsgerechten Dimensionierung von Breitschlitzwerkzeugen für thermisch instabile Thermoplaste insbesondere PVC-hart. Thesis at the RWTH Aachen (1968)
72. Schönewald, H.: Auslegung von Breitschlitzwerkzeugen für die Folien- und Plattenextrusion. Kunststoffe 68 (1978) 4, pp. 238–243
73. Kirchner, H.: Dimensionierung von Verteilerkanälen für Breitschlitzwerkzeuge. Unpublished Diploma thesis at the IKV, Aachen (1976)
74. Matsubara, Y.: Geometry design of a coat-hanger die with uniform flow rate and residence time across the die width. Polym. Eng. Sci. 19 (1979) p. 3
75. Fritz, H. G.: Extrusionsblasformen. In: Hensen, F., Knappe, W. and Potente, H. (Eds.): Handbuch der Kunststoffextrusionstechnik. Vol. II: Extrusionsanlagen. Hanser, München (1989)
76. Standke, R.: Entwicklung eines Programms zur Simulation des Verteilverhaltens von Kleiderbügelverteilern. Unpublished Studienarbeit at the IKV, Aachen (1988)
77. Helmy, H.: Aspects of the design of coathanger dies for cast film and sheet applications. Adv. Polym. Eng. 7 (1987) p. 1

78. Röthemeyer, F.: Bemessung von Extrusionswerkzeugen. Maschinenmarkt 76 (1979) 32, pp. 679–685
79. Fischbach, G.: Internal lecture (Promotionsvortrag) at the IKV, Aachen (1988)
80. Wortberg, J.: Werkzeugauslegung für Ein- und Mehrschichtextrusion. Thesis at the RWTH Aachen (1978)
81. Procter, B.: Flow analyses in extrusion dies. SPE J. 28 (1972) 2, pp. 34–41
82. McKelvey, J. M. and Ito, K.: Uniformity of flow from sheeting dies. Polym. Eng. Sci. 11 (1971) p. 3
83. Vergnes, B. et al.: Berechnungsmethoden für Breitschlitz-Extrusionswerkzeuge. Kunststoffe 70 (1980) p. 11
84. Kral, V. et al.: Modelling of flow in flat film extrusion heads. In: Progress and Trends in Rheology II. Springer, Berlin (1988) pp. 397–400
85. Schiedrum, H. O.: Auslegung von Rohrwerkzeugen. Plastverarbeiter (1974) 10, pp. 1–11
86. Kress, G.: Auslegung von Schlauchköpfen, Kühl- und Flachlegevorrichtungen. In: Extrudieren von Schlauchfolien. VDI-Verl., Düsseldorf (1973)
87. Kreth, N. and Seibel, S.: Rohrextrusion. In: Bührig-Polaczek, A.; Michaeli, W.; Spur, G. (Eds.): Handbuch Umformen. Hanser, München (2013)
88. Kleindienst, U.: Fließmarkierungen durch Dornhalterstege beim Extrudieren von Kunststoffen. Kunststoffe 63 (1973) 7, pp. 423–427
89. Harnischmacher, R.: Längsmarkierungen in extrudierten Kunststofferzeugnissen – Experimentelle Untersuchungen mit einer Modellflüssigkeit. Thesis at the University of Stuttgart (1972)
90. Caton, J. A.: Extrusion die design for blown film production. Br. Plast. (1971) 4, pp. 140–147
91. Worth, R. A., Bradley, N. L., Alfrey, T., Jr. and Maack, H.: Modifications to weld lines in extruded thermoplastic pipe using a rotating die system. Polym. Eng. Sci. 20 (1980) p. 8
92. Schenkel, G.: Rheologische Formgebung einfacher Spritzwerkzeuge für Extruder. Kunststoffe 49 (1959) 4, pp. 201–208 and 49 (1959) 5, pp. 252–256
93. Daubenbüchel, W.: Auswirkungen der Konstruktion von Speicherköpfen von Blasformmaschinen auf die Artikelqualität. Kunststoffe 66 (1976) 1, pp. 15–17
94. Boes, D.: Praxisgerechte Auslegung von Schlauchköpfen für Blasformmaschinen. Kunststoffe 67 (1977) 3, pp. 122–125
95. Schiedrum, H. O.: Der Siebkorb-Rohrkopf in der Polyolefin-Großrohr-Produktion. Plastverarbeiter 26 (1975) 9, pp. 515–518
96. Ast, W.: Schlauchfolien- und Rohr-Extrusionswerkzeuge mit Wendelverteiler. Kunststoffe 66 (1976) 4, pp. 186–192
97. Fischer, P. and Wortberg, J.: Neue Konstruktionen aktueller denn je, Plastics Special (1999) 10, pp. 50–55
98. Wortberg, J.: Neue Wendelverteiler-Werkzeuge, Kunststoffe 88 (1998) 2, pp. 175–180
99. Ast, W. and Pleßke, P.: Wendelverteiler für die Rohr- und Schlauchfolienextrusion. In: Berechnen von Extrudierwerkzeugen. VDI-Verl., Düsseldorf (1978)
100. Rotter, B.: Rheologische und mechanische Auslegung von Radialwendelverteilerwerkzeugen, Dissertation at RWTH Aachen (2002)
101. Backmann, M. and Sensen, C.: Blasfolienextrusion. In: Bührig-Polaczek, A.; Michaeli, W.; Spur, G. (Eds.): Handbuch Umformen. Hanser, München (2013)
102. N. N.: Pipe heads for high throughput. Mod. Plast. Int. 54 (1976) 9, pp. 18–20
103. Dobrowsky, J.: Wendelverteilerwerkzeuge in der Rohrproduktion. Kunststoffe 78 (1988) 4, pp. 302–307
104. Anger, A.: Verstellbare Rohrspritzwerkzeuge. Kunststoffberater 22 (1977) 10, pp. 502–504

105. De Zeeuw, K.: Untersuchungen zur Qualitätssicherung von Kunststoffrohrschweißnähten. Thesis at the RWTH Aachen (1978)

106. N. N.: Neue Zweifach-Werkzeuge für die Extrusion von Hart-PVC-Druckrohren. Kunstst. Plast. 16 (1969) 3, p. 91

107. Liters, W.: Strangpreßköpfe für die Herstellung von Rohren und Schläuchen. Gummi Asbest Kunstst. 22 (1969) 11, pp. 1196–1202 and 22 (1969) 12, pp. 1336–1346

108. N. N.: K'86: Trends – Folienextrusion. Plastverarbeiter 37 (1986) 12, pp. 124–125

109. Ast, W. and Hershey, G. F.: Entwicklungstendenzen für die Werkzeuggestaltung und Nachfolgevorrichtungen unter besonderer Berücksichtigung der Aufwicklung bei der Extrusion von Schlauchfolien. Plastverarbeiter 29 (1978) 1, pp. 5–10

110. Predöhl, W.: Untersuchungen zur Herstellung von Blasfolien aus Polyethylen niedriger Dichte. Thesis at the RWTH Aachen (1971)

111. Kress, G.: Streifenbildung bei Schlauchfolien – Einfluß von Fließkanaloberflächen. Kunststoffe 65 (1975) 8, pp. 456–459

112. Hensen, F. and Augustin, G.: Neue Hochleistungs-Anlage zum Herstellen von Schlauchfolien. Kunststoffe 59 (1969) 1, pp. 2–8

113. Plajer, O.: Schlauchkopfgestaltung beim Extrusionsblasformen (Teil I–III). Kunststofftechnik 11 (1972) 11, pp. 297–301; Kunststofftechnik 11 (1972) 12, pp. 336–340; Kunststofftechnik 12 (1973) 1/2, pp. 18–23

114. Schenkel, G.: Kunststoff-Extrudertechnik. Hanser, München (1963)

115. Kraemer, H. and Onasch, J.: Coextrusion: Märkte, Verfahren, Werkstoffe. Kunststoffberater 33 (1988) 7/8, pp. 62–67

116. Fritz, H. G. and Maier, R.: Zur Konstruktion von Speicherköpfen für Blasformmaschinen. Kunststoffe 66 (1976) 7, pp. 390–396

117. Schneiders, A.: Extrusions-Blasformen. Kunststoffe 67 (1977) 10, pp. 598–601

118. N. N.: Leaflet of Battenfeld-Fischer Blasformtechnik GmbH, Troisdorf-Spich, Germany

119. Rao, N.: Strömungswiderstand von Extrudierwerkzeugen verschiedener geometrischer Querschnitte. In: Berechnen von Extrudierwerkzeugen. VDI-Verl. Düsseldorf (1978)

120. N. N.: Kunststoff-Verarbeitung im Gespräch. 3: Blasformen. Company paper of BASF, Ludwigshafen

121. Tusch, R. L.: Blow moulding die design. SPE J. 29 (1973) 7, pp. 632–636

122. Plajer, O.: Werkzeuge für die Extrusions-Blastechnik. Plastverarbeiter 18 (1967) 10, pp. 731–737

123. Maier, R.: EP-OS 0 118 692 A2 (1984) Battenfeld-Fischer Blasformtechnik (European patent)

124. Esser, K.: Automation beim Blasformen – aus der Sicht des Maschinenherstellers. In: Menges, G. and Recker, H. (Eds.): Automatisierung in der Kunststoffverarbeitung, Hanser, München (1986)

125. Völz, V.: Zusammenspiel von axialer und radialer Wanddickenregulierung bei der Herstellung hochwertiger Hohlkörper. Tech. Conference "Fortschrittliche Blasformtechnik," Süddeutsches Kunststoffzentrum SKZ, Würzburg (1985)

126. Völz, V. and Feuerherm, H.: Produktionssichere Hohlkörperfertigung durch optimale Vorformlinggestaltung (Teil 1). Kunststoffberater 28 (1983) 1/2, pp. 17–22

127. Pickering, J.: The blow moulding of PVC bottles. PRI, PVC Conference (1978)

128. Junk, P. B.: Betrachtungen zum Schmelzeverhalten beim kontinuierlichen Blasformen. Thesis at the RWTH Aachen (1978)

129. Coen, G.: Konstruktion von Düsenwerkzeugen mittels CAD. Techn. Mitt. Krupp Werksber. 43 (1985) 3, pp. 119–122

130. Wyeth, N. C.: Pre-entered pressure coating die. SPE Tech. Pap. 11 (1965) II/2 pp. 1–2 and SPE J. 21 (1965) 10, pp. 1171–1172

131. N. N.: Wire 84–9. Internationale Drahtausstellung. Kautsch. Gummi Kunstst. 37 (1984) 9, pp. 790–791

132. N. N.: Kabel und isolierte Leitungen. VDI-Gesellschaft Kunststofftechnik, Düsseldorf (1984)

133. Okazaki, N.: Absolute non-eccentric cross head of screw extruder. Wire (1966) 4, pp. 574–578; 625–627

134. Mair, H. J.: Fernmeldekabel-Isolierung mit Polyethylen auf HochgeschwindigkeitsAnlagen. Kunststoffe 59 (1969) 9, pp. 535–539

135. Boes, D.: Schlauchköpfe für das Blasformen, Gestaltungskriterien für den Dornhalterkopf. Kunststoffe 72 (1982) 1, pp. 7–11

136. Masberg, U.: Betrachtungen zur geometrischen Gestaltung von Dornhalterköpfen. Kunststoffe 71 (1981) 1, pp. 15–17

137. Jacobi, H. R.: Berechnung und Entwurf von Schlitzdüsen. Kunststoffe 17 (1957) 11, pp. 647–650

138. Plajer, O.: Praktische Rheologie für Kunststoffschmelzen. Plastverarbeiter 23 (1972) 6, pp. 407–412

139. Fenner, R. T. and Nadiri, F.: Finite element analysis of polymer melt flow in cable-covering crossheads. Polym. Eng. Sci. 19 (1979) p. 3

140. Horn, W.: Auslegung eines Pinolenverteilersystems. Unpublished Diploma thesis at the IKV, Aachen (1978)

141. Müller, T.: Programm zur Betriebspunktsimulation von Wendelverteilerwerkzeugen. Unpublished Studienarbeit at the IKV, Aachen (1988)

142. Carley, J. F. et al.: Realistic analysis of flow in wire-coating dies. SPE Tech. Pap. 24 (1978) pp. 453–461

143. N. N.: Kunststoff-Verarbeitung im Gespräch 2: Extrusion. Brochure of the BASF, Ludwigshafen, 2nd ed. (1982)

144. Bäcker, F.: Weiterentwicklung eines Programmsystems zur Theologischen Auslegung von Extrusionswerkzeugen. Unpublished Studienarbeit at the IKV, Aachen (1989)

145. Schiedrum, H. O.: Extrudieren von PVC-Profilen am Beispiel des Fensterprofils. Kunststoffe 65 (1975) 5, pp. 250–257

146. Limbach, W.: Extrudieren von Profilen. Kunststoffe 71 (1981) 10, pp. 668–672

147. Lyall, R.: The PVC window: Materials and processing technology. Plast. Rubber Int. 8 (1983) 3, pp. 92–95

148. Schiedrum, H. O.: Profilwerkzeuge für das Extrudieren von PVC. Ind. Anz. 90 (1968) 102, pp. 2241–2246

149. Limbach, W.: Werkzeuge und Folgeaggregate für das Extrudieren von Profilen. In: Extrudieren von Profilen und Rohren. VDI-Verl., Düsseldorf (1974)

150. Fischer, P.: Herstellen von Profilen. VDI-Lehrgang Extrudieren von thermoplastischen Kunststoffen. Düsseldorf

151. Ehrmann, G. et al.: Rheometrie hochpolymerer Schmelzen – kritische Betrachtung und Ergebnisse. Kunststoffe 64 (1974) 9, pp. 463–469

152. Vinogradov, G.K and Malkin, A. Ya.: Rheological properties of polymer melts. J. Polym. Sci., Polym. Phys. Ed. 4 (1966) pp. 135–154

153. Almenräder, A.: Simulation des Vernetzungsprozesses bei Einsatz eines Kabelummantelungswerkzeuges mit Scherspalt. Unpublished report at the IKV, Aachen (1977)

154. Matsushita, S.: Recent progresses of profile extrusion dies. Jpn. Plast. 8 (1974) 4, pp. 25–29

155. Krämer, A.: Herstellen von Profilen aus PVC-hart. Kunststoffe 59 (1969) 7, pp. 409–416

156. Kaplin, L.: Berechnung und Konstruktion der Eingangszone in Extrusionswerkzeugen. Plast. Massy (1964) 1, pp. 39–46

157. Extrusion dies. Plastics (1969) 5, p. 520–521; 536
158. N. N.: Anwendungstechnisches Handbuch. Brochure of Rehau, Rehau, Germany
159. Kumar, R. and O'Brien, K.: Profile extrusion of PVC through rectangular dies. Adv. Polym. Technol. 4 (1985) p. 3
160. Schwenzer, C.: Finite Elemente Methoden zur Berechnung von Mono- und Coextrusionsströmungen. Dissertation at the RWTH Aachen (1988)
161. Barth, H.: Extrusionswerkzeuge für PVC-Hartschaumprofile. Kunststoffe 67 (1977) 3, pp. 130–135
162. Steigerwald, F.: Erfahrungen über die Herstellung von Profilen aus PVC-hart-Strukturschaumstoff. Plastverarbeiter 26 (1975) 10, pp. 3–7
163. Bush, F. R. and Rollefson, G. C.: The ABC of coextruding foam-core ABS pipe. Mod. Plast. Int. 59 (1981) pp. 38–40
164. Breuer, H.: Maschinen und Anlagen zur Herstellung von geschäumten PS-Folien. Plastverarbeiter 27 (1976) 10, pp. 539–545
165. Han, C. D. and Villamizer, C. A.: Studies on structural foam processing. I: The rheology of foam extrusion. Polym. Eng. Sci. 18 (1978) 9, pp. 687–698
166. Rapp, B.: Schlauchfolien aus Polystyrol. In: Extrudieren von Schlauchfolien. VDI-Verl., Düsseldorf (1973)
167. Lauterberg, W.: Dickentoleranzen bei Polyethylenschaumfolien. Plaste Kautsch. 25 (1978) 5, pp. 294–295
168. N. N.: DE-OS 2 249 435 (German patent)
169. N. N.: DE-OS 2 359 282 (German patent)
170. Barth, H.: Extrudieren und Spritzgießen von PVC-Weichschaum. Kunststoffe 67 (1977) 11, pp. 674–680
171. N. N.: Cellidor. Company paper of Bayer AG, Leverkusen, Germany
172. Lüers, W.: Verfahren und Vorrichtungen zum Herstellen von Rohren und Schläuchen aus Kautschuken oder Kunststoffen mit Verstärkungseinlagen. Kautsch. Gummi Kunstst. 24 (1971) 1, pp. 75–79 and 24 (1971) 4, pp. 179–184 and 24 (1971) 5, pp. 240–243
173. Siebel-Achenbach, J.: Querspritz-Verschweißkopf ummantelt Schläuche. Maschinenmarkt 90 (1984) 67, pp. 1515–1516
174. N. N.: The Netlon story. Shell Polymers 6 (1982) 3, pp. 79–83
175. N. N.: Leaflet of Netlon, Blackburn, England, GB
176. Morohashi, H. and Fujihara, T.: Recent operational experiment with screw-die extruder. Jpn. Plast. 4 (1970) 7, pp. 21–26
177. N. N.: Leaflet of Ikegai Iron Works, Tokyo, Japan
178. Ishida, M. and Ito, K.: Theoretical analysis on flow of polymer melts in a screw die. Int. J. Polym. Mater. 6 (1977) pp. 85–107
179. Han, C. D. and Charles, M.: A method for determining the melt elasticity in capillary flow. Polym. Eng. Sci. 10 (1970) p. 3
180. Harris, H. E.: Closed loop digital control of extrudate thickness and profile. SPE Tech. Pap. (1973) pp. 39–42
181. Prall, G.: Minimizing weld line effects in blown PE film. Mod. Plast. 47 (1970) 11, pp. 118–120
182. Plajer, O.: Praktische Rheologie für Kunststoffschmelzen. Zechner & Hüthig, Speyer (1970)
183. Wagner, H. M.: Zur numerischen Berechnung von Strömungsfeldern in Maschinen und Werkzeugen der Kunststoffverarbeitung. Kunststoffe 67 (1977) 7, pp. 400–403

# 6 Coextrusion Dies for Thermoplastics

Many products made from plastics cannot meet the requirements imposed on them when produced from a single material. In such cases, the solution may be to combine different materials in a multilayer structure and, thus, create a new product that combines the positive properties of the individual components. Examples include

- multiple-layer flat sheets (films) and tubular film,
- multiple-layer hollow bodies,
- cables with multiple-layer insulation,
- profiles with jackets or soft lips,
- and many more.

Some of such products could be produced by multiple-step extrusion. In such a case, one material would be processed first, in order to prepare a supporting layer or to coat a wire with a smooth layer. Subsequently, after a full or partial cooling of the first semifinished product, the application of the second or several additional layers, as the case may be, would follow (compare also Sections 5.3.2.4 and 5.7.1). In most cases, however, the coextrusion is more suitable because of economic and technological efficiency [1–10]. In doing so, two or more melts are extruded with a single die in the three following ways:

- Completely separately
- First separately and then together
- Together

# 6.1 Designs

## 6.1.1 Externally Combining Coextrusion Dies

These dies are most frequently made with only two exit slits (dual-slot dies) because more slits would make the design and manufacture of the die extremely complicated [11].

As shown in Fig. 6.1 (left) for a flat sheet die, the two melt streams flow through two fully separated channels through the die and are brought together only after emerging from the orifice.

The two melt layers can be brought together with the help of a pressure roll (Fig. 6.1, left). Such a device is particularly necessary when a third layer is added or if the extrusion speed is very high; in these cases, the air enclosed between the layers is entrained and may lead to defects in adhesion between layers [11]. The distribution of the two melts in the die can be controlled independently. The combining of the layers is simpler in the tubular film die shown in Fig. 6.1 (right).

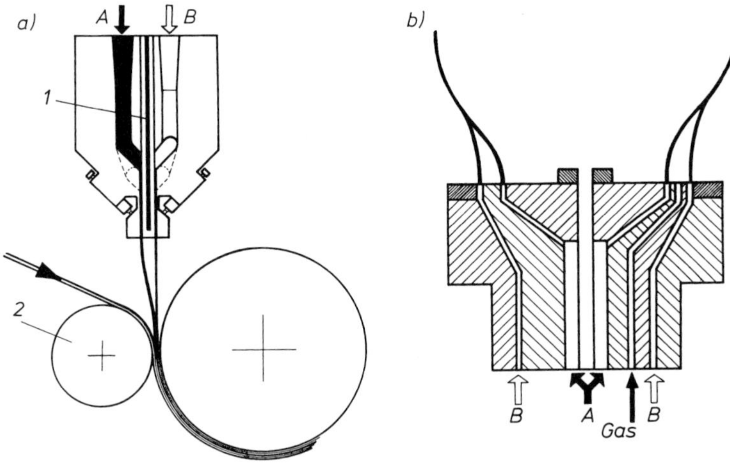


**Figure 6.1** Dual-slot coextrusion dies: (a) flat slit die: 1: heat separation, 2: pressure roll; (b) blown film die

To improve the adhesion of the layers, a stream of gas activating their surfaces can be blown into the space between the melts after they leave the die [12]. In this case, the flow regulation is possible only for the outside layer by the sliding movement of the die lip.

With dual-slot dies, it is usually very simple to thermally insulate the flow channels from each other, such as by using an air gap. This arrangement allows the processing of melts with very different viscosities and at different temperatures.

The disadvantages are that two orifices must be adjusted, the difference in neck-in of both layers must be minimal, and the formation of fumes between layers can lead to organoleptic problems. Moreover, thin layers often cool down too rapidly, causing adhesion problems [13].

### 6.1.2 Adapter (Feedblock) Dies

A multilayer extrudate can be produced by conventional wide slit dies in flat film extrusion when an adapter is implemented prior to the wide slit die. The adapter gathers the individual melt streams and feeds them to the die inlet. The multiple layers then stream together through the die and leave it as a coextrudate.

The advantage of this procedure is that any number of individual layers can be combined. The disadvantage is that all the materials must have almost identical flow behavior and processing temperatures [14]. In spite of this, most coextrusion lines today use feedblock systems [14,16].

There are in essence three types of adapters, which will be described below.

*Fixed Adapter* (Dow System)

In this adapter, as shown in Fig. 6.2 in connection with a flat sheet die, the gauges of the individual layers are determined by the cross section of the feed channels and the set mass flow rates. It is important that the individual melts meet in the die with approximately equal velocities [14–16].

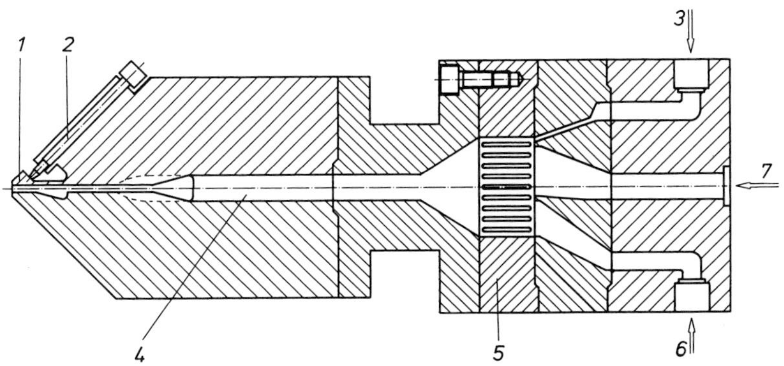


**Figure 6.2** Feedblock (Dow system) for bringing melt streams together before the flat slit die: 1: flex lip, 2: pressure bolt, 3: cover layer material, 4: melt channel with a flow restrictor, 5: adapter, 6: base layer material, 7: main layer material

*Slide Adapter* (Reifenhäuser System)

In the so-called slide adapter (Fig. 6.3), a slide serves the purpose of allowing the mass streams to meet at equal velocity. Also, it can, in a similar fashion as the choker bar, regulate the melt distribution across the width. For that purpose it either has a profile or is made in sections on the side of the melt. When assembled or exchanged, the slides can be inserted or removed from the side as a cassette without the need to take apart the connection between the extruder and the die [14].

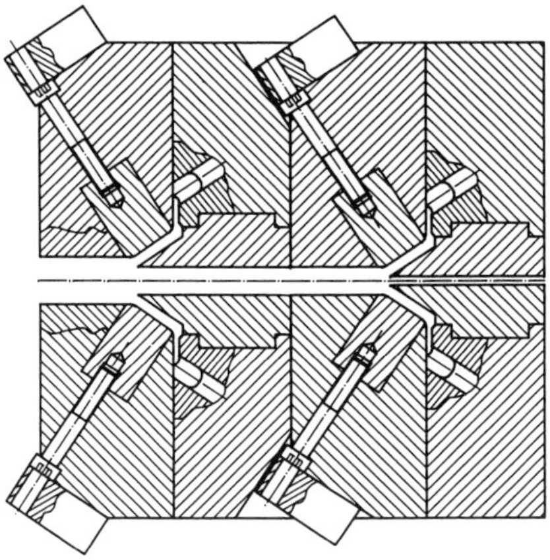

**Figure 6.3** Sliding adapter (Reifenhäuser system)

Another version of a nine-layer slider adapter is shown in Fig. 6.4. The system consists of nine distribution areas with individually exchangeable inserts to control layer thicknesses and local melt velocities.

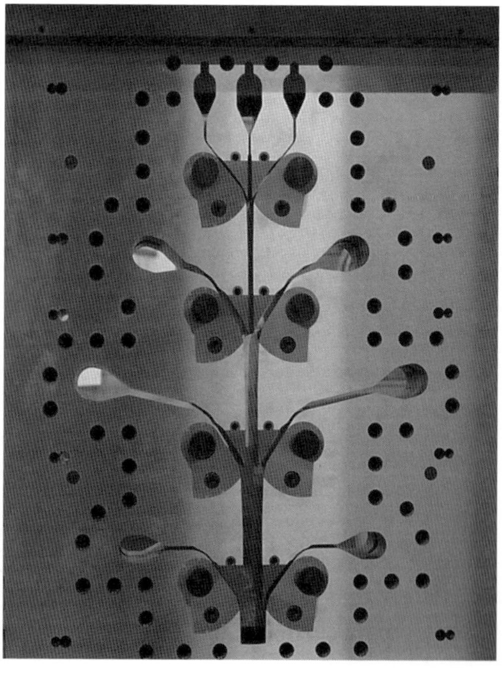

**Figure 6.4** Nine-layer sliding adapter (Reifenhäuser system)

*Vane Adapter* (Cloeren System)

The vane adapter (Fig. 6.5) differs from the slide adapter in that the elements of the system that can control the process of joining the streams are in the shape of vanes that can be turned. Since this can happen even during the operation, there is a choice either to set a certain position or let the setting be established by the melt streams [16]. Also, the vanes can have profiles in order to control the distribution of the flow over the width of the channel [16]. Because of these advantages, this system is often used.

Beyond that, adapters can also be classified as internal and external devices that can be connected in series and combined so as to attain the required configuration of layers [3].

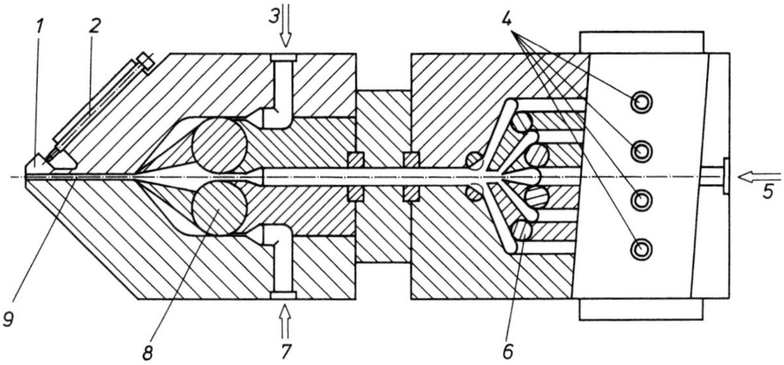


**Figure 6.5** Flat slit die with a vane adapter (Cloeren system): 1: flex up, 2: pressure screw, 3: cover layer material, 4: possible entry for other materials (adhesive, barrier material), 5: main layer material, 6: vane adapter, 7: base layer material, 8: vane adapter for cover and base layers, 9: flow restriction zone

In an inner layer adapter (Fig. 6.6, left), one melt is extruded into the other. This is a suitable method if a thin layer is embedded between two thick ones or if the internal layer should have a minimum contact with the walls [3].

The outer layer adapter is used, as a rule, to apply a new layer from the outside onto other layers (Fig. 6.6, left).

In order to be able to change the order of the layers simply and rapidly, an exchangeable distributor (feed) block that defines the sequence of the layers can be used [16]. When the arrangement of the layers is being changed by an extruder being stopped, one or more layers are eliminated. It is advisable to install check valves to prevent the return flow of the melt into the idle extruder [3].

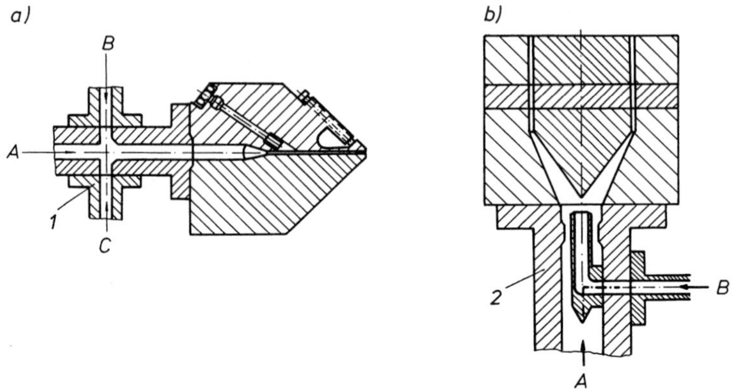


**Figure 6.6** Adapter dies for inside and outside layers: (a) flat slit die (three layers), 1: outside layer adapter; (b) blown film die (for two layers with a mandrel support, 2: inside layer adapter

### 6.1.3 Multimanifold Dies

In this type of die, each melt is first fed separately and distributed into the desired form, and then these partial streams are combined just before the die exit (). Each melt stream flows through a separate and optimized distribution unit (e.g., coathanger die for flat films, spiral mandrel die for blown films) and can be adjusted individually, provided that appropriate adjustment is available. An adjustment of the lips or die orifice allows for control of the overall gauge of the coextrudate. The method of combining the melts inside the die under pressure also improves the mutual adhesion of the layers.

Another advantage is that, with multimanifold dies, materials with vastly different flow behavior and different melt temperatures can be processed [14]. The thermal insulation of the individual channels from each other is very difficult to solve, considering the already complicated design of the die. As a result, this type of die for a combination of more than four layers becomes very complex and costly because a large building space is required and for each melt a distribution system needs to be manufactured.

### 6.1.4 Layer Multiplication Dies

With this new development layer, compositions with nearly unlimited numbers of layers can be created. Two variants can be distinguished: an interfacial surface generator (a Dow system) and a nanolayer feedblock system (a Cloeren system).

With the interfacial surface generator (ISG) system, an initial mono- or multilayer flow is duplicated once or several times. The layer multiplication is created by splitting up the melt flow into two or more streams. Afterward, each stream is widened and recombined in the die. Figure 6.7 shows the general principle of the flow channels in an ISG: a two-layer melt enters the ISG and is split up into two streams that flow parallel at first. The streams are widened separately and afterwards put onto each other. In such a splitting of the volume flow, the number of layers is always doubled. Several options of parallel or sequential splitting of the melt are imaginable, and by this the number of layers can be increased with no limit in theory.

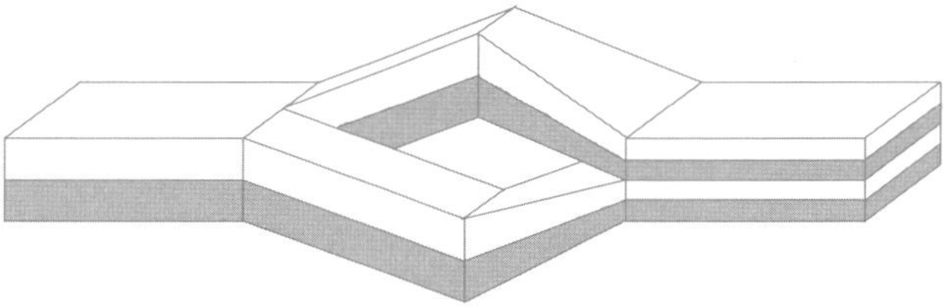

**Figure 6.7**
One stage of an interfacial surface generator [Dow]

In practical use, the number of layers is restricted, however, especially by the die length and the resulting thermal decomposition of material due to very long residence times in the die as well as the very high pressure loss.

The nanolayer feedblock system uses a sequential layer composition. In a feedblock adapter the melt is split sequentially multiple times, and then the streams are put onto each other. Additionally, top layers are conventionally added in the feedblock. The whole composition is delivered afterwards to a wide slit die and formed into a film. Figure 6.8 shows a feedblock with nanolayer technology.

**Figure 6.8** Sequential layer creation in a feedblock with nanolayer technology [Cloeren]

A comparison of the two technologies is given in Table 6.2.

**Table 6.2** Comparison of Two-Layer Multiplication Technologies

| Interfacial Surface Generator | Nanolayer Technology |
|---|---|
| Many thin layers (>100) | Fewer thin layers (10–30) |
| Modular and flexible die design with expansion possible | No modular design and no expansion possible |
| Fixed splitting up of the film, which is defined by the die design | Each layer thickness can be controlled individually and flexibly |
| Long joint flow paths | Short joint flow paths |
| Danger of layer instabilities | Less danger of layer instabilities |

With these layer compositions of up to several hundred layers, for example, unique mechanical and optical properties can be created. For example, the reflection and transmission properties of films can be customized by the multilayer composition.

## 6.2 Applications

Coextruded composite structures are today state of the art in many areas of polymer processing. In such cases, the different melt streams can either envelop each other or can be arranged on top of each other or flow side by side. In the most frequent and significant applications, either enveloping (such as for cable insulation or blown film) or an arrangement of layers on top of each other is frequently used. Because of that, several typical representative designs of extrusion dies for these applications will be discussed in the following sections.

### 6.2.1 Film and Sheet Dies

Adapter dies are used in most cases for the production of coextruded flat film and sheet. If the materials differ greatly in their flow behavior, multimanifold dies are employed. The melt distribution across the width is accomplished by the use of the melt distributors (manifolds) discussed in detail in Section 5.2.

The individual melts in both types of die generally flow in layers over each other; however, there are cases where one layer is enveloped by the other (melt encapsulation). This is done, for example, so the trim is made from one material only, which can be reground and then reused for extrusion [17], or to avoid wasting an expensive barrier material by trimming, or if the inside layer has to be protected from the environment.

For the packaging of food and beverages, coextrudates with seven or even more layers are common now. For secondary packaging, such as stretch films, up to five layers of different polyolefin types are standard in the industry.

**Melt Encapsulation**

In flat film extrusion, the width of a film is reduced after the die exit because of the neck-in effect. This means thick sections are generated at the edges of the film that need to be trimmed in most cases in order to stay within the thickness tolerances of the film. Another problem in coextrusion is that a low-viscosity melt envelopes a higher viscosity melt at the edges of a film (see Section 6.4). This flow behavior leads to an irregular distribution of single-layer thicknesses. For multilayer films, this edge trim is waste because the multimaterial trim cannot be led back into the production process directly. A mechanical decomposition of the several layers would cause an enormous effort. With a melt encapsulation, the average amount of waste due to edge trim can be reduced from 15 to 20% (depending on the film width) to less than half that.

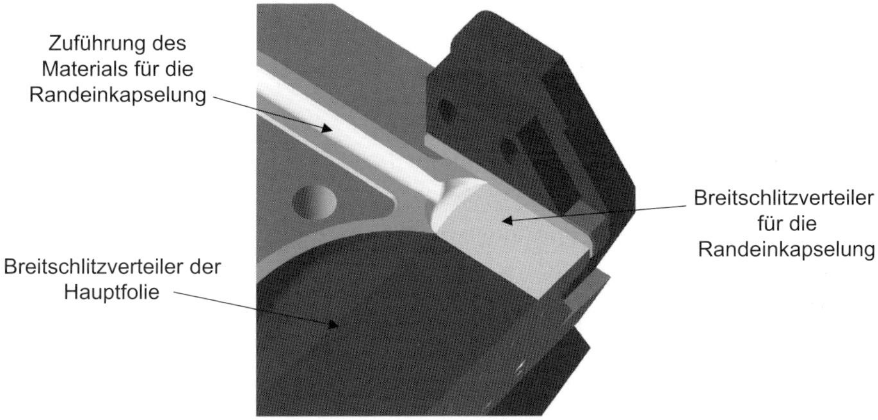


**Figure 6.9** Flow channel with melt encapsulation in a die [Verbruggen N. V.]

With a melt encapsulation, a less costly and virgin material is added in the boundary area of a flat film die. The virgin material of a first edge trim can be led back completely as pellets into the production process. With a second edge trim, the multimaterial film is reduced to a much smaller strip, resulting in a smaller fraction of expensive barrier material in the trimmed strip.

Another reason for melt encapsulation of a flat film with gas barrier functionality is connected to the characteristic of such barrier materials. Typically, EVOH is used as barrier material in coextrusion. In a layer-wise setup, each layer has direct contact with the flow channel surfaces. However, EVOH tends to stick at metallic surfaces. By enveloping EVOH using melt encapsulation, a sticking of the barrier layer is prevented. In addition, an encapsulation helps to avoid a migration of other media into the barrier layer, as the barrier functionality of EVOH against oxygen would greatly decrease with increasing moisture content [18].

For high extrusion speeds, melt encapsulation has a stabilizing effect at the film edges.

In an extrusion die, the encapsulation material can be added either in the adapter system or in the flat film die close to the die exit. In Fig. 6.9 a direct feed-in of the encapsulation material is depicted for the edge layer via additional melt inlets at the die back side. Only shortly before exiting the die does the encapsulating material come in contact with the core materials.

When using melt encapsulation, it is observed that highly viscous materials as encapsulating material are more beneficial for a homogeneous average layer thickness in the transition area than low-viscosity materials. Hence, for low-viscosity encapsulating materials, their processing temperature should be set rather low in order to increase their viscosity.

One drawback of melt encapsulation is higher die manufacturing costs, due to the additional flow channels for the encapsulating material.

### 6.2.2 Blown Film Dies

Multimanifold dies have been used for years for the coextrusion of blown film. The distribution of the melt along the circumference in these dies is accomplished by a spiral mandrel [19–21]. Figure 6.10 shows a typical representative of this type: it is a die for the production of a three-layer film. This coextrusion die, too, is equipped with a revolving mechanism and internal cooling, which are both commonly used in blown film equipment. The most common method of bringing the melts together for general multilayer structures is shown on the left half of the die. The right half of the die in Fig. 6.10 shows the case where the core is a thin tie layer or a barrier [20].

Blown film made with a minimum of five up to 11 layers for use in food packaging is state of the art today [21,22,43,47]; technical feasibility has already been shown for 13 layers.

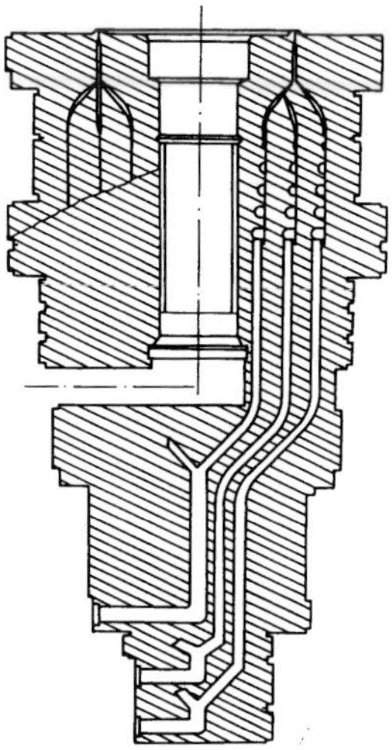

**Figure 6.10** Three-layer coextrusion die for blown film; left half: conventional layer composition, right half: addition of a tie or barrier layer [Battenfeld]

### 6.2.3 Dies for the Extrusion of Parisons for Blow Molding

Coextrusion has also become commonplace in the continuous extrusion of parisons for blow molding in many areas of application. Parisons with up to seven layers are extruded and blow-molded into hollow parts used for food packaging [23,24]. For the distribution of melt around the circumference, concentrically arranged hollow mandrels with heart-shaped and coathanger manifolds are used. Figure 6.11 shows such a head with seven heart-shaped curves for the extrusion of a six-layer parison (layer 1 is distributed with two heart-shaped curves) [24,25]. Nowadays, parison dies with axial spiral distributors are also available on the market [26,44]. In contrast to side-fed dies, no weld lines occur. In addition, the more compact buildup of spiral mandrel dies offers advantages for multilayer parisons because potentially more parisons can be extruded in parallel or the distance between parisons can be reduced, which corresponds with easier flow balancing. Recently, a radial spiral mandrel die was examined in blow-molding [45].

Coextrusion accumulator heads exist currently only as prototypes or are described in patent applications [24,27–29]. The individual accumulator chambers consist, for example, of concentrically arranged circular channels, from which the melts are ejected by individual annular rams [27,28] or as described in [29] of a single accumulator chamber in which the individual materials are arranged in a laminar form and are forced out by an outside piston and an inside piston. Design and technological problems have prevented this type of head from seeing wider use up to now.

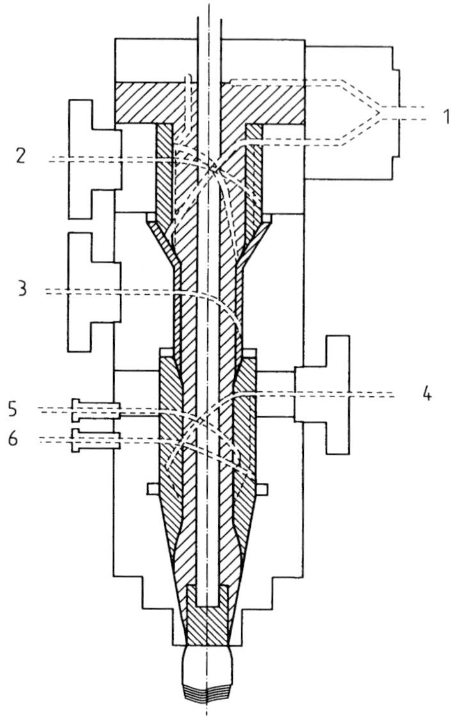


**Figure 6.11** Coextrusion die for one to six layers [KruppKautex]

## 6.3 Computations of Flow and Design

When designing coextrusion dies, a distinction must be made between adapter and multimanifold dies.

The distribution channels for the individual melts in a multimanifold die correspond to the distributors for single-layer extrusion. One extrusion stream, a parallel stream of several different melts through a common flow channel, occurs only in the exit region of the die.

On the other hand, in the adapter die, the melt streams are first brought together and then distributed together in a single manifold die. This means that the die manifold used in the adapter coextrusion must be designed for the flow of multiple layers. In addition, problems such as flow instabilities, rearrangement of the layers, and instabilities at the boundaries (interfaces) of the layers occur frequently in coextrusion. From that it can be concluded that not only the pressure loss of the multilayer flow, but also the position of the boundary (or boundaries) of the layers, the velocity, the temperature, and the shear stress profiles have to be taken into account for the purpose of design.

When computing flow of multilayer streams, the procedure is essentially the same as for the computation of a single-layer flow. However, it has to be taken into account, as a special boundary condition, that the material properties change at the boundaries of the different melts. The location of the boundaries in the flow is not known. It can be determined from the relationship of the volumetric flow rates and the viscous properties of the individual materials. Moreover, the location of the maximum velocity in the stream is not known either, perhaps with the exception of a symmetrical multilayer flow. The procedure of the computation of the flow for a flat multilayer stream is demonstrated below.

The majority of actual coextrusion flows can be either computed as the flow in a flat slit or at least approximated by it.

The following assumptions are made for the computation (for the coordinate system see Fig. 6.12):

- Wall adhesion of the melts
- Adhesion at the boundaries of the melt layers
- Laminar, incompressible flow
- Volume forces negligible when compared to friction forces
- Isothermal flow
- Neglecting the elastic behavior of the materials

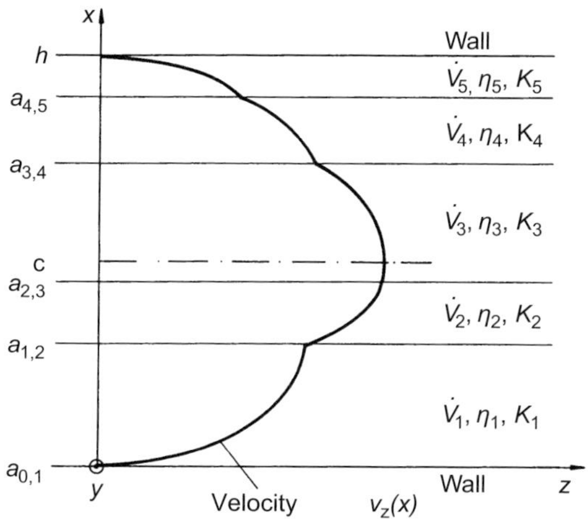


**Figure 6.12** Velocity profile in a multilayer flow

Starting with a simplified momentum balance,

$$\frac{\partial \tau}{\partial x} = \frac{\partial p}{\partial z} \tag{6.1}$$

the relationship between the shear stress and the pressure loss is obtained by integration:

$$\tau(x) = \frac{\partial p}{\partial z}(x - c), \tag{6.2}$$

where the boundary condition $\tau = 0$ at the velocity maximum $x = c$ was used. Introducing the relationship between shear stress and shear rate into Equation (6.2), we obtain

$$\eta(\dot{\gamma}(x)) \cdot \dot{\gamma}(x) = \frac{\partial p}{\partial z}(x - c). \tag{6.3}$$

and with the relationship between the flow velocity and shear rate:

$$\eta\left(\frac{\partial v_z(x)}{\partial x}\right) \cdot \frac{\partial v_z(x)}{\partial x} = \frac{\partial p}{\partial z}(x - c). \tag{6.4}$$

The boundary conditions for the solution of Equation (6.4) are

- the adhesion to the wall

$$v_z(x = 0) = 0 \tag{6.5}$$

and

$$v_z(x = h) = 0 \tag{6.6}$$

- the adhesion at the boundary layers

$$v_{z,m}(a_{m,n}) = v_{z,n}(a_{m,n}) \tag{6.7}$$

with

$1, m, n$ Indices of different layers

$v_m, v_n$ Flow velocity in layers $m$ and $n$

$a_{m,n}$ Position of boundary between layers $m$ and $n$

and the relationship between the flow velocity and the volumetric flow rate:

$$\dot{V}_m = \int_{a_{1,m}}^{a_{m,n}} v_{z,m}(x)\,dx. \tag{6.8}$$

A closed solution of Equation (6.4) is usually not possible because the relationship between the viscosity and shear rate is different in the individual layers.

Depending on which material law is chosen for the description of the relationship between viscosity and shear rate, different numerical procedures must be applied:

- The solution under the assumption of Newtonian flow of the individual melts and thus of constant viscosity between the boundary layers is truly simple, because now it is possible to integrate Equation (6.4) analytically and hence to compute the velocity profile of the flow [30]. A simple way to estimate the characteristics for two simple forms of flow, namely symmetrical three-layer flow and two-layer flow, this method of solution will be discussed more accurately below (Fig. 6.13).

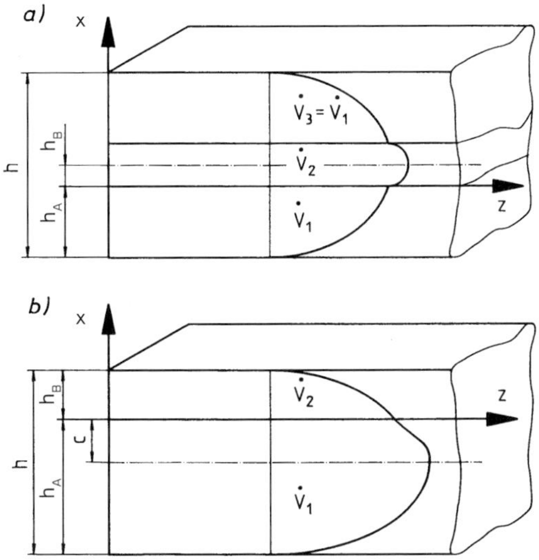


**Figure 6.13** Simple multilayer flows: (a) symmetrical three-layer flow, (b) two-layer flow

- Also when applying the power law by Ostwald and de Waele, an analytical integration of Equation (6.4) is possible. In principle, any multilayer flow can be described using this method [31–33].
- When applying the finite difference method (FDM) or the finite element method (FEM) for Equation (6.4), the use of any viscosity law at any layer arrangement is possible [30,34,35]. The FDM-based procedure, too, will be described in more detail below.

### 6.3.1 Computation of Simple Multilayer Flow with Constant Viscosity

When neglecting the dependence of viscosity on the shear rate, Equation (6.4) can be integrated:

$$v_z(x) = \frac{\partial p}{\partial z} \frac{1}{\eta} \left( \frac{x^2}{2} - cx + k \right) \tag{6.9}$$

with

$k =$ constant of integration

**Symmetrical Three-Layer Flow**

The next step in the process depends on the form of the flow under consideration. In the symmetrical three-layer flow, a core layer is enclosed by two identical surface layers that have an equal volumetric flow rate (Fig. 6.13(a)).

From symmetry, it follows that the maximum velocity is located in the middle of the flow channel. For further derivation, a coordinate system with the origin at the layer boundary was found to be of advantage. Because of symmetry, it is sufficient to study only one-half of the channel.

With the condition of the adhesion to the wall of the channel, the velocity profile for the boundary layer is

$$v_1\left(-h_{A_1}\right)=0$$
$$v_1(x)=\frac{\partial p}{\partial z}\frac{1}{\eta_1}\left[\frac{x^2}{2}-\frac{h_B}{2}x-\left(\frac{h_{A_1}{}^2}{2}+h_{A_1}\frac{h_B}{2}\right)\right]. \tag{6.10}$$

The velocity profile of the core melt follows from the condition of the adhesion at the boundary:

$$v_1(0)=v_2(0)$$
$$v_2(x)=\frac{\partial p}{\partial z}\frac{1}{\eta_2}\left[\frac{x^2}{2}-\frac{h_B}{2}x-\frac{\eta_2}{\eta_1}\left(\frac{h_{A_1}{}^2}{2}+h_{A_1}\frac{h_B}{2}\right)\right]. \tag{6.11}$$

To determine the still-unknown layer thicknesses $h_{A_1}$ and $h_B$ and the pressure gradient $\frac{\partial p}{\partial z}$, the relationship between the flow velocity and the volumetric flow rate can be utilized:

$$\dot{V}_1=B\cdot\int_{-h_{A_1}}^{0} v_1(x)\,dx, \qquad \text{with } B = \text{width of the slit} \tag{6.12}$$

$$\dot{V}_2=B\cdot 2\int_{0}^{\frac{h_B}{2}} v_2(x)\,dx. \tag{6.13}$$

After integration:

$$\dot{V}_1=B\cdot\frac{\partial p}{\partial z}\cdot\frac{h_{A_1}{}^3}{\eta_1}\left(-\frac{2}{3}-\frac{h_B}{4h_{A_1}}\right), \tag{6.14}$$

$$\dot{V}_2 = B \cdot \frac{\partial p}{\partial z} \frac{h_B{}^3}{8 \cdot \eta_2} \cdot \frac{1}{3} \left[ -2 - \frac{\eta_2}{\eta_1} \left( 3 \left( \frac{2h_{A_1}}{h_B} \right)^2 + 6 \frac{2h_{A_1}}{h_B} \right) \right]. \tag{6.15}$$

The ratio of the layer thicknesses can be determined from the ratio of the volumetric flow rates:

$$\frac{\dot{V}_2}{\dot{V}_1} = \frac{\left( \frac{h_B}{h_{A_1}} \right)^3 + 6 \frac{\eta_2}{\eta_1} \frac{h_B}{h_{A_1}} \left( \frac{h_B}{h_{A_1}} + 1 \right)}{\frac{\eta_2}{\eta_1} \cdot \left( 8 + 3 \frac{h_B}{h_{A_1}} \right)}. \tag{6.16}$$

From a given ratio of flow rates and viscosities, the ratio of the layer thicknesses can be determined from Equation (6.16) by iteration.

The location of the boundary follows from the ratio of the layer thicknesses and the following geometric relation:

$$2h_{A_1} + h_B = h. \tag{6.17}$$

The pressure loss is then obtained, for example, by rearrangement of Equation (6.14):

$$\frac{\partial p}{\partial z} = \frac{\dot{V}_1 \eta_1}{B \cdot h_{A_1}{}^3 \left( -\frac{2}{3} - \frac{h_B}{4h_{A_1}} \right)}. \tag{6.18}$$

### Two-Layer Flow

When studying a two-layer flow (Fig. 6.13(b)), an additional unknown appears during the computation of the location of the velocity maximum. Here it is also advantageous to place the origin of the coordinate system at the layer interface.

From the condition of adhesion to the wall of the channel it follows that

$$v_1(-d_1) = 0, \qquad v_1(y) = \frac{\partial p}{\partial z} \frac{1}{\eta_1} \left[ \frac{y^2}{2} - cy - \left( \frac{d_1^2}{2} + cd_1 \right) \right], \tag{6.19}$$

$$v_2(d_2) = 0, \qquad v_2(y) = \frac{\partial p}{\partial z} \frac{1}{\eta_2} \left[ \frac{y^2}{2} - cy - \left( \frac{d_2^2}{2} - cd_2 \right) \right]. \tag{6.20}$$

The condition of the adhesion at the layer interfaces leads to the relationship of the ratio of layer thicknesses, the ratio of viscosities, and the location of the maximum velocity.

$$v_1(0) = v_2(0),$$
$$c = \frac{d_1}{2}\frac{-\frac{\eta_2}{\eta_1}+\frac{d_2^2}{d_1^2}}{\frac{\eta_2}{\eta_1}+\frac{d_2}{d_1^2}}. \tag{6.21}$$

The still unknown quantities ratio of layer thickness and pressure drop can also be determined here from the relationship between the flow velocity and the volumetric flow rate:

$$\dot{V}_1 = B \cdot \int_{-h_a}^{0} v_1(y)\,dy, \tag{6.22}$$

$$\dot{V}_2 = B \cdot \int_{0}^{h_b} v_2(y)\,dy, \tag{6.23}$$

$$\dot{V}_1 = B \cdot \frac{\partial p}{\partial z}\frac{1}{\eta_1} d_1^2 \left(-\frac{d_1}{3}+\frac{c}{2}\right), \tag{6.24}$$

$$\dot{V}_2 = B \cdot \frac{\partial p}{\partial z}\frac{1}{\eta_2} d_2^2 \left(-\frac{d_2}{3}+\frac{c}{2}\right). \tag{6.25}$$

By combining Equations (6.21) with (6.24) and (6.25), the following results are obtained:

$$\dot{V}_1 = B \cdot \frac{\partial p}{\partial z}\frac{d_1^3}{\eta_1}\frac{(-1)}{12}\frac{\left(3\frac{d_2^2}{d_1^2}+4\frac{d_2}{d_1}+\frac{\eta_2}{\eta_1}\right)}{\frac{\eta_2}{\eta_1}+\frac{d_2}{d_1}}, \tag{6.26}$$

$$\dot{V}_2 = B \cdot \frac{\partial p}{\partial z}\frac{d_2^3}{\eta_2}\frac{(-1)}{12}\frac{\left(\frac{d_2^2}{d_1^2}+4\frac{\eta_2}{\eta_1}\frac{d_2}{d_1}+3\frac{\eta_2}{\eta_1}\right)}{\frac{d_2}{d_1}\left(\frac{\eta_2}{\eta_1}+\frac{d_2}{d_1}\right)}. \tag{6.27}$$

The relationship between the flow rate ratio, the viscosity ratio, and the layer thickness ratio is as follows:

$$\frac{\dot{V}_2}{\dot{V}_1}=\frac{\left(\frac{d_2}{d_1}\right)^4+4\frac{\eta_2}{\eta_1}\left(\frac{d_2}{d_1}\right)^3+3\frac{\eta_2}{\eta_1}\left(\frac{d_2}{d_1}\right)^2}{3\frac{\eta_2}{\eta_1}\left(\frac{d_2}{d_1}\right)+4\frac{\eta_2}{\eta_1}+\frac{d_2}{d_1}+\left(\frac{\eta_2}{\eta_1}\right)^2}. \tag{6.28}$$

The ratio of the layer gauges can be determined by iteration from Equation (6.28). The location of the boundary layer follows from the layer thickness ratio using the geometric relationship

$$d_1+d_2=h. \tag{6.29}$$

The pressure drop is obtained, for example, by rearranging Equation (6.26):

$$\frac{\partial p}{\partial z}=\frac{12\dot{V}_1\eta_1\left(-\frac{\eta_2}{\eta_1}-\frac{d_2}{d_1}\right)}{\left(3\left(\frac{d_2}{d_1}\right)^2+4\frac{d_2}{d_1}+\frac{\eta_2}{\eta_1}\right)d_1^3\cdot B}. \tag{6.30}$$

The procedure demonstrated above allows the computation of the velocity profiles, the location of the interface, and pressure losses in a symmetrical three-layer flow as well as in a two-layer flow in a flat slit at constant viscosities of the melts.

For the computation of the viscosity, for example, the method of representative viscosity can be applied [34] (see also Chapter 3).

The representative shear rate is determined from the total volumetric flow rate; for the computation of the representative viscosity for the individual melts, the shear rate found above enters the formula for viscosity:

$$\dot{\gamma}_{\text{rep}}=\frac{6\cdot\dot{V}_{\text{total}}}{B\cdot h^2}\cdot e_{\square}, \tag{6.31}$$

$$\eta_{\text{rep}}=\eta\left(\dot{\gamma}_{\text{rep}}\right). \tag{6.32}$$

This procedure is appropriate as long as the viscosities of the melts involved do not differ significantly. Otherwise, it is better to determine the shear rate for each individual layer. The following procedure is recommended:

- Determination of the initial values of viscosity from the representative data

- Computation of the velocity profile, computation of the shear rate profile:

$$\dot{\gamma}(x) = \frac{dv_z(x)}{dx} \tag{6.33}$$

- Determination of the shear rate at the wall $\dot{\gamma}_w$ (two-layer flow, outside (cover) layer of the three-layer flow) or of the shear rate at the interface (core layer of the three-layer flow)
- Multiplication of the shear rates with the correction factor $e_{\square}$:

$$\dot{\gamma}_{rep} = \dot{\gamma}_w \cdot e_{\square}. \tag{6.34}$$

- Determination of the corrected representative viscosities using the representative shear rates determined above
- Repeating the computation of the flow. This iterative procedure is repeated until stable values of the viscosity are found. The described simple computation can also be done with a programmable pocket calculator.

However, when the study is done on a multilayer flow with more than two different melts or volumetric flow rates (e.g., if $\dot{V}_1 \neq \dot{V}_3$ in Fig. 6.13(a)) or when temperature profiles are also to be determined, the solution can only be obtained by elaborate numerical procedures.

### 6.3.2 Computation of Coextrusion Flow by the Explicit Finite Difference Method

When describing a coextrusion flow, the continuity, momentum, and energy equations are first simplified by considering the boundary conditions and then solved; the process is similar to that for single-layer flow [34].

Since the effect of the temperature field on the velocity field is considerably weaker than vice versa, the decoupling of both fields is possible.

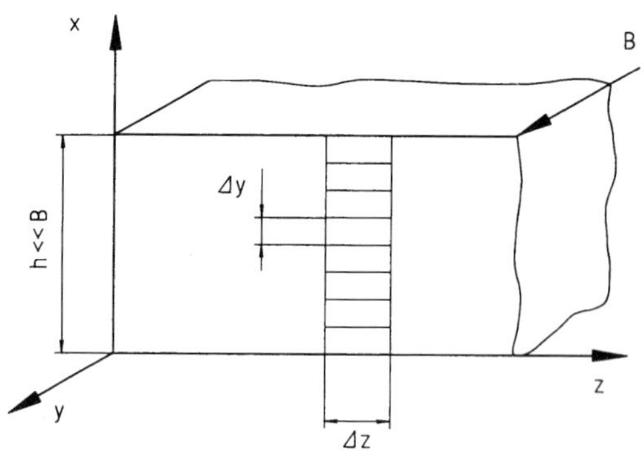


**Figure 6.14** Discretization of a flow channel for computations with the finite difference method

First, the velocity profile is computed from the temperature profile extrapolated from the preceding computational step. Second, the new temperature profile is computed taking into consideration the velocity profile obtained earlier [36].

Because of that, the computations of velocity and temperature will be done separately in the following sections.

The differential equation for the one-dimensional flow from a flat die was developed in Section 6.3.1:

$$\eta(\dot{\gamma})\cdot\dot{\gamma} = \frac{\partial p}{\partial z}(x-c). \tag{6.35}$$

When solving this equation, the location of the maximum velocity, the location of the interface(s), the pressure drop, and the shear rate profile must be determined by iteration. Then the velocity profile can be determined from the shear rate profile by integration. The course of such an iterative process is shown in Fig. 6.15.

When applying the finite difference method, the flow channel is divided into individual discrete layers (Fig. 6.14). Across each of these discrete layers, both the shear rate and the viscosity are taken as being constant locally. This way, Equation (6.35), which is the same as Equation (6.3), can be solved by iteration when the pressure drop and the position of the maximum velocity are given. From the shear rate profile, the velocity profile, and from it the flow rate, can be determined. The computation starts at one wall with a specified velocity at the wall (generally equal to zero). Then the discrete flow rates are determined layer by layer and then added. This sum is then compared to the specified flow rates for the individual melts.

As soon as the specified flow rate of the first melt is used up, the computation has reached the first interface. From here on the material data of the second melt are used for the solution of Equation (6.35). This way all the discrete layers of the flow channel are gradually subjected to the computation process. This procedure saves the otherwise necessary determination of the location of the boundary layers [30].

As a boundary condition for the determination of the location of the maximum velocity, the last computed velocity and the specified velocity at the wall are compared. The pressure loss is determined from the comparison of the computed and the specified total flow rate. In symmetrical coextrusion flows, the maximum flow velocity lies in the middle of the flow channel; moreover, it is sufficient to compute one-half of the flow channel because of symmetry.

Initial values for the pressure drop and the location of the maximum velocity can be determined from the computation of constant viscosities (see Section 6.3.1). The computation of temperature, which follows the above, is performed by using the simplified energy equation:

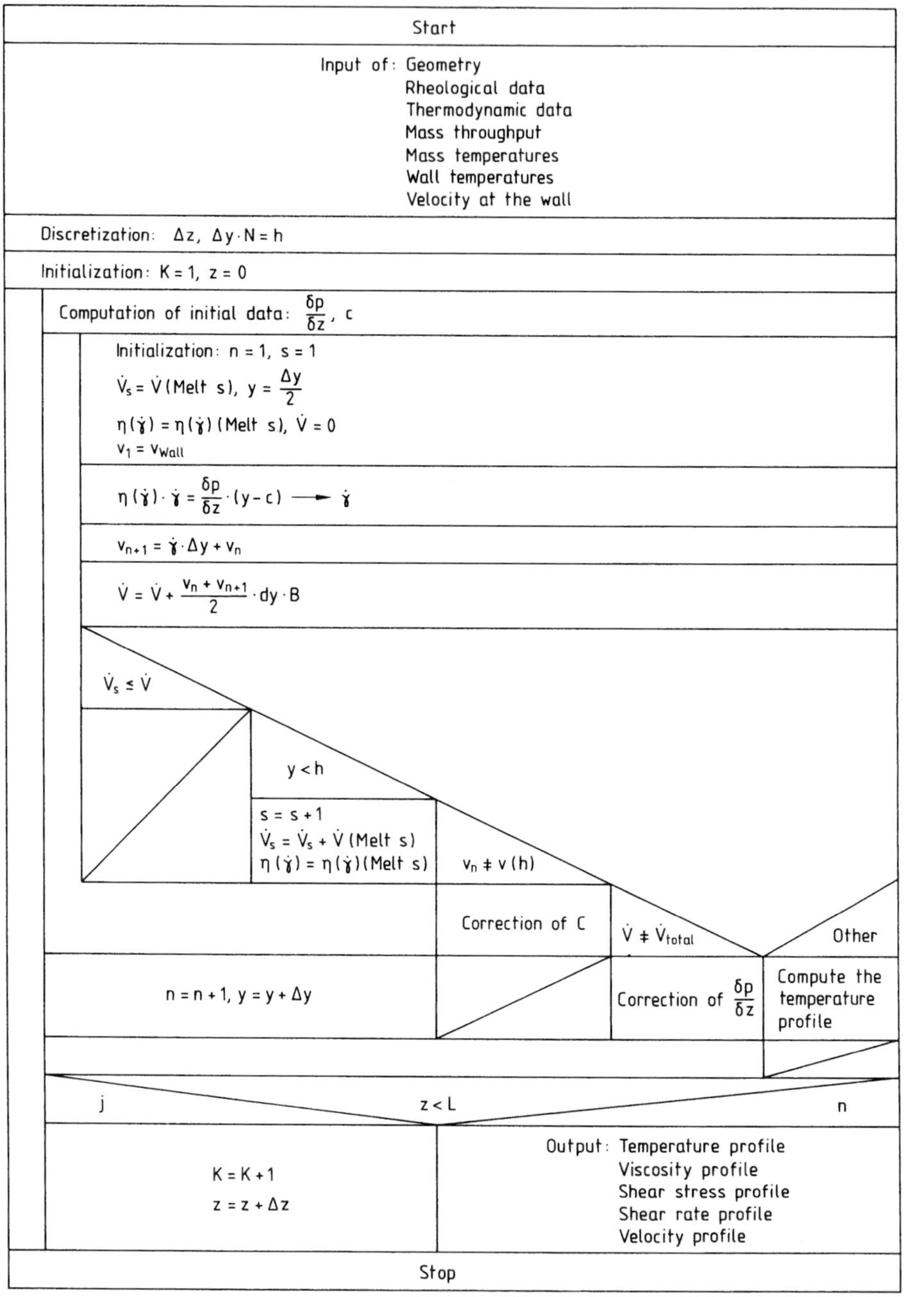


**Figure 6.15** Flow chart of a finite difference program for the computation of the distributions of velocities and temperatures in multilayer flows

$$\underbrace{\varrho \cdot c_{\mathrm{p}} \cdot v_{\mathrm{z}}(x) \frac{\mathrm{dT}}{\mathrm{d}z}}_{\mathrm{I}} = \underbrace{\lambda \frac{\mathrm{d}^2 T}{\mathrm{d}x^2}}_{\mathrm{II}} + \underbrace{\eta \cdot \dot{\gamma}^2}_{\mathrm{III}}. \tag{6.36}$$

Term I describes the convective energy transport in the flow direction, Term II the heat conduction in the direction of shear, and Term III the dissipative heating.

For the computation of flow in the normal direction, half a step size is used, and in the direction of flow a full step size is used; the step size results from a criterion of stability in difference calculations [35].

The flow chart of the described finite difference program is shown in Fig. 6.15.

### 6.3.3 Computation of Velocity and Temperature Fields by the Finite Difference Method

Velocity and temperature fields [34] are computed by the method outlined in Section 6.3.2. In the following sections some of the results will be shown (see also the example in Section 4.4.4).

Figure 6.16 shows how the viscosity ratio of both melts flowing together in the channel affect the velocity and shear rate profiles of a symmetrical three-layer flow through a slit. The dashed line represents the typical velocity profile of a pseudoplastic melt, whereas in the case of multilayer flow a block-shaped viscosity profile is formed when the low-viscosity melt flows near the wall. A lowering of viscosity in this layer leads to an additional reduction of the maximum velocity; however, the shear rate at the wall increases. The boundary layer shifts only a little with the change in viscosity. It is primarily defined by the flow rate ratio (here, the ratio of mass flow rates). The velocity profiles in the nonsymmetrical two-layer flow have a completely different appearance (Fig. 6.17). Depending on the flow rate ratio–at a constant total mass flow–the flow velocity maximum is located in the more viscous melt (i. e., if $m_2/m_1 = 0.25$ and 0.11, respectively). If $m_2/m_1 = 1$, then the maximum is in the less viscous one. The mean velocities in the layers differ more with a greater difference in viscosities and with a greater proportion of the material with a lower viscosity. The position of the maximum flow velocity depends on the viscosity ratio.

Besides the velocity profile, the thermodynamic material data and the thermal boundary conditions at the wall of the die affect the temperature profiles in the flow channel. Figure 6.18 shows the development of the temperature profile along the flow channel (wall temperature $T_w = 200\,°C$) for two velocity profiles selected from Fig. 6.17. For Melt 1 thermodynamical material data of PE-HD and for Melt 2 those of PS were used. If the maximum velocity lies in the more viscous melt, pronounced temperature maxima appear near the wall. In the case of a velocity maximum in the low-viscosity layer, a pronounced maximum appears at the interface due to a shear heating (high shear rate) magnified by a low heat transfer to the neighboring melt layers.

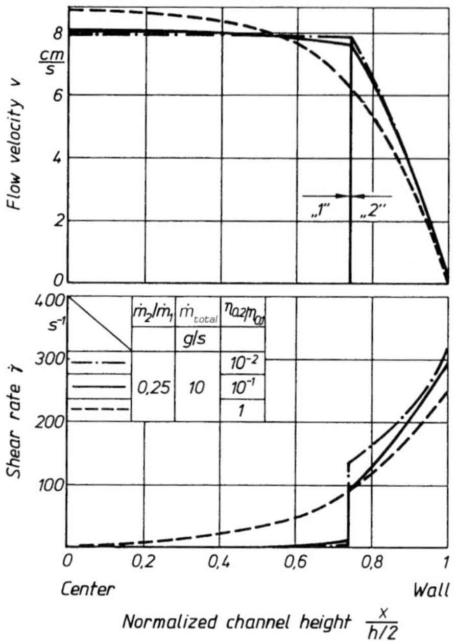


**Figure 6.16** Effect of viscosity ratio on the velocity profile of the symmetrical multilayer flow through a slit [34]

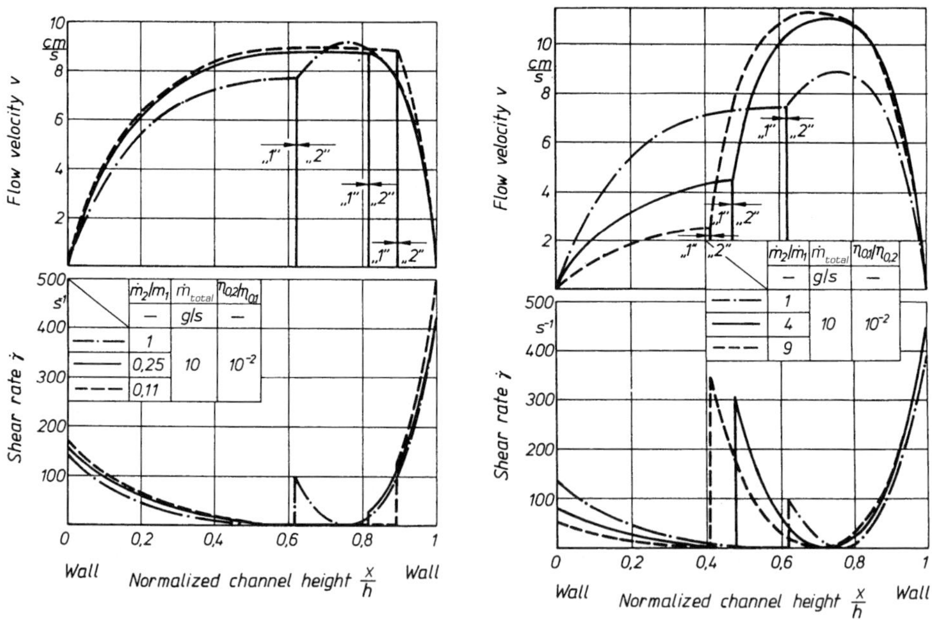


**Figure 6.17** Effect of the ratio of mass flow rates on the velocity profile in an asymmetrical multilayer flow through a slit [34]

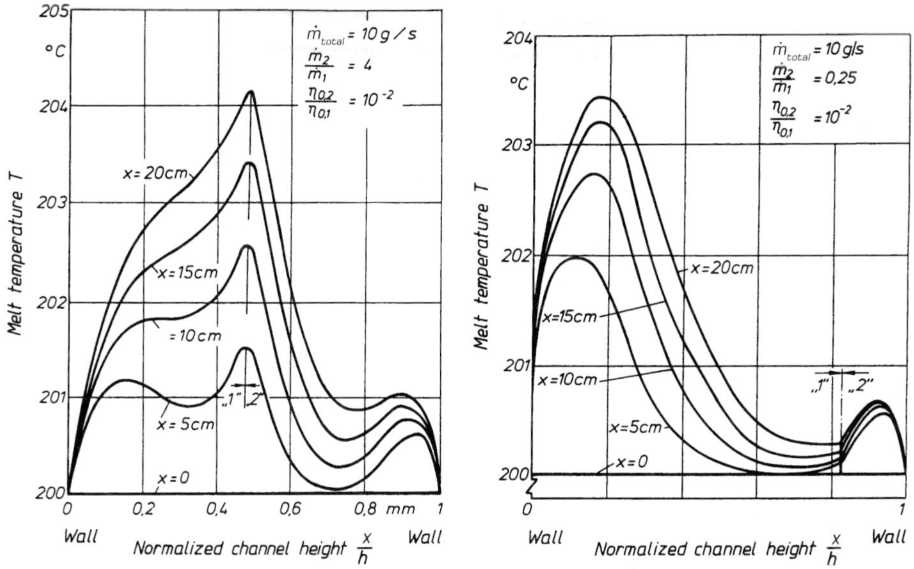


**Figure 6.18** Temperature profile in the flow direction in an asymmetrical multilayer flow through a slit [34]

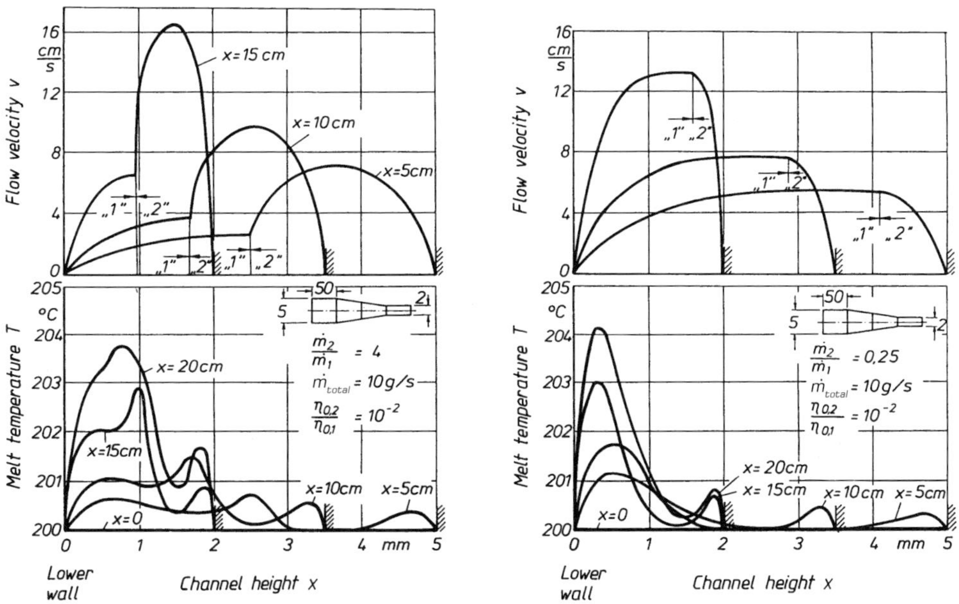


**Figure 6.19** Velocity and temperature profiles in an asymmetrical multilayer flow through a convergent slit [34]

Figure 6.19 shows the development of the velocity and temperature profiles in a convergent slit-shaped flow channel for two similar cases. Here the shape of the velocity profile is preserved, but the temperature profile exhibits its maximum values near the wall at the interface.

### 6.3.4 Computation of Velocity Fields in Coextrusion Flows by FEM

The FEM computation of flows opens the possibility of two- to three-dimensional simulation of multilayer flows (see also Section 4.4.5). This computation differs from the one for a single layer, in that one additional algorithm is necessary. This shifts the FE network in such a manner that the boundary between two melts lies always on the boundary of an element. The special problems involved in computations of multilayer flow by FEM will not be discussed here (it is recommended to consult [37] on this matter). Rather, to demonstrate the possibilities offered by FEM, a comparison of computations by FEM and flow experiments will be described in the following.

The studies of the region in which the layers in a multilayer flow merge were performed in an experimental modeling station that allows visual observation of the experiment [38]. Within the scope of the experiments, the formation of the interface and its location as well as other flow phenomena were evaluated. In the next step it was attempted to simulate a two-layer flow with a 2D-FEM computation [37]. The photographs comparing the actual flow and the computed pathways for different volume flow conditions are in Fig. 6.20. It is noteworthy that not only the shape and the position of the interface are well described, but also that the FEM computation produces the size and location of dead spots in the region where the layers merge, dependent on the flow ratio (the computation is based on purely viscous behavior of the material, described by the Carreau model).

The explanation of the formation of the dead spot is in the sizable increase of speed in the lower stream. The faster upper layer drags the slower layer in the direction of flow and thus causes a rising pressure at the lower wall in the direction of flow. A local pressure maximum is formed at the lower wall a short distance behind the point where the layers merge. This pressure field causes a reverse flow or a stagnation zone. (This effect is comparable to the one occurring in wire coating; see Section 5.3.3.4)

A check of the merging zone of coextrusion dies for the formation of the abovementioned dead spots is possible without problems. This way the maxima in the residence time and the related stress on the melt in this region can be eliminated.

This example shows that even complex flow processes can be accurately described by FEM.

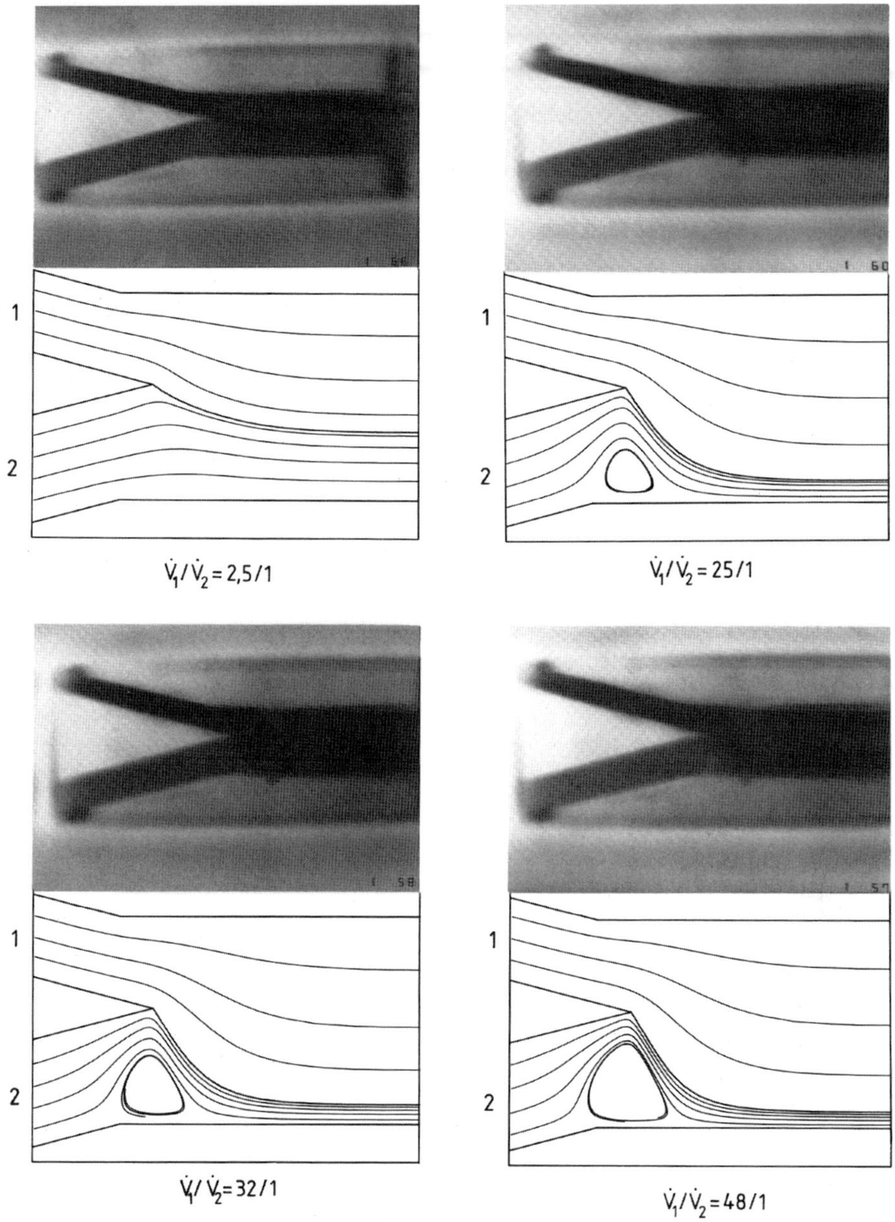


**Figure 6.20** Comparison between the results of a two-layer flow experiment and the corresponding FEM computation

## 6.4 Instabilities in Multilayer Flow

In addition to the common flow instabilities typical for single layers, such as unstable inlet angle, stick-slip, melt fracture, and wall slip, in multilayer flow two additional phenomena occur: encapsulation and interfacial instability.

The encapsulation and interfacial instabilities are two phenomena that have to be considered separately from the cause, mechanism, and consequences. The difference is clear from the comparison of Figs. 6.21 and 6.22. Two melt streams with an equal flow rate are forced through a round capillary side by side. The location of the interface was determined on thin slices of the frozen two-layer strand [39]. It is found in the coextrusion of two different grades of polystyrene (Fig. 6.21) that the melt with lower viscosity always tries to encapsulate the one with the higher viscosity.

As the results from measurements show, the tendency to encapsulate can reverse, depending on extruding conditions, when extruding two melts with viscosity curves that cross each other.

From that it can be deduced that encapsulation is a phenomenon resulting from the flow behavior of the materials.

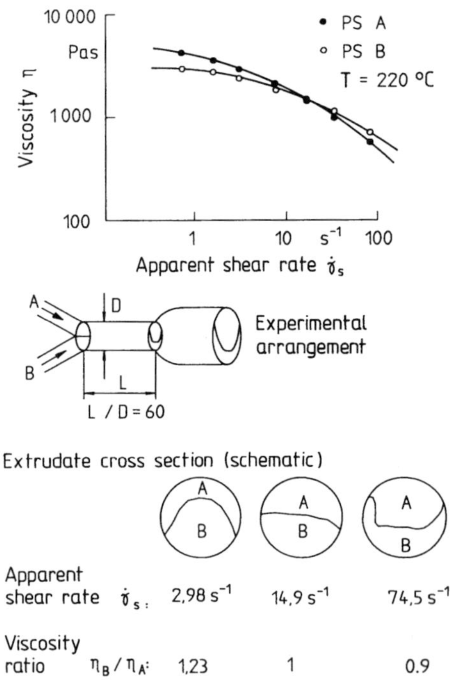


**Figure 6.21** Rearrangement of the interface in the coextrusion of two different grades of PS [39]

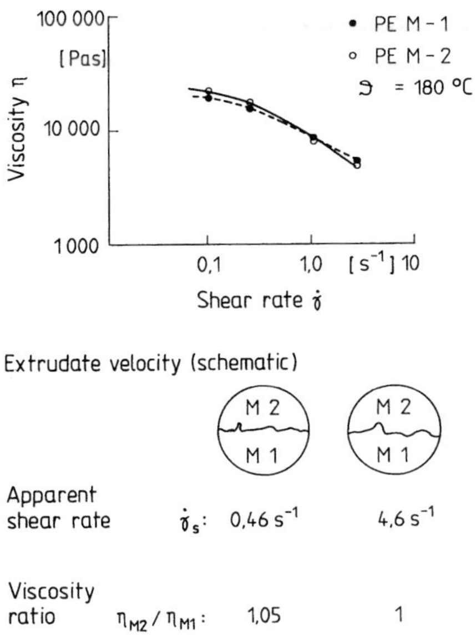


**Figure 6.22** Rearrangement of the interface in the extrusion of two different grades of PE-Typen [39]

The explanation of why it occurs is given by the principle of minimizing of energy. Both melts try to arrange themselves in such a way that the pressure loss of the flow is at a minimum. This happens only when the low-viscosity melt flows in contact with the wall and thus forms a slip film for the high-viscosity melt in the middle of the channel.

This provides the explanation of the phenomenon of encapsulation, but a prediction (by computation) of how the interface forms in a multilayer flow in dependence on the original location at the point where the melts merge, on the flow distance, melt properties, and volumetric flow rate is not possible yet. One approach to this is the three-dimensional FEM computation of coextrusion flow [37]. The channel under study and the computed location for the interface for different viscosity conditions after a length of flow that is triple the height are shown in Fig. 6.23. The definition of the boundary layer is very good; however, a quantitative experimental verification of the results is still forthcoming.

When extruding two different grades of PE (Fig. 6.22), it was observed that a very rough and irregular interface formed between the two melts [39]. This instability of the interface is not directly related to encapsulation. A typical formation of interfacial instability in the coextrusion of films is explained in Fig. 6.24.

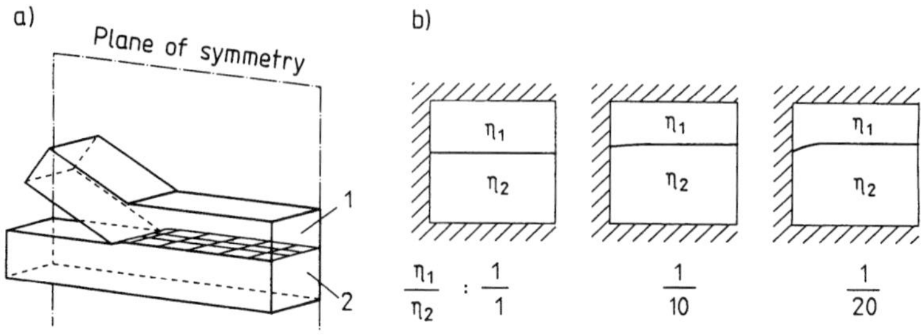


**Figure 6.23** Results from an FEM computation for the prediction of the rearrangement of the interface: (a) two-layer coextrusion die, (b) position and shape of the interface in the cross section of the channel for three viscosity ratios

A model explaining the occurrence of the interfacial instability ideally must explain the following characteristics:

- Interfacial instabilities occur, depending on the combination of materials, in thin middle layers as well as in thin outside layers.
- The disturbances in the interface vary in frequency and amplitude.
- Interfacial instabilities occur even in the coextrusion of identical materials.
- Interfacial instabilities can occur directly in the merging zone but also after a longer, stable parallel flow.

So far, two basically different approaches for the explanation of the occurrence of interfacial instabilities have been chosen.

1. Utilizing the mathematical stability analysis of the multilayer flow in the region of the boundary layer, a study was conducted to determine if the disturbance of the location of the interface is reinforced or suppressed by the flow [32]. By this analysis, the viscosity ratio in the region of the interface was found to be the criterion for the delineation between stable and unstable flows. This criterion has been verified in a great number of coextrusion experiments. When plotting the viscosity ratio in the interface region against the ratio of the layer thickness, stable and unstable operating points can be distinguished (Fig. 6.25). This has been confirmed in experiments, and also first FEM simulations have been performed in order to predict whether a certain material and layer setup potentially leads to interfacial instabilities [46].

   The main argument against the viscosity ratio as criterion for the occurrence of the instability in the boundary layer is that the instability occurring in coextrusion of identical materials cannot be explained by this model.

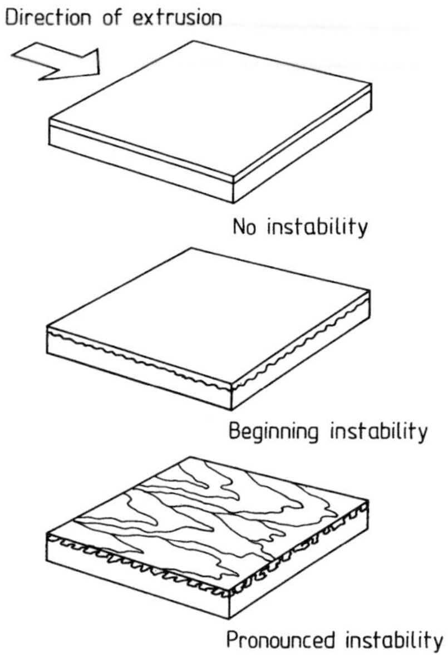


**Figure 6.24** Buildup of instabilities in the interface of the coextruded films

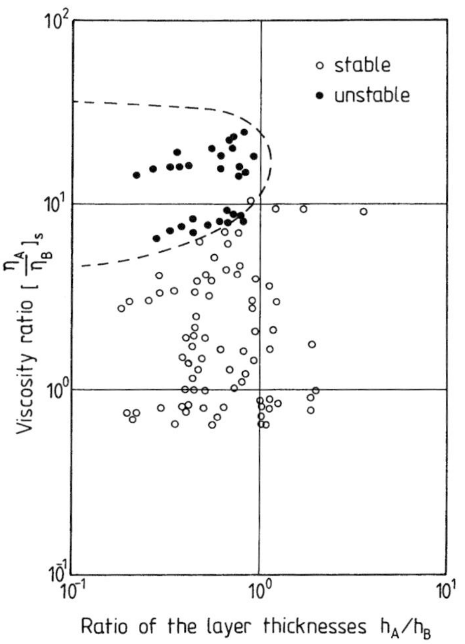


**Figure 6.25** Viscosity ratio at the interface as a critical factor

2. The alternative to the stability analysis is a model that explains the appearance of interfacial instabilities by the forces of flow acting on the interface.

   The interface is a weak spot in the bond of two adjacent melts. This should be explained by Fig. 6.26, which shows schematically the molecular events during the merging of two polymeric melts [40]. In the region of single flows, the macromolecules are, on one hand, attached to the walls of the flow channel due to the microroughness of its walls (Stokean condition for adhesion) and, on the other hand, entangled with molecules in the middle of the flow channel. Hence, the macromolecules become oriented at the walls of the flow channel parallel to the flow direction. In the direction of the flow channel walls, no entanglements can be formed. When both polymeric melts merge (II), the oriented and unattached regions meet and form the interface. The orientation remains preserved over the duration of the orientation relaxation; considerably longer times are necessary for the formation of the entanglements, as shown by statistical studies of segmental motions [41].

This last region shown (III), in which the orientation disappears completely and in which the entanglements exist across the entire boundary layer, is not reached during the usual residence times in a die.

Because of the weak spot in the interface, both melts can slide past each other much more easily than the individual layers within the material.

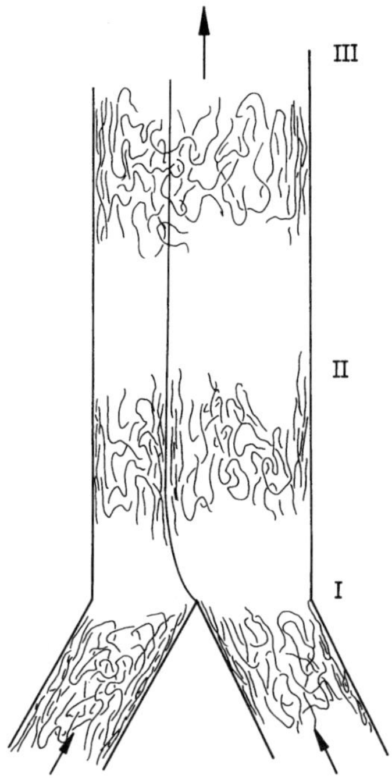


**Figure 6.26** Model for the description of of the adhesion at the interface: I: orientation of the layers near the wall; II: no entanglements, orientation exists in the interface; III: entanglements across the interface and beyond, no orientation in the interface

This hypothesis was confirmed by experimental studies, and it explains the occurrence of the interfacial instabilities during the extrusion of identical materials. The flow around the spider legs in a mandrel support die and the flow marks resulting from it represent effects similar to coextrusion of identical materials [30].

The theory of the failure of the material at the interface links the instability of the interface with the stresses acting on it.

In a one-dimensional flow the interface is subjected to shear. From this it follows that it must be possible to find a critical shear stress in the interface for each combination of materials, beyond which an instability of the interface appears [30,32,33,42].

This also was possible for a number of dies and combinations of materials. An evaluation of shear stress in the interface is possible only when the instability occurs in the one-dimensional flow region. Critical shear stress values for one-dimensional isothermal two-layer flow were determined (Table 6.3).

**Table 6.3** Values of Critical Shear Stress for a Two-Layer Extrusion

| Combination of Materials | $\tau_{crit}$ (Pa) |
|---|---|
| PE-LD 1810 H and PE-LD 1810 H | 30,000 |
| PE-LD 1810 H and PP 1060 F | 20,000 |
| PE-HD 6021 D and PP 1060 F | 24,000 |

Since the critical shear stress depends on the surface forces of the melts acting along the interface, it must be determined in an experiment for each combination of materials. A simple transfer of the critical stresses found in a one-dimensional experiment to two-dimensional flows (e. g., in divergent or convergent flow channels) is not possible. This is because, in addition to the shear stress resulting from a shear flow, there are normal stresses acting on the interface resulting from elongational flows. This can have either a stabilizing or destabilizing effect on the flow. In flow experiments, it was found that divergent flow channels with an increasing height have a strong destabilizing effect [42].

The fact that the interfacial instabilities generally do not occur in the land region of the die, although here a very high shear rate is present because of the small height of the flow channel, provides evidence that convergent flow channels have a stabilizing effect on the flow [31]. There are still two problems with the application of the model of material failure in the interface at this point:

- The effect of temperature on the critical shear stress has not been studied as yet.
- The determination of a critical shear stress in highly elastic melts is not possible with the current flow simulation programs because they are applicable only to purely viscous materials.

## Symbols and Abbreviations

| | |
|---|---|
| $a_{m,n}$ | Boundary layer between layers *m* and *n* |
| $B$ | Channel width |
| $c$ | Location of the maximum flow velocity |
| $c_p$ | Specific heat capacity |
| $e_{\square}$ | Correction factor for a rectangular cross section |
| $h$ | Channel height |
| $h_i$ | Channel height of individual layers *i* |
| $K$ | Consistency factor |
| $k$ | Integration constant |
| $m$ | Mass |
| $n$ | Viscosity exponent |
| $p$ | Pressure |
| $T$ | Temperature |
| $T_w$ | Temperature at the wall |
| $\dot{V}$ | Volumetric flow rate |
| $\dot{V}_{total}$ | Total volume flow rate |
| $\dot{V}_m$ | Volumetric flow rate of the layer *m* |
| $v$ | Velocity |
| $v_m$ | Flow velocity in the layer *m* |
| $v_n$ | Flow velocity in the layer *n* |
| $\dot{\gamma}$ | Shear rate |
| $\dot{\gamma}_{rep}$ | Representative shear rate |
| $\dot{\gamma}_W$ | Shear rate at the wall |
| $\eta$ | Viscosity |
| $\eta_{rep}$ | Representative viscosity |
| $\lambda$ | Thermal conductivity |
| $\rho$ | Density |
| $\tau$ | Shear stress |
| $\tau_{crit}$ | Critical shear stress |

## ■ 6.5 References of Chapter 6


1. Fischer, P.: Herstellung, Eigenschaften und Anwendung von coextrudierten Tafeln. Plastverarbeiter 23 (1972) 10, pp. 684–688
2. Fischer, P.: Maschinentechnische Lösungen für die Coextrusion. Ind.-Anz. 95 (1973) 35, pp. 725–729
3. Nissel, F.: Coextrusion: An advanced feed-block method. Plast. Rubber Int. 3 (1974) 3, pp. 117–119
4. Schrenk, W.J. and Veazey, E.W.: Films, multilayer. Eng. Polym. Sci. Eng. 7 (1987) pp. 106–127
5. von Ness, R.T. and Eidmann, R.A.L.: Practical coextrusion coating. TAPPI Conference, Miami Beach (1972)
6. Meyer, L.J.: Co-extrusion for economy performance properties and products. SPE Tech. Pap. (1972) 2, pp. 722–725
7. Thomka, L.M. and Schrenk, W.J.: Flat-die extrusion. Mod. Plast. Int. (1972) 4, pp. 56–57
8. Rahlfs, H.K. and Ast, W.: Coextrusion von Folien, Tafeln und Beschichtungen mit Mehrkanal-Breitschlitzwerkzeugen. Kunststoffe 66 (1976) 10, pp. 538–541
9. Johnson, J.E.: Grundzüge der Coextrusion von Platten und Folien. Kunststoffberater (1976) 10, pp. 538–541
10. Dragoni, L.: Vergleichende Betrachtungen bei der Herstellung von coextrudierten und kaschierten Verbundfolien. Plaste Kautsch. 25 (1978) 12, pp. 701–703
11. Mainstone, K.A.: Internal and external combining systems for slot die coextrusion. Paper, Film & Foil Converter (1978) pp. 65–68
12. Predöhl, W.: Herstellung und Eigenschaften extrudierter Kunststoffolien. Postdoctoral thesis at the RWTH Aachen (1977)
13. Auffermann, L. and Hub, H.H.: Aufbau und Eigenschaften von Coextrusionsfolien. In: Fortschritte bei der Folienproduktion und -Verarbeitung. IK-conference proceedings, Darmstadt (1988)
14. Reitemeyer, P.: Coextrusionswerkzeuge zum Herstellen von Flachfolien für den Verpackungsbereich. Kunststoffe 78 (1988) 5, pp. 395–397
15. N.N.: Multilayer film from a single die. Plast. Eng. (1974) pp. 65–68
16. Predöhl, W.: Rationellere Folienproduktion. In: Fortschritte bei der Folienproduktion und -Verarbeitung. IK-conference proceedings, Darmstadt (1988)
17. N.N.: Coextrusion takes a giant step into the future. Mod. Plast. Int. 61 (1983) pp. 14–18
18. Roppel, H.-O.: Verbesserung der Barriereeigenschaften beim Blasformen. In: Fortschritte beim Blasformen von Thermoplasten. IK-conference proceedings, Darmstadt (1987)
19. Hensen, F. et al.: Entwicklungsstand bei der Coextrusion von Mehrschichtfolien und Mehrschichtbreitschlitzfolien. Kunststoffe 71 (1981) 9, pp. 530–538
20. Fischer, P. and Wortberg, J.: Konzept und Auslegung von Coextrusionsblasfolienanlagen. Kunststoffe 74 (1984) 1, pp. 28–32
21. Wright, D.W.: Straight talk on five-layer blown film dies. Plast. Technol. 30 (1984) pp. 79–82
22. Caspar, G. and Halter, H.: Mehrschicht-Blasfolienextrusion. Kunststoffe 73 (1983) 10, pp. 597–598
23. N.N.: Bei der Coextrusion besonders wirtschaftlich. Plastverarbeiter 38 (1987) 4, pp. 110–112
24. Hegele, R.: Coextrusion blow molding of large hollow articles. Coex Europe ’86 (1986) pp. 355–388
25. Eiselen, O.: Konzepte für Coextrusionsblasformanlagen. Kunststoffe 78 (1988) 5, pp. 385–389
26. Onasch, J. and Kraemer, H.: Coextrusion: Märkte, Verfahren, Werkstoffe. Kunststoffberater 33 (1988) 7/8, pp. 62–67

27. Iwawaki, A., et al.: Coextrusion blow molding for gas tanks and industrial parts. Mod. Plast. Int. 55 (1977) pp. 15–17
28. IP-PS 906 128 (1975) IHI, Japan (Japanese patent)
29. DE-OS 3620144 Al (1987) Bekum Maschinenfabriken GmbH, BRD (German patent)
30. Cremer, M.: Untersuchungen zum Auftreten von Fließinstabilitäten in Mehrschichtströmungen. Unpublished thesis at the IKV, Aachen (1988)
31. Schrenk, W. J.: Interfacial flow instability in multilayer coextrusion. Polym. Eng. Sci. 18 (1978) p. 8
32. Han, C. D.: Multiphase Flow in Polymer Processing. Academic Press, New York (1981)
33. Kurrer, H.: Durchführung und Auswertung von Versuchen mit einem Coextrusions- Modellwerkzeug. Unpublished thesis at the IKV, Aachen (1988)
34. Wortberg, J.: Werkzeugauslegung für Ein- und Mehrschichtextrusion. Thesis at the RWTH Aachen (1978)
35. Hilger, H.: Berechnung von Mehrschichtenströmungen unter Berücksichtigung viskoelastischer Fließeigenschaften. Unpublished thesis at the IKV, Aachen (1986)
36. Peuler, M.: Berechnung von Fließ- und Deformationsvorgängen in Extrusionswerkzeugen. Unpublished thesis at the IKV, Aachen (1976)
37. Schwenzer, C.: Finite Elemente Methoden zur Berechnung von Mono- und Coextrusionsströmungen. Thesis at the RWTH Aachen (1988)
38. Strauch, Th.: Ein Beitrag zur Theologischen Auslegung von Coextrusionswerkzeugen. Thesis at the RWTH Aachen (1986)
39. Southern, J. H. and Ballmann, R. L.: Additional observations on stratified bicomponent flow of polymer melts in a tube. J. Polym. Sci., Polym. Chem. Ed. 13 (1975)
40. Meier, M.: Ursachen und Unterdrückung von Fließinstabilitäten bei Mehrschichtströmungen. Proceedings of 14. IKV-Kolloquium, Aachen (1988), pp. 47–54
41. Janeschitz-Kriegl, H.: Polymer Melt Rheology and Flow Birefringence. Springer, Berlin (1983)
42. Skretzka, D.: Untersuchung der Strömung von Modellfluiden in einem Modellversuchsstand. Unpublished thesis at the IKV, Aachen (1988)
43. Backmann, M. and Sensen, K.: Blasfolienextrusion. In: Handbuch Urformen. Bührig-Polaczek, A.; Michaeli, W.; Spur, G. (Eds.), Hanser, Munich (2014)
44. Thielen, M.: Extrusionsblasformen. In Handbuch Umformen. Bührig-Polaczek, A., Michaeli, W., Spur, G. (Eds.), (2014) Hanser, Munich, pp. 667 ff.
45. Hopmann, Ch., Funk, A.: Radialwendelverteiler für den Einsatz im Blasformen, Blasformen & Extrusionswerkzeuge (2013) Vol. 9, 5, pp. 12–18
46. Michaeli, W. and Windeck, C.: Limits to instabilities? Kunststoffe Int. 4 (2010) 11, pp. 41–44
47. Ederleh, L.: Barrierefolien – Produkte und ihre Prozesse. Proceedings of IKV-conference Folienextrusion – Trends bei Rohstoffen, Verarbeitung, Anwendung, Aachen, 11/12 November (2014)

# 7 Extrusion Dies for Elastomers

## 7.1 Design of Dies for the Extrusion of Elastomers

Because of the diversity of products made from this class of polymers, many of the dies common in the processing of thermoplastics are found again in the extrusion of materials based on elastomers (rubber). The dies for thermoplastics were discussed in principle in the previous chapters. A major difference in the processing of elastomeric materials is that the mass flow supplied by the extruder must be within the temperature range that allows for safe processing, that is, where no premature onset of vulcanization (so-called scorch) occurs on the extruder or in the die assembly [1].

For the extrusion of profiles from rubber compounds, mostly very simple dies have been used until recent years. They usually consisted of a steel disk with the desired profile shape cut into it. Nowadays, die designs are becoming more complex as design rules are adapted from thermoplastic extrusion dies. This is enabled by a more frequent use of numerical flow simulation techniques in order to analyze the flow behavior in a die. For hollow chambers, the necessary cores are supported by bridges. For the multiple extrusion of complicated shapes, duplex and triplex dies are employed (Fig. 7.1) [2–6]. The individual mass flows in the die are combined through a profiled disk, the shape of which is determined by the profile contour, and subsequently vulcanized (crosslinked). These dies have been used successfully in the production of tire treads.

The simple dies with a discontinuous transition lose their process dependability when extrusion speeds are increased. To increase the performance of these dies, they must be made with continuous transitions in the flow channel (Fig. 7.2). An example is the so-called roller head die. The flow channel here is very similar to that of a wide slit die. The addition of choker bars to the channel allows an effective flow adjustment, such as when compounds with altered flow properties have to be extruded. When the required gauge of the extrudate changes considerably in, for instance, roller dies, the lips of the roller head die have to be exchanged.

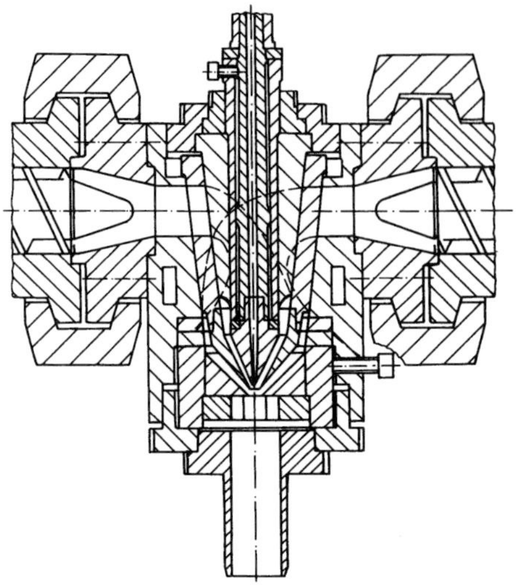

**Figure 7.1** Duplex crosshead [2]

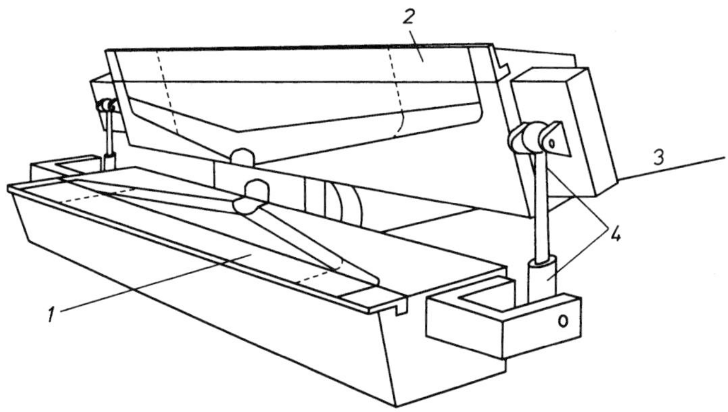


**Figure 7.2** Extrusion die for elastomers with a continuously changing flow channel [2,7]: 1: extrudate width changed by die adapters, 2: replaceable lip, 3: extruder convection, 4: hydraulic opening and shutting of the die

Dies with hydraulic opening and automatic ejector systems, which, for example, eject the contents of a wide slit die, are also widely used in the extrusion of elastomeric (rubber) compounds, such as for semifinished products for tire treads [2]. When the cleaning and the exchange of the lips and inserts are completed, the die can be closed hydraulically and clamped [2,7]. The extrusion of profiles with one smooth side and width greater than height is sometimes done by single roller die systems, so-called roller-lip systems [8]. Furthermore, side-fed mandrel dies are used for the extrusion of hoses and in cable coating.

The widespread empirical design of the flow channel is common to these dies. A prediction of pressure losses or an estimation of the distribution effect is still hardly possible with this approach. Also, the danger of scorch (onset of cure) in the die cannot be quantified. An adequate consideration of effects of viscoelasticity on the profile geometry and its surface is only estimated based on experience.

## ■ 7.2 Fundamentals of Design of Extrusion Dies for Elastomers

When considering the first chapters of this book, the question arises: What possibilities are offered here for the design of dies for elastomers, and what can be used from plastics technology, which, incidentally, has progressed considerably farther than the elastomer technology, especially with regard to the particular behavior of the elastomeric material? An analogy is possible only in cases where no scorch occurs during the flow. This, however, is the basic requirement for the correct design of an extrusion die for elastomers. As a result, this assumption is presumed valid in the following sections.

### 7.2.1 Thermodynamic Material Data

When comparing the average densities of elastomers to those of thermoplastics, they are often higher because a high proportion of fillers are added to the compounds for their practical use. However, they lie in the typical range of densities for thermoplastics. Specific heat capacity, thermal conductivity, and thermal diffusivity are similar in order of magnitude for both groups of materials (see Table 7.1) [9].

**Table 7.1** Range of Values of Thermodynamic Material Data for Thermoplastics and Elastomers [9] (Data Can Vary with Temperature and Pressure)

| Group of materials | Density $\rho$ [g/cm$^3$] | Thermal conductivity $\lambda$ [W/(m·K)] | Specific heat capacity $c_p$ [kJ/(kg·K)] |
|---|---|---|---|
| Thermoplastics | 0.7 to 1.4<br>PE-LD/PP: 0.7 to 0.9<br>PVC: 1.3 to 1.4<br>PS: 0.95 to 1.05 | 0.15 to 0.45<br>PE-LD/PP: 0.25 to 0.45<br>PVC: 0.19<br>PS: 0.16 | 1 to 3 (in the temperature range for crystallization a mostly considerable increase with a subsequent reduction to the level shown) |
| Elastomers | 1.1 to 1.7 (depending on the type and amount of filler) | Unfilled: 0.1 to 0.2<br>Highly filled: 0.2 to 0.4 | 1 to 2 (change in data in the range of typical curing temperatures) |

Since the processing temperatures of elastomers are well above their glass transition temperature ($T_g$), it can be expected that various changes in temperature during the flow through the extrusion die lead to corresponding changes in the values of material data [10]. This coupling of temperature with the thermodynamic properties can be, as a rule, neglected in the computations, especially when the data were measured at average processing temperatures. It has to be pointed out that the thermodynamic material data for elastomeric compounds have been published in a limited number of cases as each manufacturer uses individual compound recipes. Furthermore, data acquisition often requires complicated and expensive measuring techniques and, therefore, represents a fertile field for research.

### 7.2.2 Rheological Material Data

When comparing elastomers to thermoplastics in their viscosity function, it is found that both clearly exhibit a shear-dependent viscosity (Fig. 7.3).

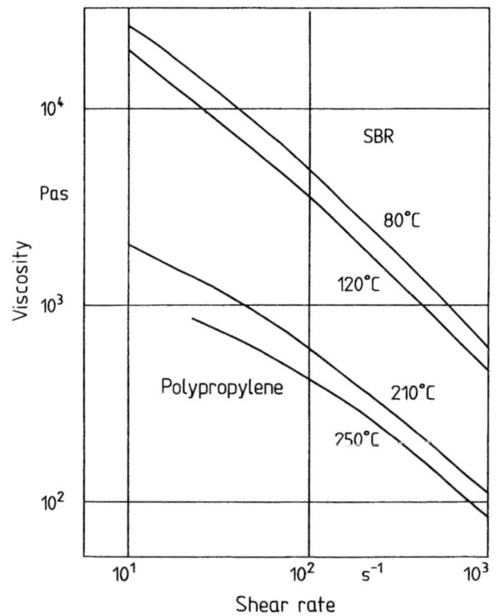


**Figure 7.3** Viscosity functions of a thermoplastic material (polypropylene) and of an elastomer (SBR)

The effect of temperature on the flow behavior of both materials can be expressed quite commonly by the Arrhenius equation or the WLF equation [11] (see also Section 2.1.1.3). Also, as to the shape of the functions, both groups of materials are comparable and can be characterized in their range of processing rates by the power law equation. The only difference is that the level of viscosity of elastomers is mostly shifted toward higher values [9]. Nevertheless, the following material-specific peculiarities must be considered:

*(a) Dissipation Model:* Since elastomers are processed mainly in the form of filled compounds and are thus processed as multiple phases, the effect of fillers on the distribution of shear rates in the flow channel must be taken into consideration. The so-called dissipation model [12] has been developed for thermoplastics and has been confirmed valid by several investigations [13,14]. It has been applied to elastomers in [15]. In this model, a differentiation is made between the flowable and solid phases. This means that there is a two-phase flow in the channel (see Fig. 7.4), while the total shear and hence the dissipation take place in the elastomeric portion, that is, between the filler particles. When measuring with a rheometer, an integral value of viscosity $\eta_{nf,int}$ and the corresponding value of the integral shear rate $\dot{\gamma}_{nf,int}$ are determined, meaning that no fillers (nf) are included (case (a) in Fig. 7.4). These values are suitable for the computation of pressure losses since they describe the flow behavior of the entire system. However, when being applied to the energy equation, corrections are necessary. As explained by Fig. 7.4, the shear rate in the elastomeric region is higher by the factor $K$ with regard to the integral parameter. As shown in [10], the total dissipated energy in the polymeric phase resulting from the above consideration is

$$P_{diss,\,total} = (1-\phi) \cdot V_{total} \cdot \eta_{nf,int} \cdot K \cdot \dot{\gamma}^2_{nf,int}. \tag{7.1}$$

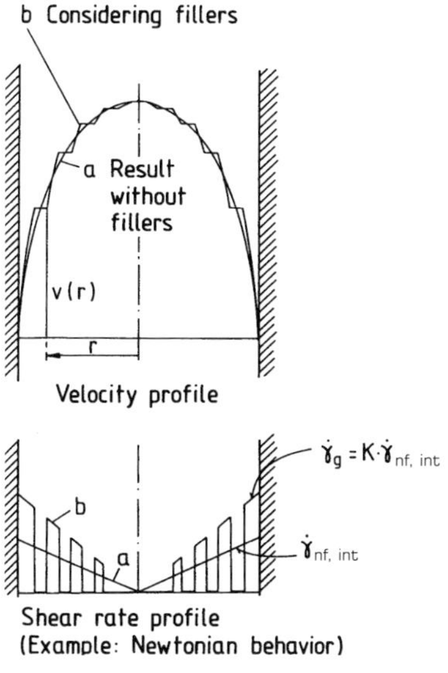


**Figure 7.4** Dissipation model [10]

This value of dissipated energy is, when compared to the integral parameters ($\eta_{nf,int}$, $\dot{\gamma}_{nf,int}$), increased by the factor

$$F_{diss} = (1-\phi) \cdot K. \tag{7.2}$$

From this formula, it can be seen that in high-viscosity rubber compounds the rise in temperature due to shear is not only by the high viscosity alone, but also by the presence of fillers. When measurements are done by capillary rheometer, it is observed that the experiments at higher throughput are no longer isothermal. Therefore, the results must be corrected. Then, a viscosity function, independent of the geometry, can be found by considering the increase in the shear rate [15]. An iterative procedure for the determination of the occurring peak temperatures is proposed in [15]. In [10,16] a formula for the computation of the factor $K$ (related to the excessive shear rate increase) from the volume filler contents $\varphi$ is given, based on a two-dimensional relationship, in which fillers are assumed to be cut spheres of a circular area $\pi \cdot d^2/4$:

$$K = \left[1 - \frac{6 \cdot \varphi}{\pi} + \frac{1}{2} \cdot \left(\frac{4 \cdot \varphi}{\pi}\right)^{\frac{3}{2}}\right]^{-1} . \tag{7.3}$$

*(b) Yield Stress:* Besides the rise in the local shear rate, interaction between individual filler particles can lead to the so-called yield stress [17]: below a certain shear stress value $\tau_0$ (yield stress), the material will not flow and it behaves as a solid body. This can be observed at times in natural rubber [10]. The flow profile of such a material can be subdivided into a region of shear flow and a region of plug flow. As shown in Fig. 7.5, the plug flow takes up a smaller part with increasing ratio of wall shear stress to the yield stress. So, the plug/shear flow model is then relevant when the shear stress at the wall is low, that is, at small volumetric flow rates or large cross sections of flow. The corresponding law of flow is

$$\tau = \tau_0 + \phi \cdot \dot{\gamma}^n \tag{7.4}$$

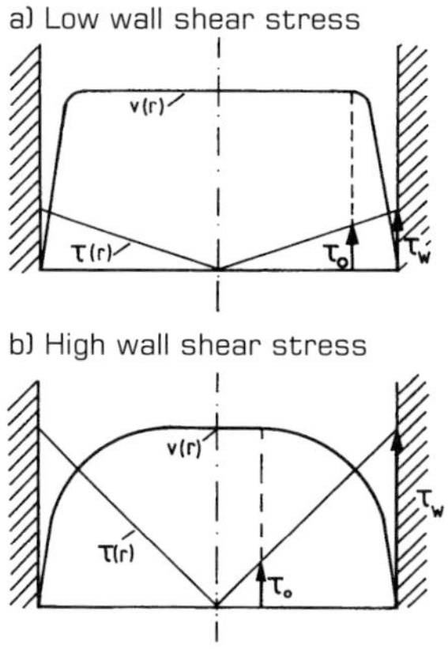


**Figure 7.5** Block/shear flow model [10]: (a) low shear stress at the wall, (b) high shear stress at the wall

This is the so-called Herschel-Bulkley material model (see also Section 2.1.1.2) [18,19]. The possibilities for the determination of yield stresses using capillary rheometers with circular and slit shaped capillaries are discussed in [10].

*(c) Wall Slip* The problem of wall adhesion (where the velocity of the layer at the wall is zero) for elastomers has been scientifically researched only selectively [20], because it is rather difficult to prove and hence to investigate and detect resulting effects. Studies conducted on NR, SBR, and CR compounds [10] as well as with injection molding of elastomers [9] suggest the conclusion that wall slip in the die can be neglected for many compounds. This, however, is only then allowable when there are no instabilities in the viscosity function, or if no changes in flow functions occur when changing the geometry of the capillary during their evaluation in a capillary rheometer. Surely, among the at least 10,000 rubber compounds used in Germany today, there will be some that exhibit wall slip in the die. But to consider these special cases in the general die design would increase their complexity and cost. Many phenomena attributed to wall slip today can also be explained by dissipation effects in the region of high shear rate at the wall of the die or by the presence of a yield stress [10].

### 7.2.3 Computation of Viscous Pressure Losses

The pressure loss at the given volumetric flow rate is an important criterion in the design of dies. The total pressure consumption, which should be as low as possible, indicates how well a die performs. An extrusion die used, for example, for spreading the mass flow from the extruder in the production of slabs should produce equal pressure loss on each flow path under the assumption of a constant exit velocity across the width. On the other hand, the determination of the shear stresses at the wall, included in the total pressure consumption, should supply the information if there is any part of the die where the yield stress was not reached, i.e., if at that point the mass stopped moving. A study examining if the computation methods developed in the plastics area are transferable to the area of elastomers was done on basic shapes, such as circle, slit, and annular gap, and is reported in [10]. In that work, both isothermal and nonisothermal computations and experiments were performed.

#### 7.2.3.1 Formulas for Isothermal

The assumption in all isothermal computations is a constant mass temperature throughout the entire flow channel. In contrast to nonisothermal computations, this assumption minimizes the amount of computing. As pointed out in the previous chapter, elastomers exhibit wall slip only as an exception. Therefore, for the isothermal computation of pressure losses, the formulas for the materials with wall adhesion behaving according to the power law $\eta = k \cdot \dot{\gamma}^{n-1}$ can be applied. If yield stresses occur, then the viscous material behavior is described by the *Herschel-Bulkley* law. Additional assertions pertaining to the computation of viscous pressure losses can be found in [21–23]. They were experimentally verified with rubber compounds in [10].

Flows in a pipe and slit are relatively simple to describe [24,25]; there are many analytical formulas for the isothermal computation available (see Tables 3.1–3.6 and 7.2).

**Table 7.2** Equations for the Calculation of Pipe and Slit Channels (Herschel-Bulkley Material Law) [10]

| | Pipe channel | Slit channel |
|---|---|---|
| Pressure loss | $\Delta p = \left(\frac{\dot{V}}{\pi \cdot R^3 \cdot K*}\right)^n \cdot k \cdot \frac{2L}{R}$ <br> $K^* = \frac{n}{n+1} \cdot \left(1 - \frac{\tau_0}{\tau_W}\right)^{\frac{1}{n}+1} \cdot \left(\frac{\tau_0}{\tau_W}\right)^2 + \left(\frac{2n}{1+2n}\right) \cdot \left(1 - \frac{\tau_0}{\tau_W}\right)^{\frac{1}{n}+2} \cdot \left(\frac{\tau_0}{\tau_W}\right) + \left(\frac{n}{3n+1}\right) \cdot \left(1 - \frac{\tau_0}{\tau_W}\right)^{\frac{1}{n}+3}$ | $\Delta p = \left(\frac{\dot{V} \cdot \left(\frac{1}{n}+2\right)}{K^{**} B \cdot \left(\frac{H}{2}\right)}\right)^n \cdot k \frac{2L}{H}$ <br> $K^{**} = \left(1 - \frac{\tau_0}{\tau_W}\right)^{\frac{1}{n}+2}$ |
| Wall shear rate | $\dot{\gamma}_W = \frac{\dot{V}}{\pi R^3 \cdot K^*}$ | $\dot{\gamma}_W = \frac{\dot{V} \cdot \left(\frac{1}{n}+2\right)}{K^{**} \cdot B \cdot \left(\frac{H}{2}\right)^2}$ |

Furthermore, it is possible to describe such flow by the method of representative viscosity [22–24] (see also Section 2.1.2). Here, advantage can be taken of the following circumstance: in the flow channel, there is always at least one streamline where the same shear rate will result whether the flow is Newtonian (i.e., $\eta \neq f(\dot{\gamma})$) or pseudoplastic.

The location *r*, normalized to the radius of the flow channel (i.e., $e_o = r/R$) for various constitutive equations, is obtained from the comparison of the equation $\dot{\gamma} = f(r)$ for Newtonian and pseudoplastic fluids [23]. The procedure for the flow through a slit is similar. The locations for the representative shear rate for the flow through a pipe and through a slit as well as for the power law and Herschel-Bulkley material flow are shown in Figs. 7.6 and 7.7.

The advantage of this method is that the flow problems can be described by the simple equations for Newtonian fluids. In the corresponding equations, the representative shear rate $\dot{\gamma}_{rep} = e_{\bigcirc} \cdot \dot{\gamma}_{Newton}$ and the representative viscosity $\eta_{rep} = \eta\left(\dot{\gamma}_{rep}\right)$ are introduced into the corresponding equations. The value is taken from the true flow curve, which can be represented by the usual viscosity functions (Carreau, power law, Vinogradov, and so on). As shown in Figs. 7.6 and 7.7, the values for $e_{\circ}$ and $e_{\square}$ lie in narrow limits over a wide range of flow exponents so that, for example, the mean value of $e_{\circ} = 0.815$ can be taken for the computation with the maximum error of 5% (for $0.2 < n < 0.6$). This independence of flow exponents over a wide range can be found in practical applications and also for Herschel-Bulkley fluids. The following mean values can be obtained from Figs. 7.6 and 7.7:

*Power law:*

$$\begin{aligned} e_{\bigcirc} &= 0.815 \left(0.2 < n < 0.6\right) \\ e_{\square} &= 0.772 \left(0.2 < n < 0.6\right) \end{aligned} \tag{7.7}$$

*Herschel-Bulkley law:*

$$\left.\begin{aligned} e_{\bigcirc} &= 0.815 \left(0.2 < n < 0.6\right) \\ e_{\square} &= 0.772 \left(0.2 < n < 0.6\right) \end{aligned}\right\} \frac{\tau_0}{\tau_w} < 0.33 \tag{7.8}$$

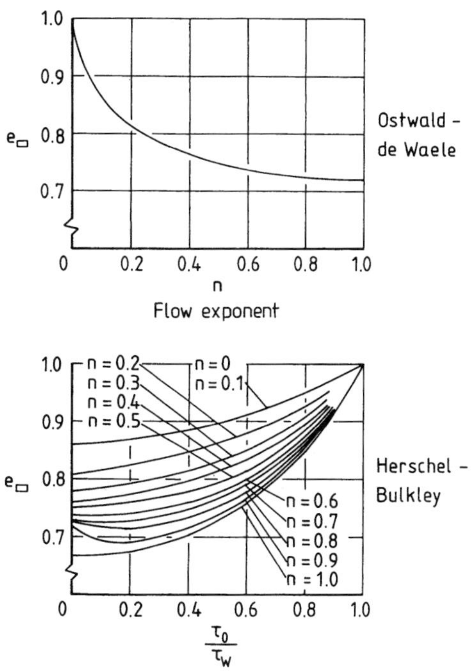


**Figure 7.6** Representative locations of the flow through a slit [23]

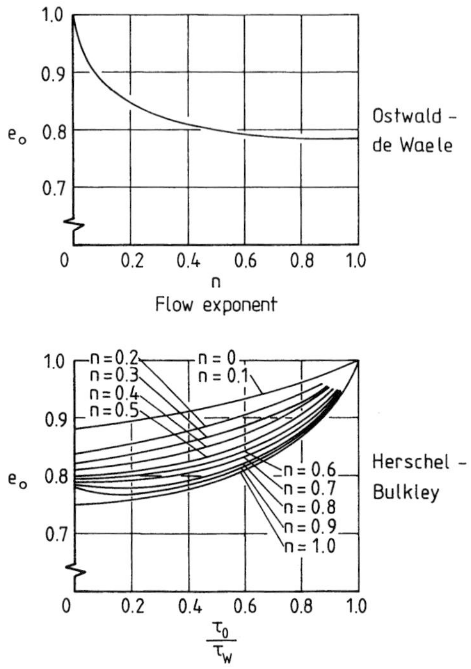


**Figure 7.7** Representative locations of the flow through a circular channel [23]

In contrast to the above-mentioned types of flow, the shear stress in an annular channel does not have a linear profile (see Chapter 3). The equations used for calculation of annular flow are, as a rule, more complicated. Therefore, it is recommended in cases where $R_o - R_i \ll R_i$ to consider the annular gap as a flat slit. The corresponding formulas can be taken from Chapter 3.

### 7.2.3.2 Approaches to Nonisothermal Computations

A nonisothermal flow in the forms of flow discussed so far can be computed with relatively simple means (difference procedure). In this case, the velocity and temperature fields are coupled. Since the problems are not specific to elastomers, the subject can be dealt with using the treatment discussed thoroughly in Chapter 4.

As shown in [10], the increased dissipation due to the increase in the shear rate in the case of nonisothermal flow can be taken into consideration, mainly to make the peak temperatures realistically visible in the mass flow. In the case of small differences between wall and mass temperatures, the effect of the peak temperature on the pressure consumption in the extrusion dies for elastomers is minimal. In this case, the computation of the pressure loss can, therefore, be done by assuming isothermal flow. Since this is usually the case in practice, the isothermal formulas represent a good resource for a practical computation of extrusion dies. Here, the power law for pseudoplastic materials according to *Ostwald and de Waele* should be used, since the yield stress in practice is important only in special cases. If the analysis of results from rheometer studies indicate the occurrence of a yield stress, an analysis of critical points in the die for the actual operation conditions should follow. In this analysis, it should be established whether the shear stress at the wall is greater than the yield stress [10]. If it turns out that the yield stress cannot be neglected, the computation of the pressure consumption should be performed, if possible, by the method of representative viscosity.

### 7.2.4 Estimation of the Peak Temperatures

A high pressure drop in the die and the correlating shear load on the melt always lead to an increased melt temperature dissipation. In order to estimate the chance of scorch in many practical applications, not only the pressure drop but also the peak temperatures in the die are of interest.

Figure 7.8 shows as an example the temperature and velocity profiles in an annular channel resulting from a nonisothermal computation. A pronounced local peak temperature can be seen that, in practice, could lead to scorch of the material.

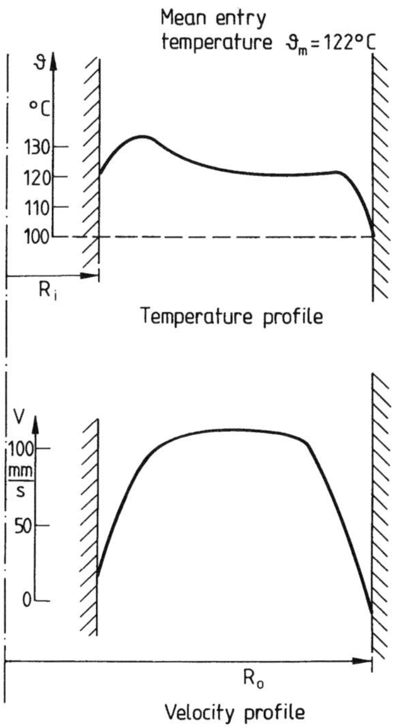


**Figure 7.8** Temperature and velocity distributions in an extrusion die (die entry) [10]

Since in a case applicable to practice (wall temperature = mass temperature) the isothermal formulas are satisfactory for the computation of the pressure loss, the processor has the most important information about the extrusion die (pressure consumption, peak temperatures) handy without the use of elaborate computational methods. In conclusion, it should be mentioned that these computational methods for the computation of nonisothermal flow are also suitable for the study of crosslinking and vulcanization reactions. This is shown in [26] for the crosslinking of polyethylene by peroxide in a continuous vulcanization in a cable jacketing line. The effect of the time-dependent viscosity on the velocity and temperature profiles of a rubber compound in an injection-molding machine is shown in [27]. The change of the viscosity with time was obtained from vulcameter curves.

### 7.2.5 Consideration of the Elastic Behavior of the Material

The viscoelastic properties of melts described in Section 2.1.2 can be observed very clearly in elastomeric compounds. The swelling of the material due to elastic effects, that is, recovery of the reversible deformations in the polymer, which are a function of deformation and relaxation history, is an important factor in the choice of size of the die cross section. Besides that, the rearrangement of the velocity profile in the region to a plug profile outside the die can lead to local elongations and compressions in the emerging extrudate and, thus, to a change in its cross section. The studies of extrudate swelling are not only limited to homogeneous melts and solutions of plastics but also to rubber compounds [28,29].

The physical cause for the extrudate swelling and the mathematical formulation pertaining to it have been interpreted in different ways because of the variety of phenomena observed [1,30–35]. Further relevant details can be found in Sections 2.1.3 and 4.6.

A complete computational description has been very difficult so far and can be done only by three-dimensional computation methods with the use of appropriate material laws [36,37]. The application of such methods is very demanding and time consuming, and it is often limited because of material data that are difficult to measure or because of the complex material laws. An alternative method, which requires easily obtainable material data and can be accomplished by the application of a $2\frac{1}{2}$-dimensional FEA, will be introduced in Section 7.4.2.

In die inlets and sudden changes in cross section as well as in convergent zones of the flow channel, not only shear but also elongational deformation occur, which cause energy transformation and therefore an additional pressure drop. There is very little published information on this subject. Therefore, the die designer often works with data obtained from experience. There are several factors determining if a given elastomeric compound behaves predominantly as either a viscous or elastic material. The factors are, for example, type and amount of filler, the elastomer content and type (chemical nature), and the melt temperature. By replacing the viscoelastic polymer with inorganic fillers, the elastic properties of the compound can be changed markedly.

## 7.3 Design of Distributor Dies for Elastomers

The extrusion of flat and circular semifinished products from elastomers requires the use of a distributor system that is capable of shaping the full strand of mass supplied by the extruder in a desired fashion. The concepts and methods of design of dies for the extrusion of thermoplastics are thoroughly discussed in Chapter 5. Some of the formulas derived from the processing of thermoplastics have been applied, after a critical evaluation, to elastomers. They have been successfully used in practice.

This is particularly true for side-fed mandrel dies of the type independent of the operating conditions for the extrusion of hose and cable jacketing. Since rubber processing plants experience a great deal of batch-to-batch variability in the properties of the material, it is advantageous to design the dies to be independent of the operating conditions and the material processed. This can offset the variability of the material. Side-fed mandrel dies designed and evaluated for different compounds were described in [10]. The design is done according to the relationships described in [25,38] and also in Chapter 5, and has been found to be effective as to the melt distribution in all cases. A wide slit die with a width of 200 mm was designed following the method published in [24]. This die also confirmed the independence of its melt distribution from the operating conditions and the material in the extrusion of rubber compounds. Therefore, these design principles can be considered applicable also to the design of extrusion dies for rubber [10,39]. If a design independent of the operating conditions is not possible, as is the case for a fishtail manifold, for example, it should be made at least insensitive to operating conditions. Dies insensitive to operating conditions are those that have the flow exponent constant over a large viscosity range in the channel and die land zones [40]. This means the flow channel has to be designed in such a way that the occurring representative shear rates are considerably smaller, or in contrast considerably larger, than the shear rate at the transition of the viscosity curve. In this case, the slope of the logarithmic viscosity curve is nearly constant, and by this the relative change of viscosity is also nearly constant in all flow channels.

The computations on which the above examples are based are model formulas for the computational design of such mass distribution systems, assuming that there is a pipe flow in the manifold and a flat slit flow in the flow resistance zones, which, at least theoretically, lead to independence of material and which have been proven in practice. Strictly speaking, in some sections of the channel there is a multi-dimensional flow that can be described completely only by numerical tools like FEM [41].

Elastomers exhibit at times a strong pseudoplasticity in their flow, which is expressed in their flow exponent. It is therefore reasonable to test the invariance of a die as to the material processed by comparing the flow exponent for the shear rate in a pipe and in a slit at the upper and lower limits of the operating field.

As shown in Chapter 5, the design of a wide slit die, in addition to the specified height of the land and the die width, includes two free parameters $R_0$ and $y_0$ which are related to each other through pressure. The shear rates in the die land zone and in the distributor channel are determined from the following relationships:

$$\dot{\gamma}_{\text{rep, pipe}} = e_{\bigcirc} \cdot \frac{4 \cdot \dot{V}}{\pi \cdot R_0^3} \tag{7.9}$$

$$\dot{\gamma}_{\text{rep, slit}} = e_{\square} \cdot \frac{6 \cdot \dot{V}}{B \cdot H^2} \tag{7.10}$$

Changes in the flow rate affect the linear change of both shear rates. The invariance is guaranteed as long as a shift in the shear rates causes a similar change in viscosity. The more both shear rates lie in the linear region of the viscosity curve, the less effect a change in operating conditions will have. In an ideal case, both resulting shear rates are identical. In such a case the die would be fully independent of the operating conditions. The disadvantages here may be a large-size die land, which may cause a space problem, high pressure consumption, and high clam-shelling forces [42].

Considering the correct length of the channel and its shape (e.g., simple rectangular cross sections or rectangular cross sections with one semicircular side) [43] improves the computation of dies for rubber compounds considerably [44].

# 7.4 Design of Slotted Disks for Extrusion Dies for Elastomers

## 7.4.1 Computation of Pressure Losses

A detailed study of the complex flow in slotted disks is possible only with extremely elaborate means [37,45]. Therefore, simpler, but effective, means that allow an integral analysis of the flow process are desirable. The inlet pressure losses are of central importance in this analysis. The total pressure loss of a profile die for rubber consists of the elastic component, which is used up in the inlet zone of the orifice, and the viscous component, which is used up in the orifice alone:

$$p_{\text{total}} = p_{\text{entry}} + p_{\text{orifice}} \tag{7.11}$$

When one knows the losses, it is possible to use them in the die design and to make the local flow distances so that there is a constant velocity at the exit of the die. Different methods for estimation of the inlet losses have been evaluated [46,47]. A practical test has shown [10] that the method outlined in [46] is the most suitable for the estimation of the inlet pressure losses. The minimum inlet pressure loss for a circular cross-sectional transition (e.g., the capillary of a capillary rheometer) is obtained from the following relationship:

$$\Delta p_{\text{entry}} = \frac{4 \cdot \sqrt{2}}{3 \cdot (n+1)} \cdot \dot{\gamma}_0 \cdot (\eta \cdot \mu)^{0.5} . \tag{7.12}$$

The quantity $\mu$ is called the elongational viscosity by Cogswell. This elongational viscosity, similar to shear viscosity, has an exponential correlation with the elongation rate $\dot{\varepsilon}$ according to studies performed on several rubber compounds:

$$\mu = \mu_0 \cdot \dot{\varepsilon}^{b} . \tag{7.13}$$

The relationship between elongational stress and the inlet pressure loss is as follows:

$$\sigma_E = \frac{3}{8} \cdot (n+1) \cdot \Delta p_{\text{entry}} \tag{7.14}$$

and the relationship between elongational stress and elongational rate is

$$\sigma_E = \mu \cdot \dot{\varepsilon} . \tag{7.15}$$

After rearrangement, the equation for the inlet pressure loss is obtained:

$$\Delta p_{\text{entry}} = \left[ \dot{\gamma}_0^{\frac{n+1}{2}} \cdot \left[ \frac{4 \cdot \sqrt{2}}{3 \cdot (n+1)} \cdot \sqrt{k} \cdot \sqrt{\mu_0^{\left(1-\frac{b}{b+1}\right)} \cdot \left[ \frac{3}{8} \cdot (n+1) \right]^{\frac{b}{b+1}}} \right] \right]^{x} \tag{7.16}$$

with

$$x = \left( 1 - \frac{b}{2 \cdot (b+1)} \right)^{-1}. \tag{7.17}$$

The constants of the elongational viscosity ($\mu_0$, $b$) can be determined from measurements by a capillary rheometer. Therefore, the inlet pressure losses were determined by the application of the Bagley correction (see Chapter 3). For this purpose, the pressure loss (at different tube lengths, at constant orifice diameter and constant flow rate) is plotted against the length of the die. The value of the pressure loss at the given flow rate is obtained as the value where the extrapolated line, connecting the points, intersects with the abscissa. This value can be inserted into Equation (7.14) so that with the known shear rate function $\eta = f(\dot{\gamma})$ the values of $\mu$ and $\dot{\varepsilon}$ are obtained directly by applying Equations (7.10), (7.12), and (7.15).

Plotting the values of $\mu$ against elongation rates $\dot{\varepsilon}$ finally leads to the function $\mu = f(\dot{\varepsilon})$, and from it $\mu_0$ and $b$ can be determined.

The above procedure has been tried in [10] on different slotted disks (rectangular and circular openings). It turned out that Equations (7.12) and (7.17) are well suited for the calculation of pressure losses due to elasticity. This is documented in Table 7.3, in which measured and calculated values from different sets of measurements are compared.

**Table 7.3** Measured and Calculated Values from Experiments with Slotted Disks [10]

| Mass throughput, g/min | Inlet pressure, measured bar | Inlet pressure, calculated bar | Error % | Disk orifice |
|---|---|---|---|---|
| 410 | 46.9 | 42.9 | 9.3 | rectangle, height = 4 mm, width = 5 mm |
| 550 | 50 | 46.9 | 6.6 | |
| 655 | 53.2 | 49.4 | 7.6 | |
| 380 | 43.3 | 43 | 0.8 | circle, 5 mm diameter |
| 550 | 46.3 | 48 | 3.7 | |
| 670 | 49.5 | 51 | 3.1 | |
| 520 | 51.7 | 52 | 0.5 | circle, 5 mm diameter |
| 655 | 58.3 | 55.7 | 4.6 | |

In another set of experiments involving more complex geometry, the formula proposed in [47] was tested. It is based on the premise that cross sections of equal area produce equal inlet pressure losses, which can be calculated from the determination of entrance pressure losses of a circular channel. As shown in Fig. 7.9, an estimation of the inlet pressure losses is possible because cross sections of equal area do not produce any appreciably different readings. To calculate the pressure loss in an orifice, a concept is proposed in [45,48] that allows estimation of the pressure gradient for each subsection of the geometry by FEM. This will be explained with an example of a rubber welting profile (Fig. 7.10).

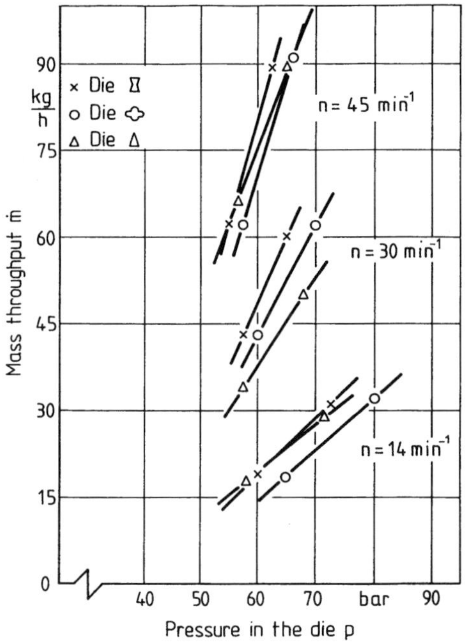


**Figure 7.9** Inlet pressure losses of slotted disks with equal cross-sectional areas [10]

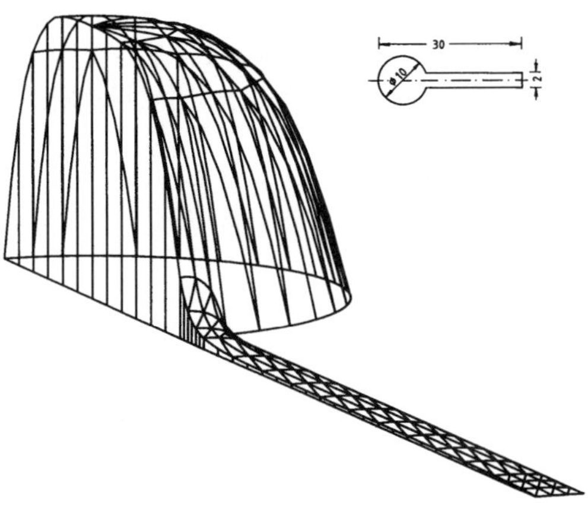


**Figure 7.10** Velocity distribution in the parallel die land zone in example beading strip die [45]

It consists of two subsections, a circular body (diameter 10 mm) and a narrow, long beading (length 20 mm, height 2 mm). The mean velocity is 10 m/min (= 167 mm/s).

Now, the flow analysis is done for the entire profile while the pressure gradient is varied within reasonable limits. The resulting values (pressure gradient, mean velocity) are plotted for each subsection of the profile in the form of a characteristic line (Fig. 7.11).

One has to make sure that the results from each subsection of the profile are calculated in the range of the desired mean velocity. Then a pressure gradient for each subsection can be determined graphically by a line going through the desired velocity and running parallel to the pressure axis. This pressure gradient corresponds to the desired velocity (Fig. 7.11).

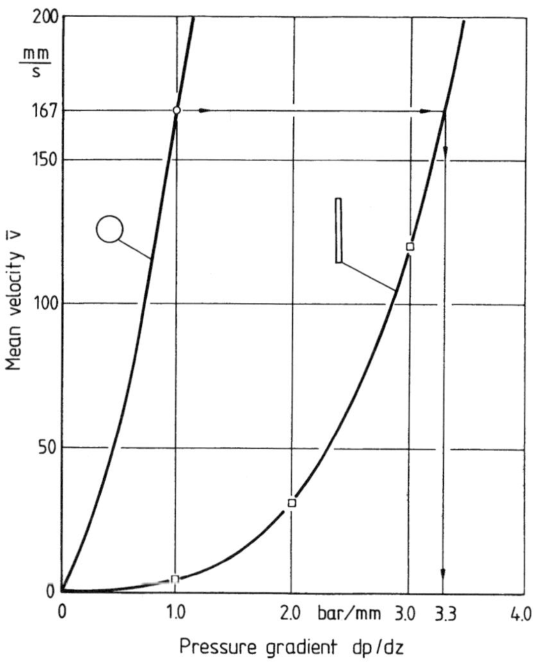


**Figure 7.11** Characteristic lines for the two sections of the die [45,48]

### 7.4.2 Extrudate Swelling (Die Swell)

The pseudoplastic polymer melts do not convert the total energy of deformation imparted by flow by dissipation into heat, but they store a part of it elastically [1,30–32]. When the melt leaves the orifice, the form constriction ceases and the melt strand expands. This condition is defined as extrudate swelling or die swell, and it marks the effect referred to as viscoelastic flow behavior. The elastically stored energy is released with time due to the relaxation process.

The theoretical grasp and modeling of viscoelastic behavior is still a subject of current research [e.g., 36,37,45]. A method applicable to practical die design is presented below. It allows one to consider the die swell in the design of relaxation zones in the die with the tools available today. This method starts with the experimental determination of the extrudate swelling for a round capillary of a capillary rheometer. Almost as a byproduct of the capillary rheometry, the emerging strand is measured mechanically or optically, and the amount of swelling of the material is determined (Fig. 7.12). The die swell is calculated from the ratio of the cross sections of the round strand and the capillary:

$$S_{swell} = \left(\frac{D}{D_0}\right)^2 = \left(\frac{R_0 + a}{R_0}\right)^2 \tag{7.18}$$

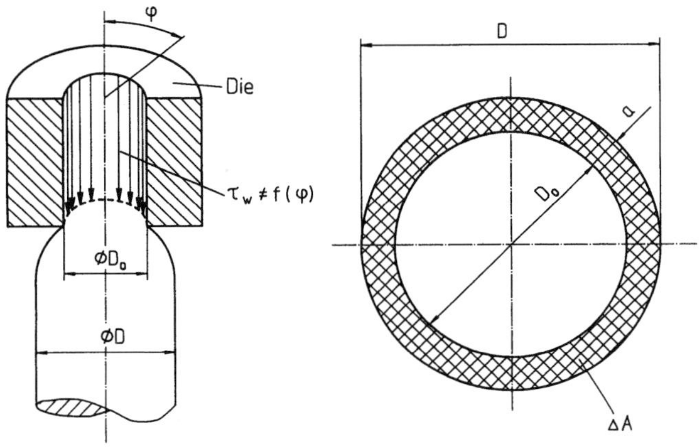


**Figure 7.12** Die swell

Figure 7.13 depicts the evaluation of the results from the experiments in the capillary rheometer. The dependence of the die swell $S_{swell,II}$ on the shear stress at the wall is shown. In many materials this correlation is almost linear [29,49]. The ratio between the length of the orifice $L$ and its diameter $D_0$ was chosen as a parameter.

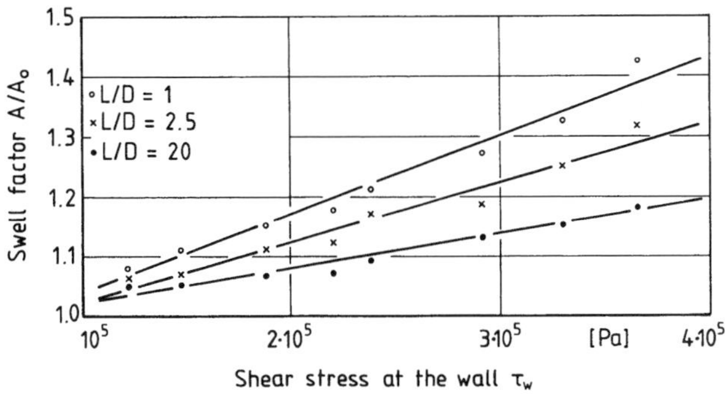


**Figure 7.13** Correlation between the swell factor and the wall shear stress in the linear case [50]

The slope of this line varies with the length of the orifice. It is the measure of the residence time of the melt in the orifice, and it indicates the amount of imposed deformation (e.g., from the inlet zone) still stored in the melt or, conversely, how much has been dissipated.

Only those variables that are defined in the die play a role. The extrudate swell $S_{swell}$ is described by two formulas: one of them takes into account the flow history of the polymer and the other the exit of the melt from the orifice:

$$S_{swell,I} = f\left(\text{residence time}\right) \tag{7.19}$$

$$S_{swell,\,II} = f\left(\text{shear stress at the wall}\right). \tag{7.20}$$

In [29,51] an exponential equation is given that expresses the time dependence of the magnitude of $S_{swell,I}$ (Fig. 7.14). Test results from capillaries with noncircular cross sections using a laboratory extruder (screw diameter 60 mm) can be taken from Fig. 7.14 [51]. The change in swelling of the material with time is shown by a solid line.

The correlation between die swell and shear stress at the wall is connected in [50] to the so-called criterion of swell at the outlet, which is shown here for the linear dependence.

$$\Delta A = K_{prop} \cdot \tau_w . \tag{7.21}$$

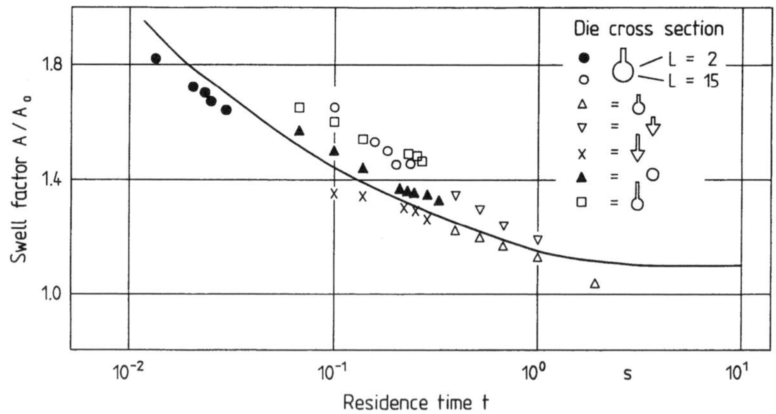


**Figure 7.14** Swell factor versus residence time [51]

In the case of a circular cross section, there is an equal shear stress at the wall along the entire circumference. The increase of area $\Delta A$ can be expressed by the distance $a$ from the circular strand to the diameter of the orifice (from Fig. 7.12):

$$\Delta A = f\left(\pi, D_0, a\right). \tag{7.22}$$

Thus, a condition of proportionality is defined:

$$\frac{a}{\tau_{\mathrm{W}}} = K^{*}_{\mathrm{prop}} \tag{7.23}$$

This condition represents a relationship for the swelling behavior of the cross-sectional areas. For the practical die design computations, the problem is reversed and the expanded area of the extrudate cross section is taken as the target value. From that it follows that the cross section of the die orifice must be reduced in such a way that the target value is reached exactly after the swelling of the extrudate. An extension of the definition from Equation (7.23) allows the computation of any desired cross-sectional shape:

$$K_{\mathrm{prop,\,W}} = \frac{\overline{a}}{\overline{\tau}} = f\left(\Delta A_{\mathrm{W}}, L, \tau\right). \tag{7.24}$$

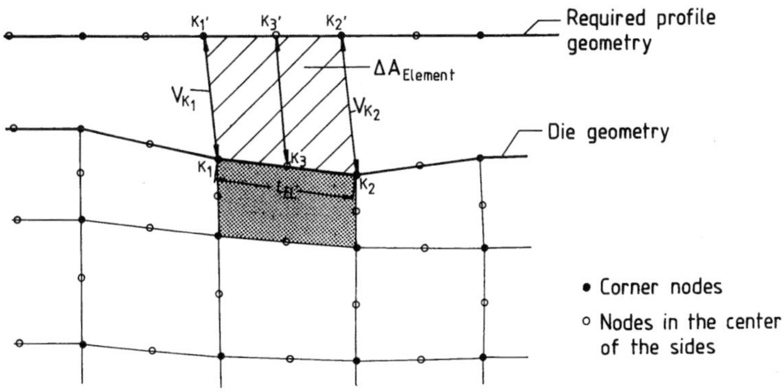


**Figure 7.15** Criterion for swelling [50]

As in the computations of pressure for arbitrary die cross sections, the cross section of the orifice is divided into finite elements [50]. Figure 7.15 shows a section of an FE mesh. The border line of the elements represents the exit contour of the die. Next to it the contour of the target geometry is shown. The numbering of the nodes of the target geometry corresponds to the numbering of the die geometry. This way the integral values of the swell obtained from the measurements in a capillary rheometer can be transfered to differential sections of any desired die geometry.

A computer program developed for problem solving [45,52,53] allows an automatic optimization of flow channels. It is based on the combination of an optimization procedure (the so-called evolution strategy [54–58]) and an FE program [50]. Examples of the capabilities of this method will be given in the next chapter.

### 7.4.3 Simplified Estimations for the Design of a Slotted Disk

A simple and fast way to obtain information about the velocity and the shear rate profile in the parallel zone of an orifice with arbitrary cross section profile is to calculate the single available velocity component with the assumption of viscometric flow. This velocity component is always perpendicular to the cross section under consideration. Therefore, each cross-sectional area corresponds to an isobaric plane, and all pathways are straight lines. The computed distributions therefore are, strictly speaking, valid only for a fully developed flow in an "infinitely" long land region. Nevertheless, the calculated values can approximate actual values quite well even for relatively short land regions, provided that the land length is the same at any point [45].

Figure 7.16 shows a profile from EPDM and the velocity distribution existing at the die exit as a 3D surface plot. A height is assigned to each node of the plane FE network, and the surface is created through these points. The computed velocity distribution shows clearly that the noses and the shank, standing out on the left, lag behind and hence are greatly stretched during the takeoff. Here, the cross section upstream has to be widened in order to increase the velocity (see also Section 4.4.5).

It is known from experience that the velocity distribution depends on the flow behavior of the melt, so the flow channel can be employed only at a certain operating point and for one material. This dependence can be determined analytically for simple cross sections [59]. For arbitrary cross sections, the dependence of the velocity distribution on the flow behavior of the material and the extrusion pressure can be quantified by numerical tools like FEM.

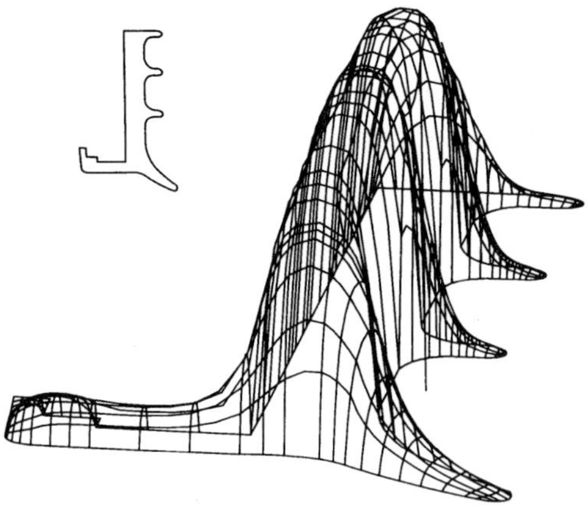

**Figure 7.16** Velocity distribution in the parallel die land of a die for EPDM [45]

For the flow channel design, the question is how to change the channel in order to obtain as uniform a flow as possible in all areas of the profile.

One possibility is to change the shape of the profile such that the mean velocity is constant at equal land length. For that purpose, the profile has to have the same wall thickness everywhere.

Another procedure involves matching the lengths of the individual land regions in such a way that a constant mean exit velocity is attained. This is shown in [59] on an example of a profile divided into simple subsections.

In [45,48] a concept is suggested for arbitrary profiles that allows an estimate of the lengths of the parallel zones for each zone of the cross section using FEM. This is shown on a rubber beading strip in Fig. 7.10.

As expected, without the correction of the flow length, the strip lags behind the circular profile. The mean velocity should be 10 m/min (167 mm/s). The behavior of the material is described by the Carreau law with the following coefficients: $A = 78{,}846$ Pa·s, $B = 2.11$ s, and $C = 0.687$.

The pressure gradient, at which there is an equal velocity in each subsection, is obtained from Fig. 7.11.

Assuming that the pressure $p$ at the inlet to the parallel zone is constant everywhere, the following relationship is true:

$$p = p_i' \cdot L_i \tag{7.25}$$

with

$p_i' = \dfrac{dp_i}{dl}$ Pressure gradient in the subsection of the profile $i$

$L_i$ Length of the subsection being sought

$i$ 1, 2, … number of subsections

Note: There is no summation by $i$ in Equation (7.25).

Denoting the round part of the profile with the index “1” and the strip with “2”, the relationship for the case above is

$$L_1 = p / p_1', \qquad L_2 = p / p_2'. \tag{7.26}$$

From Fig. 7.11:

$$p_1' = 1\,\text{bar} / \text{mm}, \qquad p_2' = 3.3\,\text{bar} / \text{mm}.$$

With a given pressure loss $p$ in the parallel zone of 50 bar, the result is

$$L_1 = 50\,\text{mm}, \qquad L_2 = 15\,\text{mm}.$$

In this design, the cross-flows in the inlet region and the entrance pressure losses were not taken into consideration. When entrance pressure losses occur, the total pressure loss used above has to be reduced by the entrance pressure loss (see Section 7.4.1). The land lengths will change accordingly. In order to calculate the cross-flow, a rather complex three-dimensional computation of flow has to be performed. Experimental verification of this procedure on slotted disks for the extrusion of rubber profiles has shown that the procedure above provides a good estimation of the flow distances. This is particularly true when the subchannels can be kept separated until just before the die exit. Whether this is possible depends on the ability of the melt to fuse sufficiently in the remaining distance to the die exit.

In the next step the die exit cross section is determined. This depends on three problem areas (Fig. 7.17) [50]:

- Zones of material stagnation
- Surface quality
- Extrudate swelling

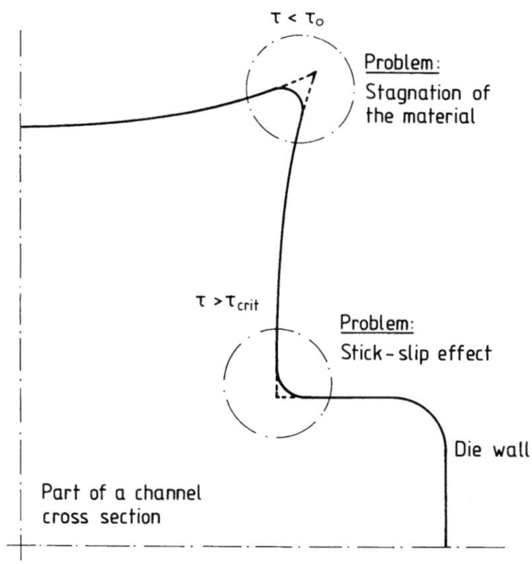


**Figure 7.17** Problem areas in the die design [50]

### *Problem Area: Material Stagnation*

Figure 7.18 shows a cross section of a die. Because of symmetry only a quarter segment is necessary.

The shear stress at the wall drops to zero in the corners and, therefore, under the critical shear stress (yield stress, $\tau_0$), which is the line drawn into the diagram. The yield stress is strongly temperature dependent for many materials. Outside corners with poor temperature control can therefore promote the occurrence of the yield stress. The goal of an optimization is the increase in the wall shear stress in the corners. The level of the shear stress is raised to exactly the level of the yield stress by rounding the edge region uniformly. In this way, the goal of the optimization is reached (Fig. 7.19).

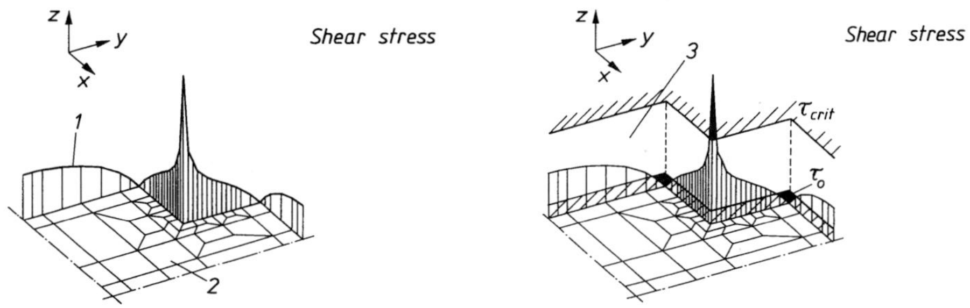


**Figure 7.18** Shear stress at the wall in a die land [50]: 1: shear stress at the wall, 2: part of the die cross section, 3: permissible processing range

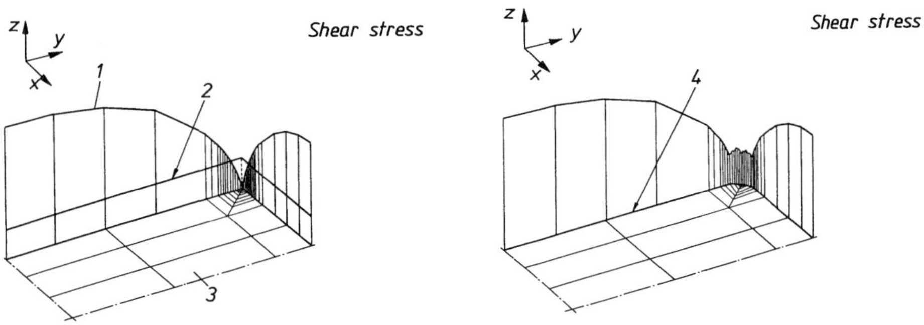


**Figure 7.19** Improvement to reach the critical yield stress [50]: 1: shear stress at the wall, 2: yield stress $\tau_0$, 3: part of a die cross section

With a larger radius in the corner, the level of the shear stress at the wall is raised even more and the overall stresses are more even, but the disadvantage would be that the finished profile is different from the required one. The advantage is that the geometry of the die is changed only as much as is absolutely necessary.

***Problem Area: Surface Quality***

Inside corners with sharp edges produce peaks in the shear stress. Figure 7.18 shows a section of a die having an inside corner susceptible to a stick-slip flow situation. The FE mesh is made finer in this location in order to increase the accuracy of the calculations. Moreover, the critical shear stress $\tau_{crit}$ is shown. When this value of shear stress is exceeded, the flow breaks down and a partial wall slip is established. The goal of the optimization is to minimize the enclosed area above the $\tau_{crit}$ line. In addition to that, it is required that the total exit area before and after the optimization remains the same. This requirement is necessary, for example, when the characteristic of the orifice must be preserved.

The relevant area of the die contour can be seen in Fig. 7.20. The optimized shape of the die is drawn, too. The difference between the two shapes in the surface contour is very small. The sharp corner is rounded, however. The regions that were displaced to the inside must be pressed outside in other places in order to keep the total area constant as required.

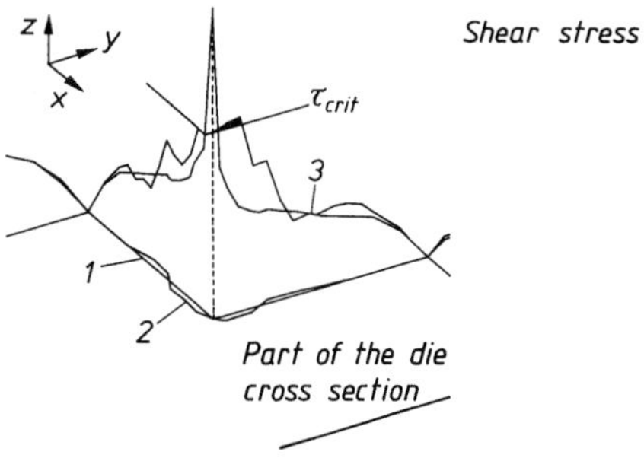


**Figure 7.20** Failure to exceed the critical shear stress [50]: 1: initial die wall, 2: optimized shape of the die wall, 3: shear stress at the wall

The resulting shear stress profile is symmetrical to the inside corner just as in the starting geometry. The crucial advantage is that the dangerous peak in the shear stress is almost completely removed by rounding off the sharp corner.

### *Problem Area: Die Swell*

Here the die exit has to be shaped in such a way that the emerging profile has the specified dimensions after die swell takes place. As an example a rectangular profile is chosen. A quarter segment is sufficient because of symmetry. The specified contour of this profile is shown in Fig. 7.21. The ratio of the height $H$ to width $B$ is 1/2.

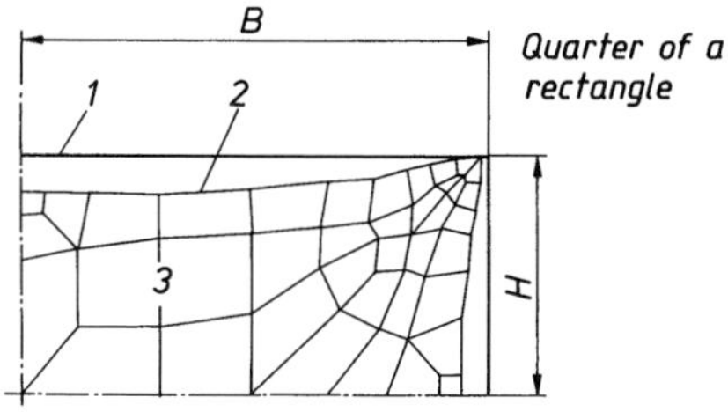


**Figure 7.21** Compensation for die swell [50]: 1: required profile geometry, 2: corrected die cross section, 3: FE mesh

The computed contour for the cross section of the die confirms the empirical values from the practical die-making.

The side edges are made concave, and the region of the corners deviates very little from the desired profile contour.

The left part of Fig. 7.22 shows the shape of several isovels (lines of equal velocity) for the optimized die contour. The flow analysis of the rectangular shape with the same area is performed with the same boundary conditions. The shape of the resulting isovels is shown on the right in Fig. 7.22 (equal pressure gradient and equal area).

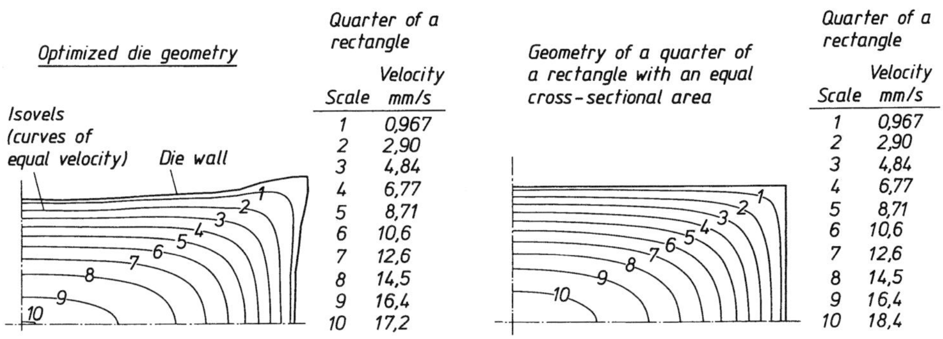


**Figure 7.22** Velocity field in a quarter of a rectangular cross section [50]

Generally, the shape of the isovels deviates from the contour of the die. From that it is not difficult to deduce that the volumetric throughput per unit area along the lines parallel to the side walls varies. Hence, too little material is supplied to the flow field in the corners. By rounding off the outside edges in the optimized die shape, this problem is corrected, however. Consequently, the isovels assume a shape gradually approaching a rectangle.

## Symbols and Abbreviations

| | |
|---|---|
| $A$ | Area |
| $A_0$ | Cross section at the die outlet |
| $\Delta A_0$ | Change of area |
| $\Delta A_W$ | Change of area at the die |
| $a$ | Difference of round strand radius and die outlet radius |
| $b$ | Exponent of extensional viscosity function |
| $B$ | Width |
| $c_p$ | Specific heat capacity |
| $D$ | Diameter |
| $D_0$ | Diameter at die outlet |
| $d$ | Diameter |
| $e_{\bigcirc}$ | Representative relative distance in a round channel |
| $e_{\square}$ | Representative relative distance in a slit channel |
| $F_{diss}$ | Magnification factor |
| $H$ | Height |
| $K$ | Factor related to the excessive shear rate increase |
| $K^*$ | Proportional factor |
| $K^{**}$ | Proportional factor |
| $K_{prop}$ | Coefficient of expansion of the cross-sectional area |
| $K^*_{prop}$ | Coefficient of expansion of the cross-sectional area |
| $K_{prop,\,W}$ | Coefficient of expansion of the cross-sectional area at the die |
| $k$ | Consistency factor of the power law |
| $L$ | Length |
| $\dot{m}$ | Mass throughput |
| $n$ | Flow exponent of the power law |
| $P_{diss,\,total}$ | Total dissipated power |
| $p$ | Pressure |
| $p'_i$ | Pressure gradient of profile subsection $i$ |
| $\Delta p$ | Pressure loss |
| $R$ | Radius |
| $R_o$ | Outer radius |
| $R_i$ | Inner radius |
| $R_0$ | Radius of the inlet of a distributor pipe |
| $r$ | Radius |
| $S_{swell}$ | Swelling factor |
| $T$ | Temperature |

| | |
|---|---|
| $T_m$ | Melt temperature |
| $t$ | Residence time |
| $V$ | Volume |
| $\dot{V}$ | Volume flow |
| $v$ | Velocity |
| $\overline{v}$ | Average velocity |
| $y_0$ | Length of the die land in the center of a distributor channel |
| $\dot{\gamma}_{filled}$ | Excessive shear velocity between two filler particles |
| $\dot{\gamma}_{rep}$ | Representative shear velocity |
| $\dot{\gamma}_{rep,pipe}$ | Representative shear velocity in a round channel |
| $\dot{\gamma}_{rep,slit}$ | Representative shear velocity in a slit channel |
| $\dot{\gamma}_{nf,int}$ | Integral shear velocity for a system without fillers |
| $\dot{\gamma}_W$ | Shear velocity at the wall |
| $\dot{\gamma}_{W,Newton}$ | Shear velocity at the wall for a Newtonian fluid |
| $\dot{\gamma}_0$ | Reference shear velocity |
| $\dot{\varepsilon}$ | Extension rate |
| $\eta$ | Viscosity |
| $\eta_{nf,int}$ | Integral viscosity of a nonfilled compound |
| $\eta_{rep}$ | Representative viscosity |
| $\lambda$ | Thermal conductivity |
| $\mu$ | Viscosity in elongation |
| $\mu_0$ | Constant of the viscosity in elongation |
| $\rho$ | Density |
| $\sigma_E$ | Elongation stress at inlet area |
| $\tau$ | Shear stress |
| $\tau_{crit}$ | Critical wall shear stress, at which wall slip begins |
| $\tau_W$ | Wall shear stress |
| $\tau_0$ | Yield stress |
| $\varphi$ | Volumetric filler content |
| $\phi$ | Fluidity |

## 7.5 References of Chapter 7


1. Dinges, K.: Kautschuk und Gummi. Reihe: Polymere Werkstoffe – Technologie 2. Vol. 3. Thieme, Stuttgart (1984)
2. Anders, D.: Roller-Head-Anlagen – Neue Entwicklungen und Einsatzgebiete. Paper of the DKG-conference, Nürnberg (1980)
3. Anders, D.: Die Co-Extrusion von Kabeln und Profilen. Kunststoffberater 28 (1983) 10, pp. 44–47
4. Anders, D.: Duplex- und Triplex-Anlagen zur Herstellung von Lauf- und Seitenstreifen. Gummi Asbest. Kunstst. 36 (1983) p. 11
5. Johnson, P.S.: Developments in extrusion science and technology. Rubber Chem. Technol. 56 (1983) p. 574
6. Gohlisch, H.J. et al.: Extrusion von Elastomeren. In: Hensen, F., Knappe, W. and Potente, H. (Eds.): Handbuch der Kunststoffextrusionstechnik. Vol. 2: Extrusionsanlagen. Hanser, München (1986), pp. 525–535
7. Gohlisch, H.-J.: Rationalisierungsmaßnahmen in der Extruder- und Kalandertechnik. Kautsch. Gummi Kunstst. 33 (1980) 12, pp. 1016–1021
8. May, W.: Das Einwalzenkopfsystem - Eine neue Technologie in der Kautschukverarbeitung. Paper of the DKG-conference, Wiesbaden (1983)
9. Masberg, U. et al.: Analogien bei der Verarbeitung von Thermoplasten und Elastomeren. Kunststoffe 74 (1984) 1, pp. 21–24
10. Limper, A.: Methoden zur Abschätzung der Betriebsparameter bei der Kautschukextrusion. Thesis at the RWTH Aachen (1985)
11. Williams, M.L. et al.: The temperature dependence of relaxation mechanism in amorphous polymers and other glass-forming liquids. J. Am. Chem. Soc. 77 (1955) 7, pp. 3701–3706
12. Geisbüsch, P.: Ansätze zur Schwindungsberechnung ungefüllter und mineralisch gefüllter Thermoplaste. Thesis at the RWTH Aachen (1980)
13. Wildemuth, C.R. and Williams, M.C.: Viscosity of suspensions modeled with a shear-dependent maximum packing fraction. Rheol. Acta 23 (1984) p. 627
14. Lyngaae-Jörggensen, J.: A phenomenological master curve for viscosity-structure: Data for two-phase polymer systems in simple shear flow. Polym. Eng. Sci. 23 (1983) p. 11
15. Menges, G. et al.: Fließverhalten von Kautschukmischungen und Modelle zur Werkzeugberechnung. Kautsch. Gummi Kunstst. 33 (1980) 4, pp. 256–260
16. Menges, G. et al.: Rheologische Funktionen für das Auslegen von Kautschuk-Extrusionswerkzeugen. Kautsch. Gummi Kunstst. 36 (1983) 8, pp. 684–688
17. Lobe, V.M. and White, J.L.: An experimental study of the influence of carbon black on the rheological properties of a polystyrene melt. Polym. Eng. Sci. 19 (1979) 9, pp. 617–624
18. Herschel, WH. and Bulkley, R.: Kolloid-Z. 39 (1926) p. 291
19. Pähl, M.H.: Praktische Rheologie der Kunststoffschmelzen und Lösungen. VDI-Verl., Düsseldorf (1982)
20. Schöppner, V. and Haberstroh, E.: Steigerung der Wirtschaftlichkeit in der Kautschukverarbeitung durch die Entwicklung schnelllaufender Kautschuk-Extrusionsanlagen. Final Report of IGF project 17545 N, (2015)
21. Röthemeyer, F.: Rheologische und Thermodynamische Probleme bei der Verarbeitung von Kautschukmischungen. Kautsch. Gummi Kunstst. 28 (1975) 8, pp. 453–457
22. Schümmer, P.: Rheologie 1. Printed lecture RWTH Aachen (1981)

23. Giesekus, J. and Langer, G.: Die Bestimmung der wahren Fließkurven nicht-newtonscher Flüssigkeiten und plastischer Stoffe mit der Methode der repräsentativen Viskosität. Rheol. Acta 16 (1977) pp. 1–22

24. Wortberg, J.: Werkzeugauslegung für Ein- und Mehrschichtextrusion. Thesis at the RWTH Aachen (1978)

25. Bird, R. B., Stewart, W. E. and Lightfoot, E. N.: Transport Phenomena, 7th ed. Wiley, London, New York (1966)

26. Franzkoch, B.: Analyse eines neuen Verfahrens zur Herstellung vernetzter Polyethylenkabel. Thesis at the RWTH Aachen (1979)

27. Menges, G. et al.: Berechnung des Formfüllvorganges beim Spritzgießen von vernetzenden Formmassen. Plastverarbeite 34 (1983) 4, pp. 323–327

28. Kramer, H.: Untersuchungen zur Spritzquellung von Kautschukmischungen. Kunstst. Gumm. 6 (1967) 12, p. 433

29. Glowania, F.-J.: Untersuchung der viskoelastischen Eigenschaften von Kautschukmischungen. Unpublished thesis at the IKV, Aachen (1987)

30. Röthemeyer, F.: Elastische Effekte bei der Extrusion von Kunststoffschmelzen. Thesis at the University of Stuttgart (1970)

31. Böhme, G.: Strömungsmechanik nicht-newtonscher Fluide. Teubner, Stuttgart (1981)

32. Zidan, M.: Zur Rheologie des Spinnprozesses. Rheol. Acta 8 (1969) 1, p. 89

33. Kannabivan, R.: Application of flow behavior to design of rubber extrusion dies. Rubber Chem. Technol. 59 (1986) pp. 142–154

34. White, J. L. and Huang, D.: Extrudate swell and extrusion pressure loss of polymer melts flowing through rectangular and trapezoidal dies. Polym. Eng. Sci. 21 (1981) 16, pp. 1101–1107

35. Röthemeyer, F.: Gestaltung von Extrusionswerkzeugen unter Berücksichtigung viskoelastischer Effekte. Kunststoffe 59 (1969) 6, pp. 333–338

36. Menges, G. et al.: Bei der Auslegung von Profilwerkzeugen wird die Strangaufweitung berechenbar. Kunststoffe 75 (1985) 1, pp. 14–18

37. Gesenhues, B.: Rechnergestützte Auslegung von Fließkanälen. Thesis at the RWTH Aachen (1985)

38. Junk, P. B.: Betrachtungen zum Schmelzeverhalten beim kontinuierlichen Blasformen. Thesis at the RWTH Aachen (1978)

39. Limper, A. and Michaeli, W.: Auslegung von Kautschuk-Extrudierwerkzeugen. In: Extrudieren von Elastomeren. VDI-Verl., Düsseldorf (1986)

40. Kaiser, O.: Auslegung und Erprobung eines Breitschlitzwerkzeuges für Kautschukmischungen. Unpublished thesis at the IKV, Aachen (1987)

41. Masberg, U.: Betrachtungen zur geometrischen Gestaltung von Dornhalterköpfen. Kunststoffe 71 (1981) pp. 15–17

42. Hartmann, G.: Auslegung von Extrusionswerkzeugen mit Rechteckkanal. Unpublished thesis at the IKV, Aachen (1986)

43. Fritz, H. G. and Winter, H. H.: Design of dies for the extrusion of sheets and annular parisons. ANTEC '84 (1984) pp. 49–51

44. Heins, V.: Auslegung eines Pinolen- und Breitschlitzwerkzeuges unter Berücksichtigung der wahren Verteilerkanallänge und -form. Unpublished thesis at the IKV, Aachen (1987)

45. Schwenzer, C. H.F.: Finite-Elemente-Methoden zur Berechnung von Mono- und Coextrusionsströmungen. Thesis at the RWTH Aachen (1988)

46. Cogswell, F. N.: Polymer Melt Rheology. George Goodwin Ltd., London (1981)

47. Ramsteiner, F.: Fließverhalten von Kunststoffschmelzen durch Düsen. Kunststoffe 61 (1971) 12, pp. 943–947

48. Imping, W.: Rechnerische und experimentelle Untersuchung der Geschwindigkeitsverteilungen in Profilwerkzeugen. Unpublished thesis at the IKV, Aachen (1987)
49. Ramsteiner, F.: Einfluß der Düsengeometrie auf Strömungswiderstand, Strangaufweitung und Schmelzbruch von Kunststolfschmelzen. Kunststoffe 62 (1972) 11, pp. 766–772
50. Lamers, A.: Profilwerkzeugauslegung mit der Evolutionstheorie. Unpublished thesis at the IKV, Aachen (1988)
51. Menges, G. and Dombrowski, U.: Gestaltung von Kautschuk-Extrusionswerkzeugen. Research report AIF Nr. 6707, Aachen (1988)
52. Poller, S.: Entwicklung eines Algorithmus zur iterativen Optimierung von Fließkanälen auf der Basis der Evolutionstheorie. Unpublished thesis at the IKV, Aachen (1987)
53. Poller, S.: Entwicklung einer mehrgliedrigen Evolutionsstrategie zur rechnerischen Optimierung von Fließkanälen mit der FEM. Unpublished thesis at the IKV, Aachen (1988)
54. Rechenberg, I.: The evolution strategy. A mathematical model of Darwinian evolution synergetics from microscopic to macroscopic. Series: Synergetics 22. Springer, Berlin (1984), pp. 122–132
55. Schwefel, H.P.: Numerische Optimierung von Computermodellen mittels der Evolutionsstrategie. Birkhäuser, Basel (1977)
56. Muth, C.: Einführung in die Evolutionsstrategie Regelungstechnik. 30 (1982) 9, pp. 297–303
57. Rechenberg, I.: Evolutionsstrategie, Optimierung technischer Systeme nach den Prinzipien der biologischen Evolution. Frommann-Holzborg, Stuttgart (1973)
58. Rechenberg, I.: Bionik, Evolution, Optimierung. Naturwiss. Rundsch. 26 (1973) 11, pp. 465–472
59. Schenkel, G. and Kühnle, H.: Zur Bemessung der Bügellängen Verhältnisse bei Mehrkanal – Extrudierwerkzeugen für Kunststoffe. Kunststoffe 73 (1983) pp. 17–22
60. Paar, M.: Auslegung von Spritzgießwerkzeugen für vernetzende Formmassen. Thesis at the RWTH Aachen (1984)
61. Benfer, W.: Rechnergestützte Auslegung von Spritzgießwerkzeugen für Elastomere. Thesis at the RWTH Aachen (1985)

# 8 Heating of Extrusion Dies

The temperature distribution in an extrusion die critically affects the viscous and elastic properties of the streaming melt, and thereby the flow pattern, the pressure loss inside the die, and the elastic properties determining the swelling of the extrudate at the exit from the orifice.

A homogeneous temperature distribution in the flow channel would be ideal for a rheological design of the flow channel. However, because of the rise in temperature due to dissipation in the melt stream, a temperature profile is formed in the flow channel that becomes more pronounced as the channel area becomes smaller toward the exit. The residence times of the melt in efficient dies are too short to allow the equalization of the temperature differences considering the relatively low heat conductivity of polymers.

This results in temperature nonuniformities across the thickness of the extrudate at the die exit even if the temperature of the walls of the flow channel is homogeneous. The objective of temperature control of the extrusion dies is, therefore, to provide a temperature distribution as uniform as possible across the thickness of the extrudate at the die exit, thus ensuring the ultimate goal of the rheological design, namely a uniform exit velocity of the melt.

The close coupling of the rheological and thermal processes in the die because of the temperature dependence of viscosity and the viscous heat generation require a simultaneous solution of the basic rheological and thermodynamic equations in the melt channel and the die body for a complete computational design. Because of the pronounced nonlinearity of the melt properties, the coupled description of these processes can be done only numerically [1–9]. The currently available solution procedures even with the use of the most modern computers are so time consuming that they are not yet suitable for a practical application.

Therefore, in the following sections the thermal processes in extrusion dies will be discussed as an uncoupled problem.

## 8.1 Types and Applications

The dies are divided into those indirectly heated by fluids and those directly heated by electricity.

### 8.1.1 Heating of Extrusion Dies with Fluids

In the processing of elastomers, dies with indirect heating are preferred.

The significant feature that distinguishes heating with fluids from electrical resistance heating is the ability to remove heat from the die independently of the surrounding conditions. This turns out to be an advantage at the typical relatively low temperatures used for the processing of elastomers.

Due to the small differences between the temperatures of the die and the ambient air, the variations in air temperature have a considerably stronger effect on the intensity of the convective cooling of the die surface than with the higher temperatures used for the processing of thermoplastics. However, the dies can be heated more homogeneously with fluids than with electrical heating elements.

In rubber processing, particularly, the considerable cost of additional equipment, such as recirculating pumps and heat exchangers, is not that pronounced because they form part of the fluid heating system of the extruder itself.

Occasionally, heating with fluids is found in the processing of thermoplastics, particularly when there are long connecting lines for melt flow between the extruder and die in large extrusion lines. In this case, on the one hand, the high losses of heat by convection and radiation (large surface) have to be compensated for; on the other hand, the dissipation heat from the long flow lines has to be taken away. In such cases (large surfaces, small cross sections) the liquid heating requires less instrumentation and maintenance than the direct electrical heating with corresponding numerous heating and control zones.

Fluid heating is also found occasionally in relatively small dies for thermoplastics. When, purposely, heat has to be supplied to or removed from a certain region, such as from the exit zone of the mandrel of blown film dies, in order to obtain a required surface appearance of the extrudate, oil heating is used [10].

The assembly of the dies, when changing or maintaining them, is usually more complicated than with direct electrical heating because of the emptying and refilling of the die and the piping with the heating fluid.

A more serious disadvantage of fluid heating is the technical complexity and cost in cases where there are different die zones that have to be maintained at different temperatures (such as in large dies for thermoplastics for the correction of the melt flow [11–14]), particularly in film and sheet extrusion. According to current technology, for each temperature zone a separate heating unit (tank, pump, heating, cooling) is required. In this context, with the development of a control valve that is not affected by temperature and with a linear control characteristic, the concept of the regulation of the flow rate becomes interesting [15,16].

In this concept, several heating zones with different temperature levels are supplied by the heating medium from only one heating unit. Usually, the amount of flow of the heating medium through the different zones, and not its temperature, is regulated.

### 8.1.2 Electrically Heated Extrusion Dies

Extrusion dies for the processing of thermoplastics are usually heated with electrical resistance heaters. Less frequently electroinductive heating elements are used. As far as their construction and shape, they are differentiated into mica- or ceramic-insulated heater bands and cartridge heaters.

In most extrusion dies, the band heaters are clamped around the body of the die on the outside or cover the surface as flat heating plates.

An essential condition for the homogeneous heating of a die, besides the uniform distribution of the heat source in the heating band, is a good physical contact between the heater and the surface of the die body [17]. In order to satisfy this requirement, the heaters are made enclosed in aluminum shells that can be fitted accurately to the contour of the die by machining them to the required shape and then fastened with screws [18]. With a satisfactorily uniform heat delivery, the attainable power densities (heating power per unit of the area of contact) for mica-insulated heaters range from 2 to 3.5 W/cm$^2$ and for ceramic and aluminum heaters from 5 to 8 W/cm$^2$ [18–20].

The significant advantages of the heating of the die *surface* are

- relatively uniform heating of the wall of the channel (local differences in the temperature on the surface of the die from heating power generally even out because of the thick die wall) and
- easy dismantling of the heating elements (when it is necessary to take the die apart, such as for cleaning).

The relatively large distance between the heat source and the flow channel surface can be a disadvantage

- when the local temperature should be increased intentionally (affecting adjacent die zones) and
- when controlling different heating zones of large dies (coupling of the control circuits, which can lead to their unstable periodic oscillation).

An additional disadvantage of the heating of the surface of the die is its relatively low energy efficiency [19]. The reason is large heat loss to the surroundings, mainly due to the much higher surface temperature when compared to that of the die. Thermal insulation of the surface could possibly reduce this energy loss, but it would adversely affect the control because the cooling rate of an "overheated" die is considerably reduced. Another consequence is, in spite of a relatively high installed heating power, the heating of the die during a startup of the line is relatively slow. Therefore, particularly for wide slit dies, heating cartridges are frequently inserted into boreholes in the *inside* of the die body, close to the flow channel. The efficiency of this internal heating of the die is higher than that of surface heating [11]. The regulation is also better, especially from the point of view of the decoupling of the regulating circuits and of the local heating of different die zones.

The internal heating of a die requires particular care in placement of the heating elements, because if the heating elements are too close to the channel and too far apart, the temperature distribution on the surface of the channel would be in the shape of a wave. Examples of this are shown at the end of Section 8.2.

The installation of heating cartridges into narrow and long boreholes is rather cumbersome. Moreover the boreholes require an extremely good surface in order to achieve a good physical contact of the cartridge with the wall. The supply of electrical energy for the heating is through a heat-resistant plug and connecting socket located at the die, which allow a quick change of the die or its parts.

### 8.1.3 Temperature Control of Extrusion Dies

For the temperature control, both on/off and continuous controls are used. For dies heated with fluid, the controls frequently have three modes (heat, off, cooling), while direct electrical heating controls have two modes (heat and off, the latter representing a natural cooling of the die surface by natural convection).

The controllers sometimes exhibit a PID (proportional integral differential) characteristic, which allows elimination of the offset error but tends to oscillate when reacting to upsets.

The regulators with PD (proportional differential) characteristics, used in the past, do not show this overshoot, but they have a lasting deviation when the operating conditions are changed.

A combination of the advantages of both concepts is in the PD/PID control system. By switching over to the desired control characteristics in response to the offset, these systems operate largely without control overshoot and offset [21].

A procedure for the setting of advantageous characteristic values for the controls resulting from simple die heating experiments is described in [22] and [23]. The measurement of temperature for the process control and the determination of actual values as a feedback for the temperature controls are generally done by Fe-Co thermocouples or by a platinum resistance thermometer (Pt 100) installed in the die. When heating by fluids, the temperature sensors are placed also in the circulation system of the heating medium (near the die) so that ultimately not the temperature of the die but that of the fluid is regulated.

The resistance thermometers are somewhat more accurate in the relevant temperature range between 50 and 400 °C (±0.5 °C to ±2.3 °C) than the Fe-Co thermocouple (±3 °C) [22] and they do not age. Thermocouples, on the other hand, have more favorable dynamic characteristics in the measurement (smaller mass and hence faster response), are more sturdy mechanically, and are less expensive [24]. The sensors are generally located in a borehole and fastened to the die by a quick-release bayonet (quarter-turn) lock that is attached to a cable.

A spring is used to ensure adequate thermal contact. This attachment system requires unplugging the sensors when the die is being changed, but it has the advantage of a strong cable connection between the sensor and the control unit. This eliminates the risk of measuring errors due to dirty or wet cable connectors.

Quick-release temperature sensors can easily get dirty and are subjected to corrosion at the point of contact, causing serious errors in measurement [22]. Therefore, these areas have to be cleaned regularly during the operation of the die. Large dies for thermoplastics often have several separately controlled heating zones. Dies with an annular exit gap have a separate heating zone for the die orifice and one or more additional zones in the die body.

Wide slit dies are generally divided across their width into several heating zones, which can be 200 to 330 mm wide [11,17]. An odd number of zones is frequently chosen so that the center of the die is controlled separately and the remaining zones are arranged symmetric to the center.

The width of the heating zones depends on the distance between the heating element and the flow channel. Consequently, the zones for surface heating are wider than those for internal heating. Very narrow zones tend to exhibit excessive swings in control behavior because of the stronger thermal coupling.

The temperature sensors are placed as close to the flow channel as possible because the measured temperatures are the most decisive factor for the flow process. When doing so, it is important to position the sensor as close as possible to the center of the effective heating zone of the heating element associated with it. The minimum distance between the tip of the borehole (well) for the sensor and the channel is determined by the strength of the remaining wall of the flow channel, and its size should be of the same order as the die diameter of the well (also depending on the local pressure in the flow channel) [25].

## 8.2 Thermal Design

For the design of the heating system of the die, it is important to have a clear idea about the degrees of freedom in thermal die design. Many theoretical possibilities are often restricted by outside realities, such as limitations in design and manufacturing.

### 8.2.1 Criteria and Degrees of Freedom for Thermal Design

The designer has the following degrees of freedom available for the optimum thermal design of an extrusion die:

- The *geometrical* form of the die does have a decisive effect on its thermal behavior. Once the shape of the channel, determined by the rheological design, is set, the degree of remaining freedom is only in its outer shape (outside contour, round, angular or slotted, wall thickness, symmetry, mechanical design).
- The choice of *material(s)* for its construction has an effect on the critical thermal properties, such as thermal conductivity, thermal diffusivity, and on the actual thermal design (insulations and special alloys with high thermal conductivity) [26].
- The spatial arrangement of the *heat supply* (location, number, and the rated power of the heating elements) can vary between uniformly divided and concentrated locally.
- The *location* of the temperature sensors and the choice of control *characteristics* affect the static (lasting deviations) and dynamic (response to disturbances and swings in temperatures) behavior of the control system.

The quality of the thermal design is measured by different criteria. These criteria can be directed toward *specific* properties of the die or the extrudate required for the given application. Therefore, the following shows *general* criteria as they apply to the thermal design of dies:

- The *thermal homogeneity* of the surface of the flow channel in production is in most cases the crucial criterion that results in a uniform exit speed of the extrudate from dies with a correct rheological design.
- The *steadiness* of the temperature, that is, its constancy with time, is the criterion against which the performance of the control system of the die is measured. The speed at which outside disturbances (changed surrounding conditions, air draft), variation in temperature of the incoming melt,or change in operating conditions (such as throughput) are controlled, or the natural temperature oscillation of the control system affect the quality of the extrudate and its fluctuation.

- The *mechanical strength* of the die, particularly in the case of wide slit dies, influences the dimensional stability of the product and typically requires dies with thick walls that are sluggish thermally.
- The *cost of producing* the die is very important. Since the rheological design resulting from computations does not always represent the final shape of the die, further machining is required. Also, it has to be taken apart relatively easily. The same is true about frequent cleaning of the finished die and in some cases surface treatment of the flow channel, such as chrome plating or polishing. The requirement of easy dismantling often has to be reconciled with the requirements from the thermal design (the arrangement of the heating elements). Thin, long boreholes (for heating channels) with a good surface quality and narrow dimensional tolerances (for heating rods, pipes [26], and so on) are difficult and very expensive to make, especially in the proximity of the flow channel.
- The *energy consumption* of the die heating systems is small in comparison to that of the entire line, but it still is a consideration in the optimization of the thermal design (external or internal heating, insulation).
- The *startup behavior* of a die is important for cost-effective operation of the entire line when frequent die changes are required (for example, for custom profiles). Thick-walled dies, although advantageous from the point of view of temperature regulation, require a long time to bring up the temperature to the required level during the startup when working with heating elements of low-rated power. Therefore, often special auxiliary heaters for the startup or continuously working control elements are added [27].
- A *special* criterion, often applicable in the extrusion of sheets and films, is the requirement for the control of the surface temperature of the extrudate at the exit from the die in order to influence the surface quality (gloss). In this case, an additional heating zone is placed at the exit that is thermally decoupled from the remaining body of the die [11].
- An example of a *special* criterion is the need for separately heated segments of the die either along its circumference or its width (blown film and wide slit dies, respectively) to control gauge distribution (thermal decoupling of segments by corresponding design measures, multichannel or computer-based controls).

The criteria listed above, at times contradictory, often lead inevitably to dies that are far from thermally optimal. Nevertheless, these dies designed with the conflicting boundary conditions operate successfully.

### 8.2.2 Heat Balance of the Extrusion Die

The estimation of the amount of required heat energy results from a heat balance of the die. This value is then used as a basis for the determination of the minimum required total heat to be supplied by the installed heating system. To accomplish this, it is necessary to balance the individual incoming and outgoing heat flows over the entire surface of the die.

The surface of the area that is attached to the last heating zone of the extruder through the connecting flange has the peculiar feature that the amount and direction of the heat flow going through it depend on the temperature gradient between the extruder and the die. If this flow, for example, becomes too large, because of great temperature differences (setting of controls) or due to an excessively large contact area die/extruder (compared to other heat exchanges in the die), a stable temperature control of the die is practically impossible, and the (thermally strongly coupled) controls for the die or the extruder tend to oscillate.

For that reason, in practical applications, the die temperature is set equal to or slightly higher than the temperature of the last heating zone of the extruder (since the extruder has a better cooling, such as by a blower, the heat flow from the die to the extruder can be controlled more effectively than heat flow in the opposite direction).

In the following considerations, the premise is that the temperatures of the die and the extruder barrel are equal; then, the heat flow across the area of contact can be neglected. With this assumption, the following heat fluxes must be taken into account for the heat balance (see Fig. 8.1):

$\dot{Q}_{\mathrm{M,In}}$ = heat flux entering the die with the melt
$\dot{Q}_{\mathrm{M,Out}}$ = heat flux leaving the die with the melt
$\dot{Q}_{\mathrm{CONV}}$ = heat flux removed by convection on the outer die walls
$\dot{Q}_{\mathrm{RAD}}$ = heat flux removed from the die by radiation exchange
$\dot{Q}_{\mathrm{DISS}}$ = energy dissipated in the die per unit of time
$\dot{Q}_{\mathrm{H}}$ = heat flux supplied to the die by the heating system (heating elemnts)

The general form of the heat balance is then

$$\begin{matrix}\text{Heat fluxes} & & \text{Heat fluxes} & & \text{Heat generated} & & \text{Heat stored}\\ \text{entering} & - & \text{leaving the} & + & \text{in the system} & = & \text{in the system}\\ \text{the system} & & \text{system} & & \text{per time unit} & & \text{per time unit}\end{matrix} \tag{8.1}$$

With reference to the entire die shown in Fig. 8.1, this means

$$\left(\dot{Q}_{\mathrm{M,In}}+\dot{Q}_{\mathrm{H}}\right)-\left(\dot{Q}_{\mathrm{M,Out}}+\dot{Q}_{\mathrm{CONV}}+\dot{Q}_{\mathrm{RAD}}+\dot{Q}_{\mathrm{DISS}}\right)=\frac{\partial}{\partial t}\left(m_{\mathrm{d}}\cdot c_{\mathrm{pd}}\cdot\vartheta_{\mathrm{d}}\right). \tag{8.2}$$

In the steady state of operation of the die, the right-hand side of Equation 8.2 becomes zero, which means that $\vartheta_d$, the die temperature, does not change any more.

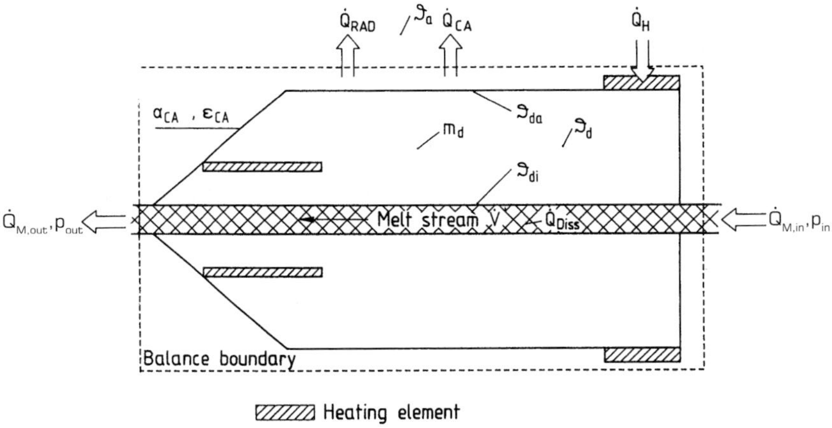


**Figure 8.1** Heat balance of an extrusion die

As explained at the beginning of this chapter,it is not generally possible to change the melt temperature drastically between the inlet and outlet of the die by varying the temperatures in the die and meet the requirement for a largely uniform temperature of the emerging extrudate.

The dissipation heat liberated in the inside of the melt channel

$$\dot{Q}_{DISS} \cong (p_{in} - p_{out}) \cdot \dot{V} \tag{8.3}$$

leads to a temperature distribution shown qualitatively in Fig. 8.2, case (b) for the case of a heated wall of the channel (heated to the melt temperature). The dissipation heat is conducted to the channel wall, where it is removed from the melt stream. The maximum temperature is close to the channel wall. Because of a pronounced dependence of the viscosity on temperature, the maximum in the shear rate shifts from the channel wall (starting with an isothermal flow, case (a)) to the region of the maximum temperature. The shear rate at the wall decreases due to the high viscosity in the proximity of the wall, when compared to the isothermal flow, assuming a constant throughput. This redistribution of the shear rates has an effect on the velocity profile in the way shown, namely that the region of the lower flow velocities near the wall becomes greater.

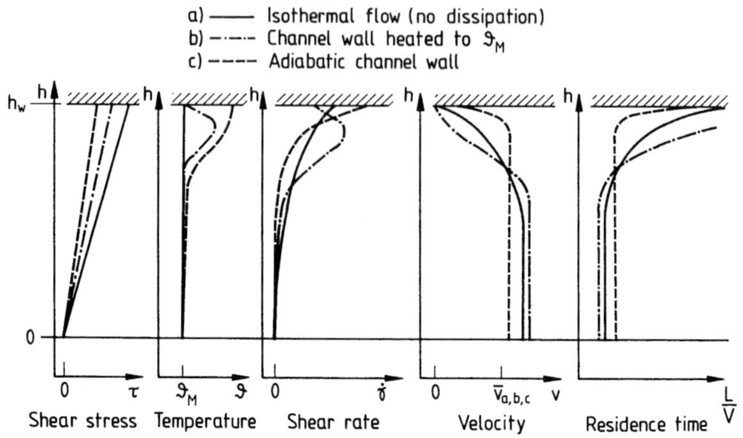


**Figure 8.2** Flow through a slit at different thermal boundary conditions and an equal throughput $\left(\dot{V}_a = \dot{V}_b = \dot{V}_c\right)$

Case (c) describes the conditions at an adiabatic channel wall. Here, the temperature maximum lies directly on the wall, where the highest shear rate occurs, which is higher than that in the isothermal case (a).

Consequently, the zone of the lower flow rates near the wall is smaller in case (c) than in case (a) [29]. The way the three different thermal boundary conditions affect the residence time spectrum is also evident from Fig. 8.2.

The residence time is determined from the ratio of the length of the channel section under consideration, $L$, and the flow velocity $\nu$ (compare Chapter 3). Especially in processing sensitive polymers, which tend to degrade or cross-link when subjected to pressure, elevated temperature, or shear for extended periods of time, the wide residence time spectrum of case (b) is disadvantageous.

Accordingly, the design of a die should not proceed under the assumption of an isothermal wall of the flow channel, which leads away a part of the dissipative heat (b), but rather under the assumption of an adiabatic wall (c) so that the generated dissipation heat causes the increase in the melt temperature, and thus the above-mentioned stagnation effects in the region near the channel wall are eliminated.

Consequently, the following is true:

$$\dot{Q}_{DISS} = \dot{Q}_{M,Out} - \dot{Q}_{M,In}, \tag{8.4}$$

The calculated increase in temperature of the melt is

$$\Delta \vartheta_M = \frac{\dot{Q}_{M,Out} - \dot{Q}_{M,In}}{\dot{m} \cdot c_p} \tag{8.5}$$

$\dot{m}$ = mass flow, $c_p$ : specific heat capacity of the melt

From Equation (8.3) it follows:

$$\Delta\vartheta_M = \left(p_{in} - p_{out}\right)\frac{\dot{V}}{\dot{m}\cdot c_p} = \frac{p_{in} - p_{out}}{\rho\cdot c_p} \quad (8.6)$$

$\rho$ melt density

Accordingly, the increase of the melt temperature is dependent only on the pressure drop across the die (and of course the material properties of the melt). For the typical material properties, for example, $\rho \cdot c_p = 2 \times 10^6$ J/m³ K, the temperature increase will be 0.05 K per bar. In order to eliminate the stagnation effects described in Fig. 8.2 resulting from too-low wall temperatures, the die temperature should be higher by $\Delta\vartheta_M$ than the temperature of the incoming mass, $\vartheta_E$. Combining Equations (8.4) and (8.2) and solving for $\dot{Q}_H$, the condition of the thermal steady state (constant die temperature), the following results:

$$\dot{Q}_H = \dot{Q}_{CONV} + \dot{Q}_{RAD}, \quad (8.7)$$

which is equivalent to the statement that the heating power must be exactly equal to the sum of the radiation and convection losses to the surroundings. The convective heat flow to the surrounding air is

$$\dot{Q}_{CA} = A_{da}\cdot\alpha_{CA}\cdot\left(\vartheta_{da} - \vartheta_a\right). \quad (8.8)$$

where $A_{da}$ is the surface area of the die that is exchanging heat with the surrounding air with temperature $\vartheta_{da}$, $\vartheta_a$ is the room temperature, and $\alpha_{CONV}$ is the heat transfer coefficient for natural convection, assumed to be approximately 8 W/m² K [30].

The radiative heat flux to the surroundings $Q_{RAD}$ is given by

$$\dot{Q}_{RAD} = A_{da}\cdot\varepsilon\cdot C_R\left(\left(\frac{T_{da}}{100}\right)^4 - \left(\frac{T_a}{100}\right)^4\right). \quad (8.9)$$

Here $\varepsilon$ is the emission coefficient: for smooth steel surfaces $\varepsilon = 0.25$ [31] and for oxidated steel surfaces $\varepsilon = 0.75$ [32]; $C_R$ is the blackbody radiation number: $C_R = 5.77 \times 10^{-8}$ W/m² K⁴.

Equation (8.9) can be written in a form similar to Equation (8.8):

$$\dot{Q}_{RAD} = A_{da}\cdot\alpha_{RAD}\left(\vartheta_{da} - \vartheta_a\right). \quad (8.10)$$

Here $\alpha_{RAD}$ is defined as the heat transfer coefficient for radiation.

In dies with a high tendency to dissipate heat (high pressure losses and melts with low $\rho \cdot c_p$), it is often necessary to extract heat from the flow channel. The design goal then is not the adiabatic but the isothermal die wall. The heat flux between the melt and the die wall is described in an equation similar to (8.8):

$$\dot{Q}_{CM} = A_{di} \cdot \alpha_{CM} \left(\vartheta_{di} - \vartheta_{M}\right) \tag{8.11}$$

The heat transfer coefficient $\alpha_{CM}$ according to [34] can be determined from Nusselt laws by a complex procedure.

Here, $A_{di}$ is the area of the inside surface of the die with the temperature $\vartheta_{di}$, and $\vartheta_M$ is the melt temperature.

Equation (8.7) for this case becomes

$$\dot{Q}_{H} = \dot{Q}_{CA} + \dot{Q}_{RAD} - \dot{Q}_{CM}. \tag{8.12}$$

The heating power is reduced by the amount of the heat taken away from the melt (or is increased by that amount, depending on if ($\vartheta_M - \vartheta_{di}$) is positive or negative). The heating power determined from the above is the minimum required for heating the die. In order to ensure a reserve that allows the control unit to operate in a favorable range, the actual nominal load should be about double that of the calculated minimum heating power. This means that the control unit operates at approximately 50% of its control range with the ratio of switch-on times of 1:2. Newer controls have adjustable points of operation so that the heating power can be oversized even to a greater extent [22]. When the oversize is a factor of four, for example, the resulting operating point is 25% in which the control can operate symmetrically. The advantage of oversizing the nominal heating capability is a reduced heat-up time during the startup.

The heat-up time $t_H$ can be calculated from the above formula [19]:

$$t_{H} = \frac{m_{d} \cdot c_{pd} \cdot \Delta\vartheta_{d}}{n \cdot \dot{Q}_{Hmax}} \tag{8.13}$$

where $m_d$ is the mass of the die, $c_{pd}$ the specific heat capacity of the die material, $\Delta\vartheta_d$ the targeted temperature increase during the heat-up, and $\dot{Q}_{Hmax}$ the installed nominal heating load.

It is clear that with the increasing nominal heating capacity, the heat-up time is shortened. The efficiency is given in [19] as a rounded value of 0.5. It takes into account the heat losses from the die surface, which is becoming hotter during the heat-up periods. The efficiency depends on the type of heating (it is small for external and large for internal heating) and on the oversizing of the heating power (which generally increases with increased oversizing).

### 8.2.3 Restrictive Assumptions in the Modeling

The main goal of the thermal design, as stated, is to place the heating elements throughout the die geometry in such a way that under given design boundary conditions a uniform temperature distribution is achieved along the surfaces of the flow channel.

The heating elements include all surfaces of the die (including the inner surfaces that are in contact with the heating cartridges) through which heat flows and also surfaces that permit heat transfer to the surroundings by convection and radiation.

The overall consideration of the heat transport processes as done in the previous section with the balance is not sufficient to judge the homogeneity of the temperature distribution before starting the actual design. Rather, a method has to be applied that allows the simulation of the processes and considers the most important dimensions and thermal boundary conditions.

In the following section, two simulation methods are introduced, and the models based on them and the simplified assumptions involved will be briefly discussed.

### 8.2.4 Simulation Methods for Thermal Design

#### *Electrical Analogy Model*

Starting with a steady operation of the die, the established temperature distribution can be considered constant with time (steady state). (The dynamic behavior of the control system is neglected.)

The heat flow in the direction of width or circumference can often be neglected for the reason of symmetry in wide flow channels or in flow channels with rotational symmetry (wide slit dies, heads for pipes, solid rods, and blown film extrusion). Then, the heat transfer process can be approximated by a two-dimensional procedure (besides the symmetry of the flow channel, a correspondingly symmetrical outside shape and placement of heating elements are the additional conditions).

Given the above conditions, the *steady-state two-dimensional* heat transfer can be described by the following differential equation:

$$\frac{\partial^2 \vartheta}{\partial x^2} + \frac{\partial^2 \vartheta}{\partial y^2} = 0. \tag{8.14}$$

The two-dimensional electrical transmission in a flat conductor with resistance is described with a similar equation:

$$\frac{\partial^2 U}{\partial x^2} + \frac{\partial^2 U}{\partial y^2} = 0. \tag{8.15}$$

From both these equations the analogy between the temperature $\vartheta$ and voltage $U$ is obvious. The heat flux $\dot{Q}$, which is proportional to the temperature gradient, corresponds to the electrical current $I$ [35].

Because of this analogy with a flat electrical conductor, a steady-state temperature field in the cross section of a die can be simulated. This means a two-dimensional analogy model can be formed from a flat conductor (or resistor) by cutting out the outline of the die cross section from a resistance paper (paper coated with graphite) and by drawing in the outline of the flow channel and of the heating elements to scale with an electrically conductive paint ("conductive silver") [12,30,36].

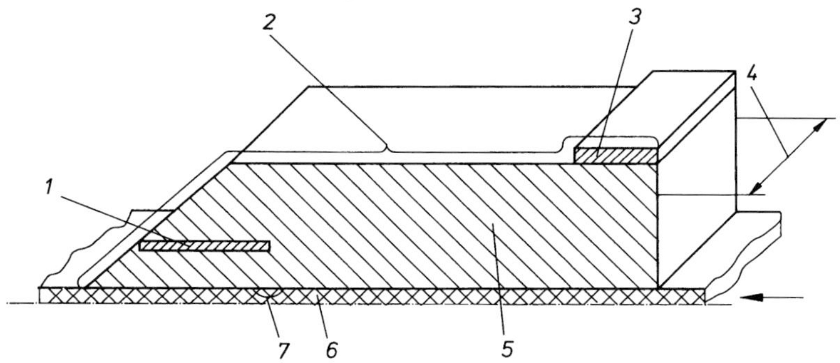


**Figure 8.3** Section through a flat slit die with the thermal boundary conditions shown: 1: cartridge heater (heating power $P_p$); 2: convective heat transfer $\alpha = 16$ W/m²K, $\vartheta_a = 20\,°C$; 3: band heater (heating power $P_B$); 4: die width ($T = 1$ m); 5: body of the die; 6: melt ($\vartheta_M = 220\,°C$); 7: heat transfer coefficient ($\alpha = 75$ W/m²K)

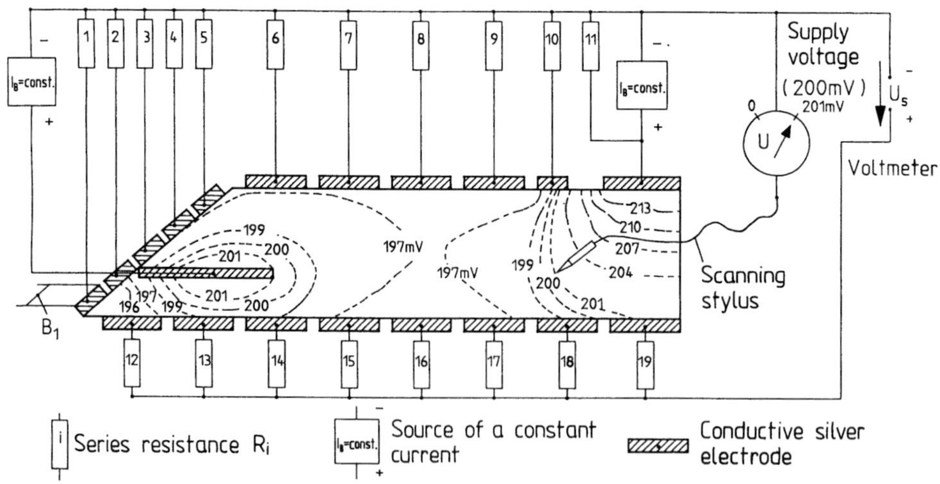


**Figure 8.4** Electrical analogy model to a flat slit die

The electrical analogy model for a partial section through a wide slit die (see Fig. 8.3) with heating elements and pertinent thermal boundary conditions is shown in Fig. 8.4.

The surfaces with the convective heat transfer are segmented so that possible inhomogeneous heat loss from the die surface can be approximated. The individual surface segments are connected through series resistances $R_i$ to a voltage source (outside segments are connected to the minus terminal $\triangleq$ at lower temperatures, and the inside segments of the channel surface to the plus terminal $\triangleq$ at higher temperatures).

The sources of energy in the heating elements are simulated by the sources of the direct current. The analogy of the relationships is summarized in Fig. 8.5. Assuming the temperature difference between the melt and the surroundings is 200 K, for example, with $n = 1$ mV/K, the supply voltage of the model will be $U_s = 200$ mV. The current intensity resulting from the source of constant current at the nominal heating load can be calculated from the formula below with the square resistance of the paper $R = 2$ kΩ [38] and the thermal conductivity of the steel from which the die is made, $\lambda = 40$ W/mK ($\lambda \cdot R = k = 80$ W kΩ/mK) for the die width $T_{die}$ of 1 m and the heating power $P_{nom}$ (for equations see Fig. 8.5):

**Analogous relationships**

| Model | | Die |
|---|---|---|
| Voltage (difference) | $\Delta U = n \cdot \Delta\vartheta$ | Temperature (difference) |
| Current | $I = \frac{n}{k} \cdot \dot{q}'$ | Heat flow (based on the die width T) |
| Voltage dropping resistance (on a surface segment i with a width $B_i$) | $R_i B_i = k \cdot m \cdot \frac{1}{\alpha}$ | Heat transfer coefficient |

**Model parameters**

| | | |
|---|---|---|
| Voltage / Temperature ratio | $n = \frac{\Delta U}{\Delta\vartheta}$ | $\left[\frac{V}{K}\right]$ |
| Conductivity ratio | $k = \lambda \, R^{\square}$ | $\left[\frac{W\Omega}{mK}\right]$ |
| Model scale | $m = \frac{l_{Model}}{l_{Die}}$ | $[\cdot/\cdot]$ |

$R^{\square}$: Square resistance of the resistive conductor (resistance paper) used

**Figure 8.5** Definitions for the electrical analogy model

$$I_{nom} = \frac{P_{nom}}{T_{die}} \cdot \frac{1\,\frac{\text{mV}}{\text{K}}}{80\,\frac{\text{Wk}\Omega}{\text{m}\cdot\text{K}}} \triangleq P_{nom} \cdot 12.5\,\frac{\mu\text{A}}{\text{kW}}. \tag{8.16}$$

The series resistance at one outside segment (e.g., $i = 1$ in Fig. 8.4) is calculated below for the model scale 1:1 ($m = 1$), a segment width $B_1 = 2$ cm, and the local heat transfer coefficient of 16 W/m² K:

$$R_1 = \frac{80}{16 \cdot 0.02} \text{k}\Omega = 250\,\text{k}\Omega \tag{8.17}$$

The voltage on each desired point of the model can be measured by an appropriately connected voltmeter using a metal stylus. The measured voltage (in millivolts) then corresponds to a temperature difference between the given point and the surroundings (thus 208 mV corresponds to 228 °C at the ambient temperature of 20 °C). The lines of equal voltage (potential) as shown in Fig. 8.4 (dashed lines) correspond to isotherms, lines of equal temperature, which represent a good description of the temperature distribution (Fig. 8.6).

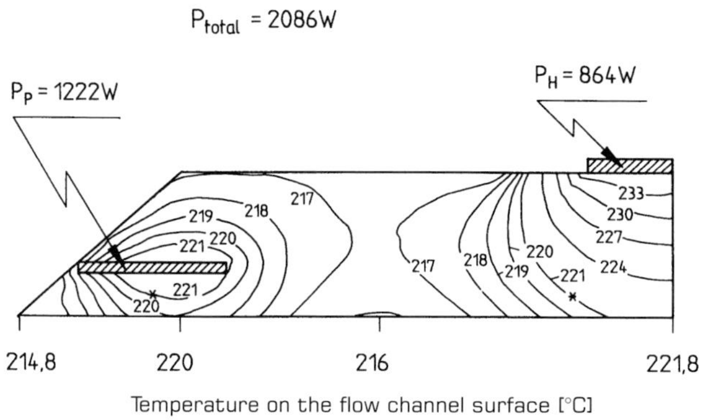


**Figure 8.6** Results from the electrical analogy model

The temperature distribution in a section of the upper half of the die is shown in Fig. 8.7 [31]. The heat fluxes within the die in the flow direction and across the edges on the sides of the die to the surroundings (insulation) were neglected. The heat transfer to the surroundings and to the melt was taken into account using the heat transfer coefficients described in Fig. 8.7. The liberated heating power in this example is simulated by sources of a constant voltage on the heating cartridges, which corresponds to the specified temperature of 240 °C.

It is noticeable that the isotherms lie much closer together on the side toward the channel than on the side against the air. This can be explained by a more intensive heat transfer to the melt.

Already in half of the distance between the heating cartridges and the wall of the flow channel, the isotherms are almost straight lines and run parallel to the surface. This indicates that the distance between the heating cartridges and the surface of the channel can be reduced without causing a wavy temperature profile, this being an important guide for setting the minimum distance between the heating cartridge and the flow channel.

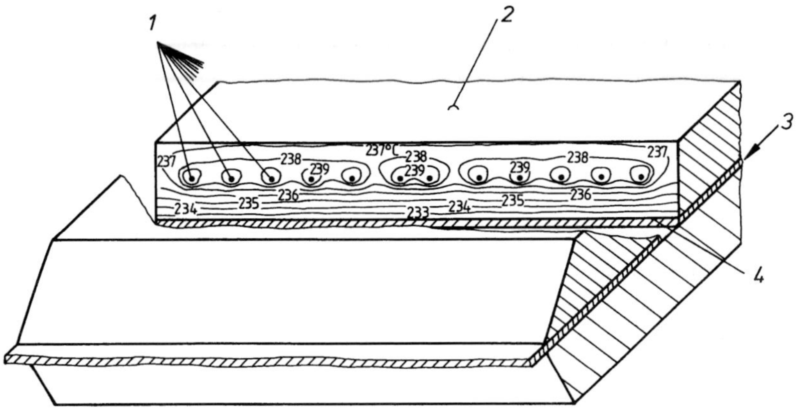


**Figure 8.7** Results in a section perpendicular to the main flow direction: 1: heating cartridges (240 °C), 2: convective heat transfer to the air ($\alpha = 8$ W/m$^2$K), 3: flow channel (melt temperature 220 °C), 4: convective heat transfer to the melt ($\alpha = 75$ W/m$^2$K)

This simulation procedure requires some investment in the instrumentation (sources of constant current and constant voltage) and certain skill by the user. However, once the outline of the die is in place and the boundary conditions installed, the different positions of the heating cartridges can be simulated quite fast by properly shaped electrodes that can be moved around [37–39]. However, the determination of lines of equal voltage is very time consuming because all points have to be searched by hand.

### *Numerical Model*

With the rapid developments in microelectronics during the last several decades, modern high-performance computers are available at relatively affordable costs. These make numerical simulation models for the computation of rheological and thermal processes under varied applicable geometrical and physical boundary conditions possible (compare also Chapter 4). Corresponding programs have also been further developed and simplified so that they can be used effectively even by someone who is not a computer specialist. Among the numerical simulation methods, the *finite element method* (FEM) stands out because of its flexibility with applicable geometries.

In this method the geometry under study is constructed from individual finite elements whereby the geometrical details can be considered very accurately.

Equation (8.14) is then solved for the entire region, completely considering all thermal boundary conditions and interactions [40].

Figure 8.8 depicts the steady-state temperature distribution in the die that was studied under identical boundary conditions as the electrical simulation shown in Fig. 8.4. Here, the values of the heating power are computed by an automatic iterative procedure in such a way that the temperature at certain points (marked with an asterisk in the pictures) adjusts itself close to the temperature set by the control circuit (melt temperature = 220 °C).

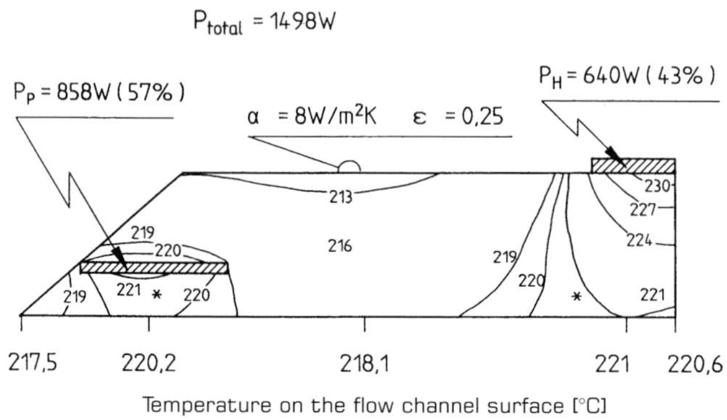


**Figure 8.8** Result of an FEM simulation (temperatures in °C)

The control of the heat delivered by the heater is here included in the simulation by a simple P-type controller. The points marked with an asterisk are the positions of the temperature sensors.

The heating loads determined in the above procedure are placed in the illustration; they pertain to a die of 1 meter width (see Fig. 8.3).

The qualitatively different shapes of the isotherms in Figs. 8.8 and 8.6 result from the segmentation of the boundary conditions in the electrical analogy model. With a finer division of the boundary region, closer results would be obtained, but the cost of the necessary electrical model would be considerably higher. The temperature distributions along the surface of the flow channel are qualitatively equal, but the differences in temperatures are greatly overestimated by the electrical model.

In the computer simulation, the heat transfer by radiation from the surface of the die was first considered for a smooth surface. The established steady-state temperature distribution for a corroded die surface at identical boundary conditions is shown in Fig. 8.9. The temperature of the surface of the flow channel becomes distinctly inhomogeneous, and the heating power requirements for the entire die increase by approximately 66 %.

The increase in the heating power requirement is not uniformly distributed over the installed heating elements. While the supplied power to the heating strip increases only by 25%, a 197% increase is required for the heating cartridge, practically double that of the heating power requirement of the new die with a smooth surface. The reason for this change in distribution of the heating power is that the part of the losses from the surface, which have to be compensated for by the cartridge heater, are considerably higher than those by the strip heater.

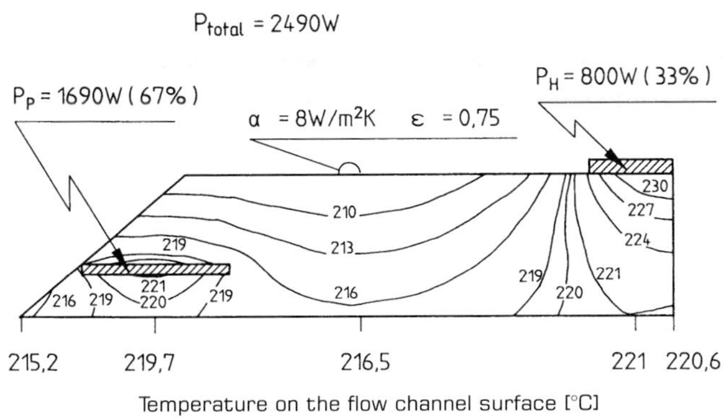


**Figure 8.9** Temperature distribution in a die with corroded surface (FEM) (temperatures in °C)

The thermal boundary conditions change during the operation, and this may cause a considerable shift in the operating points of the controllers. This has to be taken into account in the thermal design (with adjustment of the controller during operation).

The steady-state temperature distribution for a possible design alternative in the placement of heating elements is shown in Fig. 8.10 (smooth surfaces). The comparison to Fig. 8.8 makes clear that the sum of the heating loads remains practically equal, but the cartridge heater has to supply a higher share (67 % instead of 57 %). For that reason the temperature of the cartridge heater is approximately 6 °C higher. The maximum temperature of the surface of the flow channel increases about 2 °C, while the minimum temperature remains practically unchanged. This, of course, manifests itself as an inhomogeneous temperature distribution on the surface of the flow channel.

These examples illustrate clearly that the temperature distribution along the surface of the channel, starting with a melt temperature of 220 °C at the entrance (right), increases slightly, reaches a minimum in the middle region of the die, and finally (under the cartridge heater) reaches the maximum. Then, toward the orifice the temperature drops once again, and at the exit it reaches an absolute minimum, which is below the temperature of the melt.

This is a disadvantage when compared to the "ideal" channel wall temperature as discussed at the outset (Section 8.2.2). A steadily increasing channel wall temperature would be advantageous here. The surface appearance of the extrudate is frequently affected by the die temperature at its exit. This is why the die ring or lips have their own separate temperature control. Figure 8.11 shows how the temperature distribution can be affected favorably for this purpose by the addition of a strip heater in the exit region.

Because of the higher average surface temperatures, the total heating power is approximately 24 % higher than that shown in Fig. 8.8.

In the entry and core zones of the die, the surface temperatures of the flow channel are more uniform and increase markedly in the exit region. This pronounced rise in temperature is caused by an unfavorable position of the left temperature sensor, which obviously is located too far from the exit-region band heater.

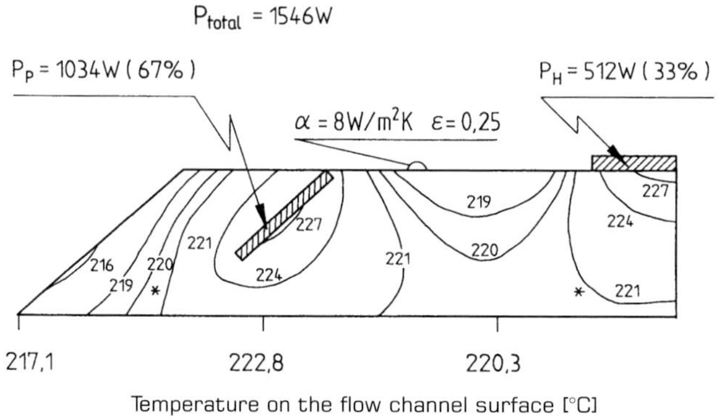


**Figure 8.10** Temperature distribution with changed position of the cartridge heaters (FEM) (temperatures in °C)

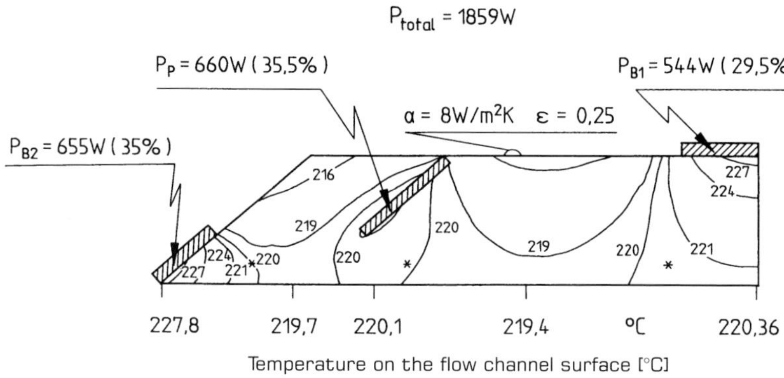


**Figure 8.11** Temperature distribution with three heating element groups (FEM) (temperatures in °C)

With the numerical simulation procedures discussed earlier, three-dimensional computations are possible. However, a considerably increased time is required to set up the model (generating the FE mesh, personnel time) and to perform the computations, making this application excessively expensive. In many cases, the knowledge gained from the simplified two-dimensional studies is sufficient for a good thermal "tune-up" of the die.

### Symbols and Abbreviations

| | |
|---|---|
| $A_{da}$ | Area of the outer die surface that exchanges heat with the ambient air |
| $A_{di}$ | Area of the inner die surface that exchanges heat with the melt |
| $B$ | Die width |
| $c_p$ | Specific heat capacity of the melt |
| $c_{pd}$ | Specific heat capacity of the die |
| $C_R$ | Radiation coefficient of the blackbody |
| $h$ | Die height |
| $I$ | Electrical current |
| k | Conductivity ratio (thermal conductivity/electrical conductivity) |
| $L_i$ | Length of the partial region *i* to be found |
| $l$ | Length |
| m | Model scale |
| $m_d$ | Die mass |
| $\dot{m}$ | Mass flow rate of the melt |
| $n$ | Proportionality factor |
| $p$ | Pressure |
| $p_{in}$ | Inlet pressure |
| $p_{out}$ | Outlet pressure |
| $P$ | Heating power |
| $P_{total}$ | Total heating power |
| $P_H$ | Heating power of a heating tape |
| $P_{nom}$ | Nominal heating power of the die |
| $P_P$ | Heating power of a heating cartridge |
| $\dot{Q}$ | Heat flux |
| $\dot{Q}_{CA}, \dot{Q}_{CM}$ | Heat flux removed from the die by convection (index A = air, M = melt) |
| $\dot{Q}_{DISS}$ | Dissipated energy per unit time |
| $\dot{Q}_H$ | Heat flux supplied to the die by the heating system (heating element) |
| $\dot{Q}_{Hmax}$ | Installed nominal heating power of the die |
| $\dot{Q}_{M,In}$ | Heat flux entering the die with the melt |
| $\dot{Q}_{M,Out}$ | Heat flux leaving the die with the melt |
| $\dot{Q}_{RAD}$ | Heat flux removed from the die by radiation |
| $R$ | Electrical resistance |
| $R_i$ | Series resistance |
| $t$ | Time |
| $t_H$ | Heat-up time of the die |
| $U$ | Voltage |
| $U_S$ | Source voltage of the model |

| | |
|---|---|
| $\dot{V}$ | Volume flow rate of the melt |
| $v$ | Velocity |
| $\alpha$ | Heat transfer coefficient |
| $\alpha_{CA}$ | Heat transfer coefficient for the convective heat exchange between die and ambient air |
| $\alpha_{CM}$ | Heat transfer coefficient for the convective heat exchange between die and a melt |
| $\alpha_{RAD}$ | Heat transfer coefficient for radiation |
| $\dot{\gamma}$ | Shear rate |
| $\varepsilon$ | Emission coefficient |
| $\eta$ | Ratio of stress/temperature |
| $\eta$ | Degree of efficiency |
| $\vartheta$ | Temperature |
| $\vartheta_a$ | Temperature of the ambient air |
| $\vartheta_d$ | Die temperature |
| $\Delta\vartheta_d$ | Increase of the die temperature during heat-up |
| $\vartheta_{da}$ | Temperature of the outer die surface in contact with the air |
| $\vartheta_{di}$ | Temperature of the inner die surface in contact with the melt |
| $\vartheta_H$ | Temperature of the heating system |
| $\vartheta_M$ | Melt temperature |
| $\rho$ | Melt density |
| $\tau$ | Sher stress |

## 8.3 References of Chapter 8


1. Winter, H. H.: Temperaturänderungen beim Durchströmen von Rohren. In: Praktische Rheologie der Kunststoffe. VDI-Verl., Düsseldorf 1978
2. Winter, H. H.: Ingenieurmäßige Berechnung von Geschwindigkeits- und Temperaturfeldern in strömenden Kunststoffschmelzen. In: Berechnen von Extrudierwerkzeugen. VDI-Verl., Düsseldorf 1978
3. Masberg, U.: Einsatz der Methode der finiten Elemente zur Auslegung von Extrusionswerkzeugen. Thesis at the RWTH Aachen 1981
4. Pittman, J. F.T. and Nakazawa, S.: Analysis of Melt Flow and Heat Transfer Using Finite Elements. Series: Polymer Engineering Reviews. Vol. 4, No. 3. Freund, London 1984
5. Mitsoulis, E. and Vlachopoulos, J.: The finite element method for flow and heat transfer analysis. Adv. Polym. Technol. 4 (1984) p. 2
6. Ben-Sabar, E. and Caswell, B.: A stable finite element simulation of convective transport. Int. J. Numerical Methods Eng. 14 (1979) pp. 545–565

7. Kelly, D. W. et al.: A note on upwinding and anisotropic balancing dissipation in finite element approximations to convective diffusion problems. Int. J. Numerical Methods Eng. (1980) pp. 1705–1711
8. Vergnes, B. et al.: Berechnungsmethoden für Breitschlitz-Extrusionswerkzeuge. Kunststoffe 70 (1980) 11, pp. 750–752
9. McKelvey, J. M.: Polymer Processing. Wiley, New York 1962
10. Mandrel cooler for higher film output. Europlastic Monthly (1972) June, pp. 101–102
11. Predöhl, W.: Herstellung und Eigenschaften extrudierter Kunststoffolien. Postdoctoral thesis at the RWTH Aachen 1977
12. Menges, G. et al.: Temperaturbeeinflussung des Durchflußverhaltens in Breitschlitzwerkzeugen. Plastverarbeiter 26 (1975) 7, pp. 361–367
13. Schumacher, F.: Fließregulierung in Breitschlitzdüsen durch Temperatursteuerung. Kunststoffe 75 (1985) 11, pp. 798–801
14. Feistkorn, W and Sensen, K.: Automatik-Schlauchfolienwerkzeuge für coextrudierte Folien. Kunststoffe 78 (1988) 2, pp. 1147–1150
15. Menges, G. et al.: Flüssigkeitstemperiersysteme für Kautschukextruder – Optimieren und Auslegen. Maschinenmarkt 90 (1984) p. 67
16. Menges, G. et al.: Prozeßregelungskonzepte in der Extrusion. Kunststoffe 78 (1988) 10, pp. 936–941
17. Barney, J.: Design requirements for PVC film and sheet dies. Mod. Plast. 46 (1969) 12, pp. 116–122
18. Catón, J. A.: Extrusion die design for cast film production. Brit. Plast. 44 (1971) 3, pp. 95–99
19. Dalhoff, W.: Systematische Extruderkonstruktion. Krausskopf, Mainz 1974
20. Hensen, F.: Anlagenbau in der Kunststofftechnik. Postdoctoral thesis at the RWTH Aachen 1974
21. Fischer, P.: Stand der Regelungs- und Steuerungstechnik bei Extrudern und Extrusionsanlagen. In: Rechnergesteuerte Extrusion. VDI-Verl., Düsseldorf 1976
22. Ällerdisse, W.: Überprüfung und Angleichen von Thermofühlerleitungen und Reglern an gegebene Regelstrecken. In: Messen an Extrusionsanlagen. VDI-Verl., Düsseldorf 1978
23. Schwab, E. et al.: Nach Faustregeln berechnet - Anhaltspunkte zur Ermittlung günstiger Reglereinstellwerte. Elektrotechnik 59 (1977) Mai, pp. 12–13
24. Handbuch der Temperaturmessung. Firmenschrift der Linseis GmbH
25. Schiedrum, H. O.: Erfahrungen mit Temperaturreglern und Regelstrecken. Kunststofftechnik 12 (1973) 3, pp. 57–60
26. Noren, D.: New approach aids uniform heating and cooling of extrusion cooling. Plast. Technol. 28 (1982) 2, pp. 75–79
27. Menges, G. et al.: Temperaturverhältnisse beim Anfahren von Breitschlitzwerkzeugen. Kunststoffe 65 (1975) 7, pp. 432–436
28. Upmeier, H.: Prozeßoptimieren, Steuern und Regeln an Blasfolienanlagen. In: Rechnergesteuerte Extrusionsanlagen. VDI-Verl., Düsseldorf 1981
29. Woodworth, C. L.: Computer simulation of steady-state nonisothermal melt flow in tubes. Adv. Polym. Technol. 6 (1986) 3, pp. 251–258
30. Wiibken, G.: Thermisches Verhalten und thermische Auslegung von Spritzgießwerkzeugen. Research report, AIF Nr. 2973, IKV, Aachen 1976
31. Gemmer, H. et al.: Auslegung der Werkzeugtemperierung und Einfluß auf die Eigenschaften von Spritzgußteilen. Ind. Anz. 93 (1971) 53, pp. 1300–1306; 62, pp. 1601–1603
32. Hottel, H. C. and Sarafim, A. F.: Radiative Transfer. McGraw-Hill, New York 1967
33. Renz, U.: Grundlagen der Wärmeübertragung. Lecture at the RWTH Aachen 1981

34. Vlachopoulos, J. et al.: Numerical studies of non-Newtonian flow and heat transfer. Conference: Heat Transfer in Plastics Processing, University of Bradford 1974
35. Hackeschmidt, M.: Elektrisch leitendes Papier für Potentialfeld-Ausmessungen. Elektrotechnik 46 (1968) 5, pp. 92–98
36. Menges, G. and Wübken, G.: Einfaches elektrisches Analogmodell zur Optimierung der Kühlkanalanordnung in Spritzgießwerkzeugen. Plastverarbeiter 23 (1972) 6, pp. 394–395
37. Jung, P. B.: Zweckmäßige Werkzeugkühlung für die Hohlkörperherstellung. Research report, AIF Nr. 3022, 1976
38. Dierkes, A. et al.: Auslegung von Extrusionswerkzeugen. 9. Kunststofftechnisches Kolloquium of IKV, Aachen 1978, pp. 61–88
39. Michaeli, W.: Zur Analyse des Flachfolien- und Tafelextrusionsprozesses. Thesis at the RWTH Aachen 1976
40. User Handbook for the program package MICROPUS, Vers. 2.1, IKV, Aachen 1989

# 9 Mechanical Design of Extrusion Dies

The mechanical design of extrusion dies, which involves calculation of the forces and deformations arising during the operation of the die, is important for two reasons: first, to ensure that the die will not be damaged during operation; second, to see that the distribution channel, particularly in distributor (manifold) dies, retains the geometry established by the rheological design during its operation.

The important applications of the mechanical design are

- the design of screw joints and sealing surfaces with respect to the internal pressure,
- the design of wall thicknesses for a permissible deformation by the internal pressure,
- and the design of systems for the adjustment of the geometry of the die in the exit region (automatic dies and blow-molding dies with an adjustable outer ring for the programming of the wall thickness of the extrudate).

The mechanical design is always closely linked to the rheological design. First of all, in the rheological design, the geometry of the flow channel is defined. Then the pressure distribution has to be estimated conservatively by computation of the flow for the die within the expected operating range, with the material of the highest viscosity, at the lowest mass temperature and the highest mass throughput. With that, the isotropic pressures and shear stresses at the wall existing in the die are known. The forces acting in the die can be calculated from the stresses and the areas of the flow channel walls. The weight of the die also has to be considered in the design of medium-sized and large dies.

A universally valid method for the mechanical design of extrusion dies does not exist. Different ways to proceed in the mechanical design of extrusion dies will be demonstrated here on three different examples.

# 9.1 Mechanical Design of a Breaker Plate

In the mechanical design, the pressure losses and the deformations resulting from them have to be considered not only for the screen packs (Fig. 9.1(b)), but also for the supporting breaker plate (Fig. 9.1(a)). The deflection of the screens $f_s$ results from the pressure loss across the screen pack, and it must not exceed the minimum specified value in the unsupported area. For the calculation of the deflection of the breaker plate, both the pressure losses across the screen packs and the pressure losses across the breaker plate itself must be taken into account.

### *Strength Calculations for the Breaker Plate*

According to the geometrical conditions, the breaker plate can be considered as a flat circular plate subjected to bending.

Based on Kirchhoff's theory of plates, the maximum deflection of a rigidly fixed round plate with a field of holes due to pressure (see Fig. 9.1) is determined from

$$f_{max} = \frac{\Delta p \cdot R^4}{64N}. \tag{9.1}$$

For loads and fixing conditions different from the ones shown in Fig. 9.1, [2] provides additional formulas.

The stiffness of the plate is characterized by the following relationship:

$$N = \frac{Eh^3\alpha}{12\left(1-\nu^2\right)} \tag{9.2}$$

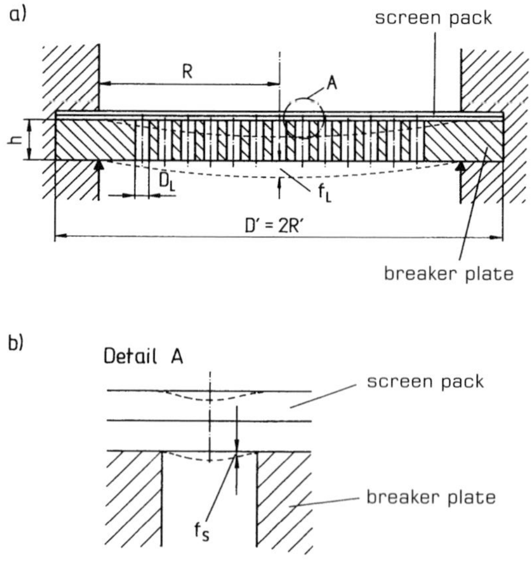


**Figure 9.1** Symmetrically stressed circular plate with a perforated area [1]: (a) $f_L$ = deflection of the breaker plate, (b) $f_S$ = deflection of the screen pack

In the above two relationships, $\Delta p$ is the pressure difference across the breaker plate and the screen pack, $E$ is the modulus of elasticity, $\nu$ is Poisson's ratio, and $\alpha$ is the attenuation coefficient. The latter describes the ratio of $N/N_0$ of the drilled plate to an undrilled one and is expressed as follows:

$$\alpha = 1 - \frac{2R_L}{t} \quad \text{with the hole radius} \quad R_L = D_L / 2 \tag{9.3}$$

The hole spacing $t$ is equal to the distance between the holes in the plate. If it appears necessary to calculate the stress for each crosspiece between the holes separately, pertinent information is contained in [2]. In practice, the lowest value of $\alpha$ is 0.12; lower values are meaningless.

Figure 9.2 shows the comparison of experimental data from [2] and the ratio $N/N_0$ as a function of $\alpha$ (the attenuation coefficient). A detailed analysis by structural mechanics, published in [13], concluded that this formula does not provide reproducible results for very thin plates and very thick plates. The bending calculated for the thin plates is too great and for the thick ones too little. By that finding, the region of validity of Equation (9.3) is limited to

$$0.7 \leq \frac{h}{t} \leq 3 \tag{9.4}$$

For the calculation of the thickness of the plate, the corresponding clamping conditions have to be taken into account. Because of the maximum elastic bending stress in the center of the plate, the plate thickness is calculated from the following formula:

$$h = B_p \cdot D \cdot \left( \frac{p \cdot S}{\alpha \cdot \sigma_p} \right)^{0.5} . \tag{9.5}$$

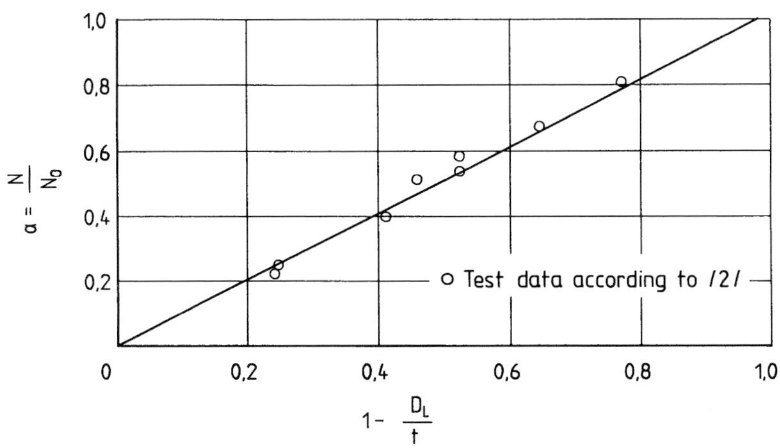


**Figure 9.2** Flexural strength of plates with perforated area [2]

In this relationship, the calculation coefficient $B_p$ represents different clamping conditions of the plate at its rim; its values vary between 0.454 for freely resting and 0.321 for a rigidly fixed plate. The thickness of breaker plates made from several parts should be calculated in a similar way but with the coefficient increased by 10%. Figure 9.3 gives suggestions for the choice of the calculation coefficient. Equation (9.5) also contains the attenuation coefficient $\alpha$ and the plate diameter, which for the purpose of the calculation should be set to the mean diameter of the seal ($2R$). The safety factor $S$ usually lies between 1.5 and 1.8, and $\sigma_p$ is the maximum permissible bending stress.

Two types of pressure spikes occurring suddenly in front of the breaker plate have to be considered in its design:

1. Pressure spikes $\Delta p_p$, which occur suddenly during the startup and which can be double or triple that of the pressure difference during the steady-state operation ($S_p = 2$ to 3).
2. The increase in pressure in front of the breaker plate due to plugging or obstruction of the supported screen pack. As recommended by [3], an additional pressure loss $\Delta p_{sa} = 50$ bar should be assessed along with the usual pressure loss of the screen pack $\Delta p_s$, since at this pressure increase the screen pack is usually changed. For safety reasons, this value should be multiplied by a factor of 1.2 to 2 before it is used for the calculation.

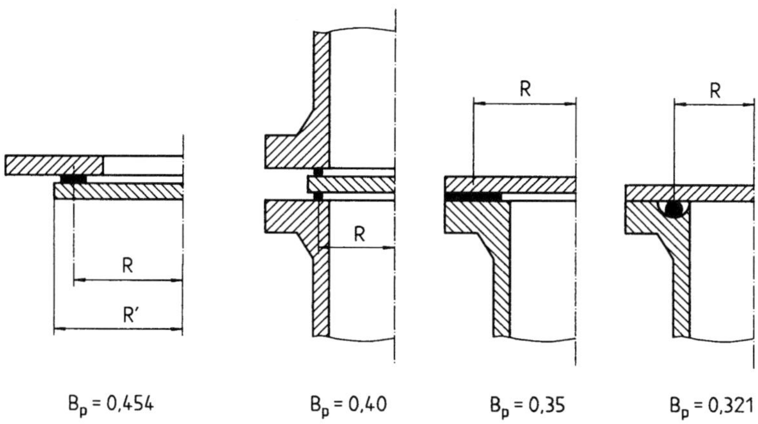


**Figure 9.3** Coefficient $B_p$ for circular plates with different types of clamping [1]

The pressure loss Δp to be used in Equation (9.4) is the maximum value from the addition of all pressure drops across the screen pack and the breaker plate:

$$p = \left(S_p \Delta p_p + (\Delta p_{sa} + \Delta p_s) \cdot S\right) \qquad \textit{with } S_p = 2-3 \textit{ and } S = 1.2-2. \quad (9.6)$$

### *Determination of the Bending of the Screen*

Screens or screen packs are supported mechanically against the differential pressure by the breaker plate. The actual supporting function is done by the spaces between the boreholes of the plate. In the regions of the holes, the screen or the screen pack is subjected to bending stresses. The diameter of the borehole in the plate has to be chosen so that the maximum allowable deflection of the screen or the screen pack is not exceeded. The maximum deflection depends on the maximum permissible tension stress of the wire used in the screen. The geometrical data required for further study are presented in Fig. 9.4.

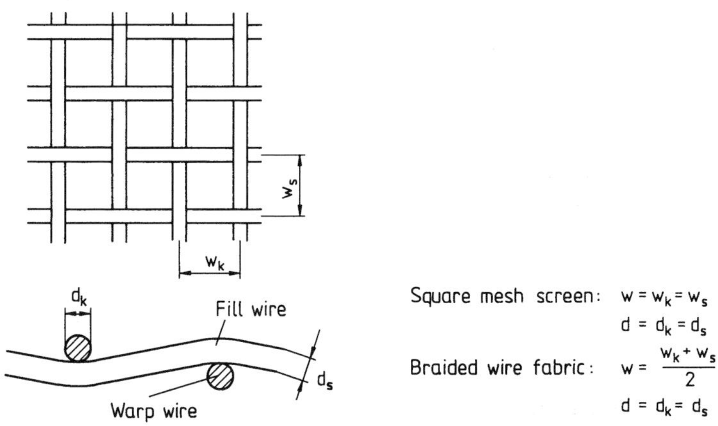


**Figure 9.4** Structure of screen and braid webs and designations [1]

The same procedure for calculating the breaker plate is applicable for the deflection of the screen part. The basis here, again, is the plate theory by Kirchhoff. In contrast to the breaker plate, the rim of the screen in the region of the borehole is considered to be movable and not clamped tightly. The maximum deflection of the screen under the differential pressure $\Delta p_s$ is obtained as follows:

$$f_{max} = \frac{p_s R_L^4}{64 N_S} \cdot \frac{5+v}{1+v}. \tag{9.7}$$

The radius of the borehole $R_L$ is one of the constituents of the above equation. In case there is a countersink or a profile in the geometry of the inlet, the maximum radius $R_{L\,max}$ and the screen stiffness $N_s$ have to be used. The latter is given by

$$N_S = \frac{E h_S^3 \alpha_S}{12\left(1 - v^2\right)} \tag{9.8}$$

with the attenuation coefficient for screens $\alpha_S$ and the screen thickness $h_S$; the latter takes into consideration the crimping of the screen wire, corresponding to

$$h_S = d_S \left(1 + \frac{d_S}{w}\right). \tag{9.9}$$

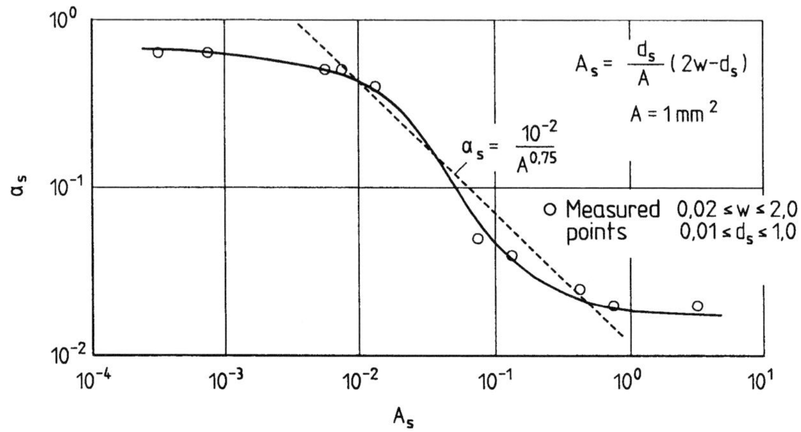


**Figure 9.5** Attenuation coefficient $\alpha_S$ for the computations of screens [1]

The attenuation coefficient describes the ratio $N/N_{0S}$ of a plain flat plate of thickness $h_S$ to the steel screen used. Figure 9.5 shows the dependence of the attenuation coefficient $\alpha_S$ on the standardized proportion of the area of the wires in the screen:

$$A_S = \frac{d_S}{A}(2w - d_S) \qquad A = 1\,\text{mm}^2, \tag{9.10}$$

where $A_S$ is formed with the wire diameter $d_S$ and mesh width $w$ corresponding to Fig. 9.4. As a result of the more complex geometrical relationships in screens and screen fabric, $\alpha_S$ cannot be expressed in a simple linear fashion. For an estimation of $\alpha_S$ that is sufficient in many cases, the following relationship can be applied (see dashed curve in Fig. 9.5):

$$\alpha_S = \frac{10^{-2}}{A_S^{0.75}} \tag{9.11}$$

Single screens are in practice not used; packs made up of three to five screens of different mesh apertures are common. For the calculation of the deflection of these screen packs, the sum of all stiffness values for single screens and the total pressure drop $\Delta p_{S,\,tot}$ are used in Equation (9.7).

The stiffness of the pack, mentioned above, is calculated as follows:

$$N_{S1} = \sum_{i=1}^{n} N_{Si} = \frac{E}{12\left(1-v^2\right)} \sum_{i=1}^{n} \left(h_{Si}^3 \cdot \alpha_{Si}\right) \tag{9.12}$$

In order to avoid a blowout of the screens in the area of the borehole, the permissible stress has to be checked. After [4] the tensile stress for the wire used in the screen is obtained from the following relationship:

$$\frac{\sigma_t}{S_S E} = \frac{1}{2}\left(\frac{f_{max}}{R_L}\right)^2 . \tag{9.13}$$

This tensile stress has to meet the conditions $\sigma_{t,p} > \sigma_t$. The values of $\sigma_{t,p}$ are available from individual tables. As an additional precaution against the wire break, the safety factor $S_s$ can be increased, thus $1.2 < S_s < 1.5$.

The design of a breaker plate requires alternating mechanical and rheological considerations because the number of boreholes and their diameter and the thickness of the plate affect the pressure loss and thus the forces acting on it as well as the mechanical strength.

## 9.2 Mechanical Design of a Die with Axially Symmetrical Flow Channels

The die to be considered is the center-fed die, which, as shown in Fig. 9.6, has a flow channel that is deliberately kept simple. This flow channel can be thought of as consisting of five sections. The mandrel is held by spider legs (their number being $n$) in Segment III. The whole mandrel support plate is constructed in one piece and can be thought of as being rigidly clamped from the outside by bolts (their number being $m$).

Furthermore, Fig. 9.6 shows the pressure drop in the flow channel, which can be calculated, as shown in Chapters 3 and 4, for a given operating point with the rheological material data for the channel geometry. Two cases are shown. In Case 1, the pressure drop in Segments I to II is very slight and, as a first approximation and in contrast to Case 2, can be set equal to zero.

While the melt is flowing through the die, viscous forces $F_Z$ as well as compressive forces $F_p$ (in three directions) act on the mandrel and on the walls of the flow channel. The internal pressure $p_i$ causes the die to open up radially (Fig. 9.7). (The forces acting on the spider legs are disregarded for the time being. The dimensions of the spider legs are determined below.)

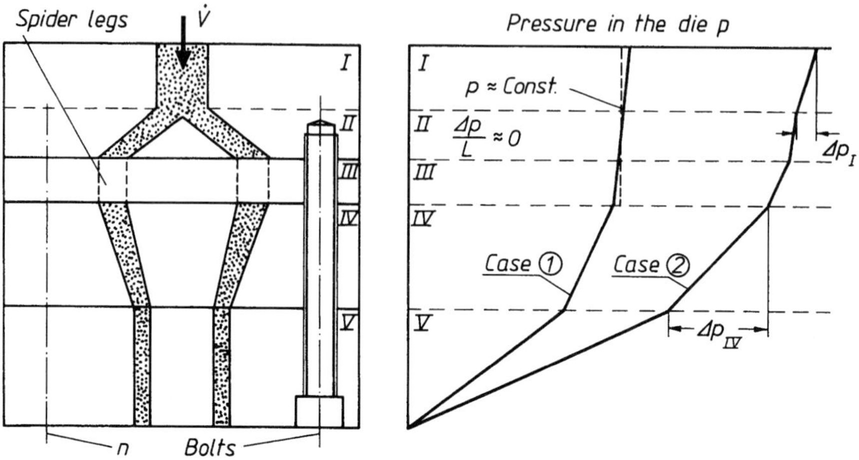


**Figure 9.6** Center-fed die: qualitative pressure distributions in the die

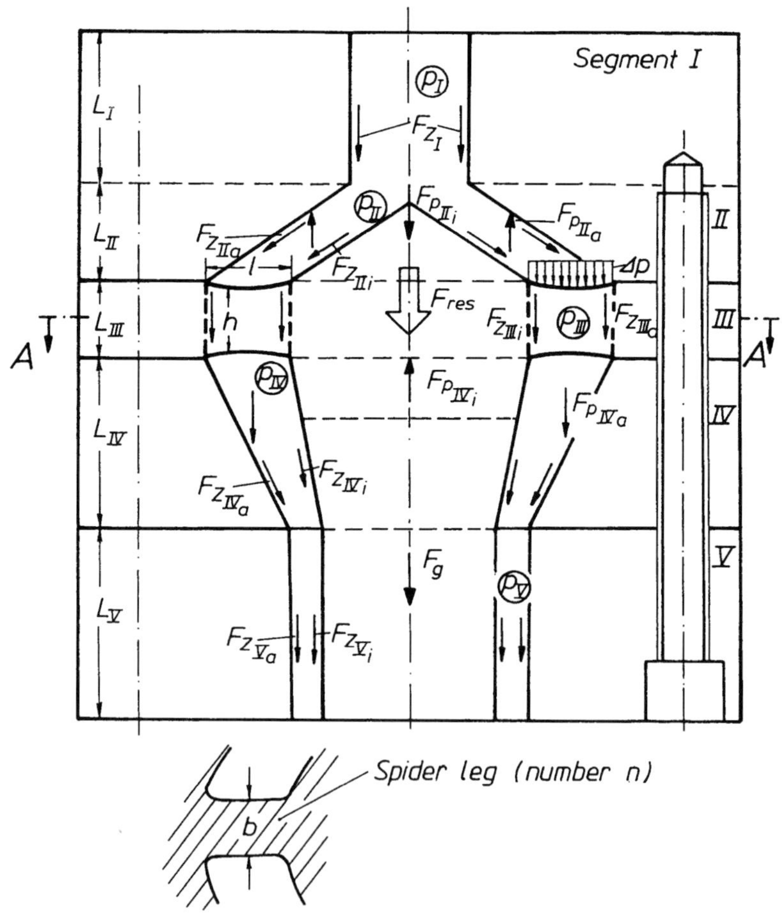


**Figure 9.7** Forces acting in the flow channel of a center-fed die

First of all, when determining the dimensions of large dies, it is necessary to check if the gravitational force $F_g$ (own weight) has to be considered. (When there are no guide pins or registered centerings in addition to the bolts, its own weight has to be considered even for medium-sized dies when determining the size of the bolts.) The forces of gravity will be neglected from now on.

For instructional reasons, the forces of the individual segments of the die are not considered in the sequence of the segments.

### Segment I

(a) Forces at the flow channel wall due to viscosity.

The shear stress $\tau_W$ acts at the wall of the flow channel. For the pipe flow existing here, it follows from Table 3.1 that

$$\tau_W = \frac{\Delta p}{2L} R \rightarrow \tau_{W_I} = \frac{\Delta p_I}{2L_I} R_{i_I}. \quad (9.14)$$

with

$$F_Z = \tau_W \cdot A \quad \left(A = 2\pi R_I L_I\right) \quad (9.15)$$

the following relationship holds:

$$F_{Z_I} = \Delta p_I \pi R_{i_I}^2. \quad (9.16)$$

For Case 1: $\Delta p_I \approx 0 \rightarrow F_{ZI} \approx 0$.

(b) Opening up under internal pressure $p_i$.

Segment I represents a circular, thick-walled hollow cylinder that is opened up by the internal pressure by the amount $f_i$ given by the following general formula [7]:

$$f_i = \frac{p_i R_i}{E} \left( \frac{R_a^2 + R_i^2}{R_a^2 - R_i^2} + \mu \right). \quad (9.17)$$

where $\mu$ is Poisson's number (Poisson's ratio $m = \frac{1}{\mu}$) for steel: $\mu \approx 0.33$ [6,8]. The maximum stresses occur at the interior wall, that is, at the flow channel. The following equation holds for the biaxial state of stress to be set up there $\left(c \triangleq \textit{circumference}, r \triangleq \textit{radial}\right)$:

$$\varepsilon_c = \frac{1}{E}\left(\sigma_c - \mu \sigma_r\right) = \frac{f_i}{R_i}. \quad (9.18)$$

According to [7]:

$$\sigma_{c_{max}} = p_i \frac{R_a^2 + R_i^2}{R_a^2 - R_i^2}, \tag{9.19}$$

$$\sigma_{r_{max}} = -p_i. \tag{9.20}$$

When selecting die dimensions, every effort should be made to keep $f_i$ below 0.05 mm.

In Case 1, $p_i$ is constant and can therefore be used directly for the determination of the dimensions. If the pressure in the segment drops, as it does in Case 2, the dimensions of the die should be selected so as to accommodate the maximum pressure at the inlet for safety reasons.

(As already assumed in this calculation, normal stresses caused by deformation can be neglected when determining the dimensions of the extrusion dies.)

### *Segment V* (9.8)

(a) Viscous forces at the flow channel wall.

If the channel height $H_V$ is small when compared to $R_{iV}$, the following equation can be used according to Table 3.1:

$$\tau_w = \frac{\Delta p}{2L} H \rightarrow \tau_{w_V} = \frac{\Delta p_V}{2L_V} H_V. \tag{9.21}$$

with $A_V = \pi(R_{aV} + R_{iV})L_V$, it follows that $F_{ZVa}$ and $F_{ZVi}$ are given by

$$F_{Z_{Va}} \approx F_{Z_{Vi}} = \tau_{w_V} A_V = \frac{\pi \Delta p_V}{2} H_V \left(R_{a_V} + R_{i_V}\right). \tag{9.22}$$

If, however, $H_V$ is large in comparison to the radius of the annular slit, the following equation has to be chosen according to Table 3.1:

$$\tau = \frac{R_a}{2} \frac{\Delta p}{L} \left[\left(\frac{r}{R_a}\right) - \frac{1-k^2}{2\ln\frac{1}{k}} \left(\frac{R_a}{r}\right)\right] \tag{9.23}$$

with $k = R_i/R_a$.

For $\tau_{WVi}$ and $\tau_{WVa}$, $r$ is set to equal $R_i$ and $R_a$, respectively. Then $F_{ZVi}$ and $F_{ZVa}$ result.

(b) The opening up under internal pressure is calculated in a similar fashion as for Segment I.

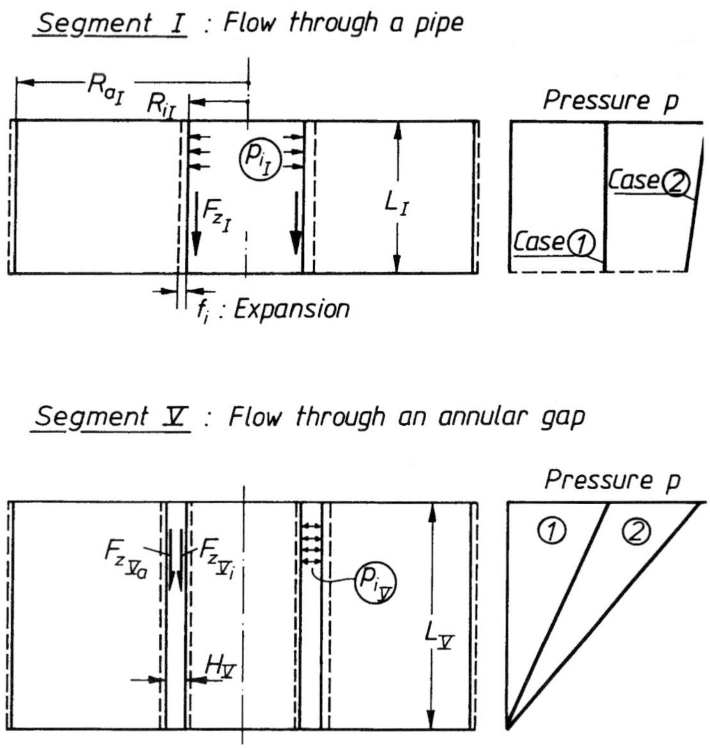


**Figure 9.8** Forces acting in a flow through a pipe and in a flow through an annular gap

It should be noted, however, that the inner mandrel is also compressed in the radial direction because of the internal pressure. In order to calculate this deformation, it is necessary to consider the multiaxial condition of stress in the mandrel in the circumferential, radial, and axial directions.

For a circular, thick-walled cylinder under an external pressure $p_a$, the following general equations can be used according to [5,9]:

$$\sigma_c = -p_a \frac{R_a^2 + \frac{R_a^2 R_i^2}{r^2}}{R_a^2 - R_i^2}, \tag{9.24}$$

$$\sigma_r = -p_a \frac{R_a^2 - \frac{R_a^2 R_i^2}{r^2}}{R_a^2 - R_i^2}, \tag{9.25}$$

$$\sigma_a = -p_a \frac{R_a^2}{R_a^2 - R_i^2}. \tag{9.26}$$

For the deformation, the following equations hold:

$$\varepsilon_c = \frac{1}{E}\left(\sigma_c - \mu\left(\sigma_r + \sigma_a\right)\right), \tag{9.27}$$

$$\varepsilon_r = \frac{1}{E}\left(\sigma_r - \mu\left(\sigma_c + \sigma_a\right)\right), \tag{9.28}$$

$$\varepsilon_a = \frac{1}{E}\left(\sigma_a - \mu\left(\sigma_c + \sigma_r\right)\right). \tag{9.29}$$

If the mandrel has no inner borehole, then $R_i$ equals zero and consequently follows $\sigma_c = \sigma_r = \sigma_a = -p_a$. During the calculation, it should be noted that $p_{iV}$ acts as an external pressure ($\rightarrow p_a$) on the mandrel. Then, in the calculation of the deformation, $p_{i\,max}$ (here $P_{V\,max}$) should be used. The deformation of the exterior part of the die and the deformation of the interior part then give the total deformation.

***Segment II*** **(Fig. 9.10)**

(a) Forces at the flow channel wall due to viscosity.

As in the procedure for calculating the pressure loss in a conical region of a die, it is proposed to replace this region by channel sections with parallel walls (tubular or annular sections) and to calculate the shear stresses at the walls in these sections. According to the relationships given for Segment V, the forces at the inner and outer walls of the channel can be calculated for each section of length $L^*$. The sum of all the forces so calculated gives the respective total force due to the melt viscosity at the inner and outer walls of the flow channel $\left(F_{Z_{III}} = \sum F^*_{Z_{III}}\right)$.

Since pressure $p$ is approximately constant for Case 1, finite forces are obtained only for Case 2.

(b) Axial compressive forces.

As a result of the melt pressure, compression forces act in the axial direction on the tip of the mandrel support (Area 2) as well as opposite to the direction of extrusion on Area 1.

For Case 1, where $p_{II}$ = constant, these forces can be determined easily by means of the projected stress area.

For Area 1:

$$F_{p_{IIa}} = p_{II}\pi\left(R^2_{a_{II}} - R^2_{i_I}\right). \tag{9.30}$$

For Area 2:

$$F_{p_{IIi}} = p_{II}\pi R^2_{i_{II}}. \tag{9.31}$$

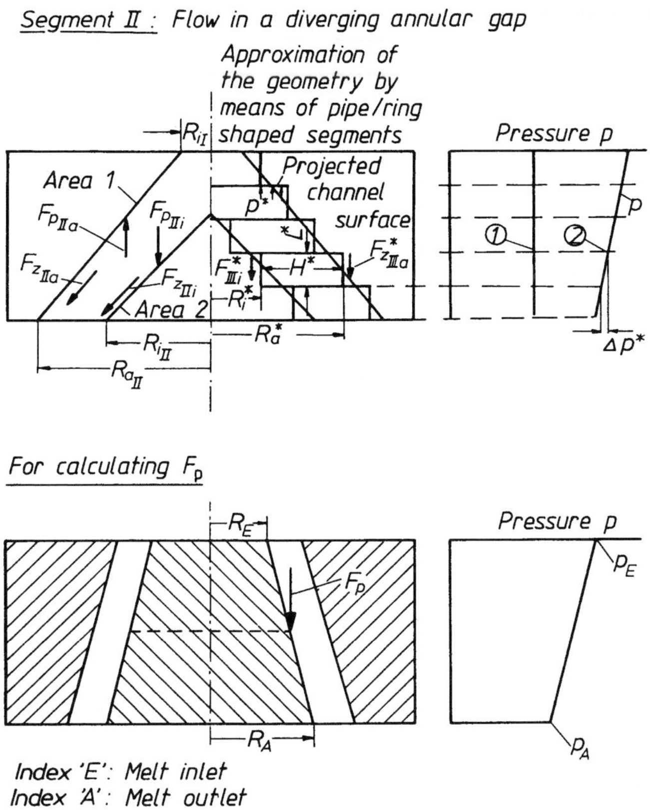


**Figure 9.10** Forces acting in a flow through an annular gap with a changing mean diameter

In considering Case 2, it is necessary to differentiate between a nonlinear and linear pressure drop in the region under consideration.

If the pressure does not decrease linearly, the calculation must again be carried out in individual steps. As an approximation, the pressure is considered to be constant in each individual segment and to act on the projected area of the section of the wall. The sum of the forces in the individual sections thus obtained is then the total acting force.

However, if the pressure changes linearly, the total force acting can be calculated directly, $p_E$ being the pressure at the beginning and $p_A$ at the end of the segment under consideration (see Fig. 9.10):

$$F_p = \pi \cdot p_E \cdot \left(R_A^2 - R_E^2\right) + \pi \cdot \left(p_E - p_A\right) \cdot \left[R_E \cdot \left(R_A + R_E\right) - \frac{2}{3} \cdot \frac{R_A^3 - R_E^3}{R_A - R_E}\right] \qquad (9.32)$$

where $R_E$ and $R_A$ are the respective radii of the channel at the beginning and the end of the segment.

(c) Opening up under internal pressure.

The determination of dimensions can be carried out as for Segment I. The maximum pressure, acting on the point of the least wall thickness, should be checked. Because of the conicity of the flow channel, the forces $F_{\text{pIIa}}$ and $F_{\text{pIIi}}$ act to compress the segments. The compression deforms the segment in the radial direction. This deformation is not considered further here, other than to refer to equations in [5,6]. As in Segment V, pressure $p_{\text{II}}$ also compresses the top of the mandrel in the radial direction.

### *Segment IV*

The compression and viscous forces acting on this segment, as well as the opening up, are calculated in a similar fashion as in Segment II. However, here the varying height of the flow channel must be taken into consideration.

### *Segment III* (Fig. 9.7) (with determining the dimensions of the spider legs)

For this region of the die, the effective viscous forces as well as the radial opening up under the internal pressure can be calculated as for Segment V. If only a few spider legs are present, they can be disregarded.

The correct choice of dimensions of the spider legs, which support the mandrel, is particularly important in this region of the die. Figure 9.7 shows the forces acting on the mandrel. These can be combined into the resulting forces $F_{\text{res}}$. If the number of spider legs is $n$, then each spider leg must absorb the force $F$:

$$F = \frac{F_{\text{res}}}{n} \tag{9.33}$$

The case, illustrated by Fig. 9.11, can be used as an approximation for describing the bending deformation of the spider leg due to this force. The spider leg may not be regarded only as a slender beam; the deformation due to shear must also be taken into consideration (see Fig. 9.11).

If the pressure drop in the mandrel support region and also the projected area of the spider legs (especially when offset spider legs with a supporting ring are used) are large, this additional stress on the spider legs must be taken into consideration for bending deformation as well as for shear deformation (see Fig. 9.11). However, as a rule, the pressure drop in the mandrel support region is slight, so it can be disregarded.

The total deformation of the spider leg $f_{\text{total}}$ and therefore the axial displacement of the mandrel are given by the sum of the individual deformations (see Fig. 9.11).

The moments of inertia required for the calculation are given in Fig. 9.12 [5,6] for some cross-sectional shapes that can be used as an approximation of the cross section of the spider legs.

The strength should be checked in the stressed cross section of the spider leg at the loads that occur. The hypothesis of the change in shape energy provides the equivalent stress $\sigma_{eq}$, which must be smaller than the allowable strength $\sigma_p$ of the material. For the shear and bending stress at the spider leg, the following expression can be used [6]:

$$\sigma_{eq} = \sqrt{\sigma_b^2 + 3\tau^2} < \sigma_p. \tag{9.34}$$

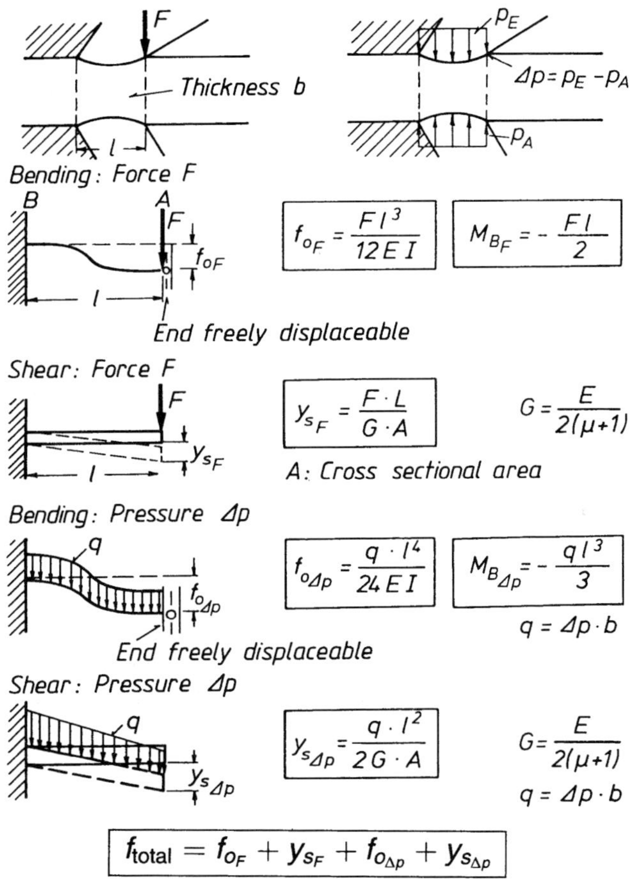


**Figure 9.11** Deformation of the spider leg (equations from [5,6,8])

For the bending stress $\sigma_b$ the following relationship holds:

$$\sigma_b = \sigma_{b\max} = \frac{\Sigma M_{bi}}{W_x} = \frac{M_{bF} + M_{b\Delta_p}}{W_x} \tag{9.35}$$

and for the shear stress:

$$\tau = \tau_{\max} = \frac{\Sigma F_i}{A} = \frac{F + \Delta pbl}{A}. \tag{9.36}$$

For $M_{bF}$, $M_{b\Delta p}$, $A$, and $W_X$ see Figs. 9.11 and 9.12.

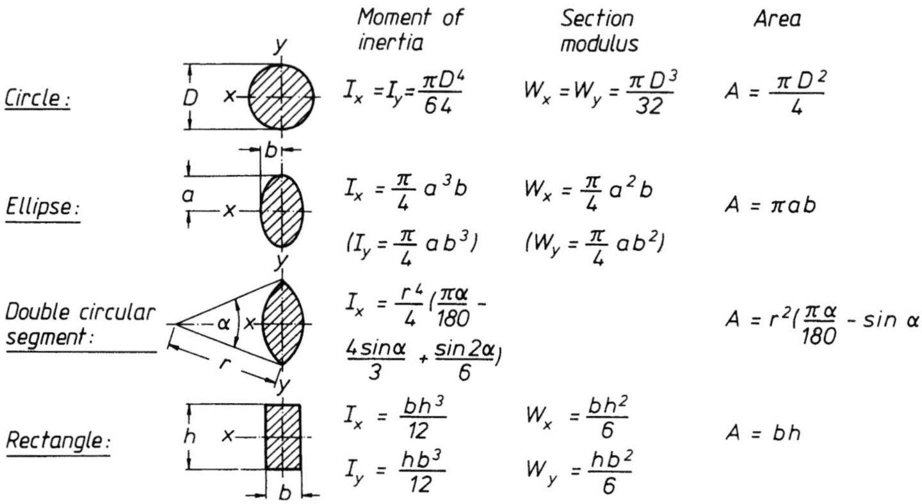


**Figure 9.12** Determining the dimensions of spider legs (moments of inertia, section moduli, areas [5,6])

If the allowable material strength is exceeded under the operating conditions chosen ($\sigma_{eq} > \sigma_p$), the dimensions of the die parts must be changed or the flow channel must be redesigned; a solution with a lower pressure consumption must be found.

**Further Comments Regarding the Determination of Dimensions:**

The forces that must be absorbed by the bolts (their number being $m$) can be calculated by means of the forces shown in Fig. 9.7, the direction of these forces being taken into account. It should be noted that for the design selected for the die, the total load on the mandrel must be absorbed by the bolts. The standard literature dealing with the calculation of the bolt dimensions should be consulted [5,6].

Furthermore, the sealing surfaces between the individual segments must be tested for surface pressure and attention paid to screw pretension. This pretension is to be selected so that the sealing surfaces always remain pressed together.

If there are large temperature differences in the die body, which is usually not the case (the body should have approximately the temperature of the melt), it may be necessary to check for the thermal stresses in the critical regions of the die. A procedure similar to that in this chapter was chosen in [9,10] to design a servo drive that adjusts the mandrel in extrusion blow molding. In this calculation, efforts were made to have equal and opposite viscous and compressive forces acting on the adjustable mandrel. This can be achieved in only a few actual cases, however.

## 9.3 Mechanical Design of a Slit Die

The geometry of the manifold of a slit die, designed for a given operating point according to [11], is shown in Fig. 5.32. The pressure distribution in the flow channel of this die, as calculated from the equations in Section 5.2.2 and in Chapter 3, is shown in Fig. 5.28. As illustrated by Fig. 5.28, the greatest pressure is in the center of the die. The axial extension of the flow channel is also the greatest here. The pressure relationships existing here, which are critical for the mechanical design, are shown once again in Fig. 9.13.

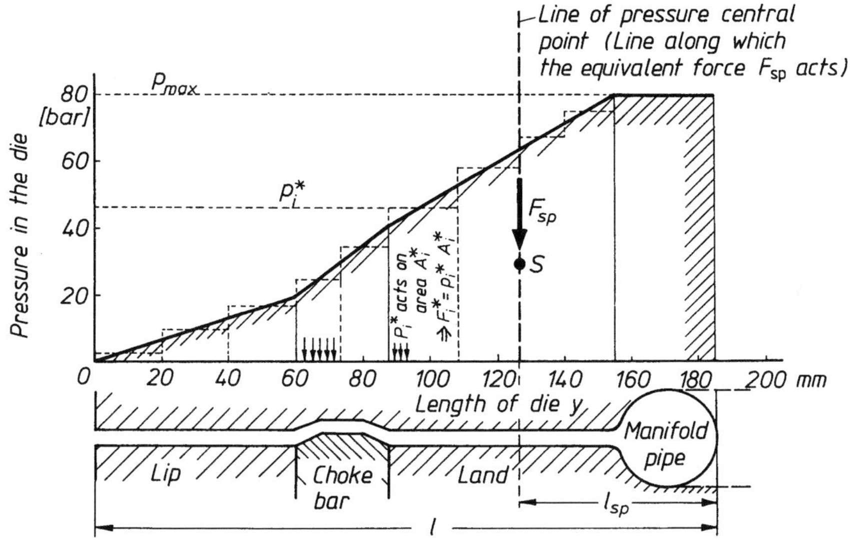


**Figure 9.13** Determining the dimensions of a flat slit die (pressure drop in the center of the die)

There are several ways to determine the dimensions of the die wall thickness from the mechanical point of view, but all of them must take into account the bending and shear deformations. From a thermal point of view, the wall of the die must be sufficiently thick that, when a heating cartridge is used, for example, a constant temperature of the flow channel wall is ensured (see Chapter 8).

(a) Method of calculating die deformation (using $p_{max}$).

Only section A–A in Fig. 9.14 is considered for the determination of dimensions. The maximum pressure in the flow channel $p_{max}$ (Fig. 9.13) should act on the area $A = B \cdot l$.

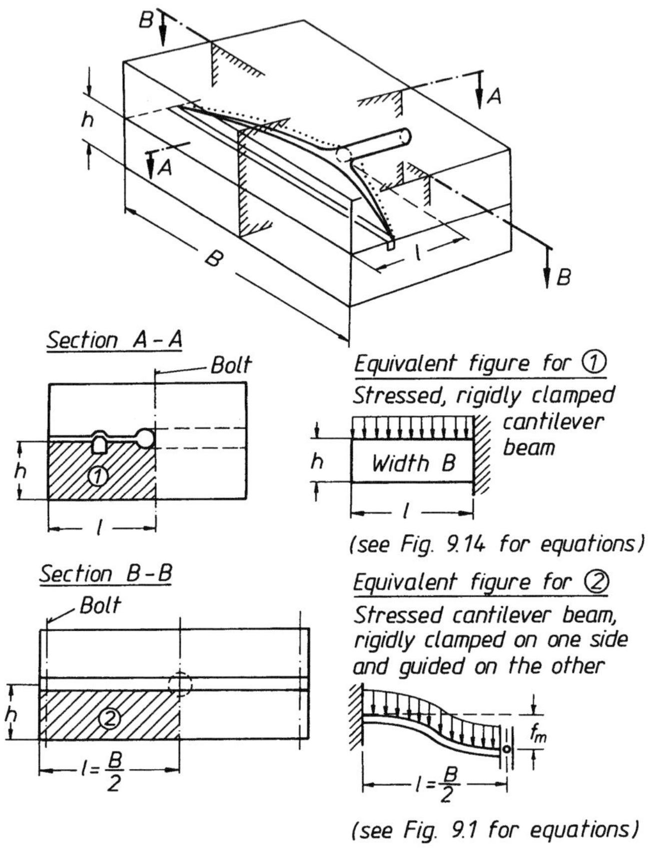


**Figure 9.14** Determining the dimensions of flat slit dies

This means that the coathanger shape of the flow channel is disregarded. In this way, the die in section A–A can be regarded as a cantilever beam with a constant area load $p_{max}$ clamped on one side. The resulting deflection is shown in Fig. 9.15. Because of the high load applied ($p_{max}$), the calculation already contains safety factors. A similar method is chosen for the shear deformation (Fig. 9.15).

(b) Method of calculating die deformation (using $F_{sp}$).

The center of gravity of the area under the pressure curve in Fig. 9.13 can be determined graphically (the procedure is described in detail in [5,6]). The so-called line of pressure concentration, in which a dummy force $F_{sp}$ (the total force on the flow channel surface) is applied, can be drawn through this center of gravity. The following equation applies (see Fig. 9.13):

$$F_{sp} = \sum F_i^* = \sum p_i^* A_i^* . \tag{9.37}$$

The individual forces $F_i^*$ can be determined by approximating the pressure curve by a step function.

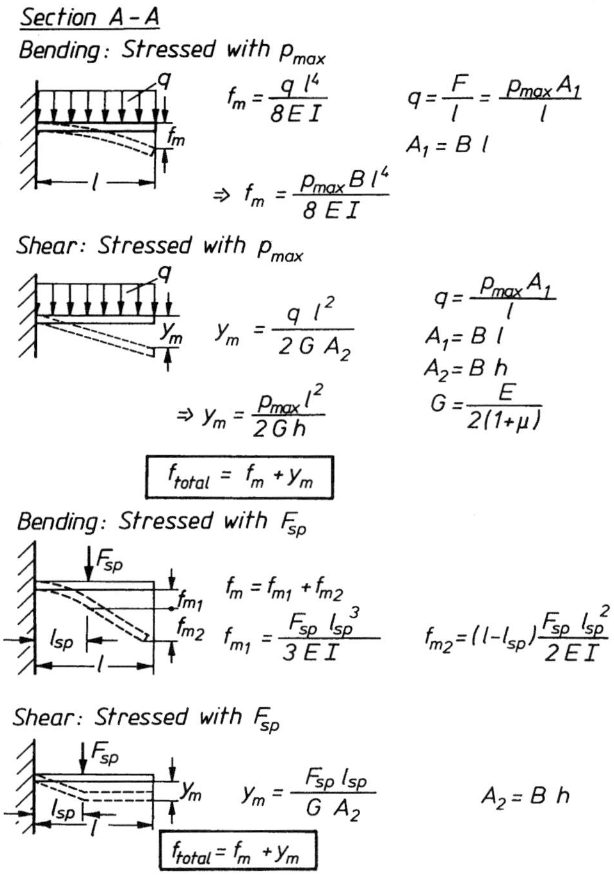


**Figure 9.15** Determining the dimensions of a flat slit die (considered as a cantilever beam)

(c) Method of calculating die deformations (superimposing deformations in sections A–A and B–B).

If a section is taken in the plane B–B of the slit die, as shown in Fig. 9.14, and if it is assumed that the die is not clamped together by bolts, the deformation of this resulting new cantilever beam, which is rigidly clamped at one end and free to move at the other, can be calculated from the equations given in Fig. 9.11. As shown in [12] for an injection mold, the total deformations from the two cases of loading under consideration (sections A–A and B–B), assuming the same load in each case, can be superimposed.

The equation below then applies:

$$\frac{1}{f} = \frac{1}{f_{\text{total}_{A-A}}} + \frac{1}{f_{\text{total}_{B-B}}}. \tag{9.38}$$

If the superimposition principle of Equation (9.38) is used, the total deformation $f$ is always smaller than the smallest individual deformation. This means that an analysis of the particular case that leads to the smallest deformation is sufficient to select the dimensions of the die safely.

Even a slit die should be designed with such dimensions that the maximum deformation at the orifice remains less than 0.05 mm.

(d) Selection of bolt sizes for a slit die.

The forces acting on the bolts holding the two halves of the slit die together can be determined by balancing the moments about the rear edge of the die (Fig. 9.16). This balancing calculation can be carried out in different sections similar to section A–A. The force $F_{sp}$ acting in the respective center of gravity $l_{sp}$ must then be recalculated. The analysis of the forces acting only in the middle region of the die assuming that they exist over the whole width of the die offers a reliable approach to selecting the size of the screws. The equations in Fig. 9.16 illustrate that the bolts should lie as close to the flow channel as possible. In addition, the distance between the flow channel and the rear edge of the die should be chosen to be relatively large.

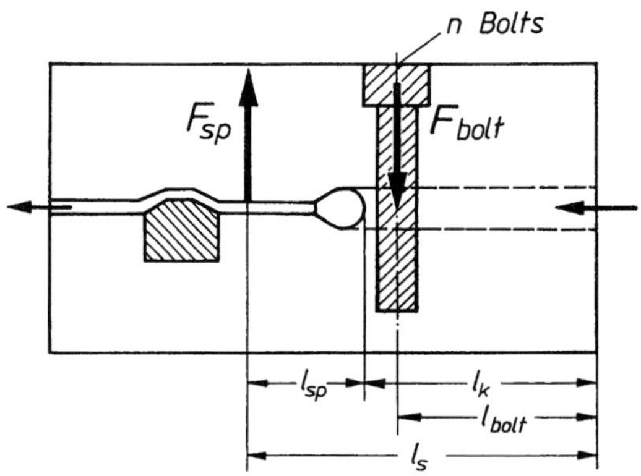


$$n \cdot F_{bolt} \cdot l_{bolt} = F_{sp} \cdot l_s$$

$$F_{bolt} = \frac{F_{sp} \cdot l_s}{n \cdot l_{bolt}}$$

$$F_{bolt} = \frac{F_{sp} \cdot (l_{sp} + l_k)}{n \cdot l_{bolt}}$$

**Figure 9.16** Calculating bolt forces for a flat slit die

## ■ 9.4 General Design Rules

Some basic principles must be considered in the design and configuration of extrusion dies. These are emphasized again here, independent of the comments in Chapters 3 through 8.

In summary, the following principles should be taken into consideration in the design of dies:

- A die should consist of as few individual parts as possible, in order to minimize the time needed for assembly and cleaning. In this connection, care must be taken to center the parts of the die accurately relative to each other [14]. The die is to be manufactured with close fits [15].
- A small number of individual parts also leads to few joints in the die body in the region of the flow channel. Not only is there a possibility that leaks will develop at these joints because of inadequate sealing, but there is also a risk that the material will degrade there. The number of joints should therefore always be kept as few as possible [16]. These joints should be placed in advantageous cross sections [15] to simplify the cleaning of the die.
- The sealing surfaces, since they cannot be avoided, should be as flat and small as possible, in order to ensure a uniform distribution of sealing forces over the entire sealing surface [15,16]. At this point, the surface pressure should be checked.
- The gap between stationary and moving parts of the die (e.g., between the choker bar and the die body in a slit die) can be sealed by inserting a gasket cord or an oversized packing strip in a groove (rectangular or semicircular) in the stationary part of the die. Soft metal (like aluminum) and temperature-resistant plastics (like PTFE) are available as materials for the sealing element [17].
- The die should be held together by a few heat-resistant bolts (instead of many small ones); the service life of larger diameter bolts is longer. The bolts should be easily accessible (e.g., from above for slit dies [18]) without having to dismantle the strip heaters [15].
- A hinged flange or a quick-connect seal should be provided for the die. Large dies are to be mounted on adjustable or movable supports [15].

In order to achieve a better distribution of load, the die orifice should be adjusted by several bolts, such as when centering a blown film die.

- The dimensions of the die body must be adequate for keeping pressure deformations within justifiable limits (see Sections 9.1 and 9.2). Since the body of the die is weakened by boreholes for screws, cartridge-type heaters, and pressure and temperature probes, these holes must be taken into consideration [18].
- The thermal expansion must be taken into consideration when individual parts of the die have different temperatures.

When establishing the configuration of the flow channel of a die, attention should be generally paid to the following items:

- When possible, the melt should be supplied to the die centrally [15].
- There must not be any dead spots or corners in the flow channels (sites of melt stagnation). This means that sharp, sudden transitions in cross section or changes in direction must be avoided. Furthermore, all radii in the channel must be greater than 3 mm [17].
- Zones with a large cross section, which thus are zones in which the residence time of the melt is long because of the low flow velocity, generally lead to degradation of thermally sensitive compounds, such as rigid PVC. For such materials, the principle of a minimum flow channel volume should be observed [16]. In slit dies, the necessary minimizing of the distribution channel (manifold and die land) on which the pressure is acting leads to dies that are short in the axial direction and that more or less fulfill this principle directly.
- Flow lines always lead to lower quality of extruded products. Their formation should be avoided or diminished and their number reduced by a proper design of the flow channel (see Section 5.3.1.1).
- Depending on the semifinished product produced and on the material processed, a parallel die land should be provided so that reversible deformations in the melt can relax at the end of the flow channel. It should be possible to control the temperature of this region separately.
- The surface of the flow channel should be polished or even honed [19] and, if necessary, chrome plated [19]. The surface roughness should be less than 0.2 μm [18]. The chrome plating reduces the tendency of the melt to adhere to the surface of the flow channel and therefore leads to a reduced residence time of the melt in the die and to rapid and simple cleaning of the die [16]. The chromium layer wears, however, and must be renewed after a certain operating time. Materials, such as polyesters and polyamides, which exert high tension on the die wall on cooling, may even detach this layer [18]. Therefore, corrosion-resistant steel is increasingly being used in extrusion dies (see Section 9.4). The flow channels of such dies are then only hardened and polished (honed and lapped) [16,18,19].

## 9.5 Materials for Extrusion Dies

The materials for extrusion dies have to meet the following requirements:

- Can be readily machined (by cutting or erosion)

- Are resistant to pressure, temperature, and wear
- Have sufficient strength and toughness
- Have sufficient surface hardness
- Can be readily polished to a satisfactory surface (without porosity)
- Respond adequately to a simple heat treatment
- Have minimum tendency to distortion and change in dimensions during heat treatment
- Are resistant to (corrosive) chemical attack
- Offer possibilities for surface treatment (e.g., chromium plating, nitriding)
- Have a good thermal conductivity
- Are stress-free

These requirements are not specific to extrusion and are mentioned in the literature pertaining to the processing of polymers [15,18,20–24], where they are more or less modified for each specific process. However, these requirements cannot be met by a single material. So, according to [20], the following questions must be asked when selecting the material:

- What type of compound will be processed (processing temperature range, corrosion, increased wear, possibly by the fillers present)?
- What is the nature and the magnitude of the mechanical stresses (see Sections 9.1 to 9.3)? The bending stresses present are of significant and crucial importance for the selection of the material. Because of their brittleness, fully hardened steels cannot be used for larger dies [23].
- By what process is the die manufactured? Machining methods are possible for materials with strength up to approximately 1500 N/mm$^2$ [22]; however, the most advantageous machining conditions exist at strengths of 600–800 N/mm$^2$.
- What heat treatment is required and does it tend to cause distortions or dimensional changes?

Besides the occasional few nonferrous materials, the following steels are used for the manufacture of extrusion dies [20–22,24]:

- Case-hardened steels
- Nitrided steels
- Fully hardened steels
- Quenched and tempered steels (to be used as supplied)
- Corrosion-resistant steels

### *Case-Hardened Steels*

Case-hardened steels have proven their value for the manufacture of dies because they are most likely to fulfill the necessary requirements. They are readily machined and, after case hardening or heat treatment, have a very wear-resistant hard surface with a tough core of high compression strength [20,22]. These steels have a very low carbon content (less than 0.2%). The carbon content can be enriched by carburization (up to a carbon content of about 0.8% with a carburizing depth of 0.6 to 2.0 mm), very wear-resistant carbides being formed. The regions of the die that are not to be hardened are covered during the process. The literature should be consulted for information about the relevant heat-treatment processes [25–27].

A list of case-hardened steels for plastics processing is given in [22]. Primarily, the following grades are used [24]: material nos. 1.2162 (21 MnCr 5), 1.274 (X19 NiCrMo 4), 1.2341 (X 6 CrMo 4).

#### *Nitrided Steels*

Nitrided steels are steels whose alloying additives—chromium, aluminum, molybdenum, and vanadium—form hard nitrides in the presence of nitrogen (e. g., as gas (ammonia) or in a cyanide bath) at temperatures between 480 and 580 °C and at case-hardening times of 10 to 100 hours [20–22]. The nitrides lead to a high surface hardness with a tough material core. In this process, there is no quenching after case hardening (nitriding), so no distortion is expected. Nitrided steels are also supplied in the stress-free, annealed state.

In nitriding, the greatest hardness is achieved at about 0.03 to 0.08 mm below the surface [20]. This represents an oversize that must be later worked off [22].

In the so-called ion-nitriding, on the other hand, the already finished products are hardened at the surface by a high voltage discharge in an atmosphere containing nitrogen at about 350 to 580 °C, during which the nitrogen diffuses into the surface. Apart from polishing the flow channel and sealing surfaces, no other finishing work is necessary. If very wear-resistant surfaces are to be attained without increasing the hardness of the surface greatly, the nitriding can be carried out in a salt bath (temperature approx. 500–550 °C, case-hardening time 10 minutes to 2 hours). This is called the soft nitriding or the Tenifer process [20,21], and it is used successfully today for the bearing surfaces of rotating parts, such as choker bar bolts in slit dies.

Nitrided steels are not completely corrosion resistant. To improve the corrosion resistance, large amounts of chromium have to be added.

The following are some of the nitride steels [22,24]: material nos. 1.2852 (33 AlCrMo 4), 1.2307 (29 CrMoV 9), 1.2851 (34 CrAl 16). The increase in hardness of these grades of steel is due to the formation of martensite by precipitous cooling after heating. The resulting mechanical properties depend to a large extent on the attainable rates of cooling [22].

#### *Fully Hardened Steels*

Fully hardened steels are very hard and show a good wear resistance. However, their toughness is small when compared to that of case-hardened as well as quenched and tempered steels. Because they are susceptible to crack formation and distortion, they are used only infrequently and then only for small dies [20,22], primarily when there are high compressive stresses [24]. Some fully hardened steels in polymer processing are [24] material nos. 1.2344 (X 40 CrMoV 51), 1.2367 (X 32 CrMoV 53), 1.2080 (X 210 Cr 12), 1.2379 (X 155 CrVMo 12.1), 1.2767 (X 45 NiCrMo 4), and 1.2842 (90 MnCrV 8). Additional particulars can be found in [22].

#### *Quenched and Tempered Steels*

Quenched and tempered steels, which are used as supplied, are employed whenever there is the possibility that a heat treatment will result in distortion and dimensional changes, such as in the case of large dies.

In the manufacturing process, these steels are annealed after being hardened. This results in a decrease in hardness and strength and an increase in toughness and elasticity. The relatively low strength after quenching and tempering (hardening and annealing) leads to surfaces of an inferior wear resistance and a lesser ability to be polished. This condition, however, can be improved by nitriding or chrome plating the surfaces [20].

Such quenched and tempered steels are, for example [24], material nos. 1.2312 (40 CrMnMoS 86), 1.2347 (X CrMoVS 51), and 1.2711 (54 NiMoV 6). Additional particulars are available in [22].

#### *Corrosion-Resistant Steels*

These are steels containing more than 12% chromium and are used whenever chemically corrosive materials (e.g., hydrochloric acid) are released during plastics processing and when it is not feasible, for technological reasons, to provide a chrome-plated or nickel-plated surface for the flow channel.

Because of the affinity of chromium for carbon (which must be present in small amounts so that the steel can be hardened), there is a danger that chromium carbide will be formed above 400 °C, taking chromium locally away from the area surrounding the carbon, leading to a negative change in the properties of the material and its rust resistance [20]. As a rule, these temperatures are not usually reached in extrusion.

Corrosion-resistant steels are, for example [22,24], material nos. 1.2083 (X 40 Cr13) and 1.2316 (X 36 CrMo17).

In addition to the nitriding, the following methods are used to increase the corrosion stability and wear resistance of extrusion dies:

- Protective chromium layers (hard chrome plating)
- Protective coatings of nickel or alloys with high nickel content
- Titanium carbide coatings
- Appropriate die inserts

*Hard chrome plating* produces corrosion-resistant, quite wear-resistant, highly polished surfaces, to which the melt adheres only to a slight extent and that are therefore cleaned easily [18,19]. The chromium layer is applied electrolytically (thickness 0.015–0.03 mm). The anode must be matched accurately to the contour of the flow channel in order to achieve a chromium layer of uniform thickness. An uneven thickness of the coat and sharp edges can lead to stresses in the layer, which then can easily chip or flake off. Therefore, the layer must in many cases be renewed periodically [19,20,22].

*Nickel coatings and nickel alloys* are sometimes used in dies for the extrusion of PVC; their high resistance to corrosion by hydrochloric acid is regarded as their main advantage [19,28]. However, it appears to be a problem that some lead stabilizers react chemically with nickel coatings, which can be detached completely by such a reaction, although the coatings are applied by a hard facing process to the part [19] and subsequently refinished. The coatings are relatively thick (approx. 1 mm), and several layers can be superimposed on each other. Repairs to dies can also be made using the hard-facing process.

In addition, dies made from alloys containing 95 % nickel are used for the processing of PVC. Nevertheless, the potential problem of the chemical reaction remains [19].

*Titanium carbide coatings,* which stand out because of their high wear resistance and increased corrosion resistance, are deposited on the surface of the die (at a thickness of 6–9 µm) in a reactor, in which gaseous chemical products are reacting. In this case, it is advisable for the die material to have an adequate hardness since the coating itself is very hard. Furthermore, it is necessary to pay attention to the die distortion since the reaction takes place above 900 °C [24]. The successful use of this method in wire-coating dies has been reported [29].

The wear problem of extrusion dies can also be solved by using *inserts of appropriate, wear-resistant material,* for example, hard metal or diamond inserts in cable-jacketing dies (see Section 5.3.2.4) in the exposed zones of the die.

So-called "bimetallic" slit dies are also used today. In these, the regions of the flow channels up to the exit of the die that are exposed to wear consist of a highly wear-resistant material [32].

If extremely high-grade surfaces are required in the flow channel (free of porosity), it is advisable to use steels that have been fused in the vacuum furnace and in which defects have been excluded because of the high degree of purity. Although these steels are very expensive, most sheet and flat film dies for rigid PVC are made from them today.

Because of their good thermal conductivity, dies made from aluminum [29,30] and copper-beryllium [31] are also occasionally used for the extrusion of profiles. The wear resistance of aluminum can be increased to a value above that of hard chrome plated steel by anodizing (anodic oxidation). Anodizing also improves the corrosion resistance, so it becomes adequate for PVC. At the same time, the tendency of the melt to adhere and therefore its residence time at the wall are reduced to values below those of hard chrome plating. Nevertheless, the very low strength and the high susceptibility to shock remain as distinct disadvantages of this material [28].

### Symbols and Abbreviations

| | |
|---|---|
| $A$ | Area of cross section |
| $A_S$ | Normalized share of the wire surface |
| $B_P$ | Calculation coefficient |
| $b$ | Width |
| $D$ | Diameter |
| $d_K$ | Diameter of the warp wires of a screen pack |
| $d_S$ | Diameter of the weft wire of a screen pack |
| $E$ | Elasticity modulus |
| $F$ | Force |
| $F_p$ | Pressure force |
| $F_{bolt}$ | Bolt force |
| $F_{sp}$ | Substitute force (force that acts in the pressure area center) |
| $F_Z$ | Viscous force |
| $f$ | Deflection |
| $f_{max}$ | Maximum deflection |
| $f_p$ | Deflection of the breaker plate |
| $f_s$ | Deflection of the screens |
| $G$ | Shear modulus |
| $H$ | Channel height |
| $h$ | Height |
| $h_S$ | Height of a screen pack |
| $I$ | Moment of inertia |
| $k$ | Ratio of inner and outer radius of an annular gap |
| $L$ | Length |
| $l$ | Half-width of the outlet of a flat slit die |

| | |
|---|---|
| $l_K$ | Distance of the manifold to the rear edge of the die |
| $l_{bolt}$ | Distance of the bolt to the rear edge of the die |
| $l_{sp}$ | Distance of the pressure area center to the manifold |
| $M_b$ | Bending moment |
| $m$ | Number (of pieces) |
| $N$ | Stiffness of the plate |
| $N_S$ | Stiffness of the screen pack |
| $N_0$ | Stiffness of the plate without holes |
| $n$ | Number (of pieces) |
| $p$ | Pressure |
| $p_A$ | Outlet pressure |
| $p_a$ | Ambient pressure |
| $p_E$ | Inlet pressure |
| $\Delta p$ | Pressure loss |
| $\Delta p_p$ | Pressure loss in a breaker plate |
| $\Delta p_s$ | Pressure loss in a screen |
| $\Delta p_{sa}$ | Additional pressure loss in a screen |
| $R$ | Radius |
| $R_A$ | Radius at outlet |
| $R_a$ | Outer radius |
| $R_E$ | Radius at inlet |
| $R_h$ | Radius of a hole |
| $R_i$ | Inner radius |
| $r$ | Radius |
| $S$ | Safety factor |
| $S_p$ | Safety factor of a breaker plate |
| $S_s$ | Safety factor of a screen |
| $T_{max}$ | Initial depth of a spiral (= maximum depth of a spiral) |
| $t$ | Hole spacing |
| $W$ | Resistance moment |
| $w$ | Mesh width |
| $\alpha$ | Attenuation coefficient |
| $\alpha_S$ | Attenuation coefficient for the screen |
| $\varepsilon$ | Elongation |
| $\mu$ | Coefficient of transverse contraction (Poisson's ratio m $= 1/\mu$) |
| $\nu$ | Transverse contraction ratio |
| $\sigma_b$ | Bending stress |
| $\sigma_c$ | Comparative stress |

$\sigma_t$ Tensile stress
$\sigma_p$ Permissible bending stress
$\tau_W$ Shear stress at the wall

**Abbreviations**

DLC Diamond-like carbon
PA Polyamide
PET Polyethyleneterepthalate
PTFE Polytetrafluorethylene
PVC Polyvinylchloride
TiN Titanium nitride

## 9.6 References of Chapter 9


1. Masberg, U.: Auslegen und Gestalten von Lochplatten. In: Filtrieren von Kunststoffschmelzen. VDI-Verl., Düsseldorf (1981)
2. Schwaigerer, S.: Festigkeitsberechnung im Dampfkessel-, Behälter- und Rohrleitungsbau. 4th ed. Springer, Berlin (1997)
3. Kreyenborg, U.: Ind. Anz. 106 (1973) 95, pp. 2513-2515
4. Wolmir, A. S.: Biegsame Platten und Schalen. VEB-Verlag für Bauwesen, Berlin (1962)
5. Formelsammlung. (Eds.): VGB Technische Vereinigung der Großkraftwerksbetreiber e. V. Lehrstuhl für Materialprüfung, Werkstoffkunde und Festigkeitslehre, University of Stuttgart (1981)
6. Czichos, H. (Ed.): Hütte – Die Grundlagen der Ingenieurwissenschaften., 3rd ed. Springer, Berlin, Heidelberg (2000)
7. Fehling, J.: Festigkeitslehre. VDI-Verl., Düsseldorf (1986)
8. Beitz, W. and Kütter, K.-H.: Dubbel Taschenbuch für den Maschinenbau. Springer, Berlin (1990)
9. Pritchatt, R. J.et al.: Design considerations in the development of extrudate wall thickness control in blow moulding. Conference: Engineering Design of Plastics Processing Machinery. Bradford, England (1974)
10. Parnaby, J. and Worth, R. A.: The variator mandrel forces encountered in blown moulding parison control systems: Computer-aided design. Mech. Eng. 188 (1974) 25, p. 357
11. Wortberg, J.: Werkzeugauslegung für Ein- und Mehrschichtextrusion. Thesis at the RWTH Aachen (1978)
12. Bangert, H.et al.: Spritzgerechtes Formteil und optimales Werkzeug. 9. IKV Colloquium, Aachen (1978)
13. Kempis, R. D.: Strukturmechanische Analyse von Rohrbündeltragplatten mittels FE- Methode. Thesis at the RWTH Aachen (1981)
14. Schiedrum, H. O.: Auslegung von Rohrwerkzeugen. Plastverarbeiter 25 (1974) 10, pp. 1–11
15. Hensen, F.: Anlagenbau in der Kunststofftechnik. Postdoctoral thesis at the RWTH Aachen (1974)

16. Kreft, L. and Doboczy, Z.: Die Verarbeitung thermoplastischer Kunststoffe auf Einschneckenpressen. Part XI: Spritzköpfe. Kunstst. Gummi 6 (1967) 2, pp. 50–56
17. Kauschke, M. and Michaeli, W.: Zur Konstruktion eines rechneransteuerbaren Breitschlitzwerkzeuges. Unpublished thesis at the IKV, Aachen (1975)
18. Predöhl, W.: Herstellung und Eigenschaften extrudierter Kunststoffolien. Postdoctoral thesis at the RWTH Aachen (1977)
19. Barney, J.: Design requirements for PVC film and sheet dies. Mod. Plast. 46 (1969) 12, pp. 116–122
20. Treml, F.: Stähle für die Kunststoffverarbeitung. Lecture at the IKV, Aachen, Nov. 15 (1967)
21. Seiwert, H.: Werkstoffe für Extruder und Extrusions-Werkzeuge. Kunststoffe 52 (1962) 9, pp. 552–553
22. Menges, G. and Mohren, P.: How to Make Injection Molds, 3rd ed., Hanser, München (2001)
23. Dember, G.: Stähle zum Herstellen von Werkzeugen für die Kunststoffverarbeitung. In: Kunststoff-Formenbau, Werkstoffe und Verarbeitungsverfahren. VDI-Verl., Düsseldorf (1976)
24. Höller, R.: Vergüten von Stahlwerkstoffen. In: Kunststoff-Formenbau, Werkstoffe und Verarbeitungsverfahren. VDI-Verl., Düsseldorf (1976)
25. Malmberg, W.: Glühen, Härten und Vergüten des Stahls, 7th ed. Springer, Berlin (2013)
26. Illgner, K. H.: Gesichtspunkte zur Auswahl von Vergütungs- und Einsatzstählen. Metalloberfläche 22 (1968) 11, pp. 321–330
27. DIN 17210: Einsatzstähle
28. Avery, D. H. and Csongor, D.: Wear in plastic extrusion machinery. SPE Tech. Pap. (1978) pp. 446–449
29. Verbesserte Wirtschaftlichkeit durch Oberflächentechnik. Technica 14 (1977) pp. 1034–1035
30. Zerkowski, G.: Werkzeugwerkstoff Aluminium. In: Kunststoff-Formen bau, Werkstoffe und Verarbeitungsverfahren. VDI-Verl., Düsseldorf (1976)
31. Beck, G.: Kupfer-Beryllium. In: Kunststoff-Formenbau, Werkstoffe und Verarbeitungsverfahren. VDI-Verl., Düsseldorf (1976)
32. Information from Johnson-Leesona Company

# 10 Handling, Cleaning, and Maintaining Extrusion Dies

In general, extrusion dies are expensive, high-precision parts of the extrusion line that require, besides well thought out, careful handling, a frequently underestimated amount of care. To avoid malfunctions of the die that can result in periods of unplanned and costly stoppage [1,2], it is necessary to operate the dies properly and to take preventive measures so that they are ready for use for as much time as possible.

Human errors, especially careless or wrong handling of the die during maintenance and cleaning or while adjusting it during operation, are the main causes of damage to the die. Therefore, it is necessary to provide intensive training to all personnel coming into contact with the die [1,2].

The operator must learn to understand the function of the die completely and to recognize how product quality is affected by changes in temperature and by adjustments of the orifice or the choker bar. Moreover, he must pay attention to sources of error and to boundary conditions.

The operator must also realize the importance of good contact between the thermocouple or strip heater and the die and determine whether and where malfunctions exist in the extrusion line. For example, when the extruder output pulsates, he must not continuously adjust the choker bar of a slit die, as this leads to increased wear. (As a rule, such systems are not conceived or designed for such a purpose.) He must also be able to interpret the pressure measured at the die inlet from the point of view of specified critical limits. Furthermore, he must learn not to make "sharp" adjustments between only two adjustment points at the die orifice or the choker bar, but to make appropriate changes also in the neighboring regions. Bolts are broken off, threads are destroyed, and choker bars are snapped if this rule is disregarded.

It should also be noted that melt at the die orifice should be removed only with spatulas made from a soft material (copper, brass, soft aluminum) in order to avoid notching or denting the edges of the die orifice. Hence, the gap width of a die exit should be measured only with a soft feeler gauge made, for example, from brass.

The slightest markings at the die orifice lead to striations in the extruded product, resulting in a lower quality.

Whenever it is necessary to change dies frequently, as is the case particularly when extruding flat films and sheets, it is advisable to employ special personnel. These persons should be familiar with the construction and function of each die and have the necessary descriptive literature (drawings, assembly instructions, parts lists) available. It is wise to have such persons familiarize themselves with their new die by disassembling, inspecting, and assembling it before it is used [1].

Dies should be disassembled, cleaned, serviced, and maintained at a special designated place, sufficiently far removed from rough production. Such a work area should be very clean and lined with corrugated cardboard. A small crane should be available in this work area. With a manually operated block and tackle, there frequently is the danger that the die will be damaged by the freely swinging ends of the chain. The use of hemp ropes should be considered in order to eliminate this source of damage to the die.

In the work area, there should be tools (screwdrivers, torque wrenches), soft scrapers (brass, copper, soft aluminum), cleaning and polishing materials, as well as possibly a die preheating station, which can considerably shorten the time required for changing dies and is useful for checking the heater elements. Such a preheating station consists of an appropriate number of electrical leads for the heating zones as well as one control system, which usually is sufficient. Individual controls for the die heater generally are not necessary here [1,2], but this depends on the size of the die.

If a die must be cleaned completely, the main bolts of the die should be loosened, but not removed, while the die is still attached to the extruder. Only those bolts associated with exposing the whole of the flow channel should be loosened.

Dies should be disassembled only in the heated state. It is necessary to work quickly in order to avoid premature cooling or scorch (incipient vulcanization). Rubber compounds can be usually removed from the walls of the flow channel quite easily. A thermoplastic resin in the flow channel can be removed either by pulling it away from the surface of the die (by blowing compressed air against the point of contact between the resin and the hot die) or by removing it with a soft scraper. Any resin remaining can be removed with brass or copper wool–but never with a steel object! Solvents or household detergents sometimes also help; however, their action should always be tested first at a noncritical place on the die.

When electric drills with brass brushes or the like are used, there is a risk that the chromium layer will be stripped from the surface of the flow channel or that the edges will be unintentionally rounded off slightly. Sealing surfaces of the die can be cleaned with fine, flat grindstones or with a very fine emery cloth [1,2].

In addition to manual methods of cleaning, it is possible to use salt baths (Kolene bath), in which a hot nitrate salt melt (approx. 400–500 °C) oxidatively destroys the polymeric melt residues and carries them away. However, this process cannot be used with chromium-plated surfaces [3,4]. Cleaning can also be carried out by melt evaporation in a fluidized bed of aluminum particles, heated to about 550 °C. Vacuum pyrolysis ovens are used with an increased frequency. During their operation, large remainders of material drip off into a cooled pit in which they can solidify, while the other residues are vaporized and removed into a waste water stream. In this process, it is important to find out if the use of high temperatures causes changes in the construction material of the die, leading to the reduction of its strength [3]. Also, it is important to consider the local laws pertaining to the removal of fumes and wastewater. In addition to the above, slow and gentle melt removal is possible by a boiling solvent [3]. Spinnerets are often cleaned by ultrasound [5]. It is always advisable to consult the die manufacturer for the correct cleaning method.

Before assembly, the flow channel should be polished to a high gloss. This operation should be followed by an extensive examination of the surface of the flow channel, in which the chromium layer should be checked, especially when processing PVC. At the same time, smaller, fine scratches can be removed. If the damage is more extensive, its effect on the extrusion result should be checked. More serious damage can be eliminated, for example, by buildup welding followed by machining [1,2,6].

Before the die is assembled, it is advisable to coat the flow channel with a thin film of a silicone grease; this is done to protect the flow channel during long periods of storage and to aid the passage of the resin melt, which is still relatively cold, when the die is started up [1,6].

Since it is possible that the surfaces of the flow channel may touch each other during assembly, it is advisable to place a sheet of paper or a plastic film between them [1].

During assembly, all bolts, bolt bearing surfaces, and threads must be provided with a high-temperature grease, such as molybdenum or graphite, in order to ensure ease of turning during the operation of the die and when the die is disassembled at some later time. If a die is not disassembled for a considerable time, it is advisable to check and, if necessary, remove individual bolts and grease them [2].

Bolts should be tightened after the die has been flanged to the extruder and has reached the operating temperature once again. The instructions of the die manufacturers as to the torque to be applied should be followed. Maintenance personnel should also check that there is proper contact between the heaters or thermocouples and the die. Moreover, the thermocouples must be calibrated.

It is recommended in [1] that the die be disassembled completely, cleaned, and checked approximately every six months, and all parts of the die (screws, bolts, cartridge heaters, electrical leads), which might malfunction, replaced. The time interval for such maintenance, however, depends on the material being processed. In some plants processing PVC, for example, the weekend after five production days is used for cleaning and maintaining the dies. In many cases, especially when extruding flat film and sheets, it is advantageous to keep a second, cleaned and preheated die ready for use in order to reduce production interruptions to a minimum when changing over to a new product or when repairs to the die necessitate disassembling it [1,2].

## 10.1 References of Chapter 10


1. Flanagan, J. L.: The maintenance of flat film and sheet dies. Plastics Machinery & Equipment, Aug. (1974)
2. Stone, H. N.: Dollars and sense of extrusion die maintenance. SPE J. 24 (1968) 1, pp. 57–58
3. Kress, G.: Auslegung von Schlauchköpfen, Kühl- und Flachlegevorrichtungen. In: Extrudieren von Schlauchfolie. VDI-Verl., Düsseldorf (1973)
4. Predöhl, W.: Herstellung und Eigenschaften extrudierter Kunststoffolien. Postdoctoral thesis at RWTH Aachen (1977)
5. Gollmick, H. J.: Unhörbarer Schall löst schwierige Reinigungsaufgaben. Chemiefasern (1969) 4, pp. 275–280
6. Doyle, R.: Clean out that die for smoother extrusion. Plast. Technol. 11 (1965) 12, p. 39

# 11 Calibration of Pipes and Profiles

For the extrusion of profiles, pipes, and similar semifinished products, the calibration and cooling sections are connected to the die that has shaped the melt (Fig. 11.1). The goal is to solidify the melt by the contact with the calibration (sizing) die in the calibration section to a degree sufficient for transferring the take-off forces and still maintaining the desired dimensions. The mean temperature $\bar{T}$ of the profile should drop below the solidification temperature (melting point) $T_E$ as it passes through the cooling section, so that the remelting of the already solidified layers is avoided. The whole profile, as it reaches the saw or the flying knife, should have a temperature below the melting point [1].

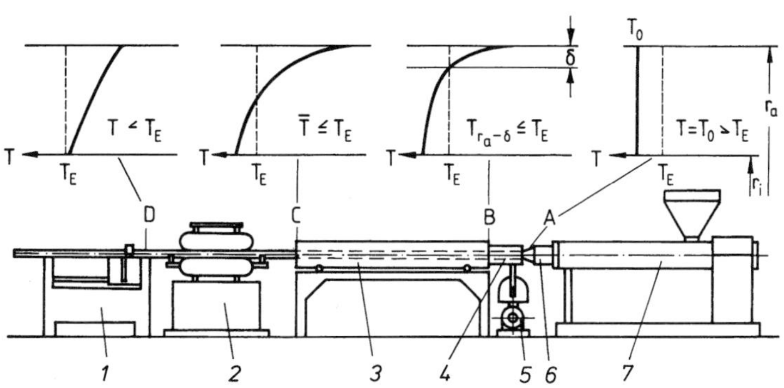


**Figure 11.1** Temperature profiles in the wall of a plastic pipe during extrusion [1], *1* Cutter (saw), *2* Take-up unit, *3* Water bath, *4* Calibrator (sizing), *5* Vacuum pump, *6* Die, *7* Extruder; $T_E$ Temperature of solidification

The calibration and cooling equipment, therefore, has the task of fixing the dimensions of the extrudate and, consequently, it represents an integral part of an extrusion die. Today, extruded plastic profiles are almost always calibrated. Only very simple profiles and profiles from plasticized PVC are occasionally not sized [2,3], but placed onto conveyor belts, where they are cooled, for example by spraying. The calibration practically always means that the extrudate is pulled through one or more so-called calibrators, made predominantly from metal (brass, steel, aluminum). The calibrator surface, which is in contact with the extrudate, is fitted to its shape. There are essentially two methods of the heat dissipation in this process: dry and wet calibration.

In the dry calibration the extrudate does not have direct contact with the cooling medium, but the heat contained in the melt is led away strictly by the contact with the metal surface of the calibrator to the cooling channels and thus from their surface to the cooling medium.

In the wet calibration the heat from the melt is transferred at least partially, depending on the method, directly by convection to the coolant. The remaining part of the heat is taken away by contact with the calibrator (or by a film of coolant on its walls).

The intensity of the indirect cooling by the calibrator is affected definitely by the thermal contact between the surfaces of the extrudate and the calibrator.

In connection with the above, a sufficiently large normal force between the surfaces is important, especially in the dry calibration. This normal force causes friction forces during the draw of the extrudate through the calibrator, which are spread along the entire length of the calibrator and are transferred onto the extrudate and have to be overcome by the take-up device. The maximum permissible force at the take-up is determined by the load capacity of the profile, which in turn depends on its cross sectional area and the temperature distribution in the extrudate.

Particularly in the inlet to the calibration section, where the temperature of the extrudate is still relatively high and the load capacity of the profile is still relatively low (because only a small portion of it is solidified), the take-up forces (still relatively low) lead to an increase in the longitudinal stretch of the strand (marked with a + in Fig. 11.2). These deformations cannot fully relax because of the subsequent cooling and therefore become frozen-in. This is often referred to as frozen-in elongations. They ultimately affect the shrinkage of the extrudate to a very high degree.

Because of the longitudinal stretch and the thermal shrinkage, the dimensions of the extrudate in the direction across the direction of the take-up are reduced (marked with a – in Fig. 11.2). This generally leads to a reduction of the normal forces in the calibrator and consequently to the reduction of friction and thermal contact.

The reduction in the dimensions in the transverse direction (wall thickness, dimensions of the contour) is often prevented by a corresponding over-sizing in the cooling section, thus obtaining a product with proper dimensions.

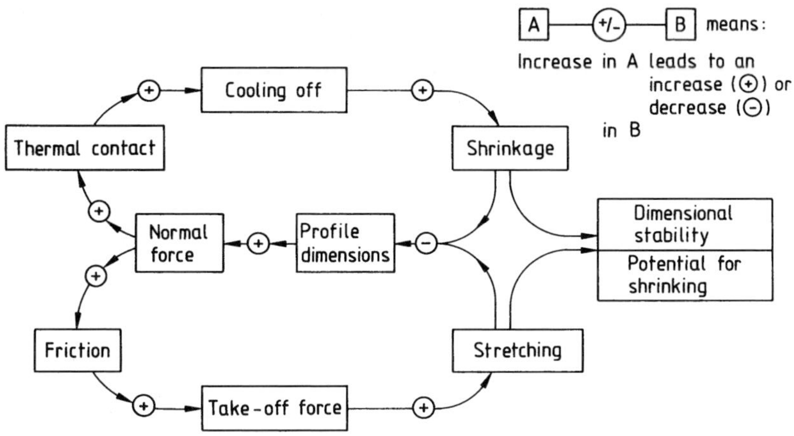


**Figure 11.2** Interdependencies in calibration

The calibration line can also be tapered (long calibrators) or can consist of several calibrators in series with gradually changing dimensions to match the dimensions of the extrudate, which is changing with the progressive cooling. Because the take-up forces generally increase with the lengthening of the calibration line (due to increasing friction of the surfaces), the maximum length of the calibration line is given by the loading capacity of the extrudate or by the maximum acceptable stretch of the extrudate.

Even when the cooling is very intensive, the extrudate must be kept in the calibrator until its internal stability is high enough to resist external stresses (gravity, take-up force) and internal stresses without an excessive change in dimensions. Finally, the minimum calibration time obtained from the above along with the maximum length of the calibration section define the attainable line speed.

In order to achieve as high line speeds as possible, the calibration section then should cool

- with a minimum friction
- as intensively as possible.

Calibration has to be adjusted to the special requirements pertaining to the shape of the extrudate and the material used. The particulars of the calibration methods used today are discussed in the following sections.

## 11.1 Types and Applications

### 11.1.1 Friction Calibration

Simple open profiles are calibrated on charts with the profile being pulled through cooled platens (Fig. 11.3) that more or less rest on it (the platen load can be reduced by springs or counterweights). Some profile stretching is unavoidable with this method of sizing.

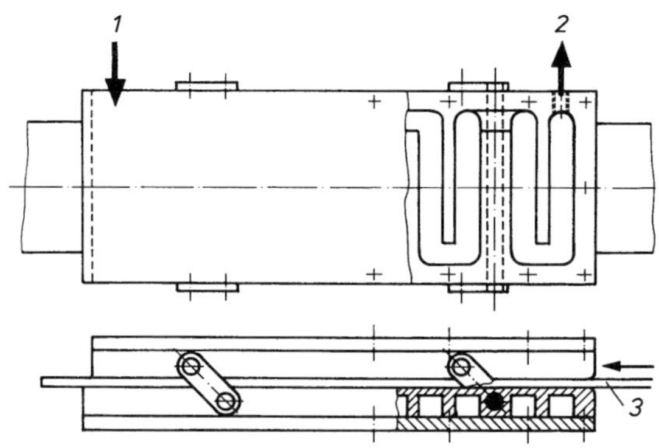


**Figure 11.3** Friction calibration [4,5], *1* Cooling water inlet, *2* Cooling water outlet

The outer contour of the profile is worked into the platens. For complicated profiles with undercuts, it may be necessary to subdivide the chutes several times in the longitudinal direction, so that each section can be folded separately [2–5].

The upper parts of the calibration equipment, lying on the profile, are generally subdivided several times into individual sizing sections.

The chutes are partially coated with PTFE dispersions which, however, wear and therefore must be frequently renewed [2,3]. With this, as with all other calibration processes, the line speeds that can be reached depend largely on the geometry of the profiles. Approximate speeds for friction calibration are 3 to 4.5 m/min for wall thickness of 1 mm and 0.5 to 0.7 m/min for wall thickness of 4 mm [2,3]. As shown in Fig. 11.3, the chutes should be cooled with water flowing against the direction of extrusion [4]. An open profile can be drawn only on one side of the friction calibration equipment. In order to avoid possible distortion of the profile, the side away from the calibration is then cooled with air. Water cooling may possibly be too intensive in this case [3].

### 11.1.2 External Calibration with Compressed Air

External calibration generally means that the external dimensions of the extruded profile are fixed by the calibration equipment. It has become universally accepted because a number of processing advantages. This is particularly true of mass-produced products, such as plastic pipes, which are standardized on the basis of the outside diameter.

External calibration with compressed air, also referred to as compressed air sizing, is used only for pipes, mainly those made from PVC with a diameter greater than 355 mm and for polyolefin pipes with a diameter greater than 90 to 110 mm [6,7]. In this process (Fig. 11.4) contact between the calibration die (also referred to as sizing sleeve) and the extrudate is established by an air overpressure (approx. 0.2–1 bar). For this purpose, compressed air is introduced into the pipe by way of the mandrel of the pipe die. The ends of smaller pipes and hoses, which can be coiled up, are closed off. Larger pipes are sealed off with floating plugs that consist of a series of circular rubber seals, and to some extent guided by rollers which are forced against the interior wall of the pipe by springs [6,7]. The floating plug is attached by a rope or chain to the mandrel, making sure that the rope or the chain do not drag along the inner wall of the pipe. The maximum take-off force must be taken into consideration in determining the dimensions of the floating plug holder. Furthermore, this possible stress should be taken into account in the mechanical design of the die (see Chapter 9).

With this calibration process, the equipment is well centered and flanged as directly as possible to the pipe die to prevent the internal pressure to expand or tear the pipe open. However, provisions must be made for effectively keeping the heat of the hot pipe die separate from the cold sizing system. This can be accomplished by an air gap and a small contact area between these two parts of the line.

It is advisable to install an air pressure controlling system in order to compensate for air losses through leaks, for example, due to a worn floating plug [6,7].

The cooling of the sizing sleeve and the pipe is done in the connecting cooling section by circulating water (see Fig. 11.4), by trickling water or by a water spray, with the latter being used increasingly for large pipes.

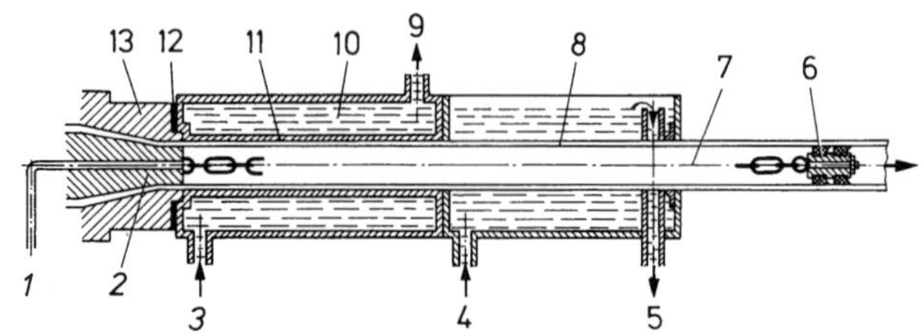


**Figure 11.4** External calibration with compressed air (overpressure calibration) [6,7], *1* Compressed air supply, *2* Mandrel, *3* Water inlet, *4* Water inlet, *5* Water outlet, *6* Floating plug, *7* Chain, *8* Plastic pipe, *9* Water outlet, *10* Cooling water, *11* Calibration (calibration die), *12* Heat separation, *13* Extrusion die

### 11.1.3 External Calibration with Vacuum

In external calibration with vacuum, the contact between the calibrator system and the extruded profile necessary for its cooling and sizing is achieved by applying a vacuum to the calibration equipment. In this process, the equipment can surround the profile completely (closed external sizing), the vacuum being applied through fine boreholes or slits in the surrounding wall (Fig. 11.5). A different sizing system is shown in Fig. 11.6 in which the profile is pulled through plates, between which vacuum is applied (plate sizing). The main advantage of this process is that there is no floating plug in the interior of the profile. It is only necessary to maintain atmospheric pressure inside the profile. For this purpose, boreholes for equalizing the air pressure are arranged in the mandrel or the cores of the profile die.

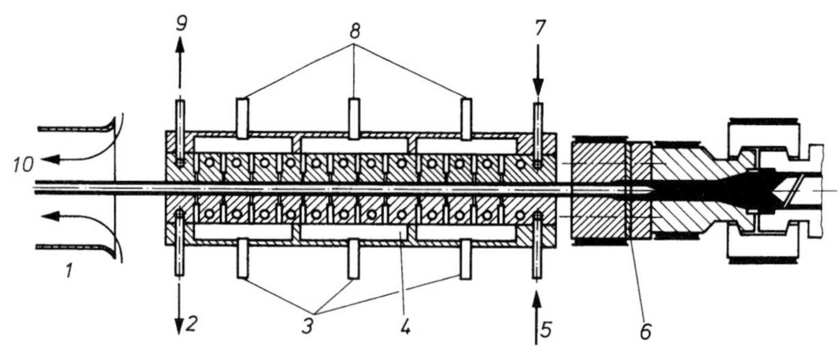


**Figure 11.5** External calibration with vacuum (closed vacuum with an air cooling channel). *1* Air cooling channel, *2* Water outlet, *3* Vacuum, *4* Vacuum calibration die, *5* Water inlet, *6* Profile die, *7* Water inlet, *8* Vacuum, *9* Water outlet, *10* Cooling air

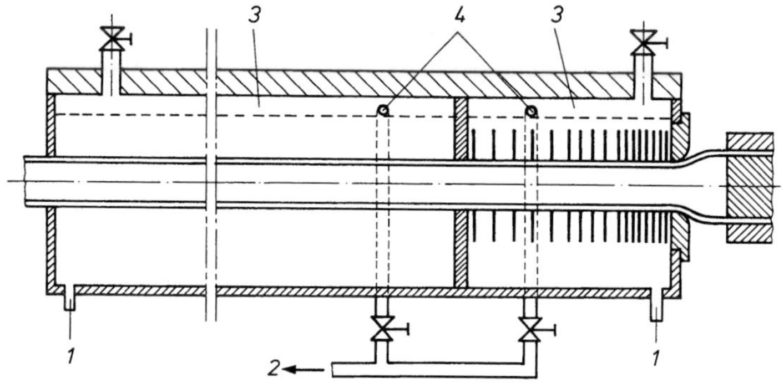


**Figure 11.6** External calibration with vacuum plate sizing, Gatto system [12], *1* Cooling water inlet, *2* To vacuum pump, *3* Vacuum, *4* Cooling water outlet

Closed external sizing (Fig. 11.5) is used for hollow sections and small pipes. For profiles, generally several sizing devices, in between which are cooling sections, are arranged in series (e.g., window profiles, 3 calibration units, each approximately 400–450 mm long), the dimensions of the respective calibration units being matched to the shrinkage (volume contraction of the cooling profile (Fig. 11.7)) [8–10]. With such an arrangement an optimum contact for cooling is achieved between the sizing system and the profile. The profile is drawn into the first sizing section. In doing so, it is deformed plastically and laid along the wall (this underdraft may amount from 5 to 30%) [2,3] and the desired profile contour, corresponding to the cross section of the sizing equipment, is formed. Lugs and bosses which, for example, are intended to form the grooves in the profile, are then left out in the later sections of the equipment in order to reduce the risk of jamming (Fig. 11.7) [8,10].

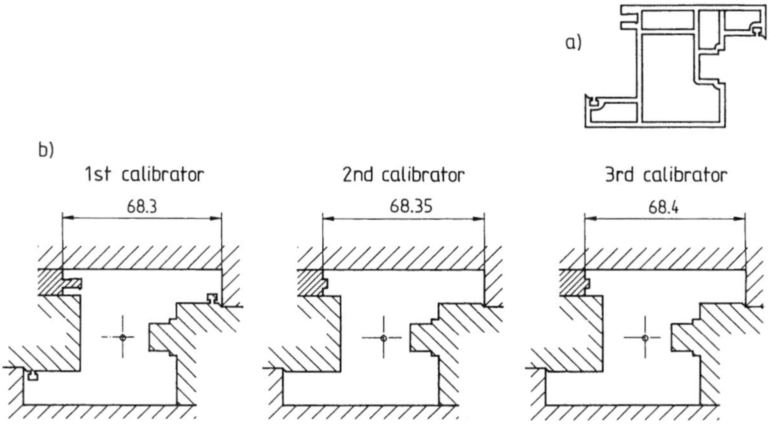


**Figure 11.7** Window profile and calibration [8], a) Deviation of the die contour from the profile contour, b) Channel cross sections of the three consecutive calibrators

With complicated profiles, it is not always possible to apply a uniform vacuum over the whole extent of the profile, so that the calibration is carried out much like friction calibration with more or less pronounced vacuum support. In order to prevent jamming of the external profile fins, the grooves in which they are cooled and partially sized are deliberately kept 10 to 30 % too high in some places. The profile may be ripped because of high friction forces; therefore, for safety reasons the vacuum in the system is frequently reduced by a so-called sniffing valve [2,3].

Figure 11.8 shows the quite involved arrangement of cooling channels in calibration equipment for the window profile illustrated in Fig. 11.7. For larger profiles that are open at the side, cooling webs can be inserted in the calibration system in order to improve the cooling effect (Fig. 11.9) [4]. In that case, it is advisable to change the exit cross section of the extrusion die as shown in Fig. 11.9 in order to make it easier to start-up the extrusion plant and to insert the cooling web [4].

Because of the need for high thermal conductivity, brass and copper beryllium are used as the construction material of such calibration sleeves. As a rule, the surfaces are hard chrome-plated to reduce wear.

Very smooth calibration surfaces, slightly roughened in part, are used for PVC profiles, while rough surfaces have proven their value for polyolefin pipes [3].

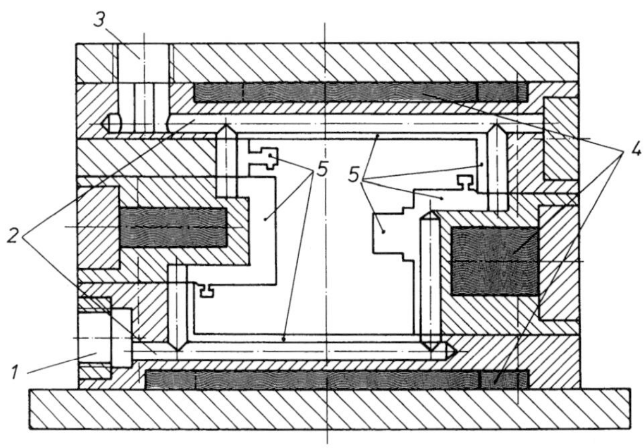


**Figure 11.8** Section through the sizing equipment for window profiles [8], *1* Vacuum connection (lower part), *2* Vacuum channels, *3* Vacuum connection (upper part), *4* Cooling channels, *5* Vacuum slits

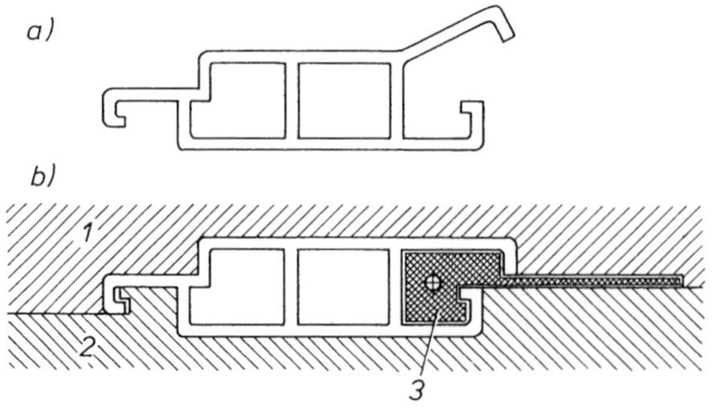


**Figure 11.9** Calibration with a cooling web, a) Construction of the cross section of the die outlet, b) Calibration with a cooling web [4], *1* Upper half, *2* Lower half of the calibrator, *3* Cooling web

Figure 11.10 shows a calibration die, that has been built using standardized components [11]. The upper part of such sizing equipment generally can be taken off in order to make start-up easier. Figure 11.11 shows a short calibration system with vacuum for simple profiles. The profiles are then cooled with a water bath, spray, or by air.

The calibration speed always depends on the profile to be sized: its shape and wall thickness as well as the material from which it is made. For closed external calibration, it is of the order of 4–5 m/min [5].

The plate sizing discussed previously (Fig. 11.6) represents one way of reducing the friction. In this process, the profile (most frequently a tube) is drawn-in much the same way as metal wires are drawn-through a series of drawing plates, arranged consecutively in a closed water bath or with interposed spray cooling [12]. The profile is in direct contact with the cooling water.

By applying vacuum by means of a suction pump (approximately 50 to 200 cm of water for pipe extrusion) [3], the plastic material running into a closed water bath can be made to lie closely against the plates. This process is therefore referred to as vacuum tank sizing. The system can be sealed at the first plate against the environment by means of an appropriate underdraft, which may be as high as 30% for hollow sections [2,3]. Friction is reduced further by a film of water between the plate and the profile. Wetting agents and lubricants can have a supporting function [13].

At the entrance to the calibration section, the plates of brass or aluminum, which are 5 to 8 mm thick and have an inlet angle of 30–45° [2,3] with a slightly decreasing internal diameter, are arranged closely together. Once the extrudate has solidified to a sufficient thickness, the axially displaceable plates are arranged increasingly further apart. The plates may also be replaced by a perforated sizing sleeve [2,3].

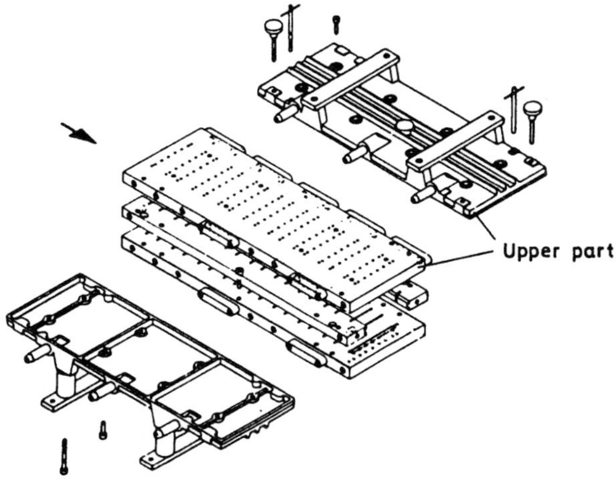


**Figure 11.10** Vacuum calibration die [11]

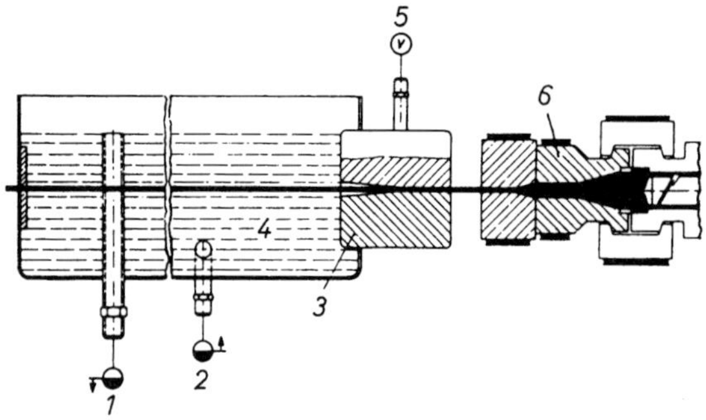


**Figure 11.11** Short calibrator with water bath [11], *1* Water outlet, *2* Water inlet, *3* Short calibrator, *4* Water bath, *5* Vacuum, *6* Profile extrusion die

In vacuum tank calibration, openings in the plates (arranged around the central sizing borehole) improve the circulation of the cooling water, which leads to considerably improved cooling [7,14].

The calibration equipment must be axially adjustable, in order to be able to produce the extremely important seal at the profile inlet for the two external sizing processes mentioned here. The distance between the die and the calibration equipment is between 10 and 100 mm [12].

The exit of the profile from the vacuum tank is sealed with a rubber gasket. Speeds of up to about 10 m/min are reached with the plate sizing [5].

Pipes with external diameters up to 630 mm are already being produced [15] by a modified plate sizing process with spray cooling between sets of plates under vacuum [12].

### 11.1.4 Internal Calibration

As the name implies, internal sizing fixes the internal dimensions of extruded hollow sections of simple cross sectional shape (Fig. 11.12). This process finds little application for extruding pipes because the external diameter of pipes is usually the basis for the classification and standardization. Pipes for pneumatic tube conveyor systems with close tolerances for the diameter represent an application of this calibration system and, at the same time, an exception [7].

In this process, the mandrel of the extrusion die is joined to a calibration mandrel that can be cooled and over which the molten tube, extruded from the die, is pulled and cooled at the same time. (There can be additional cooling from the outside from an air spray or a water bath [13].) Dies in which the melt is deflected by 90 degrees (e.g., side-fed dies) or offset dies are frequently used in this process, in order to be able to hold the sizing mandrel and the sizing cooling system securely.

By a simple profiling of the calibration mandrel, it is possible (as shown in Fig. 11.12) to convert a round tube, generally produced with a simple tube die, into a simple profiled shape (here a street marking post). This process is considerably less expensive than processes using profile dies with an appropriate exit cross section [2,3,16]. Making calibration mandrels entirely from PTFE blocks without using any cooling system has also been attempted [6].

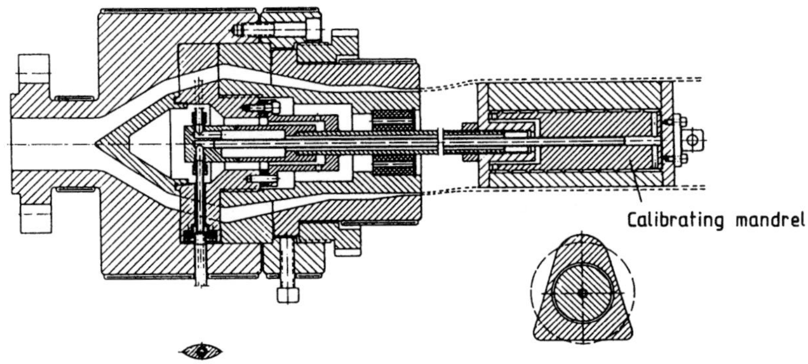


**Figure 11.12** Internal calibration [2,3,16]

If the material cools down on the calibration mandrel, considerable take-up forces may be required when using internal sizing because of the shrinkage of the material. In order to be able to introduce these take-up forces in a thin-walled profile without deforming it, roller cores are occasionally used; these are held by a rod on the sizing mandrel in the interior of the profile at the point of attack of the contact force of the take-up [3,6].

### 11.1.5 Precision Extrusion Pullforming (the Technoform Process)

In the precision extrusion pullforming process, also referred to as the Technoform process, a slight accumulation of material is sought between the extrusion die and the calibration equipment. The profile is then drawn from this accumulation–in much the same manner as in metal drawing processes–in a short calibration equipment which is strongly cooled. Further cooling takes place in a connected water bath. In the calibration equipment, a film of water may be forced between the wall of the profile and the wall of the sizing die for cooling purposes and to reduce friction. The magnitude of the melt accumulation is scanned with a sensor and kept constant by controlling the take-up speed [17,18].

Low take-up forces due to the short calibration line, high calibration speeds, and the fact that the profile dies do not have to be optimized as to their flow behavior, are advantages of this method [19].

### 11.1.6 Special Process with Movable Calibrators

If the calibrators move continuously with the extrudate, profiles with a changing cross section (in the direction of the take-up) can be produced.

Thus method is used, for example, to produce corrugated pipes (Fig. 11.13). Caterpillar take-up with special, semicircular grooved metal dies (molding chain) are used in which the molten plastic tubes are molded by pressure or by vacuum.

In order to prevent a premature cool-down or expansion (when using internal pressure) of the extrudate, it must be directed by a special heated outer die ring close to the grip of the molding chain. Since the nozzle must be very "slim" because of the proximity of the molding chain, it cannot be heated along its entire length. Therefore, it must be made from a material with exceptionally good heat conductivity (e.g., copper-beryllium alloy) [3].

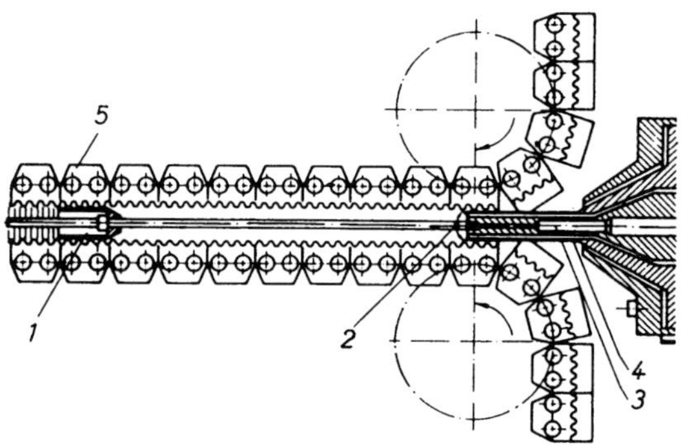


**Figure 11.13** The manufacture of corrugated pipes [4], *1* Floating plug, *2* Compressed air outlet, *3* Mandrel, *4* Die, *5* Chain with corrugated internal surface

Closed hollow bodies are also made continuously by this method, as shown in [20] on an example of a flat element for a heat exchanger.

## 11.2 Thermal Design of Calibration Lines

The profitability of a manufacturing process is closely related to the quantity of products produced. With regard to an extrusion process, this means that the output achievable is one of the factors having a decisive effect on the economic success. However, when profiles are extruded, the output achievable is usually not determined by the extruder, but by the output of the calibration and cooling equipment, which follows the extruder. The calibration and cooling equipment are not allowed to exceed a certain length as was pointed out earlier. The reason for that is that the friction forces between the profile and the calibration equipment must not become too large; it also is important to assure that the profile is introduced into the sizing equipment rapidly and in a simple manner [3].

When designing the calibration equipment, the required length of the calibration die (sizing equipment) is of interest. In calculating the length of the calibration die, it is wise to start out from the assumption, shown in Fig. 11.1, that only a thin outer layer of the extruded semifinished product must be cooled to a temperature below the softening temperature in order to ensure that the profile retains its shape and dimensions [1,21–23].

This required frozen layer $d_E$ must be sufficiently thick after leaving the calibration equipment to be able to withstand, for example, the take-up forces, or, when sizing externally with compressed air, the tangential forces due to the pressure in the interior of the profile.

It therefore appears to be reasonable to analyze the forces acting on the profile and to determine the cross-sectional area of the solidified profile or the thickness of the solidified layer required for the transfer of forces on the basis of the strength of the extruded material below the solidification temperature. This is shown in [21,22] for external sizing with compressed air. In the same article, the forces acting on the profile are discussed from the point of view of a possible deformation of the profile.

The following forces can act on the profile: the calibration pressure, the weight of the profile, the buoyancy in a cooling bath, the take-up force, frictional forces along the wall of the sizing die, as well as the compressive forces due to hydrostatic pressure in a cooling medium [21,22].

The forces acting on the extrudate can be divided as follows:

- Longitudinal forces caused mainly by the frictional forces between the extrudate and its outer surface area, which add up to the total take-up force across the length of the cooling section, and
- Transverse forces caused by pressure differences between the hollow chambers (cavities) in the extrudate and its outside surface (in pressure and vacuum calibrations) or by the effect of gravity (own weight, buoyancy in the water bath).

The transverse forces alone absorbed directly either by the calibrator wall or by suitable support (rollers) so that they do not sum up over the length of the cooling section. However, they can affect the friction forces.

The friction force between the surfaces of the extrudate and the calibrator can be expressed as:

$$F_{\text{friction}} = F_{\text{normal}} \cdot \mu \tag{11.1}$$

$\mu$ is the friction coefficient typical for the pair of the materials in question and $F_{\text{normal}}$ the normal force between the surfaces.

The friction coefficient itself is temperature dependent (Fig. 11.14) [21], while it is clear that the lowest possible contact temperature (particularly at the entry into the calibrator) leads to a reduction in the friction force.

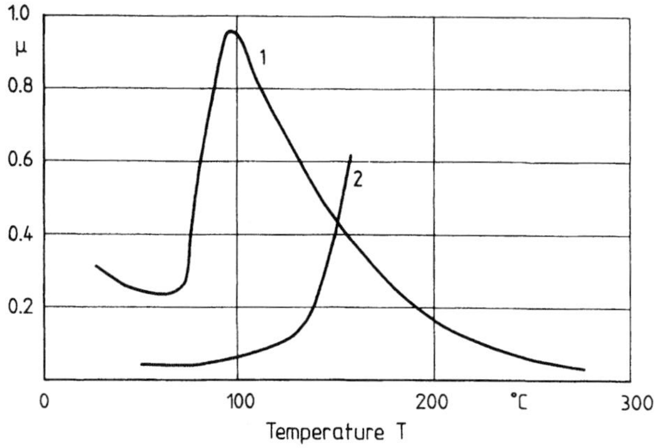


**Figure 11.14** Friction coefficients for different material combinations [21], *1* Polyethylene, *2* PVC, medium hardness

The known fact of the stickiness of polyolefins is manifested in an exceptionally high friction coefficient around 100 °C.

If the thickness of the frozen layer $d_E$ is determined or is given based on practical knowledge, the length of the calibration die necessary to achieve it can be estimated by several different methods.

### 11.2.1 Analytical Computational Model

The one-dimensional heat transfer in a plate-shaped body (Fig. 11.15) is described by the following equation

$$\frac{\partial T}{\partial t} = -\frac{\partial}{\partial x}\left(a\frac{\partial T}{\partial x}\right) \tag{11.2}$$

with $a$ being the thermal diffusivity.

The resistance to the heat transfer at the cooled surface of the said body is

$$\dot{q}'' = -\lambda \frac{\partial T}{\partial x}\bigg|_{x=D} = \alpha(T_0 - T_F) \tag{11.3}$$

with the following assumptions:

- temperature independent material data ($\lambda$, $\rho$, $c_p$)
- constant geometry and size (no shrinkage)

- constant boundary conditions ($\alpha$, $T_F$)
- a homogeneous distribution of the initial temperature ($T_M$ everywhere at t = 0)

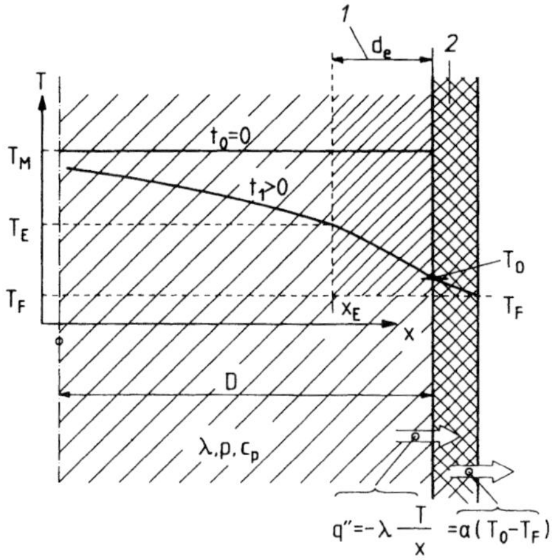


**Figure 11.15** Model for the cooling of an extrudate, *1* Thickness of the layer solidified at the time $t_1$, *2* Resistance to heat transfer, *T* Temperature, $T_M$ Temperature of the melt at the time $t_0$ = 0, $T_F$ Temperature of the cooling medium, $T_O$ Surface temperature, $T_E$ Solidification temperature, $\varrho$ Density, $c_p$ Specific heat capacity, $\lambda$ Thermal conductivity, $\alpha$ Heat transfer coefficient, $\dot{q}''$ Heat flow per area

The analytical solution to Equation (11.2) is given by [24]:

$$\frac{T(x,t)-T_F}{T_M-T_F}=\frac{2\sin\delta}{\delta+\sin\delta\cos\delta}\cdot e^{-\delta^2\frac{at}{D^2}}.\cos\left(\delta\frac{x}{D}\right), \tag{11.4}$$

$\delta$ can be obtained by iteration from the following function

$$\delta=\frac{\alpha D}{\lambda}\cot(\delta) \quad \text{with} \quad 0<\delta\leq\frac{\pi}{2} \tag{11.5}$$

- The dimensional quotient in Equation (11.5)

$$\frac{\alpha D}{\lambda}=Bi \tag{11.6}$$

is the Biot Number, which can be interpreted as the ratio between the internal ($D/\lambda$) and external ($1/\alpha$) resistance to the heat transfer.

- The dimensionless quotient in Equation (11.4)

$$\frac{a\cdot t}{D^2}=\frac{\lambda\cdot t}{\varrho\cdot c_p\cdot D^2}=Fo \tag{11.7}$$

is the Fourier Number which can be interpreted as a dimensionless cooling time

- The dimensionless quotient in Equation (11.4)

$$\frac{T(x,t)-T_F}{T_M-T_F}=\Theta(x,t) \tag{11.8}$$

is called the degree of cooling.

When observing the degree of cooling as a function of time on any location of the plate mentioned above, it is seen that when at the beginning ($t_0$ = 0), $T = T_M$, and consequently, $\Theta$ = 1, while for long cooling times, ($T = T_F$) approaches zero.

Setting the required degree of cooling of the solidified layer with

$$\Theta_E=\frac{T_E-T_F}{T_M-T_F} \tag{11.9}$$

as well as the required thickness of the solidified layer with

$$d_E=D-x_E \tag{11.10}$$

known in advance, the Equation (11.4) can be solved (after Equation (11.5) is inserted) with respect to the dimensionless cooling time $F_{0E}$ using Equations (11.6) and (11.8):

$$F_{0_E}=-\ln\left[\Theta_E\cdot\frac{\delta+\sin(\delta)\cdot\cos(\delta)}{2\cdot\sin(\delta)\cdot\cos\left(\delta\frac{D-d_E}{D}\right)}\right]\cdot\frac{1}{\delta^2} \tag{11.11}$$

with

$$\delta=Bi\cdot\cot(\delta)\quad 0<\delta\leq\frac{\pi}{2}. \tag{11.12}$$

In [21–23], different procedures to simplify Equation (11.11) are introduced and compared to each other. They are based on data derived from experience.

The procedure in [22] that yields the best results for the design of a cooling section for pipes described in it, leads to a diagram shown in Fig. 11.16. The dimensionless cooling time $Fo_E$ can be obtained from the predetermined thickness of the solidified layer and the degree of cooling and can be used for the calculation of the length of the cooling section from known line speed $v_{ab}$ and the thermal diffusivity $\alpha$ of the extrudate, thus:

$$L_E=Fo_E\frac{D^2}{a}v_{ab} \tag{11.13}$$

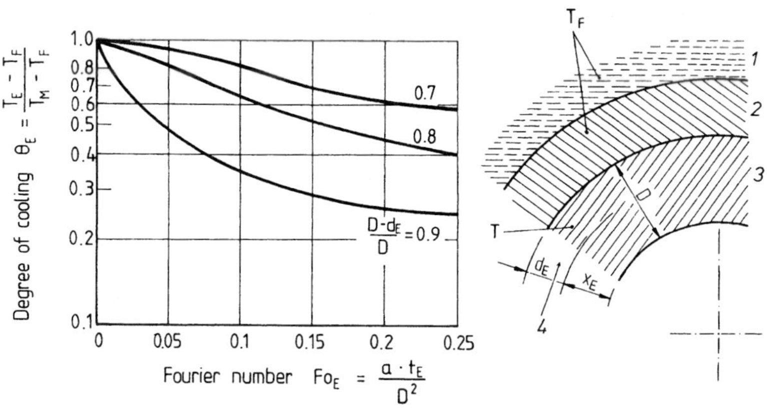


**Figure 11.16** Estimating the length of the calibration [21, 22], *1* Heating/cooling medium, *2* Calibrating sleeve, *3* Profile, *4* Solidified layer

For the sake of simplicity an ideal thermal contact between the extrudate and the cooling medium was assumed here (Biot $\rightarrow \infty$, $\delta = \pi/2$)

The relationships between the wall thickness of the extrudate $D$, the thickness of the solidified layer $d_E$, the relative cooling intensity $Bi$ or $\delta$ described respectively by Equation (11.10), and the required dimensionless cooling time $Fo_E$ is useful [26] for transferring characteristics of well designed cooling section to new cooling section for extrudates with different dimensions. The simplifications made for the analytical solution of Equations (11.2) and (11.3) allow generally only a rough estimate of the thermal events in the cooling section. The following three critical points should be mentioned here:

1. In the external calibration with vacuum, a gap will form between the sizing die and the profile as a result of the shrinkage of the material during its cooling. This gap can either act as thermal insulation or can be filled with cooling water—in such a case where the calibration sleeve is open on the side of the cooling bath. This, of course, affects the heat transfer at the surface of the extrudate considerably; it is no longer possible to assume a constant heat transfer resistance.
2. The simplification by the assumption of constant material data appears to be also critical due to a transformation of the material that occurs in the solidified layer (particularly in semicrystalline polymers).
3. In many cases, particularly in profile extrusion, the heat transfer cannot be assumed to occur in one dimension only. It is necessary then to apply procedures which consider the typical two-dimensional geometry of both the extrudate and the calibrator.

### 11.2.2 Numerical Computational Model

The solution of Equation (11.2) can be determined numerically only when considering the temperature dependence of the thermal material data specific for the given material and the time dependent thermal boundary conditions. FDM or FEM procedures are applicable here.

The typical ranges of values of the thermal resistance between the extrudate and the cooling medium for different configurations of the cooling sections are given in Fig. 11.17 [23,27–29]. The widths of the ranges given there indicate that there are hardly heat transfer coefficients typical for a given configuration. This is particularly true for the dry calibration, which [27] gives values with a spread up to 75%. The heat transfer coefficient (a characteristic value of the resistance to the heat transfer) in this type of calibration shows not only a dependence on the geometry of the cross section and the cooling time, but also a pronounced dependence on operating conditions of the line (throughput, take-up speed, temperature of the material) [29–31]. The spreads of values for the "wet" calibration process are considerably smaller (approximately 30%).

There are no generally valid equations for the determination of the heat transfer coefficients. The reason for that is that the relationships between the shaping of the profile by the friction or take-up forces, shrinkage and distortions of the profile (due to internal stresses), and the heat transfer mechanisms are exceedingly complex and therefore do not lend themselves to a general description.

Nevertheless, the time dependent temperature patterns in the extrudate can be estimated from mean values of the heat transfer. Also from this, the effects of processing variables (changes in the temperature of the cooling medium, in the length or configuration of the individual cooling sections) on these temperature patterns can be correctly calculated in the order of magnitude and trends.

Figure 11.18 shows, for example, the effect of the change in the configuration of the cooling section (from a water bath cooling to a spray cooling) or the temperature pattern in the wall of a pipe (results of an FDM computation). Taking the temperature at the end of the cooling section averaged over the wall thickness of the pipe as the basis of the design criterion, the length of the cooling section can be reduced by 70% by using a spray cooling system (at the same take-up speed).

The computation of the cooling processes in profiles with complex structure is described in [27,28]. Here the FEM procedure is applied in order to describe the heat transfer in the extrudate in two dimensions (Fig. 11.19).

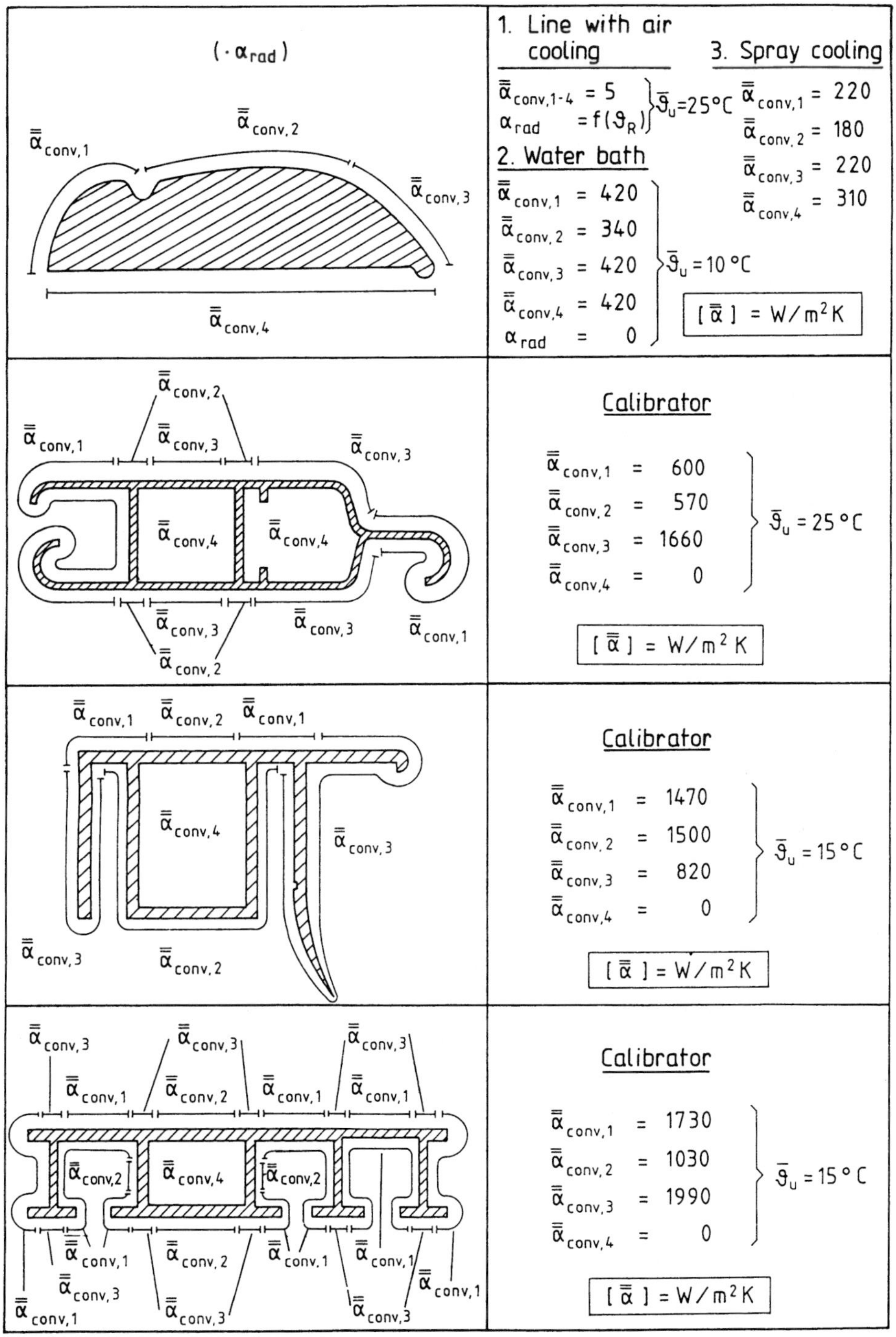


**Figure 11.17** Heat transfer coefficients for different types of profile calibration lines [27]

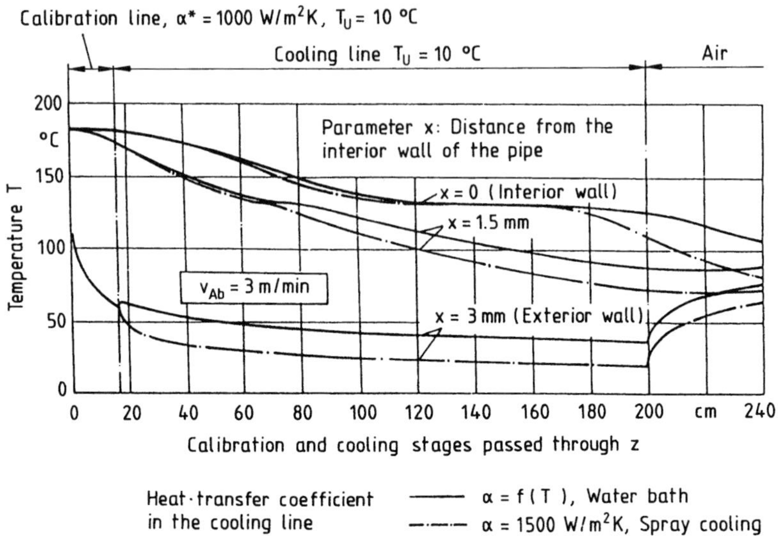


**Figure 11.18** Computed course of temperature distribution in the wall of an extruded PE-HD pipe (25 × 3) for different cooling conditions [23]

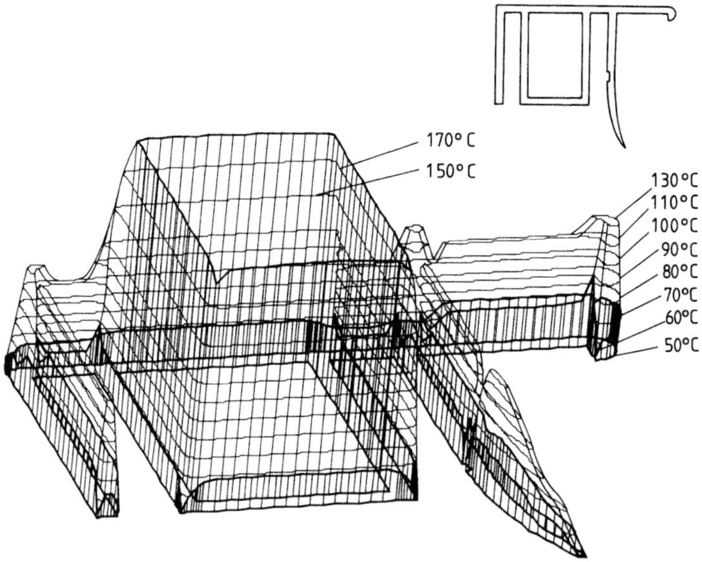


**Figure 11.19** Temperature distribution in a picture frame profile [27]

The same procedure is applied in [32] to compute the effect of different materials from which the calibrator is made on the temperature distribution on the extrudate surface (Fig. 11.21). This pertains to the calibration of a rectangular pipe as shown in Fig. 11.20.

The results show that the temperature distribution is homogeneous, except in the area of the corner of the pipe, when the calibration is done by brass components. For a calibration with steel components the position of the cooling channels had to be optimized in order to attain similarly homogeneous temperature distribution.

The computed temperature distributions with regard to time for selected points in the pipe wall (see Fig. 11.20) with the use of dry calibrators are shown in Fig. 11.22. The first calibrator is made from brass (M) and the following ones from steel (S).

The individual calibrators are arranged here with a gap so that the surface of the extrudate can be warmed up in the air gaps between the calibrators by the heat flowing from the inside (Points 4 and 5). Figure 11.23 shows conclusively that the effectiveness of the calibration line can be considerably improved by this measure. In that illustration, the graphic representation of the temperature as a function of time, for the case of calibrators with a gap, is done by a solid curve a), and the case without a gap by a dashed curve b). The origin of this functional representation is located at the entrance to the fourth calibrator. The friction forces are of the same order of magnitude in both cases, because the *total length* of the *calibration dies* remains the same (the *length* of the *cooling section* is increased in a, against b, by the length of the gaps, which is 38% here).

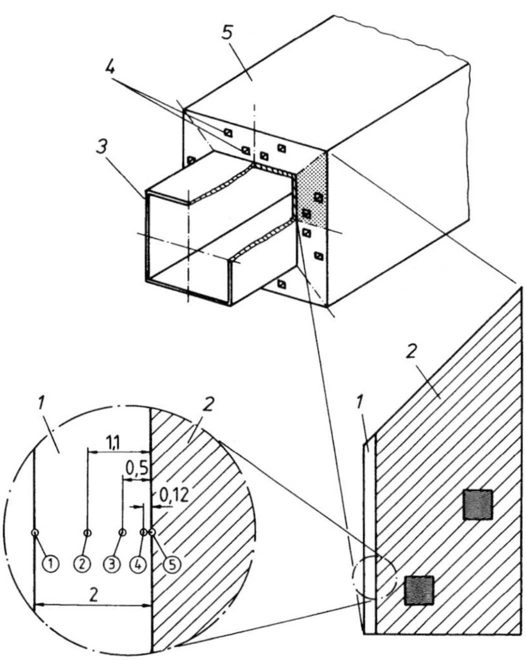


**Figure 11.20** A profile and the calibrator, *1* PVC, *2* Metal, *3* Extrudate, *4* Cooling channels, *5* Calibrator

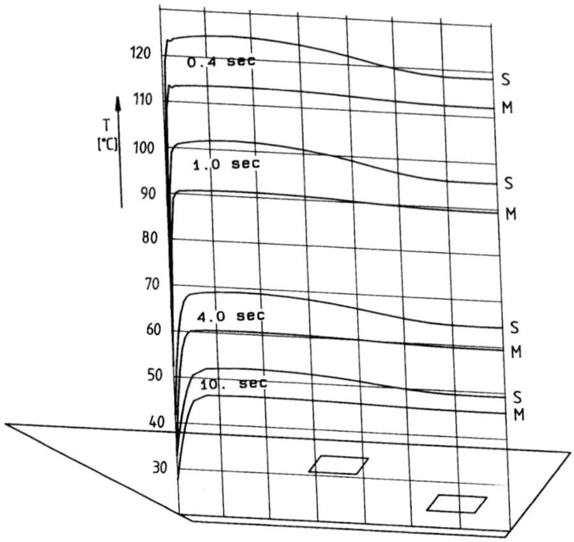


**Figure 11.21** Temperature distribution on the extrudate surface for different cooling times in a brass (M) and steel (S) calibrator

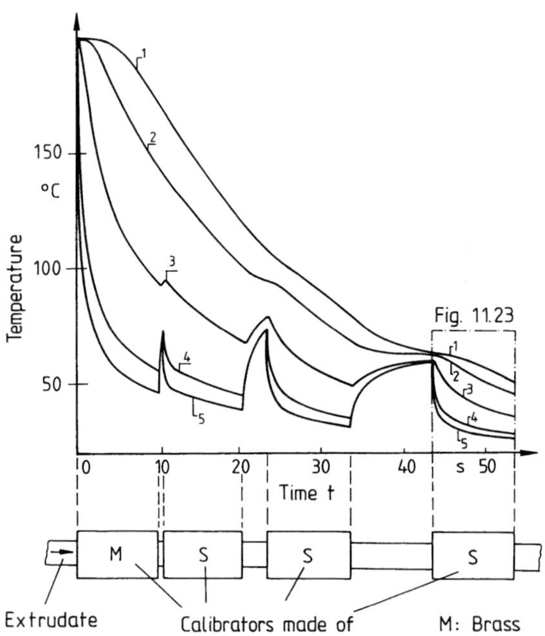


**Figure 11.22** Temperature-versus-time plot of a calibration line with an intermediate annealing

When taking the temperature of the inside surface 60 °C (Point 1) as the basis for the design, it is clear from Fig. 11.23 that the total length of the calibration in Case 1 can be shortened by approximately 13 % (5.2 seconds from the total 4 times 10 seconds) or the line speed can be increased accordingly.

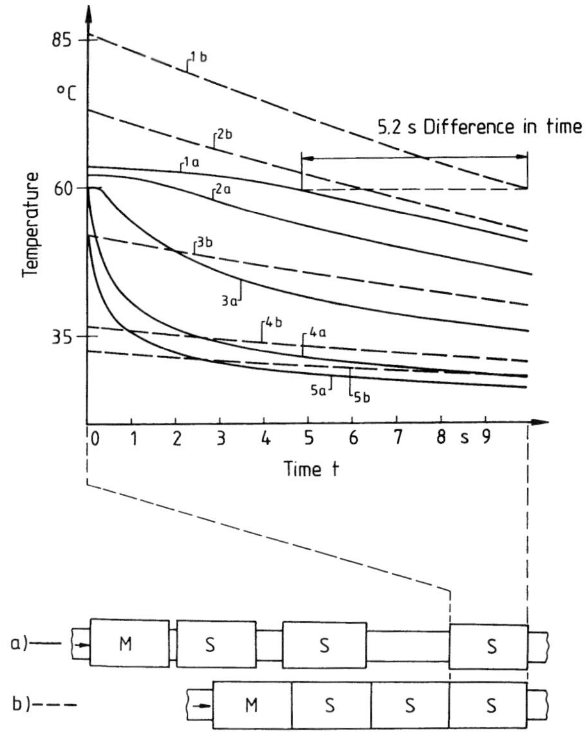


**Figure 11.23** Temperature histogram with (a) and without (b) an intermediate annealing in the above calibrator

These examples show how the cooling processes can be evaluated using theoretical considerations with regard to their effectiveness while changing individual processing parameters, the geometry, and type of construction used in the cooling sections. The materials used can also be evaluated this way. With such a simulation, it is possible to recognize the trends caused by the different variables in the cooling and to separate the important factors from the unimportant ones. Additional studies of this kind dealing with thermal after-treatment of extrudates in different dies are found in [27,33–39].

### 11.2.3 Analogy Model

The use of two-dimensional numerical procedures (Finite Difference Method or Finite Element Method) requires a much longer time when compared to the one-dimensional methods of estimation, because the respective geometry (or partial geometry, as the case may be) of the cross section of the extrudate must be first transformed into a form (FD or FE mesh) suitable for loading into the computer.

A method to estimate the change of temperature with time in basic geometric elements, which frequently occur in profile cross sections, and in which the heat transport cannot be described in one dimension, is described in [40–42]. This method does that by bypassing the demanding numerical computational procedures.

Such basic geometric elements for a typical profile geometry are shown in Fig. 11.24. These basic elements (L and T regions) often determine the cooling time because of the changed ratio of surface to volume, which means they often cool slower than the straight plate-like (and hence thermally one-dimensional) partial geometries of the extrudate cross section.

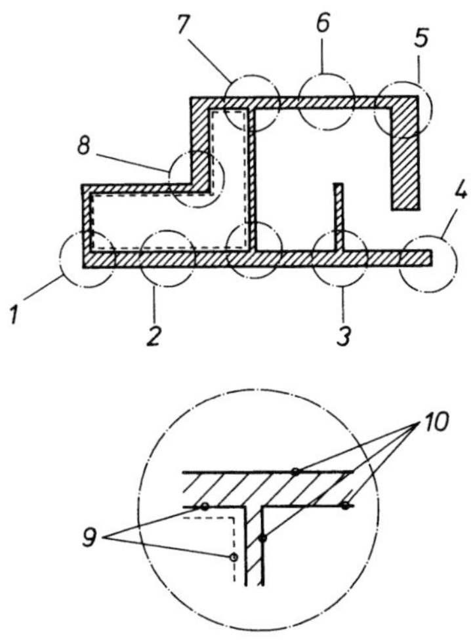


**Figure 11.24** Basic geometries of a typical profile cross section, *1* Box corner, *2* Plate (cooled on one side), *3* T-Profile, *4* End of a plate, *5* L-Profile, *6* Plate (cooled on both sides), *7* Corner rib, *8* Groove corner, *9* non-cooled surfaces (hollow chamber), *10* Cooled surfaces

The theory of this analogy is based on the assumption that the thermal processes in these basic elements can be characterized by dimensionless constants in a similar way as in a one-dimensional case.

Assuming constant thermal material data and boundary conditions in Fig. 11.25, the definitions of the characteristic numbers for the three cases of cooling in the L-region are shown in the following examples:

- L-profile (cooled on both sides),
- corner of a box (cooled outside),
- corner of a groove (cooled inside).

In the L-region, in addition, there are the following three characteristic numbers derived from the one-dimensional studies:

- relative cooling intensity (Biot Number, see also Equation (11.6))
- dimensionless cooling time (see also Equation (11.7))
- dimensionless temperature (see also Equation (11.8))

the following constant is added:

- the (dimensionless) thickness ratio $D$ of both L-shanks.

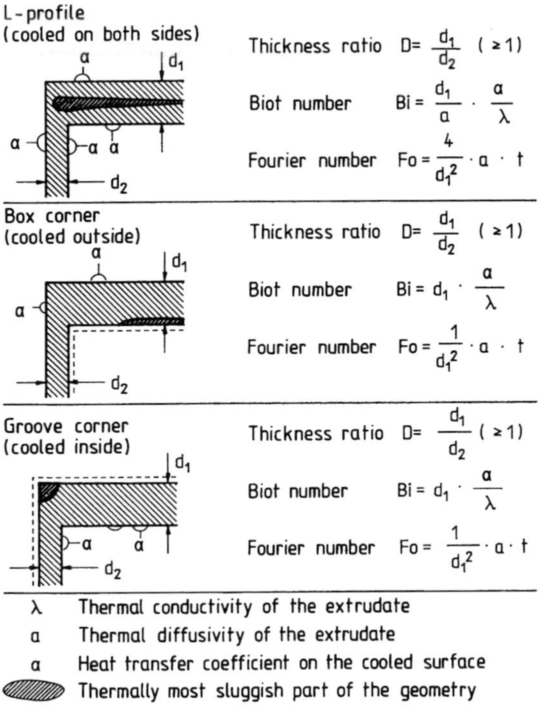


**Figure 11.25** Definitions of characteristic coefficients for different cooling conditions of the L-shaped region

When computing the two-dimensional temperature distributions as a function of time in the basic geometry for the different cases of cooling and different combinations of the constants *Bi* and *D*, applying a numerical simulation procedure, the temperature as a function of time for individual points of the geometry can be plotted as a function of the Fourier number.

To estimate the required length of the cooling section (or the cooling time), it is now not necessary, in most cases, to know the temperature as a function of time at each point of the cross section of the extrudate. On the contrary, often it is sufficient to know only the cooling behavior of the points in the geometry, which are thermally the most sluggish ones. The numerically determined temperature profile as a function of time for each of these thermally most sluggish points is then approximated in dependence on *Bi* and *D* using simple equations on a pocket calculator or plotted graphically.

Figure 11.26 shows the result of such an investigation. Finally, with this procedure, the temperature patterns in more complicated partial geometries can be determined equally quickly and under the same conditions as for a simple plate or a pipe wall.

The critical points with regard to the assumed constancy of the thermal material data and boundary conditions mentioned there are valid to the same extent when applying this analogy model.

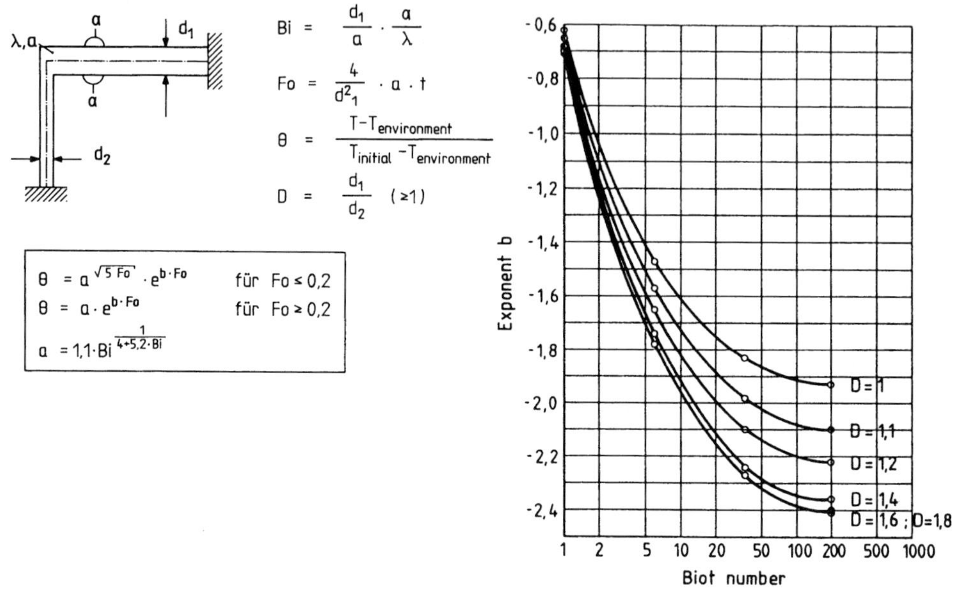


**Figure 11.26** Temperature of the thermally most sluggish point in an L-profile

### 11.2.4 Thermal Boundary Conditions and Material Data

The thermal material data of the extrudate are essential for computations leading to a meaningful thermal design. When any of the computations is based on incorrect or inaccurately determined data, the resulting cooling times are equally inaccurate. The measurement of the above data is difficult and costly, particularly for semicrystalline thermoplastics, which exhibit a pronounced temperature dependence of heat conductivity and thermal diffusivity in the region near their crystalline melting temperature $T_m$.

In addition to that, these material data can be dependent on processing conditions (cooling rate, degree of crystallinity). In many cases, the average values obtained from the manufacturers of the raw materials are the only ones available.

This possible source of errors is particularly disturbing when the thermal boundary conditions under which the extrudate is cooled are to be quantified by means of simulative computations.

Since the coefficient of heat transfer is virtually impossible to measure, as mentioned before, procedures were developed which allow its indirect determination from measurements of the temperature in dependence on time in the extrudate.

The computations for the cooling are performed in such a way that the calculated boundary conditions are subjected to iteration until the measured and computed temperature patterns are in agreement [27,29,31].

If the computations in this procedure are based on incorrect material data, the resulting heat transfer coefficients will be critically affected [27,31].

## 11.3 Effect of Cooling on the Quality of the Extrudate

The criteria used for measuring the quality of an extrudate, and which are affected by the cooling process, are:

- the condition of the surface (gloss, grooves)
- dimensional stability
- the stability against mechanical loads and influence of media
- the texture of the solidified extrudate

The gloss of the surface is produced definitely in the first cooling phase. Polished, dry entrance into the calibration line leads to a high gloss, while direct cooling by liquids produces dull surfaces. Grooves are formed due to worn calibrations or by abrasive particles brought in by the liquid in wet calibration (which often is the reason for the preferential use of dry calibration).

Under the term dimensional stability, not only reaching the required dimensions and tolerances is understood, but also a stability of dimensions in heat (e.g., shrinkage), which is affected by frozen-in mechanical stresses in the extrudate (underdraft due to friction in the calibration line, distribution of internal stresses in the cross section of the extrudate due to cooling). The level of internal stresses can be reduced by two-sided cooling (which is not always possible) [43].

The residual stresses in the extrudate also affect the stability in mechanical loading and the resistance against corrosive media. Generally, when compressive stresses are produced in the extrudate surface while being cooled, a high stretch of the solidified surface layers during cooling (e.g., by large friction and take-up forces) can lead to residual tensile stresses in the surface [23,44,52]. In this case, the extrudate is very sensitive to even miniscule injury to the surface in use or to corrosive media (environmental stress cracking).

The texture, particularly that of extrudates with thick walls, depends on the wall thickness. This is especially true about semicrystalline thermoplastics. Near the cooled surfaces, the degree of crystallization can be negligible, so that there is a boundary layer without any texture; however, inside layers, which were cooled more slowly, are highly crystalline. The thickness of the boundary layer without texture increases considerably with more rapid cooling [45,46].

## 11.4 Mechanical Design of Calibration Lines

When designing a calibration line it is important to build its fixed elements strong enough to resist the high take-up forces. In many cases, so-called calibration tables are used, which have a mounting plate with adjustable height, and located in their base is additional equipment, such as vacuum pumps as well as tanks for coolant. The mounting plates hold different kinds of calibration tooling. The take-up forces for large profiles (e.g., main profiles for windows) can be up to several tons, therefore, the equipment has to be anchored well in the floor of the production facility. The gap between the cooling section and the extrusion orifice should be easily adjusted (e.g., by pinions) in order to facilitate the start-up of the line.

Closed outside calibrators have to be made in sections for threading the extrudate, particularly for profiles with complicated cross section geometry.

The weight of individual calibrator blocks (at least their tops) should not exceed 20 kg so that they can be handled with relative ease.

The material from which the calibrator is made is chosen based on the production requirements. In most cases, brass or steel have been used successfully; occasionally aluminum is selected [47].

Brass and aluminum have advantages over steel in thermal conductivity and machining. The higher tendency of these two materials to wear requires special surface treatment (chrome-plating of brass, anodization of aluminum), which has to be repeated locally if the wear is excessive.

Wet calibrators tend to have higher wear than dry ones due to abrasive particles in the cooling fluid.

## 11.5 Cooling Dies, Process for Production of Solid Bars

Profiles with thick walls, such as solid bars up to 500 mm diameter and thick plates (slabs), are produced with cooling dies.

The tooling for this process consists of a calibration die, which is attached directly to the shaping die (without any gap) and cooled intensively (Fig. 11.27). In this setup, it is important to achieve a solid attachment of the calibrator to the die and a good thermal separation of both at the same time.

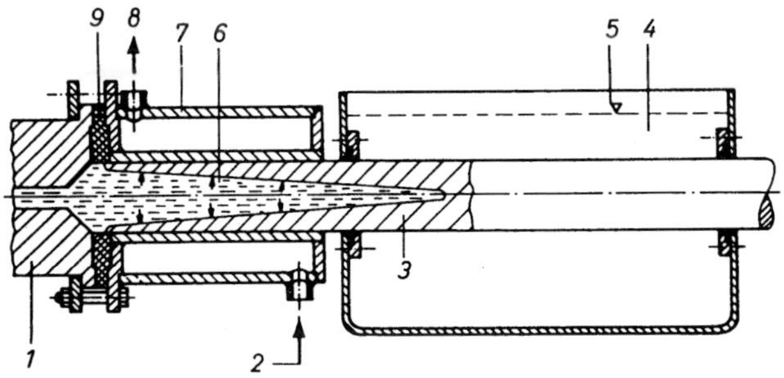


**Figure 11.27** Extrusion of solid bars [3], *1* Extrusion die, *2* Cooling water inlet, *3* Solidified layer, *4* Water bath, *5* Water level, *6* Melt, *7* Calibrating section, *8* Cooling water outlet, *9* Thermal barrier

The flow channel in the shaping die moves the melt from a relatively small flow diameter (independent of the geometry of the semifinished product [13]) under a relatively steep aperture angle (up to 45 degrees) into the transition zone and towards the cooling section. A solidified layer on the extrudate surface is formed, which becomes increasingly thicker in the direction of extrusion. The inside of the extrudate contains a melted core, shaped as a cone or a wedge, which can reach far into the calibration line.

The operating conditions of the extruder and the speed of the line have to be matched in such a way that the pressure in the melt cone to its tip is high enough to prevent the formation of voids due to thermal shrinkage of the solidified polymer [48]. These high melt pressures (of the order of 100 bar) lead to accordingly high normal and friction forces in the former between the die and the extrudate. The friction forces can be reduced by a smoothly finished surface of the die, by using lubricant (e.g., oil) or by surface coating (e.g., PTFE) [3,49]. On the other hand, the melt pressure supports the take-off of the extrudate though the projected area of the profile cross section. Depending on the ratio between the area of the profile cross section and the surface area of the extrudate being in contact with the former and subjected to friction, it can be necessary to *slow down* the profile instead of *pulling it away,* especially when the strands are thick and extruded at low speeds.

Special take-up devices (mostly caterpillar take-up) are used for this purpose that have the ability to slow down the speed of the extrudate. Special requirements are necessary for the drive of these devices with regard to the freedom from play (when changing the load).

Another way to transfer breaking forces to the extrudate as close as possible behind the exit from the calibrator is the use of spring loaded shoe breaks (see also friction calibration) with adjustable pressure onto the extrudate.

The cooling of the extrudate occurs mostly in two steps. The calibration section is designed so short so that a thick enough solidified layer is formed that is capable of resisting the internal pressure and the take-up forces. A cooling section with a water bath or spray cooling line (possibly with support rollers) generally follows the calibrator [49].

The progress of the solidification front in a PP solid profile with a height of 95 mm is shown in Fig. 11.28 as an example. After a take-up length of 800 mm the solidified outside skin reached the minimum thickness of 10 mm so that the calibration can end here and the profile can be further cooled down in a water bath [27].

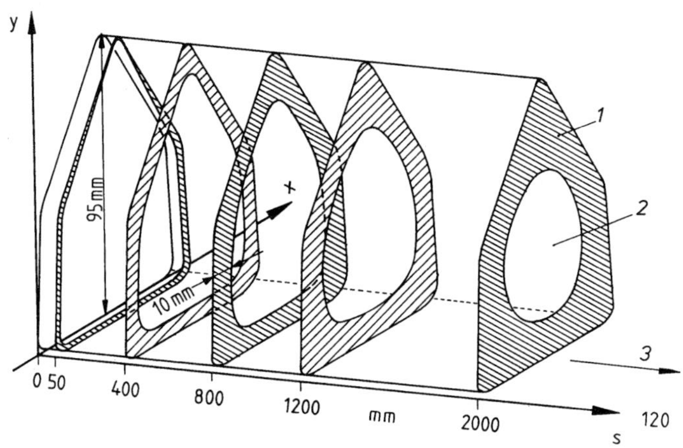


**Figure 11.28** Computed shape of isothermal planes in a block profile, *1* Already solidified, *2* Still melted, *3* $v_{let\text{-}off}$ = 0.03 m/min

The time dependent temperature profiles of different locations in the extrudate for a combined calibration and water bath cooling sections are shown in Fig. 11.29. The calibration ends here at the cooling time of 1,600 seconds (take-up speed 0.03 m/min). It is obvious that the layers of the extrudate near the surface (Points 1 and 5) warm up slightly when entering the water bath. This can be attributed to a less intensive heat transfer in this part of the cooling section.

After approximately 6,200 seconds, the thermally most sluggish point of the extrudate (4) reaches the temperature of 50 °C, which corresponds to a length of a water bath of 2.3 meters.

Because of the high internal pressure in the calibrator, it is important to pay attention to sufficient mechanical stability, particularly with regard to the location of the cooling channels.

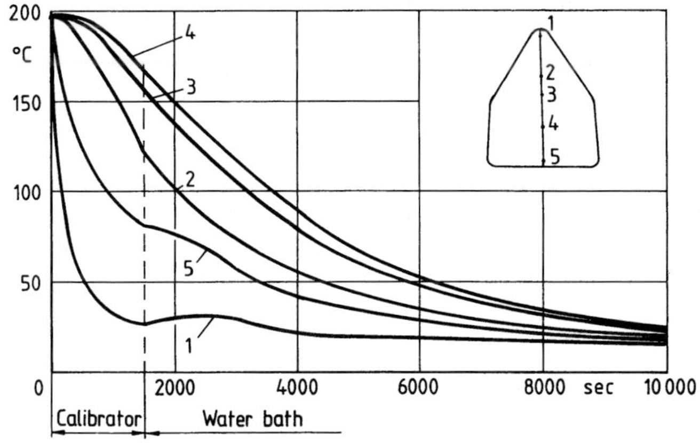


**Figure 11.29** Computed temperature-versus-time plot (histogram) of a block profile

Because of the very low cooling rates, small extruders (screw diameter 30–45 mm) [4,51], sometimes with multiple dies, are used for the extrusion of solid rods. The extrusion speeds, for example, for a PA (polyamide) round rod with a diameter of 60 mm, are about 2.5 m/hour, and with a diameter of 200 mm, about 0.5 m/hour [51].

The residual stresses generated during the cooling of the extrudate have an adverse effect on its properties, particularly if machined (distortion during the operation). Therefore, such profiles are often first annealed.

## Symbols and Abbreviations

| | |
|---|---|
| $a$ | Thermal diffusivity |
| $Bi$ | Biot number |
| $c_p$ | Specific heat capacity |
| $D$ | Thickness of the extrudate |
| $d_E$ | Thickness of the solidified layer |
| $F_{friction}$ | Friction force |
| $F_{normal}$ | Normal force |
| $Fo$ | Fourier number |
| $Fo_E$ | Dimensionless cooling time |
| $L$ | Length of the cooling zone |
| $\dot{q}$ | Heat flux density |
| $t$ | Time |
| $T$ | Temperature |
| $T_E$ | Temperature of solidification |
| $T_F$ | Temperature of the cooling medium |

| | |
|---|---|
| $T_m$ | Mass (melt) temperature |
| $T_0$ | Surface temperature |
| $T_U$ | Ambient temperature |
| $v_{ab}$ | Take-off speed |
| $x$ | Height coordinate |
| $x_E$ | Thickness of the melt layer |
| $\alpha$ | Heat transfer coefficient |
| $\Theta$ | Degree of chilling (quenching) |
| $\Theta_E$ | Degree of chilling (quenching) of the solidified layer |
| $\lambda$ | Heat conductivity |
| $\mu$ | Coefficient of friction |
| $\rho$ | Density |

## 11.6 References of Chapter 11

1. Kleindienst, U.: Einflußgrößen beim Vakuumkalibrieren von extrudierten Kunststoffrohren. Kunststoffe 63 (1973) 1, pp. 7–11
2. Schiedrum, H. O.: Profilwerkzeuge für das Extrudieren von PVC. Ind. Anz. 90 (1968) 102, pp. 2241–2246
3. Schiedrum, H. O.: Profilwerkzeuge und -anlagen für das Extrudieren von thermoplastischen Kunststoffen. Plastverarbeiter 19 (1968) 6, pp. 417–423 and 19 (1968) 8, pp. 635–640 and 19 (1968), pp. 728–730
4. Kunststoffverarbeitung im Gespräch. 2: Extrusion. Ed. by BASF, Ludwigshafen (1986)
5. Krämer, A.: Herstellen von Profilen aus PVC-hart. Kunststoffe 59 (1969) 7, pp. 409–416
6. Schiedrum, H.O: Auslegung von Rohrwerkzeugen. Plastverarbeiter (1974) 10, pp. 1–11
7. Schiedrum, H. O.: Kalibrieren und Kühlen von Rohren. In: Kühlen von Extrudaten. VDI-Verl., Düsseldorf (1978)
8. Schiedrum, H. O.: Extrudieren von PVC-Profilen am Beispiel des Fensterprofils. Kunststoffe 65 (1975) 5, pp. 250–257
9. Brinkschröder, F. J. and Johannaber, F.: Profilextrusion schmelzeinstabiler Thermoplaste. Kunststoffe 71 (1981) 3, pp. 138–143
10. Weber, H.: Extrusion von Fensterprofilen. Plastverarbeiter 30 (1979) 10, pp. 608–614
11. Limbach, W.: Werkzeuge und Folgeaggregate für das Extrudieren von Profilen. In: Extrudieren von Profilen und Rohren. VDI-Verl., Düsseldorf (1974)
12. Berger, P. and Krämer, A.: Kalibrieren von Rohren bei hohen Abzugsgeschwindigkeiten. Kunststoffe 65 (1975) 1, pp. 1–6
13. Domininghaus, H.: Einführung in die Technologie der Kunststoffe. 2. - Company document, Hoechst AG, Frankfurt
14. Gebler, H. et al.: Maschinen zur PP-Rohrherstellung. Kunststoffe 70 (1980) 5, pp. 246–253
15. Information of Fa. Plastic-Industrie. Ausrüstungs GmbH, Löhne (1979)

16. Schiedrum, H. O.: Kalibrieren und Kühlen von Rohren. Plastverarbeiter 30 (1979) 5, pp. 255–258
17. DE-OS 2532085 (1977) Reifenhäuser KG, Troisdorf und Gofini AG, CH Glarus (German patent)
18. Profilanlagen für die Extrusion von Vollprofilen nach dem Technoform-Präzisions-Profilzieh-Verfahren. Information of Reifenhäuser KG, Troisdorf (1977)
19. Barth, H.: Kalibrieren und Kühlen von Profilen. In: Kühlen von Extrudaten. VDI-Verl., Düsseldorf (1978)
20. Ehnert, M.: Kontinuierliches Fertigen in sich geschlossener Flachhohlkörper. Kunststoffe 73 (1983) 1, pp. 13–16
21. Kreft, L. and Doboczky, Z.: Berechnung und Konstruktion von Kalibrierdüsen. Kunstst. Gummi 1 (1962) 1, pp. 15–20
22. Kreft, L. and Doboczky, Z.: Konstruktion und Berechnung der Kühlsysteme für die Kalibrierung von Halbzeugen. Kunstst. Gummi 6 (1967) 5, pp. 173-177 and 6 (1967) 6, pp. 207–212
23. Kleindienst, U.: Untersuchung des Abkühlungsvorganges und dessen Einfluß auf das Eigenspannungsfeld in der Wand extrudierter Kunststoffrohre. Thesis at the University of Stuttgart (1976)
24. Grigull, U.: Temperaturausgleich in einfachen Körpern. Springer, Berlin (1964)
25. Menges, G. and Michaeli, W.: Calculation of Cooling Processes in Extrusion. Ann. Techn. Conf. SPE (1979) conference proceedings, pp. 141–145
26. Ast, W.: Einfache Methode zur Auslegung von Kühlstrecken bei der Kunststoff-Extrusion. Kunststoffe 69 (1979) 4, pp. 186–193
27. Schmidt, J.: Wärmetechnische Auslegung von Profilkühlstrecken mit Hilfe der FEM. Thesis at the RWTH Aachen (1985)
28. Gesenhues, B. and Schmidt, J.: Auslegung von Düsen und Kühlstrecken in der PVC-Extrusion. Plastverarbeiter 35 (1984) Nr. 5, pp. 86–92
29. Grünschloß, E. and Radtschenko, L.: Experimentelle Bestimmung der Wärmeübergangszahlen bei der Kühlung extrudierter Kunststoffrohe aus HD-PE im Wasserbad. Plastverarbeiter 30 (1979) Nr. 10, pp. 631–639
30. Hader, W.: Parameterstudie bei der Extrusion von Profilen. Unpublished thesis at IKV, Aachen (1986)
31. Tietz, W.: Analyse der Verteilung der Wärmeübergangswiderstände an einer Profilextrusionsanlage. Unpublished thesis at IKV, Aachen (1986)
32. Menges, G. et al.: Anwendung der Wärmeausgleichsrechnung bei der Extrusion und beim Schweißen. Kunststoffe 79 (1989) 2, pp. 182–187
33. Michaeli, W. et al.: Produktbeeinflussung durch Werkzeug und Folgeaggregate bei der Extrusion. Proceedings of 8. Kunststofftechnisches Kolloquium of IKV, Aachen (1976)
34. Franzkoch, B.: Analyse eines neuen Verfahrens zur Herstellung vernetzter Polyethylen-kabel. Thesis at the RWTH Aachen (1979)
35. Michaeli, W.: Zur Analyse des Flachfolien- und Tafelextrusionsprozesses. Thesis at the RWTH Aachen (1975)
36. Michaeli, W.: Berechnen von Kühlprozessen bei der Extrusion. In: Kühlen von Extrudaten. VDI-Verl., Düsseldorf (1978)
37. Menges, G. et al.: Entformungstemperatur und Kühlzeit bei der Herstellung extrusionsgeblasener Hohlkörper. Plastverarbeiter 24 (1973), pp. 621-623 and 24 (1973), pp. 685–690
38. Menges, G. et al.: Zum Temperaturverhalten des als Zwischenprodukt beim Extrusions-Blasformen erstellten Vorformlings. Plastverarbeiter 24 (1973) 6, pp. 333–340
39. Ast, W.: Extrusion von Schlauchfolien - Theoretische und experimentelle Untersuchungen des Abkühlvorganges. Thesis at the University of Stuttgart (1976)

40. Koch, A.: Berechnung der Temperaturfelder in Kühlstrecken unter Verwendung analytischer Lösungsansätze. Unpublished thesis at the IKV, Aachen (1985)

41. Kalwa, M.: Anwendung der Finite Elemente Methode zur Simulation von Wärmetransportvorgängen in der Kunststoffverarbeitung. Thesis at the RWTH Aachen (1990)

42. Rietmann, M.: Ermittlung von Kennzahlendiagrammen zum zweidimensionalen Wärmetransport in Elementargeometrien. Unpublished thesis at the IKV, Aachen (1988)

43. Berndtsen, N.: Extrusion von Profilen. Kunststoffe 78 (1988) 10, pp. 960–963

44. Menges, G. and Kalwa, M: Berechnung von Eigenspannungen und Verzug in Extrusionsprofilen. Final report of the DFG research project Me 272/179-1 (1988)

45. Kamp, W. and Kurz, H.-D.: Kühlstrecken bei der Polyolefin-Rohrextrusion. Kunststoffe 70 (1980) 5, pp. 257–263

46. Kurz, H.-D.: Neue Untersuchungen zum Abkühlvorgang bei HDPE- und PP-Rohren. Conference Rohrsymposium, Frankurt (1979)

47. Optimieren von Kalibrier- und Kühlverfahren. Plastverarbeiter 36 (1985) 3, pp. 50–51

48. Titomanlio, G. et al.: Analysis of Void Formation in Extruded Bars. Polym. Eng. Sci. 25 (1985) 2, pp. 91–97

49. Voigt, J. L.: Kontinuierliches Herstellen von Polyamid-Stäben. Kunststoffe 51 (1961) 8, pp. 450–452

50. Käufer, H. and Vogel, T.: Anwendung eines Bremswerkzeugs anstelle einer Kalibrierung. Information of the TU Berlin, Kunststofftechnik, Oktober (1980)

51. Hinz, E.: Erfahrungen beim Herstellen von Vollprofilen aus thermoplastischen Kunststoffen. In: Extrudieren von Profilen und Rohren. VDI-Verl., Düsseldorf (1974)

52. Menges, G. et al.: Der Einfluß von Eigenspannungen auf das mechanische Verhalten von Kunststoffen. Kunststoffberater 33 (1988) p. 10

# Index